Communications in Computer and Information Science 961

Commenced Publication in 2007
Founding and Former Series Editors:
Phoebe Chen, Alfredo Cuzzocrea, Xiaoyong Du, Orhun Kara, Ting Liu,
Dominik Ślęzak, and Xiaokang Yang

More information about this series at http://www.springer.com/series/7899

Antonia Moropoulou · Manolis Korres
Andreas Georgopoulos · Constantine Spyrakos
Charalambos Mouzakis (Eds.)

Transdisciplinary Multispectral Modeling and Cooperation for the Preservation of Cultural Heritage

First International Conference, TMM_CH 2018
Athens, Greece, October 10–13, 2018
Revised Selected Papers, Part I

 Springer

Editors
Antonia Moropoulou
National Technical University of Athens
Athens, Greece

Constantine Spyrakos
National Technical University of Athens
Athens, Greece

Manolis Korres
National Technical University of Athens
Athens, Greece

Charalambos Mouzakis
National Technical University of Athens
Athens, Greece

Andreas Georgopoulos
National Technical University of Athens
Athens, Greece

ISSN 1865-0929   ISSN 1865-0937 (electronic)
Communications in Computer and Information Science
ISBN 978-3-030-12956-9   ISBN 978-3-030-12957-6 (eBook)
https://doi.org/10.1007/978-3-030-12957-6

Library of Congress Control Number: 2019930855

This Springer imprint is published by the registered company Springer Nature Switzerland AG
The registered company address is: Gewerbestrasse 11, 6330 Cham, Switzerland

Preface

Innovative scientific methodologies and challenging projects marking future trends in the protection of cultural heritage have initiated a universal conversation within a holistic approach, merging capabilities and know-how from the scientific fields of architecture, civil engineering, surveying engineering, materials science and engineering, information technology and archaeology, as well as heritage professionals and stakeholders in cultural heritage. The advanced digital documentation permits the data fusion of interdisciplinary innovative modeling, analytical and non-destructive techniques, supporting the emergence of a transdisciplinary field for multispectral sustainable preservation and management of cultural heritage.

The project of the rehabilitation of the Holy Sepulchre's Holy Aedicule, as a pilot multispectral, multidimensional, novel approach based on transdisciplinarity, holistic digital reconstruction and cooperation in the protection of monuments, motivated the organization of this conference. As discussed in the homonymous panel with the NTUA Interdisciplinary Team (Prof. Antonia Moropoulou, NTUA School of Chemical Engineering- Coordinator, Prof. Manolis Korres, Emeritus, Academy of Athens, Prof. Andreas Georgopoulos, NTUA School of Rural and Surveying Engineering, Prof. Constantine Spyrakos and Asst. Prof. Charalambos Mouzakis, NTUA School of Civil Engineering), National Geographic Society (Fredrik Hiebert, Archaeologist), World Monuments Fund (Yiannis Avramides, Civil Engineer), European Commission (Albert Gauthier), as well as the representatives of the three Christian Communities, Guardians of the Holy Tomb and Asst. Prof. Stefanos Athanasiou, University of Bern, Faculty of Theology, the opening of the scientific community towards society, which managed to reach out to all humanity, was successfully achieved through the cooperation of National Geographic with NTUA, as characteristically demonstrated through its contribution in highlighting and transmitting the ecumenical values related to the opening of the Holy Tomb.

The conference also explored the sustainable preservation and management of monuments, sites, and historic cities as a prerequisite, which must be met in order for them to remain within the listed Monuments of World Heritage recommended by UNESCO, such as the Medieval City of Rhodes, for which the strategic management plan proposed by the NTUA was discussed in detail. In this direction, new funding tools, schemes, and mechanisms are sought, based on the re-use of heritage assets at the service of local communities and in order to enhance their value, aiming to increase social participation and awareness and at the same time generate revenue for a sustainable preservation within the framework of circular economy, as stressed by Mrs. Bonnie Burnham, President Emeritus of the World Monument Fund, Mr. George Kremlis, E.C.-D.G. Environment, Responsible for the Circular Economy, Mrs. Le Minaidis, Secr. Gen. OWHC (Org. of World Heritage Cities) – Coordinator, Prof. Ismini Kriari, Rector at the Panteion University of Social and Political Sciences, Prof. Michael Turner, UNESCO Chair in Urban Design, Conservation Studies/Bezalel

Academy of Arts and Design, Jerusalem, Mr. Teris Chatziioannou, Deputy Mayor of Rhodes, as well as NTUA Professors Antonia Moropoulou, President of the TCG General Assembly, Sophia Avgerinou-Kolonia, ICOMOS CIVVICH, President of the Board of Directors of *Eleusis 2021* - European Capital of Culture, Eleni Maistrou, Society for the Environmental and Cultural Heritage (ELLET), Council of Architectural Heritage.

In the panel "Novel Education Approach for the Preservation of Monuments," the education of both society and members of the scientific community was discussed. The Interdisciplinary NTUA Post-Graduate MSc Program "Protection of Monuments," 20 years after its founding, offered by the NTUA School of Architecture, with the cooperation of the NTUA Schools of Chemical Engineering, Civil Engineering and Rural and Survey Engineering, represented in the panel by Prof. Irene Efesiou, NTUA, Post-Graduate MSc Program "Protection of Monuments" Em. Prof. Eleni Maistrou, President of the Council of Architectural Heritage of the Society for the Environmental and Cultural Heritage, ELLET, and Antonia Labropoulou, NTUA, the Post-Graduate Program "Protection, Conservation and Restoration of Cultural Monuments Interdepartmental Program of Post Graduate Studies" offered by the Aristotle University of Thessaloniki, represented in the panel by Prof. Maria Arkadaki, the University of the Peloponnese, represented by Prof. N. Zacharias, the School of Pedagogical and Technological Education (ASPETE), represented by Prof. Anastasia Sotiropoulou and the Frederick University of Cyprus, represented by Prof. Marios Pelekanos, recognized the significant educational value of problem-based learning and the hands-on approach in educating new scientists, while the value of the holistic digital fusion of data was highlighted by Professor Peter De Vries, Delft, European Society for Engineering Education (SEFI). The significant contribution of multimedia in the education of not only scientists, but the whole of society was discussed with Marcie Goodale, Product Director of National Geographic Learning and J. J. Kelley, Emmy-nominated film-maker of National Geographic.

In order to achieve a multilateral approach for the preservation of cultural heritage, heritage stakeholders, science, and industry must be brought together. This issue was discussed in the relative panel by Amalia Androulidaki, Dir. Gen. of Anastylosis, Hellenic Ministry of Culture and Sports – Coordinator, Yuval Baruch, Israeli Antiquity Authorities, Michalis Daktylides, PEDMEDE, European Construction Industry Federation (FIEC), Kyriakos Themistokleus, Cyprus Scientific and Technical Chamber (ETEK), member of the Executive Committee, Prof. Panagiotis Touliatos, Frederick University of Cyprus, Stefano Dellatore, Politecnico di Milano, Athina Chatzipetrou, President Hellenic Archaeological Fund, Dr. Evgenia Tzannini, Law Lecturer NTUA - T.C.G., Özgür Turan Aksan, Architect, Istanbul, Dr. Ekaterini T. Delegou, NTUA, representing the Hellenic Construction Technology Platform - T.C.G., NTUA, and Elena Korka, ICOMOS, member of the International Board.

The innovative panel procedures of the conference formatted the future landscape for the preservation of cultural heritage. The First TMM_CH conference attracted researchers from all over the world, as 237 contributions were submitted for oral presentation from 22 countries. In addition, the participants of the conference, including stakeholders, industry, and academy, especially scholars and young researchers, who came from all over the world, highlighted the fact that it was an international conference, which successfully achieved bringing together a vast number of brilliant specialists who represent the aforementioned fields to share information about their latest projects and scientific progress.

Striving to ensure that the conference presentations and proceedings are of the highest quality possible, we only accepted papers that presented the results of various studies focused on the extraction of new scientific knowledge in the area of transdisciplinary multispectral modeling and cooperation for the preservation of cultural heritage. Hence, only 73 papers were accepted for publishing (i.e., 32% acceptance rate). All the papers were peer reviewed and selected by the international Steering and Scientific Committees, comprising 105 and 155 reviewers, respectively, and from over 90 institutions. Each submission was reviewed by three reviewers. We would like to express our deepest gratitude and appreciation to all the reviewers for devoting their precious time to thoroughly review the submissions and provide feedback to the authors.

Subsequently, we are happy to present this book, *Transdisciplinary Multispectral Modeling and Cooperation for the Preservation of Cultural Heritage*, which is a collection of the papers presented at the First International TMM_CH Conference held during October 10–13, 2018, at the Eugenides Foundation of Athens in Greece. This book consists of 14 chapters, which correspond to the 14 thematic areas covered during the conference, as follows: Opening Lecture; The Project of the Rehabilitation of Holy Sepulchre's Holy Aedicule as a Pilot Multispectral, Multidimensional, Novel Approach Through Transdisciplinarity and Cooperation in the Protection of Monuments; Digital Heritage; Novel Educational Approach for the Preservation of Monuments; Resilience to Climate Change and Natural Hazards; Conserving Sustainably the Materiality of Structures and Architectural Authenticity; Interdisciplinary Preservation and Management of Cultural Heritage; Sustainable Preservation and Management Lessons Learnt on Emblematic Monuments; Cross-Discipline Earthquake Protection and Structural Assessment of Monuments; Cultural Heritage and Pilgrimage Tourism; Reuse, Circular Economy and Social Participation as a Leverage for the Sustainable Preservation and Management of Historic Cities; Inception – Inclusive Cultural Heritage in Europe Through 3D Semantic Modeling; Heritage at Risk; Advanced and Non-Destructive Techniques for Diagnosis, Design and Monitoring.

We would like to acknowledge all who made the conference possible and successful, i.e., the organizers: National Technical University of Athens, Technical Chamber of Greece, National Geographic Society, World Monuments Fund; the ministries: Ministry of Culture and Sports of the Hellenic Republic, Ministry of Digital Policy, Telecommunications and Media of the Hellenic Republic, and Ministry of Foreign Affairs who collaborated; the President of the Hellenic Republic and the Hellenic Parliament and His Beatitude Archbishop Hieronymus II of Athens and All Greece and the representatives of the three Christian communities, Guardians of the Holy Tomb who blessed the conference. The conference was held within the framework of the European Year of Cultural Heritage 2018.

We would also like to acknowledge the following for the inauguration addresses of the conference: NTUA Rector Prof. Ioannis Golias, National Geographic Society Archaeologist Dr. Fredrik Hiebert, European Commission D.G. Connect Dr. Albert Gauthier, World Monuments Fund Yiannis Avramides, ICOMOS Dr. Elena Korka,

Technical Chamber of Greece (TCG) President Mr. Giorgos Stasinos, President of the Hellenic Parliament Mr. Nikos Voutsis, President of the Standing Committee of the Cultural and Educational Affairs of the Hellenic Parliament NTUA Prof. Dimitris Sevastakis, Deputy Minister of Foreign Affairs of the Hellenic Republic Mr. Markos Bolaris, Minister of Digital Policy, Telecommunications and Media of the Hellenic Republic Mr. Nikos Pappas, and the Secretary General for Telecommunications and Media Mr. Georgios Florentis.

The conference would not have been possible without the full cooperation and commitment of the NTUA professor Ioannis Golias and the NTUA Interdisciplinary Team, Prof. Manolis Korres, Emeritus, Academy of Athens, Prof. Andreas Georgopoulos, NTUA School of Rural and Surveying Engineering, Prof. Constantine Spyrakos and Asst. Prof. Charalambos Mouzakis, NTUA School of Civil Engineering, the commitment of the Technical Chamber of Greece (TCG), president and administrative council, as well as without the valuable assistance of Leonie Kunz, Aliaksandr Birukou, Amin Mobasheri, Natalia Ustalova, Miriam Costales, all from Springer, to whom we are utmost grateful. We are very proud of the result of this collaboration and we believe that this fruitful partnership will continue for many more years to come.

TMM_CH for the year 2020 has already been announced.

December 2018   Antonia Moropoulou

Organization

General Chair

Antonia Moropoulou National Technical University of Athens, Greece

Program Committee Chairs

Manolis Korres National Technical University of Athens, Greece
Andreas Georgopoulos National Technical University of Athens, Greece
Constantine Spyrakos National Technical University of Athens, Greece
Charis Mouzakis National Technical University of Athens, Greece

Steering Committee

Antonia Moropoulou National Technical University of Athens, Greece
Manolis Korres National Technical University of Athens, Greece
Andreas Georgopoulos National Technical University of Athens, Greece
Constantine Spyrakos National Technical University of Athens, Greece
Charis Mouzakis National Technical University of Athens, Greece
Kountouri Elena Directorate of Prehistoric and Classical Antiquities, Greece
Florentis Georgios Ministry of Digital Policy, Telecommunications and Media of the Hellenic Republic, Greece
Keane Kathryn National Geographic Society, USA
Hiebert Fredrik National Geographic Society, USA
Burnham Bonnie World Monuments Fund, USA
Avramides Yiannis World Monuments Fund, USA
Korka Elena International Council of Monuments and Sites (ICOMOS), France
Minaidis Lee Organization of World Heritage Cities (OWHC), Canada
Potsiou Chryssy International Federation of Surveyors (FIG), Denmark
Rodriguez-Maribona Isabel European Construction Technology Platform (ECTP), Belgium
Forest Emmanuel European Construction Technology Platform (ECTP), Belgium
Murphy Mike European Society for Engineering Education (SEFI), Belgium
De Vries Pieter European Society for Engineering Education (SEFI), Belgium

Caristan Yves	European Council of Academies of Applied Sciences, Technologies and Engineering Euro-Case, France
Prassianakis Ioannis	Academia NDT International, Italy
Massué Jean Pierre	Council of Europe, France
D'Ayala Pier Giovanni	International Scientific Council for Island Development (INSULA), France
Ronchi Alfredo	European Commission-MEDICI Framework, Belgium
Di Giulio Roberto	University of Ferrara, Italy
Ioannides Marinos	Cyprus University of Technology, Cyprus
Gravari-Barbas Maria	Paris 1 Panthéon-Sorbonne University, France
Efesiou Irene	National Technical University of Athens, Greece
Alexopoulou Aleka	Aristotle University of Thessaloniki, Greece
Tzitzikosta Aikaterini	Hellenic National Commission for UNESCO, Greece
Achniotis Stelios	Cyprus Scientific and Technical Chamber (ETEK), Cyprus
Baruch Yuval	Israel Antiquities Authority, Israel
Groysman Alec	Association of Engineers, Architects and Graduates in Technological Sciences, Israel
Nakasis Athanasios	Hellenic Section of ICOMOS, Greece
Lianos Nikolaos	Democritus University of Thrace, Greece
Avgerinou-Kolonias Sofia	National Technical University of Athens, Greece
Hadjinicolaou Teti	International Council of Museums Hellenic National Committee, Greece
Pissaridis Chrysanthos	ICOMOS, Cyprus
Mavroeidi Maria	International Committee for the Conservation of Industrial Heritage (TICCIH), Greece
Maistrou Eleni	National Technical University of Athens, Greece
Tournikiotis Panayotis	National Technical University of Athens, Greece
Corradi Marco	University of Perugia, Italy
Forde Michael	University of Edinburgh, UK
Aggelis Dimitris	Free University of Brussels, Belgium
Anagnostopoulos Christos	University of the Aegean, Greece
Asteris Panagiotis	School of Pedagogical and Technological Education (ASPETE), Greece
Athanasiou Stefanos	University of Bern, Switzerland
Balas Kostas	Technical University of Crete, Greece
Botsaris Pantelis	Democritus University of Thrace, Greece
Boutalis Ioannis	Democritus University of Thrace, Greece
Castigloni Carlo	Polytechnic University of Milan, Italy
Chetouani Aladine	University of Orleans, France
Chiotinis Nikitas	University of West Attica, Greece
Christaras Basile	Aristotle University of Thessaloniki, Greece
Coccossis Harry	University of Thessaly, Greece

D'Ayala Dina	University College London, UK
Dimakopoulos Vassilios	University of Ioannina, Greece
Distefano Salvatore	Kazan Federal University, Russia
Erdik Mustafa	Bogazici University, Turkey
Foudos Ioannis	University of Ioannina, Greece
Ioannou Ioannis	University of Cyprus, Cyprus
Kallithrakas-Kontos Nikolaos	Technical University of Crete, Greece
Katsifarakis Konstantinos	Aristotle University of Thessaloniki, Greece
Koufopavlou Odysseas	University of Patras, Greece
Koutsoukos Petros	University of Patras, Greece
Kyriazis Dimosthenis	University of Piraeus, Greece
Kyvellou Stella	Panteion University of Social and Political Sciences, Greece
Leissner Johanna	Fraunhofer Institute, Germany
Osman Ahmad	Saarland University of Applied Sciences, Germany
Mouhli Zoubeïr	Association de Sauvegarde de la Médina de Tunis, Tunisia
Loukas Athanasios	University of Thessaly, Greece
Mataras Dimitris	University of Patras, Greece
Mavrogenes John	Australian National University, Australia
Nobilakis Ilias	University of West Attica, Greece
Paipetis Alkiviadis	University of Ioannina, Greece
Papageorgiou Angelos	University of Ioannina, Greece
Pappas Spyros	European Commission, Belgium
Philokyprou Maria	University of Cyprus, Cyprus
Prepis Alkiviades	Democritus University of Thrace, Greece
Providakis Konstantinos	Technical University of Crete, Greece
Sali-Papasali Anastasia	Ionian University, Greece
Saridakis Yiannis	Technical University of Crete, Greece
Skianis Charalambos	University of the Aegean, Greece
Skriapas Konstantinos	Network PERRAIVIA, Greece
Sotiropoulou Anastasia	School of Pedagogical and Technological Education (ASPETE), Greece
Spathis Panagiotis	Aristotle University of Thessaloniki, Greece
Stavroulakis Georgios	Technical University of Crete, Greece
Tokmakidis Konstantinos	Aristotle University of Thessaloniki, Greece
Touliatos Panagiotis	Frederick University, Cyprus
Tsatsanifos Christos	International Society for Soil Mechanics and Geotechnical Engineering, UK
Tsilaga Evagelia-Marina	University of West Attica, Greece
Tucci Grazia	University of Florence, Italy
Xanthaki-Karamanou Georgia	University of the Peloponnese, Greece
Zacharias Nikos	University of the Peloponnese, Greece
Zendri Elisabetta	Ca'Foscari University of Venice, Italy

Scientific Committee

Antonia Moropoulou	National Technical University of Athens, Greece
Manolis Korres	National Technical University of Athens, Greece
Andreas Georgopoulos	National Technical University of Athens, Greece
Constantine Spyrakos	National Technical University of Athens, Greece
Charis Mouzakis	National Technical University of Athens, Greece
Abdal Razzaq Arabiyat	Jordan Tourism Board, Jordan
Abuamoud Ismaiel Naser	University of Jordan, Jordan
Achenza Maddalena	University of Calgary, Canada
Adamakis Kostas	University of Thessaly, Greece
Aesopos Yannis	University of Patras, Greece
Agapiou Athos	Cyprus University of Technology, Cyprus
Aggelakopoulou Eleni	Acropolis Restoration Service, Greece
Andrade Carmen	International Centre for Numerical Methods in Engineering, Spain
Argyropoulou Bessie	University of West Attica, Greece
Asteris Panagiotis	School of Pedagogical and Technological Education ASPETE, Greece
Athanasiou Stefanos	University of Bern, Switzerland
Avdelidis Nikolaos	University of Thessaly, Greece
Babayan Hector	National Academy of Sciences of Armenia, Armenia
Badalian Irena	Church of the Holy Sepulchre, Israel
Bakolas Stelios	National Technical University of Athens, Greece
Benelli Carla	Church of the Holy Sepulchre, Israel
Bettega Stefano Maria	Superior Institute of Artistic Industries (ISIA), Italy
Betti Michele	University of Florence, Italy
Biscontin Guido	Ca'Foscari University of Venice, Italy
Boniface Michael	University of Southampton, UK
Boyatzis Stamatis	University of West Attica, Greece
Bounia Alexandra	University of the Aegean, Greece
Bozanis Panayiotis	University of Thessaly, Greece
Caradimas Constantine	National Technical University of Athens, Greece
Cassar JoAnn	University of Malta, Malta
Cassar May	University College London, UK
Cavaleri Liborio	University of Palermo, Italy
Chamzas Christodoulos	Democritus University of Thrace, Greece
Chiotinis Nikitas	University of West Attica, Greece
Chlouveraki Stefania	University of West Attica, Greece
Correia Marianna	School Gallaecia of Portugal, Spain
Daflou Eleni	National Technical University of Athens, Greece
De Angelis Roberta	University of Malta, Malta
Delegou Aikaterini	National Technical University of Athens, Greece

Dellatore Stefano	Polytechnic University of Milan, Italy
Demotikali Dimitra	National Technical University of Athens, Greece
Dimitrakopoulos Fotios	National and Kapodistrian University of Athens, Greece
Don Gianantonio Urbani	Studium Biblicum Franciscanum in Jerusalem - Diocese of Vicenza, Israel
Doulamis Anastasios	National Technical University of Athens, Greece
Drdácký Miloš	Academy of Sciences of the Czech Republic, Czech Republic
Dritsos Stefanos	University of Patras, Greece
Economopoulou Eleni	Aristotle University of Thessaloniki, Greece
Economou Dimitrios	University of Thessaly, Greece
Exadaktylos George	Technical University of Crete, Greece
Facorellis Yorgos	University of West Attica, Greece
Farouk Mohamed	Bibliotheca Alexandrina, Egypt
Firat Diker Hasan	Fatih Sultan Mehmet Vakıf University, Turkey
Frosina Annamaria	Centre for Social and Economic Research in Southern Italy (CRESM), Italy
Fuhrmann Constanze	Fraunhofer Institute for Computer Graphics Research IGD, Germany
Ganiatsas Vasilios	National Technical University of Athens, Greece
Gavela Stamatia	School of Pedagogical and Technological Education (ASPETE), Greece
Ghadban Shadi Sami	Birzeit University, Birzeit, Palestine
Gharbi Mohamed	Institut Supérieur des Etudes Technologiques de Bizerte, Tunisia
Gómez-Ferrer Bayo Álvaro	Valencian Institute for Conservation and Restoration of Cultural Heritage, Spain
Hamdan Osama	Church of the Holy Sepulchre, Israel
Iadanza Ernesto	University of Florence, Italy
Ioannidis Charalambos	National Technical University of Athens, Greece
Izzo Francesca Caterina	Ca'Foscari University of Venice, Italy
Kaliampakos Dimitrios	National Technical University of Athens, Greece
Kapsalis Georgios	National University of Ioannina, Greece
Karaberi Alexia	National Technical University of Athens, Greece
Karellas Sotirios	National Technical University of Athens, Greece
Karoglou Maria	National Technical University of Athens, Greece
Katsioti-Beazi Margarita	National Technical University of Athens, Greece
Kavvadas Michael	National Technical University of Athens, Greece
Kioussi Anastasia	Fund of Archaeological Proceeds, Greece
Kollias Stefanos	National Technical University of Athens, Greece
Konstanti Agoritsa	National Technical University of Athens, Greece
Konstantinides Tony	Imperial College London, UK
Kontoyannis Christos	University of Patras, Greece
Kourkoulis Stavros	National Technical University of Athens, Greece

La Grassa Alessandro	Centre for Social and Economic Research in Southern Italy (CRESM), Italy
Lambropoulos Vasileios	University of West Attica, Greece
Lambrou Evangelia	National Technical University of Athens, Greece
Lampropoulos Kyriakos	National Technical University of Athens, Greece
Lee-Thorp Julia	Oxford University, UK
Liolios Asterios	Democritus University of Thrace, Greece
Liritzis Ioannis	University of the Aegean, Greece
Lobovikov-Katz Anna	Technion Israel Institute of Technology, Israel
Lourenço Paulo	University of Minho, Portugal
Lyridis Dimitrios	National Technical University of Athens, Greece
Maietti Federica	University of Ferrara, Italy
Mamaloukos Stavros	University of Patras, Greece
Maniatakis Charilaos	National Technical University of Athens, Greece
Maravelaki Pagona-Noni	Technical University of Crete, Greece
Marinos Pavlos	National Technical University of Athens, Greece
Marinou Georgia	National Technical University of Athens, Greece
Mataras Dimitris	University of Patras, Greece
Matikas Theodoros	University of Ioannina, Greece
Mavrogenes John	Australian National University, Australia
Milani Gabriele	Polytechnic University of Milan, Italy
Miltiadou Androniki	National Technical University of Athens, Greece
Mitropoulos Theodosios	Church of the Holy Sepulchre, Israel
Mohebkhah Amin	Malayer University, Iran
Neubauer Wolfgang	Ludwig Boltzmann Institute for Archaeological Prospection and Virtual Archaeology, Austria
Nevin Saltik Emine	Middle East Technical University, Turkey
Oosterbeek Luiz	Polytechnic Institute of Tomar, Portugal
Ortiz Calderon Maria Pilar	Pablo de Olavide University, Spain
Osman Ahmad	Saarland University of Applied Sciences, Germany
Pagge Tzeni	University of Ioannina, Greece
Panagouli Olympia	University of Thessaly, Greece
Pantazis George	National Technical University of Athens, Greece
Papagianni Ioanna	Aristotle University of Thessaloniki, Greece
Papaioannou Georgios	Ionian University, Greece
Papi Emanuele	University of Siena, Italy
Pérez García Carmen	Valencian Institute for Conservation and Restoration of Cultural Heritage, Italy
Perraki Maria	National Technical University of Athens, Greece
Piaia Emanuele	University of Ferrara, Italy
Polydoras Stamatios	National Technical University of Athens, Greece
Psycharis Ioannis	National Technical University of Athens, Greece
Rajčić Vlatka	University of Zagreb, Croatia
Rydock James	Research Management Institute, Norway
Saisi Antonella Elide	Polytechnic University of Milan, Italy

Santos Pedro	Fraunhofer Institute for Computer Graphics Research IGD, Germany
Sapounakis Aristides	University of Thessaly, Greece
Sayas Ion	National Technical University of Athens, Greece
Schippers-Trifan Oana	DEMO Consultants, The Netherlands
Siffert Paul	European Materials Research Society, France
Smith Robert	Oxford University, UK
Stambolidis Nikos	University of Crete, Greece
Stavrakos Christos	University of Ioannina, Greece
Stefanidou Maria	Aristotle University of Thessaloniki, Greece
Stefanis Alexis	University of West Attica, Greece
Tavukcuoglu Ayse	Middle East Technical University, Turkey
Theoulakis Panagiotis	University of West Attica, Greece
Thomas Job	Cochin University of Science Technology, India
Triantafyllou Athanasios	University of Patras, Greece
Tsakanika Eleutheria	National Technical University of Athens, Greece
Tsatsanifos Christos	International Society for Soil Mechanics and Geotechnical Engineering, UK
Tsoukalas Lefteris	University of Thessaly, Greece
Turner Mike	Bezael Academy of Arts and Design Jerusalem, Israel
Tzannini Evgenia	National Technical University of Athens, Greece
Van Grieken René	University of Antwerp, Belgium
Van Hees Rob	Delft University of Technology, The Netherlands
Varum Humberto	University of Porto, Portugal
Varvarigou Theodora	National Technical University of Athens, Greece
Vesic Nenad	ICOMOS International Scientific Committee for Places of Religion and Ritual (PRERICO), Greece
Vintzilaiou Elissavet	National Technical University of Athens, Greece
Vlachopoulos Andreas	University of Ioannina, Greece
Vogiatzis Konstantinos	University of Thessaly, Greece
Vyzoviti Sophia	University of Thessaly, Greece
Ward-Perkins Bryan	Ertegun Graduate Scholarship Programme in Oxford, UK
Yannas Simos	Architectural Association School of Architecture, UK
Zacharias Nikos	University of Peloponnese, Greece
Zachariou-Rakanta Eleni	National Technical University of Athens, Greece
Žarnić Roko	University of Ljubljana, Slovenia
Zervakis Michael	Technical University of Crete, Greece
Zervos Spyros	University of West Attica, Greece
Zouain Georges	GAIA-heritage, Lebanon

Executive Organizing Committee of the Steering Committee

Lampropoulou Antonia	National Technical University of Athens, Greece
Psarris Dimitrios	National Technical University of Athens, Greece
Directorate of European Affairs & International Relations, Department of International Relations	Technical Chamber of Greece, Greece

Organizational Support of the Steering Committee

Kolaiti Aikaterini	National Technical University of Athens, Greece
Keramidas Vasileios	National Technical University of Athens, Greece
Kroustallaki Maria	National Technical University of Athens, Greece
Alexakis Emmanouil	National Technical University of Athens, Greece
Skoulaki Georgia	National Technical University of Athens, Greece

Technical Editing

Psarris Dimitrios	National Technical University of Athens, Greece
Kroustallaki Maria	National Technical University of Athens, Greece
Kolaiti Aikaterini	National Technical University of Athens, Greece

Kind Support

Stavridi Christina	Tsomokos S.A., Greece
Georgia Vlachou	Tsomokos S.A., Greece
George Markantonatos	Tsomokos S.A., Greece

The venues of the conference are a kind offer of:

Eugenides Foundation

Sponsors

1.Aegean Airlines

2. DALKAFOUKIOIKOS

3. Sintecno

post
scriptum

4. post scriptum

5.NEOTEK

NeoMech
www.shop3d.gr

6.NeoMech

With the kind support of:

SGT SYMEON G. TSOMOKOS S.A.

Contents – Part I

Novel Educational Approach for the Preservation of Monuments

Resilience to Climate Change and Natural Hazards

Conserving Sustainably the Materiality of Structures and Architectural Authenticity

Interdisciplinary Preservation and Management of Cultural Heritage

Contents – Part II

Cross-Discipline Earthquake Protection and Structural Assessment of Monuments

Cultural Heritage and Pilgrimage Tourism

Reuse, Circular Economy and Social Participation as a Leverage for the Sustainable Preservation and Management of Historic Cities

Inception – Inclusive Cultural Heritage in Europe through 3D Semantic Modeling

Heritage at Risk

**Advanced and Non-Destructive Techniques for Diagnosis, Design
and Monitoring**

Opening Lecture

The Project of the Rehabilitation of Holy Sepulchre's Holy Aedicule as a Pilot Multispectral, Multidimensional, Novel Approach Through Transdisciplinarity and Cooperation in the Protection of Monuments

Antonia Moropoulou[1]([✉]), Manolis Korres[2], Andreas Georgopoulos[3], Constantine Spyrakos[4], Charalambos Mouzakis[4], Kyriakos C. Lampropoulos[1], and Maria Apostolopoulou[1]

[1] School of Chemical Engineering, National Technical University of Athens, Athens, Greece
amoropul@central.ntua.gr
[2] School of Architecture, National Technical University of Athens, Athens, Greece
[3] School of Rural and Survey Engineering, National Technical University of Athens, Athens, Greece
[4] School of Civil Engineering, National Technical University of Athens, Athens, Greece

Abstract. The Holy Aedicule of the Holy Sepulchre, an emblematic monument that has survived throughout the centuries, recently underwent a major and demanding rehabilitation under the responsibility of the National Technical University of Athens Interdisciplinary Team. The requirement for reinstating structural integrity to the Holy Aedicule, for preservation of the values it represents and for achieving a sustainable rehabilitation in a demanding environment, demanded a complex framework of cooperation between different disciplines and the religious and societal carriers. Innovations were developed and successfully implemented in order to assist in achieving the project goals, marking future trends in the field of monuments protection. The rehabilitation of the Holy Aedicule, through fruitful cooperation between engineering disciplines, evolved beyond a purely engineering achievement into an emblematic example of transdisciplinarity, where Holistic Innovation and Research are intertwined with Social accessibility and the Sciences of Archaeology and Theology, creating new paths into discovering the secrets of the Tomb of Christ.

Keywords: Transdisciplinarity · Holy Aedicule · Multispectral modeling · Cultural heritage protection

---

Technical Editing: Kyriakos C. Lampropoulos and Maria Apostolopoulou.

1 Introduction

The history of the Holy Aedicule [1–5] begins with the Crucifixion and Resurrection of Jesus Christ. According to the ancient scriptures, both Golgotha and the Tomb of Christ were just outside the old walls of Jerusalem. Constantine the Great and Saint Helena, excavating the Temple of Aphrodite built two centuries earlier by the Roman Emperor Hadrian, revealed the tomb that was recognized as the Tomb of Christ, as well as the site of the Crucifixion, and built the Church of Resurrection to encompass both holy sites. The Holy Aedicule, enclosing the Tomb of Christ, evolved over the centuries, the result of successive interventions, disasters and restorations.

Its current form dates back to the restoration of 1810 by Kalfa Komnenos from Mytilene, after the devastating fire of 1808. The Holy Aedicule stands in the center of the Rotunda of the Church of Resurrection, with an area of approximately 70 m^2, and a height up to its terrace corresponding to a typical two-storey residence. A dome, supported on 12 small columns, reaches even higher. The interior of the Holy Aedicule is divided into an antechamber, at its eastern part, and a tomb chamber at the west, that corresponds to the original tomb chamber, hewn in the natural rock, a significant part of which has been preserved to this day. This is actually the only preserved part of a much larger rock mass, in which the original tomb was hewn. At its exterior, the monument adopts the architectural style of classicism, however, with elements of Central European architecture, as well as of typical Constantinople architecture of the era. Two centuries later, however, the Holy Aedicule already presented serious deformations due to earthquakes and environmental issues, and was in need for rehabilitation.

2 The Exemplary Project of the Holy Aedicule Rehabilitation

In 2015 the National Technical University of Athens (NTUA), after invitation from His All Holiness, Beatitude Patriarch of Jerusalem and All Palestine, Theophilos III, signed a programmatic agreement with the Jerusalem Patriarchate and implemented an *"Integrated Diagnostic Research Project and Strategic Planning for Materials, Interventions Conservation and Rehabilitation of the Holy Aedicule of the Church of the Holy Sepulchre in Jerusalem"* [6]. Based on the results of the NTUA interdisciplinary team study, a historic common agreement was signed in 2016, by the three Christian Communities, Guardians of the Holy Tomb, and entrusted the NTUA interdisciplinary team with the implementation of the emblematic project of *"Conservation, reinforcement and repair interventions for the rehabilitation of the Holy Aedicule"* which was successfully implemented, with the scientific supervision of the NTUA interdisciplinary team [7].

This project developed and implemented Innovation in order to face two categories of challenges:

Scientific Coordination and Supervision Challenges

- The coordination of an interdisciplinary team at the highest level of technical responsibility: This entailed the coordination of a complex project in a changing environment, in conditions of risk and within a limited timeframe, with many restrictions and prerequisites (to allow the continuous pilgrimage and religious functions, to preserve and highlight the values of the monument, to implement the project within a strict timeframe - Easter 2016 to Easter 2017). It also entailed an effective decision making process through scientific support, to manage and solve arising issues during the project

- Scientific support to the integrated governance throughout all phases of the rehabilitation project
- Scientific coordination of a diverse group comprising of: the interdisciplinary NTUA team; the scientific team supporting the scientific support of the project, comprising of over 50 members (architects, civil engineers, survey and rural engineers, chemical engineers, archaeologists, etc.); the Greek teams of conservators and restorers, comprising of 12 members
- The integrated coordination of a complex project in the sensitive environment of the open city of Jerusalem, a city of two different people and three different religions

Main Scientific and Technical Challenges of the Project

- Ensure structural integrity upon the dismantling of the stone panels and during the masonry strengthening interventions
- Design and successfully implement compatible and performing restoration materials and rehabilitation interventions, addressing thermohygric, dynamic and other environmental stresses within the alive monument of the Holy Sepulchre
- Assess the effectiveness and sustainability of the implemented rehabilitation interventions
- Document and interpret the findings and reveal and highlight the values of the monument

In order to address these challenges, NTUA formed an interdisciplinary team of scientists, restorers, conservators and masons, who all worked day and night, so that in less than nine months by March 22nd 2017, the Holy Aedicule was delivered to the three Christian communities and to the whole World.

Goals of the Project
The goals of this project were the following:

- Address the critical deformations and achieving performance, with emphasis on structural integrity, by compatible and performing materials and conservation/reinforcement/rehabilitation interventions.
- To preserve and highlight the values of the monument.
- To ensure sustainability, by monitoring, controlling and dynamically adjusting the thermohygric behavior of the Holy Aedicule.

Exemplary Project
This is an exemplary project highlighting Greek and European Innovation and Expertise in the field of Cultural Heritage protection, exploiting a multilevel integrated interdisciplinary approach, because:

- The Holy Aedicule of the Holy Sepulchre is a monument of unique value for the Christian World and not only, emblematic for the values it transmits to humanity across borders.
- The project is an achievement of Greek and European Know-how, highlighting Greek and European Know-how, Innovation and Expertise in the field of cultural heritage protection.
- Through the interest & media coverage it attracted, the project demonstrates Europe's position as a world leader in the digital transformation of the Cultural Heritage "Industry".
- The project can function as a flagship for Europe in relation to objectives for digitally-driven interdisciplinary cultural heritage protection.

Through this project it was highlighted that Research and Innovation in Cultural Heritage can become a tool for mutual understanding and coexistence, even in areas of conflict. As demonstrated in the rehabilitation of the Holy Aedicule, Cultural Heritage has a potential to tear down walls, borders and stereotypes by fostering dialogue and freedom in exchange of ideas, practices and people. The Holy Aedicule rehabilitation project, as well as the social accessibility achieved, highlights the role of Cultural Heritage protection in contributing to cultural, religious and social inclusion and openness towards a multicultural and tolerant World. The Holy Sepulchre, thus, emerged as a world centre of religious, scientific, cultural interest, a centre of pilgrimage, a centre of Research and Innovation.

3 Innovations of the Holy Aedicule Rehabilitation in Transdisciplinary Multispectral Modeling and Cooperation for the Preservation of Cultural Heritage

The interdisciplinary project of the rehabilitation of the Holy Aedicule, within a holistic approach, merges capabilities and know-how from the scientific fields of architecture, civil engineering, surveying engineering, materials science and engineering, information technology, archaeometry and archaeology. Throughout the project, innovative and high-measuring technologies were applied - with emphasis on non-destructive techniques - to fully document the Holy Aedicule, assess its state of preservation, identify the causes of the observed damages, and monitor and assess all rehabilitation interventions, while preserving and highlighting the values of the monument.

3.1 Dynamic Digital Documentation as an Integrated Core Space

Although the Church of the Holy Sepulchre is well documented (historic, architectural, geometric, structural), this has not been the case with the Holy Aedicule of the Holy Sepulchre, for a variety of reasons. To a large degree this is attributed to the belief that the current Aedicule dates entirely from the 19th century, following the devastating fire of 1808 and thus, few remnants, if any, of the previous Aedicule structures remained within it. Moreover, a sensitive status quo limited past geometric documentation efforts, resulting in fragmented documentation not readily available. Combined with limited historic documentation of the earlier forms of the Holy Aedicule prior to its 19th century reconstruction, it resulted in crucial information regarding its structural layers and their state of preservation not being available prior to the recent rehabilitation project.

This was addressed in 2015–2016 through the implementation by NTUA of an integrated diagnostic study [6]. This diagnostic research project adopted a digital driven approach for Cultural Heritage preservation using interdisciplinary scientific tools. Specifically, one of the main challenges was to create an accurate and functional three-dimensional reconstruction of the monument which would provide an integrated core space, enabling and optimizing:

- Accurate and detailed (3D) design of rehabilitation interventions
- Estimation of required quantities of restoration materials
- Dynamic environment for the optimization and redesign processes throughout all stages of the rehabilitation project
- Assessment of the rehabilitation, with reference to the main goals

Geometric Documentation. The integrated geometric documentation of built cultural heritage is a complex dynamic process involving acquisition, processing, presentation and recording of the necessary data for the determination of the position and the actual existing form, shape and size of an asset in the three-dimensional space at a particular given moment in time. In this framework, a detailed geometric documentation of the Holy Aedicule was accomplished [8], prior to the initiation of the works, but also during and after the completion of the rehabilitation interventions. It involved the orthogonal projection of a carefully selected set of points on horizontal or vertical planes, in order to effectively record all geometric properties of the monument. The selection of these points was not arbitrary, but was based on historic and architectural analyses to identify parts of the Holy Aedicule that were critical for the design of the necessary interventions, and on specifications set by engineers and scientists of the project regarding the required geometric documentation products.

These products consisted high resolution three-dimensional models (Fig. 1) of the Holy Aedicule (both for its exterior and interior) and specialized highly accurate geodetic measurements. These allowed the creation of conventional 2D base material and of sections at specific positions and enabled documentation of the deformations and deviations of the structure. To achieve these, the most contemporary geomatics techniques and specialized instrumentation were applied. Specifically, an automated 3D imaging methodology based on high resolution digital images, terrestrial laser scanning and high-accuracy geodetic measurements were implemented. These data were georeferenced to an already existing local plane projection reference system from previous work of NTUA.

Digital Documentation. Digital documentation of the Holy Aedicule included not only the products of geomatics techniques and the data derived from geodetic instrumentation, but also digitalized information regarding the characteristics of the different materials comprising the structure, NDT prospection results and finite elements models. Indeed, the enrichment of the created geometric digital documentation products, could not be based only on geomatics techniques and geodetic instrumentation, due to their inherent limitation in prospecting internal layers of the surveyed structure. It should be emphasized that a limited knowledge of the internal structure of a built cultural heritage asset, as compared to contemporary structures, is very crucial, since it does not allow effective and inclusive assessment of its state of preservation nor the design of appropriate restoration materials and interventions. Regardless of how accurate – geometrically - a three-dimensional model of a historic structure may be, it is of limited use if its building materials are not known in detail, or information about the structural layers and the state of preservation of the structure are not provided. Thus, the digital documentation applied was a necessary innovation, providing the project with a useful tool, with the capability to "in-practice" assist and optimize the rehabilitation process.

Fig. 1. Three-dimensional models of the Holy Aedicule. Left: Initial state (January 2016); Right: After the implementation of the rehabilitation (March 2017) [6–8].

In the past, more than often, accurate geometric documentation was seen merely as a means to support the historic documentation of a CH asset, or to simply document its current architectural and geometric form for future reference. In this emblematic project, however, geometric documentation was seen as a vital element of an integrated, functional and dynamic digital documentation that served many needs of the rehabilitation project. Therefore, during the diagnostic study, geometric documentation was enhanced by innovative prospection tools which fused historic and architectural analyses with findings from non-destructive techniques.

Innovative Prospection. Ground penetrating radar (GPR) was a non-destructive techniques utilized in this project that enabled the prospection of the structure. Specifically, a GPR survey was implemented over the exterior and interior surfaces of the Holy Aedicule [9], in conjunction with historic and architectural analyses. It revealed remnants of the original monolithic Aedicule, embedded within the current Aedicule structure, and elucidated the correlation of the current Aedicule structure with its previous construction phases (Fig. 2). This was an important finding, of great historic significance, since the presence of remnants of past construction phases of the Holy Aedicule was only theorized prior to the NTUA diagnostic study. Moreover, the internal structural layers were revealed and accurately documented, demonstrating that the Holy Aedicule was a rather complex structure, containing materials and building elements from various construction phases and of varying states of preservation. This was critical information, as it defined "boundaries" both for the preservation of values of the Holy Aedicule and for the design of compatible and effective materials and restoration interventions, necessitating a rehabilitation approach addressing a wide range of requirements.

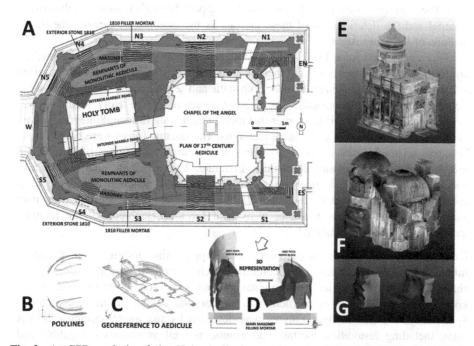

Fig. 2. A: GPR analysis of the Holy Aedicule revealing the internal layers and previous construction phases [9]. B–D: Conversion of GPR information into 3D representation of the remnants of the monolithic Aedicule [10]. E–F: 3D representation of internal structure and the georeferenced remnants of the monolithic Aedicule [6–8].

One of the innovations of this project was the mutual utilization of geometric documentation with non-destructive prospection and historic and architectural analyses. The non-destructive prospection findings were georeferenced to the local plane projection reference system, transforming the three-dimensional geometric models into truly three-dimensional models of the Holy Aedicule containing additional internal layering information.

Interrelated Geometric, Architectural, Materials and Structural Documentation. Geometric documentation and information from the prospection of the internal structure of the Aedicule were further correlated with findings from the assessment of the exterior and interior surfaces by other non-destructive techniques such as infrared thermography and portable digital microscopy, providing additional materials-related information to the surface texture of the aforementioned three-dimensional models. In parallel, the NTUA interdisciplinary team also tested core samples from the Aedicule to characterize the building materials—the mortars and stones—and to detect any evidence of decay. This was accomplished through optical and digital microscopy, scanning electron microscopy coupled with microanalysis, thermal analysis, x-ray diffraction, mercury intrusion porosimetry, total soluble salts tests, and mechanical tests. These analyses identified the mortars that support the site's masonry as the critical material and the deterioration of these mortars as the main factor causing the deformations of the Holy Aedicule.

This augmented knowledge of both the geometry of the Aedicule, its internal layering, its building materials and their state of preservation, facilitated the optimized assessment of the Holy Aedicule under both static and seismic loads via elaborate finite element modeling and analysis. The bearing structure of the Holy Aedicule was assessed in terms of the seismic forces that might threaten its structural integrity as well as the static loads from its constituent materials. The seismic forces were based on the historical seismicity of Jerusalem, according to the current provisions of Eurocode 8 and the available, international scientific literature. This structural analysis (Fig. 3), thus, identified the critical areas of the Holy Aedicule that required remedying measures and were the focus of the rehabilitation interventions. In addition, this joint analysis provided reliable data that were used for the calculation of the required quantities of restoration materials.

Knowledge Based Digital Infrastructure to Support the Design of the Rehabilitation. Based on the above, the combined architectural, historical and materials characterization, geometrical documentation and structural assessment acted as knowledge based digital infrastructure to support the design of the rehabilitation. Specifically, a series of interventions was designed by the interdisciplinary NTUA team, including restoration mortars and grouts, as well as various compatible and performing reinforcement systems. The use of the proposed materials and interventions was validated prior to the implementation of the works, by incorporating all proposed materials and interventions into the aforementioned optimized finite element model of the Aedicule. When the modified finite element model was assessed under both static and seismic loads, it became clear to the team that the proposed rehabilitation would provide adequate strength and reinstate the monument's structural integrity.

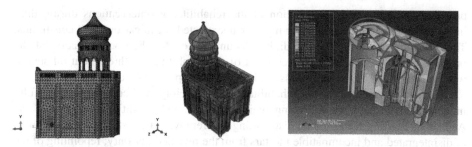

Fig. 3. Left: Finite element model used for structural analysis. Middle: Optimization of the finite model by introduction of accurate layering and materials properties. Right: Response of the model of the Holy Aedicule at its initial state, identifying critical areas [6, 7, 11].

3.2 Implementation of an Integrated Methodology for the Rehabilitation of the Holy Aedicule Through an Iterative Process of the Main Stages of Works

As previously described, the rehabilitation of the Holy Aedicule was a complex and sensitive project due to the importance of the monument, its state of preservation and the required interventions, and the site's historical and religious significance—in particular, the need to preserve the remnants of the original monolithic Aedicule including the Holy Rock embedded within the current structure. Thus, the rehabilitation of this monument, a monument of the utmost importance to Christianity, demanded an interdisciplinary cooperation and the use of our hands for its completion, validating the ancient Greek saying *"Σύν Αθηνά και χείρα κίνει"* (translated: along with your prayer to the Goddess Athena, also move your hands), i.e. *along with your prayers act as well.* These issues necessitated the organization of the worksite and the scheduling of the timetable in such a way as to accommodate various restrictions and engineering challenges. One such challenge was the requirement for the site to remain accessible to pilgrims and those engaged in the performance of religious functions. In combination with safety requirements, these requirements necessitated most of the work to be accomplished at night. Moreover, a strict deadline was set to complete the rehabilitation work within 12 months after the signature of the Common Agreement to ensure that Easter celebrations could be held at the rehabilitated Holy Aedicule; indeed the rehabilitated Aedicule was delivered on schedule, on March 22nd 2017.

These demanding requirements, without, obviously, any compromise on the quality and effectiveness of the rehabilitation materials and interventions, could only be fulfilled through an integrated methodology for the rehabilitation of the Holy Aedicule, the main stages of which would be implemented through an iterative process. All stages were carefully planned to ensure that works could be performed in parallel at different areas of the Aedicule or at laboratories set up within the Church of the Holy Sepulchre (Cleaning and Protection Laboratory, Interdisciplinary Monitoring and Documentation Laboratory). In addition, it was ensured that critical stages of the works would be completed on-time before the initiation of other stages, while maximizing utilization of human and technical resources. This addressed all project requirements, but also provided flexibility and optimization of the rehabilitation interventions during their implementation, based on new findings or unexpected technical or other constraints. The innovation of using the 3D digital documentation for the documentation of the progress of the works and the assessment of the rehabilitation, throughout all stages, was critical in this iterative process, as it provided important feed-back.

Within this framework, the rehabilitation of the Holy Aedicule was successfully implemented through the following main stages [7, 11]: Additional support to the Aedicule's existing metal frame; dismantling and removal of the stone panels; removal of disintegrated and incompatible mortars from the revealed masonry; repointing of the masonry; repair and partial reconstruction of part of the masonry; injection of grouts up to 3 m; resetting and anchoring of exterior columns; reassembly of stone panels; resetting and anchoring of stone column railing; grouting of top zone and terrace; anchoring of interior marbles; conservation interventions on the Onion Dome, the Dome of the Chapel of the Angel and the Dome of the Burial Chamber; final mortar application, pointing and finishing; cleaning and protection of interior and exterior architectural surfaces and conservation of decorative elements.

3.3 Scientific Support to the Integrated Governance of the Project

The Holy Aedicule rehabilitation project emerged as an emblematic project of innovative applications, research and education [7], in addition to its religious, cultural and political significance for the three religions and the two peoples of Jerusalem. This was successfully accomplished through the integrated governance (Fig. 4) with the responsibility of the three Christian communities and the scientific supervision and monitoring of the NTUA interdisciplinary team.

The key points in this effective cooperation of Science and Religion are the intersection of the different disciplines involved in the project and the multi-interface interactions between the scientific teams, the workers, the Common Technical Bureau of the Church of the Holy Sepulchre and the three Christian communities, all under the authority of His Beatitude, Theophilos III. The Common Agreement respecting the Status Quo of the three Christian communities, Guardians of the Holy Tomb, provided the statutory framework for the implementation of the project. Within this framework the project was initiated, became possible and was successfully implemented under an integrated governance (Fig. 4), comprising of the Project Owners Committee, the Steering Committee, adjoining the scientific and the technical directors of the project. A key decision making process was the Construction Site Management Meeting. The meetings took place on average once every two weeks and resolved all pending issues, or referred them to the Scientific Committee, thus presenting solutions in real time.

The three Christian communities trusted Science at an unprecedented level. This trust evolved into the centripetal force that bonded a diverse team of more than seventy, scientists, workers, conservators, researchers, practitioners, heritage professionals, policymakers, all cooperating to implement the challenging rehabilitation of the Holy Aedicule. In this effort, the scientific support to decision making was at the core of this integrated governance and an important achievement of the NTUA interdisciplinary team. This transdisciplinary context [12] had an analogous impact on academic production itself, shifting the academic and technical communities from the technical/

scientific rationality towards an epistemic culture, where no discipline prevailed over the others but instead all cooperated with and benefited from the achievements of each discipline.

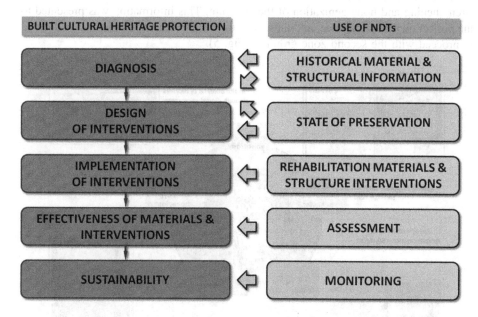

Fig. 4. Emerged methodological approach for the use of non-destructive techniques for monitoring and assessing the preservation state of built heritage.

During the implementation of the project, the innovation of using non-destructive techniques for monitoring and assessment of materials and interventions was very effective and crucial in the decision making process of the integrated governance. One example is the use of infrared thermography in the survey of the façade areas after the dismantling of the stone panels and during the works. The interdisciplinary NTUA diagnostic study [6] identified the water infiltration as the main decay factor, inflicting significant damage to the historic mortars (filling and masonry) of the Holy Aedicule. This was taken into account in the design of the restoration mortars used in the project. However, at the stage of opened facades, before and after the historic filling mortar had been removed, infrared thermography detected persistent phenomena of rising damp on the main masonry, as well the presence of incompatible materials. After repointing of the masonry, NDTs validated in-situ and in-real time the compatibility of the restoration materials used. The rising damp phenomenon, however, persisted even after repointing of the masonry, although in a less intense manner due to the use of appropriate restoration materials [13]. The effect of rising damp from the underground was highlighted as an environmental load with a negative impact, which threatened the long-term sustainability of the interventions and the Holy Aedicule in general. This initiated a comprehensive three-dimensional non-destructive prospection of the underground environment of the Holy Aedicule.

Another example is the identification of "higher temperature" areas, at the elevated parts of the Aedicule, through infrared survey [14]. These observations, indicative of internal voids and cracks within the upper parts of the Aedicule, confirmed that grouting of the upper zone of the Aedicule (heights > 3 m) was necessary for the strengthening and homogenization of the structure. This information was presented to the project owners committee, and thus, with their unanimous approval, it was decided to proceed with the second zone grouting (Fig. 5).

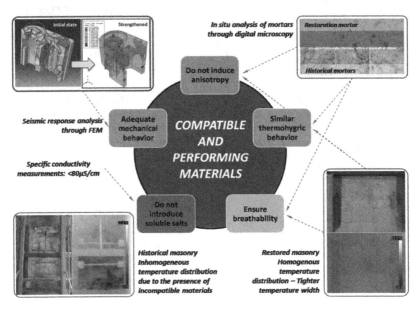

Fig. 5. Innovation in selecting and designing compatible and effective materials for the rehabilitation of the Holy Aedicule.

The use of non-destructive testing during the works was also critical in providing scientific support in the decision to open the Holy Tomb, during the grouting process. Specifically, an array of NDTs (GPR, videoendoscopy, ultrasonic tomography) was used to prospect the interior of the Tomb, in order to reveal its internal morphology, since during the diagnostic study it was evident that the remnants of the original monolithic Aedicule embedded within the current structure presented a complex configuration that could pose a threat for grout flow into the interior of the Tomb, thus covering the original burial rock surface. NDTs documented a filling layer of loose material between the top marble facing of the tomb and the burial rock surface into which grout material could protrude through cracks from the adjacent masonry and rock remnants. The results were immediately presented to the Project Owners committee and it was decided that the Holy Tomb would open and remain so throughout the 60-h grouting process in order to control and prevent grout flow into this sensitive area, thus, ensuring the preservation of the values of the Tomb of Christ.

4 Innovative Methodology to Ensure Structural Integrity, Achieving the Project's Goals

As described above, one of the main goals of the project was to ensure the structural integrity of the Holy Aedicule. Based on the finding of the diagnostic study [6] an array of interventions were designed and implemented to achieve this goal. The rehabilitation process [7, 11] began with the dismantling and removal of the stone panels of selected areas from the Aedicule's facades, since deformation was only observed at the lower parts of the structure, relating to the causes of deformation (humidity and deteriorated mortars). The disintegrated filling mortar was removed and the inner-rubble-type masonry of the Aedicule was repointed with a compatible and performing restoration mortar [14], as proposed by the interdisciplinary NTUA study [6]. Parts of the masonry around the burial chamber that presented a poor state of preservation were removed, using a specially design masonry retaining system [7, 11] to avoid collapse of the original adjacent parts of the Aedicule. A new masonry using compatible stone blocks was reconstructed, up to a height of 2 m and to a depth that followed the curvature of the embedded remnants of the original monolithic Aedicule (Holy Rock). The reconstructed masonry served to protect the embedded original parts of the monument, contributing to a large degree in the structural integrity of this area of the Aedicule.

The exterior and interior dislocated columns were then reset and diminution of the deformations was documented throughout the duration of the project (Fig. 6A) [7]. Following the resetting of columns, the homogenization of the Aedicule's masonry and the consolidation of the embedded remnants of the original structure were achieved by grouting. The grout injection tube matrix [7] was designed appropriately based on digital geometric data (Fig. 6B) [8] and on digitalize materials data and non-destructive prospection by GPR (Fig. 6C) [7, 9] of the retrofitted structure of the Aedicule. Grouting of the structure was accomplished in three zones: from floor level up to 1.5 m; the second zone from 1.5 m to 3.0 m (apex of the façades arches); upper parts of the facades and from the roof of the structure. NDTs were utilized throughout the grouting process to monitor, document and assess the effectiveness of the process.

After the completion of the grouting process, the reset columns were anchored to the strengthened masonry with the use of specially ribbed Grade 2 titanium bars. A titanium mesh was installed over the strengthened masonry to enhance the bond between the successive concrete layers. The external stone panels were then gradually reassembled in their reset position. Each stone slab was also anchored to the strengthened masonry with titanium bars and secured with adjacent units with titanium connectors. After the reassembly of a stone slab zone, a compatible and performing restoration concrete was added in the layer between the external stone slabs and the reinforced masonry [13]. The procedure continued until the façade opening was closed. It should be noted that the design of the anchoring was based on digital documentation (architectural, geometric, materials) in cooperation with the results of experimental tests and Finite-Element-Modeling (which validated the response of the anchored stone slabs and the retrofitted structure under seismic loads).

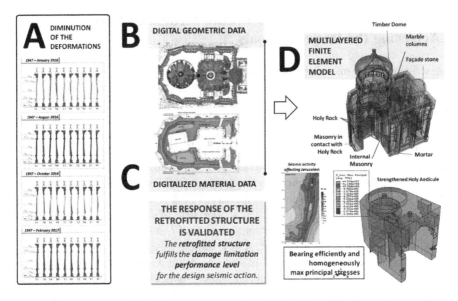

Fig. 6. Innovative methodology to ensure structural integrity, achieving project goals.

The response of the retrofitted structure was then validated, prior to the removal of the British frame surrounding the Aedicule and the additional metal support (diagonal and vertical steel beams) installed during the works [7, 11]. The assessment of the rehabilitation interventions was accomplished through an integrated interdisciplinary approach (Fig. 6). As described above, digital documentation allowed the creation of a three-dimensional model of the Aedicule using high-resolution digital images, terrestrial laser scanning, and highly accurate geodetic measurements to document the Aedicule prior to the initiation of the works. Additional measurements were performed during the works, and after the completion of the strengthening measures, to document the Aedicule after the diminution of the deformations (Fig. 6B). Digitalized materials data, regarding the applied materials (restoration mortars, concrete and stones) and implemented interventions (restoration masonry, titanium anchors network) (Fig. 6C) were then introduced, in conjunction with the updated geometric data into a modified multilayered finite element model (Fig. 6D) to assess the retrofitted bearing structure; the model confirmed that structural integrity was indeed achieved. The iron frame installed by the British was then successfully removed, "freeing" the monument after almost seventy years.

5 Innovative Methodology to Reveal and Interpret Findings to Preserve and Highlight the Values of the Monument

The rehabilitation of the Holy Aedicule, was not only a project aiming to ensure the structural integrity of the structure, but a project which aimed to achieve this goal while preserving, revealing and highlighting the values of this important monument.

The rehabilitation project provided the unique opportunity to elucidate the construction phases of the monument and reveal the history of the Tomb and the Aedicule surrounding it. Science was instrumental in this direction, providing concrete scientific data, to serve as a tool in the interpretation of historical references and architectural and archaeological theories. Thus, the scientific support provided by the NTUA interdisciplinary team exceeded the needs of the project in regards to the technical issues, and evolved into providing invaluable data of archaeological importance.

The opening of the Tomb, after almost five centuries, provided a unique opportunity to shed light onto the evolution of the burial monument, as, for the first time, at least in the recent history of the monument, the hidden internal layers of the Tomb were brought to light and studied implementing state-of-the-art equipment and techniques, revealing exciting information, until today unknown. Until recently, no indisputable data were available that could verify whether this Tomb was related to the original Tomb of Christ. Furthermore, the accounts regarding the morphology of the burial monument throughout the centuries, mainly scarce pilgrim accounts and artistic depictions, were often contradicting, thus allowing various interpretations. During the project, knowledge of the Tomb morphology was essential for its effective rehabilitation and preservation of values. Therefore, although the Tomb was opened to serve the project, the interpretation and use of innovative techniques exceeded purely technical goals and revealed the "secrets" of the Tomb.

In this context, OSL dating [15] in combination with material characterization techniques, were performed on historical materials from areas of the Aedicule of great architectural and archaeological interest in order to clarify the construction evolution of the monument. This approach revealed the covering of the original rock surface with marble cladding at the time of Constantine the Great. This was of great importance, since for the first time it was verified that the Tomb worshipped today as the Tomb of Christ is in fact the same one identified by Constantine and Helena as the Tomb of Christ. Results from OSL dating verified the presence of different construction phases within the Aedicule, dating from Constantine the Great, to the Byzantine reconstruction, the Renaissance period and up to the 19th century restoration, confirming findings revealed by other NDTs [9]. This proves that the Holy Aedicule was never reconstructed from its foundations, as many have claimed in the past (Bonifaccio da Ragusa-1555, Komnenos-1810), but instead evolved through the centuries by embedding remnants of the previous constructions phases [9, 15]. This approach is transdisciplinarity in practice.

The prospection of the underground environment of the Church of the Holy Sepulchre with innovative NDTs, revealed that the Holy Aedicule is located at the center and above a cluster of underground cisterns and natural and manmade underground voids and spaces, some of which may be correlated with tombs from various historical periods, completed the documentation of the underground area and provided new semantics, opening new chapters in archaeological research [7].

6 Innovative Methodology to Ensure Sustainability

6.1 Risk Assessment

The intense rising damp affecting the structure, as detected through NDTs during the rehabilitation project, initiated a comprehensive three-dimensional non-destructive prospection of the underground environment of the Holy Aedicule with ground penetrating radar, electric resistivity tomography, electromagnetic probe system, robotic cameras and geometric documentation systems to identify the underground risks. Findings from these analyses formed the basis for the proposal of underground interventions in the Rotunda area for the underpinning, reinforcement, water and humidity control of the Holy Aedicule [7, 11]. In addition, the Church of the Holy Sepulchre is located in an area which has experienced, in the past, significant seismic activity, thus, the earthquake response of the Church complex, and the Holy Aedicule in particular, are of importance.

6.2 Innovative Methodology to Ensure Sustainability

Underground Interventions to Reverse the Risk to Sustainability
The interdisciplinary NTUA diagnostic study [6] revealed the critical role of rising damp and moisture transfer within and around the Aedicule for the Aedicule's state of preservation prior to the initiation of the works. A comprehensive three-dimensional non-destructive prospection and geometric documentation of the underground environment of the Holy Aedicule, the Rotunda area and the Church of the Holy Sepulchre, revealed a complex morphology with an array of underground features, either natural or man-made - often interconnected with each other – and water-drainage and sewage networks in a poor state of preservation.

This complex underground environment affects significantly the foundations of the Aedicule and poses a risk to its sustainability. As part of the rehabilitation project, an initial study was implemented [7], which revealed that the foundation system of the Aedicule is at risk, since the surface of the natural rock of the quarry area below and around the Aedicule is strongly irregular, and the monument is founded on a degraded thin bedding layer of low silicate mortar or rubble of older structures that are possibly not sufficiently consolidated. A numerical analysis of the Aedicule's foundation system was performed, and revealed a differential settlement of the structure because of a long-term reduction in the rubble's stiffness.

The NTUA interdisciplinary team proposed a novel underground intervention that involves the excavation of the natural rock and the construction of a peripheral drainage and ventilation gallery, in combination with grouting of the rubble and/or removing it and replacing it with compatible and performing mortar and stonework [7]. This gallery will also allow drainage of the rising underground water through a built-in system of open canals and pipes in a space between the gallery and the natural rock. A ventilating system will be integrated within this gallery, for an optimal aeration and humidity regulation of the perimeter corridor, the cistern, and the adjacent earlier excavation site south of the Aedicule, whereas, a remote-controlled multi-sensor system will monitor humidity uptake in the Aedicule's vicinity. At the surrounding area, a new functional sewage and rainwater network will be constructed within the perimeter of the Rotunda and in the Church of the Holy Sepulchre to replace the complex and ineffective existing network.

Monitoring the Response of the Aedicule

The rehabilitation of the Holy Aedicule, actively improved its sustainability through the implemented interventions. The proposed underground interventions will further improve the Aedicule's sustainability against the prevailing risks. However, it is important to monitor critical risks, within an integrated cultural heritage management approach, in order to assess the effect of the total environment (environmental factors and natural/manmade threats) on the Aedicule structure.

As part of the rehabilitation project, a climatic monitoring and control system was designed and installed. Specifically, a wireless sensor system that monitors the microclimatic conditions (air temperature and relative humidity) was installed within the Tomb chamber and the Chapel of the Angel [7]. A ventilation and dehumidification system was installed at the roof of the aedicule, which dynamically adjusts its function levels through data acquired from the monitoring system. An additional wired sensor system for measuring materials' moisture inside the Aedicule's masonry, at selected areas of different height and orientation, was installed, since rising damp is a critical risk and the thermo-hygric response of the monument is important for its sustainability.

Structural health monitoring was the other component of the monitoring of the response of the Holy Aedicule. It regards earthquake monitoring and spatial monitoring. Specifically, a monitoring system for evaluating the structural response of the monument under seismic loads was installed [7]. The aim of this seismic response monitoring system is to record and assess the dynamic behavior of the Holy Aedicule, through the use of digital triaxial accelerometers installed at the ground surface (southeastern corner) and at the roof of the Aedicule, in order to define its translational and rotational response. This system can record accelerations continuously, offering remote data and real-time data acquisition. Spatial monitoring of the Aedicule regards static monitoring, where measurements are conducted in wide-time intervals (weekly, monthly, annual) followed by total post-processing, and dynamic monitoring, implemented at close-time intervals (hourly, etc.) followed by simpler real-time data processing. A 3D network consisting of 13 points was established for the determination of the Aedicule's displacements and the high accuracy measurements. An advanced total station (Trimble S9) was setup that provided static and dynamic monitoring of the Holy Aedicule's displacements, during the rehabilitation works and after the completion of the works. The internet access of the workstation (Total station, laptop) allows transfer of data and results to authorized users worldwide, for further processing and documentation.

7 Innovative Multilayered Data Management

As described above, an important basis for transdisciplinarity is the capability to fuse data from dynamic digital documentation within an integrated core space. An integrated Information System Platform, with data integration in five "dimensions", is developed in the framework of the Holy Aedicule's rehabilitation project. This multilayered data management system establishes and develops transdisciplinarity among relevant scientific and engineering fields, digital and non-digital layers of information and non-destructive and analytical information creation technologies. It utilizes the information created through the rehabilitation project, setting interrelationships and providing a digital infrastructure where information can be assigned spatially for further correlation with others. This system offers modular functionalities, is extendable, is applicable to other cases and is transferable. Such integration establishes transdisciplinarity. This management system can offer geo-referenced data optimal archiving, retrieval and cross-discipline data analysis for automated identification of threats and alarm triggering. This is accomplished through a spatial database formulated by the dynamic digital documentation of the Holy Aedicule (historic, architectural, geometric documentation, materials layers, rehabilitation phases, monitoring of the structure and its environment, etc.), and encompasses multi-layered management of information (NDTs, analytical testing, spatial and time-data registration, historic documentation, etc.) and big data integration. Thus, various researchers from different disciplines are able to register their data directly to the location and/or area where the measurements were made. This platform is the cornerstone for data management, knowledge acquisition and information sharing and a significant step towards a holistic digital driven infrastructure (Fig. 7).

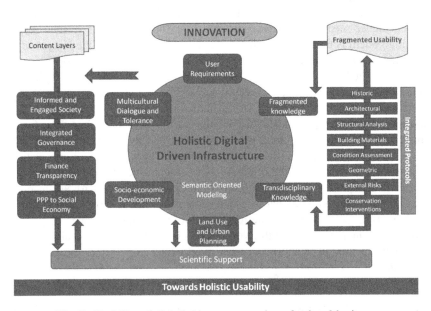

Fig. 7. Usability of digital driven preservation of cultural heritage.

8 Recognition of the Holistic Innovation Inferred by the Holy Aedicule Project to the Protection of Monuments

The research developed for the rehabilitation of the Holy Aedicule by the NTUA interdisciplinary team was implemented and completed with Innovation, marking future trends in the protection of monument, as declared in Brussels on 14.03.2017 by the representatives of the DG: Research & Development, Education & Culture, ICT/Connect. In addition, the digital aspects and general transdisciplinary methodology implemented during and after the project were presented in *"Digital solutions and multi layer data innovations in the rehabilitation of the Holy Aedicule of the Holy Sepulchre in Jerusalem"*, in Innovation and Cultural Heritage High-level Horizon 2020 conference of "The European Year of Cultural Heritage", 20.03.2018, Royal Museum of Arts and History, Brussels.

9 Transdisciplinarity

Transdisciplinarity was the research strategy implemented during and after this project to achieve the goals of this effort. This strategy covers disciplines beyond the field of engineering. The opening to Transdisciplinarity with the scientific "world" of History, Archaeology and Theology was established [16–19] through public dialogue with the Israel Antiquities Authority; through exchange of ideas with scholars at the International Conference on Art and Archaeology and the School of Archaeology at the University of Oxford and at the Institut für Christkatholische Theologie at the University of Bern, where new cooperations were established; through the participation of archaeologists from the Hellenic Ministry of Culture and Sports; through cooperation with architects of the Common Technical Bureau of the Church of the Holy Sepulchre. The fruitful cooperation of all disciplines, established Transdisciplinarity, providing an added value, both to the project of the Holy Aedicule's rehabilitation and to the post-project research, aiming and succeeding in enhancing the values of the monument and providing new insight into the history of this most important site of the Christian World and not only.

It should be noted that, although the rehabilitation project is completed, the Exhibitions are ongoing and Research continues. The social accessibility and the impact of Narration provided by the scientific data, together with Theology and Social Practice of Faith, lead to a new "Chapter" of significant interest for theologists, sociologists, in cooperation with Christian Communities.

10 Social Accessibility

The cultural impact of the project is increasing. The rehabilitation of the Holy Aedicule represents the values of intercultural and interreligious communication and the virtue of mutual understanding, between countries and people of different religious and theological backgrounds. This was expressed during the project through the fruitful coexistence and cooperation of various religious and political authorities, public and social institutions, pilgrims from all over the world and of different faiths.

A strategic alliance was formulated with the three Christian leaders and with National Geographic, which was instrumental in conveying news regarding the rehabilitation of the Holy Aedicule to the world. In accordance to the common agreement, Prof. A. Moropoulou, the chief scientific supervisor of the project was responsible for communicating the results and findings of the project. Thus, the National Geographic Society and the Media were present and disseminated to the rest of the world critical stages of the project, in conjunction with the Steering Committee Meetings.

When on October 26 2016 the burial monument was opened and its layers revealed, the broadcast by National Geographic brought two billion people, in spirit, to the burial monument to kneel in front of the Tomb of Christ. The presence of media, thus, provided the seeds for the project to develop a broader cultural, religious and educational context. Although the Holy Sepulchre and the values it represents are undeniably of utmost importance to the world of Christianity and not only, the project, though the media exposure, further enhanced the monument's accessibility to a wider audience worldwide.

In this framework, the organization of a series of exhibitions, with emphasis on new technologies and digital experiences, aims to further augment the social accessibility of the Holy Tomb and the values it represents:

The exhibition *"Tomb of Christ: The Church of the Holy Sepulchre Experience"* is organized at the National Geographic Museum, Washington DC, USA, by National Geographic, between November 15, 2017 and January 6, 2019. This exhibition offers an immersive 3-D experience, virtually transporting visitors to the City of Jerusalem and to the Church of the Holy Sepulchre, to discover its fascinating history and presents the restoration of the Holy Aedicule by the NTUA interdisciplinary team.

A digital exhibition of advanced technology *"Tomb of Christ: The monument and the project"* is organized at the Byzantine and Christian Museum, Athens, Greece, between May 21, 2018 and January 31, 2019, with the cooperation of the Ministry of Culture and Sports of the Hellenic Republic with the National Technical University of Athens and the Ministry of Digital Policy, Telecommunications and Media of the Hellenic Republic, is held under the auspices of HE the President of the Hellenic Republic, Mr. Prokopios Pavlopoulos, with the support of the Hellenic Parliament the Byzantine and Christian Museum, National Geographic, and with the blessings of His Beatitude, Archbishop Ieronymos II of Athens and All Greece, and of the Holy Synod of the Church of Greece. The data presented in the exhibition "Tomb of Christ: The monument and the project" are provided by the National Technical University of Athens Interdisciplinary Team, acquired during the project and further processed and analyzed, and data that have been bestowed to the NTUA Interdisciplinary Team by National Geographic, deriving from the exhibition 'Tomb of Christ' in the NG Museum, Washington DC, where the NTUA Interdisciplinary Team provided project data. In the first part of the exhibition, data derived from the virtual Exhibition "Tomb of Christ" of the National Geographic Museum in Washington D.C. is presented, bestowed to the NTUA Interdisciplinary Team. In the second part, data provided by the NTUA Interdisciplinary Team are presented, acquired during the project and further processed and analyzed, especially, findings that arose after the opening of the Holy Tomb for the first time after five centuries.

Other museums around the world have already expressed their interest to host the exhibition. The kind of experience offered by these exhibitions provides the opportunity to those that have not or cannot make the pilgrimage to Jerusalem and the Church of the Holy Sepulchre, to get a glimpse of the Holy Sepulchre and its values, and to recognize the importance of its rehabilitation and the cooperation behind this accomplishment.

In the Virtual Multimodal Museum - VIMM Coordination meeting, Thematic Area 1, Berlin, 12.04.2018 at the Ethnologisches Museum, Berlin, the methodology of the exhibition, its products and its values were discussed and it was recognized that the appropriate and focused application of augmented technology - as employed – is capable of stimulating emotions to the exhibition visitors, in contrast to typical digital exhibitions of purely "cold" representations. Therefore, this exhibition is an exemplary case amongst digital exhibitions as the visitors can "travel" in spirit, with the aid of virtual reality.

A series of lectures and talks regarding the project and its findings, have been presented to the Christian Communities in Greece, Italy [20] and around the world. Such presentations had a high attendance and were widely received. In addition, scientific presentations are continuous and ongoing at conferences and lectures across the world, raising awareness on the issue of transdisciplinarity.

The Holy Aedicule was delivered on March 22nd 2017 to the three Christian Communities, Guardians of the Holy Tomb, who had entrusted the NTUA interdisciplinary team with its rehabilitation. At the completion ceremony, the united Christian World embraced the messages of the Holy Tomb in an ecumenical celebration, where important representatives of all communities of the Christian world as well as representatives from authorities and governments conveyed a message of unity, mutual understanding and cooperation.

The dynamics developed through and beyond the project show that the Narration, Faith and the faithful rehabilitation of the Holy Aedicule are capable to raise awareness of all Humanity concerning united Christianity, intercultural dialogue between the different religious beliefs, between believers and non-believers, between the two people of Jerusalem and the three religions, between the people of the conflagrated Middle East, the Ecumene, with contemporary Greece at the center.

11 Conclusions

The rehabilitation of the Holy Aedicule of the Holy Sepulchre in Jerusalem was an emblematic project addressing scientific coordination and supervision challenges as well as challenges related to scientific and technical issues. The goals of the project focused on addressing the critical deformations and achieving structural integrity through the design and use of compatible restoration materials and interventions; on preserving the values of the monument; on ensuring sustainability. All these goals were achieved through an array of innovations developed. Specifically, innovations in transdisciplinary multispectral modeling and cooperation for the preservation of cultural heritage enabled the evolution of dynamic digital documentation into an integrated core space. The implementation of an integrated methodology for the rehabilitation through an iterative process of the main stages of the works, engaged 3D digital documentation providing to the project flexibility and adaptability, and provided scientific support to decision making by the integrated governance of the project. The adoption on an innovative multilayered data management approach, the introduction of transdisciplinarity at all levels of the project, as well as the enhanced social accessibility by the exhibitions and all the dynamics created through and beyond the project, are innovative contributions of the Holy Aedicule of the Holy Sepulchre in Jerusalem and proved that the transdisciplinary multispectral modeling and cooperation for the preservation of Cultural Heritage may be achieved by a culture-driven technology protection.

Acknowledgments. The authors wish to acknowledge the leaders of the Christian communities: His Beatitude the Greek Orthodox Patriarch of Jerusalem Theophilos III; His Paternity Archbishop Pierbattista Pizzaballa, who was the Custos of the Holy Land until May 2016 and is now the Apostolic Administrator of the Latin Patriarchate of Jerusalem; Father Francesco Patton, who has been Custos of the Holy Land since June 2016; and His Beatitude the Armenian Patriarch of Jerusalem Nourhan Manougian. Contributions from all over the world secured the project's funding. Worth noting Mica Ertegun's and Jack Shear's donations through World Monuments Fund, Aegean Airlines et al. Acknowledgements are attributed to the members of the Holy Sepulchre Common Technical Bureau: Th. Mitropoulos, Ph.D., the director of the Common Technical Bureau and construction site manager; O. Hamdan; C. Benelli; and I. Badalian. Acknowledgements are also attributed to everyone from the Christian Communities, the Scientific Community, the Media and the Society, who offered us the privilege to transmit and discuss the message of the monument and the project.

References

1. Coüasnon, C.O.P.: The Church of the Holy Sepulchre Jerusalem. The Schweich Lectures of the British Academy 1972. Oxford University Press, London (1974)
2. Biddle, M.: The Tomb of Christ. Gloucestershire England, Sutton (1999)
3. Lavas, G.: The Holy Church of the Resurrection in Jerusalem. The Academy of Athens (2009). (in Greek)
4. Corbo, V.C.: "Il Santo Sepolcro de Gerusalemme, Aspetti archeologici dale origini al period crociato, Parte I, II, III" Studium Biblicum Franciscanum, Collectio Maior. Franciscan Printing Press, Jerusalem (1981)
5. Mitropoulos, Th.: The Church of Holy Sepulchre – The Work of Kalfas Komnenos. European Centre of for Byzantine and Post-Byzantine Monuments (2009). (in Greek)
6. Project: Integrated Diagnostic Research Project and Strategic Planning for Materials, Interventions Conservation and Rehabilitation of the Holy Aedicule of the Church of the Holy Sepulchre in Jerusalem. Scientific Responsible: A. Moropoulou, National Technical University of Athens, 2015–2016
7. Project: Conservation, reinforcement and repair interventions for the rehabilitation of the Holy Aedicule of the Holy Sepulchre in the All-Holy Church of Resurrection in Jerusalem. Scientific Responsible: A. Moropoulou, National Technical University of Athens. Duration: 22.03.2016–22.03.2017

8. Georgopoulos, A., et al.: The role of digital geometric documentation for the rehabilitation of the Tomb of Christ. In: Digital Heritage 2018. IEEE, Elsevier (2018, in print)

9. Lampropoulos, K.C., Moropoulou, A., Korres, M.: Ground penetrating radar prospection of the construction phases of the Holy Aedicule of the Church of the Holy Sepulchre, in correlation with architectural analysis. Constr. Build. Mater. **155**, 307–322 (2017)

10. Agrafiotis, P., Lampropoulos, K., Georgopoulos, A., Moropoulou, A. 3D modelling the invisible using ground penetrating radar. In: Aguilera, D., Georgopoulos, A., Kersten, T., Remondino, F., Stathopoulou, E. (eds.) TC II & CIPA 3D Virtual Reconstruction and Visualization of Complex Architectures, 1–3 March 2017, Nafplio, Greece. The International Archives of the Photogrammetry, Remote Sensing and Spatial Information Sciences, vol. XLII-2-W3, pp. 33–37 (2017)

11. Moropoulou, A., et al.: Faithful rehabilitation. J. Am. Soc. Civ. Eng. **54–61**, 78 (2017)

12. Moropoulou, A., Farmakidi, C.M., Lampropoulos, K., Apostolopoulou, M.: Interdisciplinary planning and scientific support to rehabilitate and preserve the values of the Holy Aedicule of the Holy Sepulchre in interrelation with social accessibility. Sociol. Anthropol. **6**(6), 534–546 (2018)

13. Apostolopoulou, M., et al.: Study of the historical mortars of the Holy Aedicule as a basis for the design, application and assessment of repair mortars: a multispectral approach applied on the Holy Aedicule. Constr. Build. Mater. **181**, 618–637 (2018)

14. Alexakis, Emm., Delegou, E.T., Lampropoulos, K.C., Apostolopoulou, M., Ntoutsi, I., Moropoulou, A.: NDT as a Monitoring tool of the works progress and assessment of materials and rehabilitation interventions at the Holy Aedicule of the Holy Sepulchre. Constr. Build. Mater. **189**, 512–526 (2018)

15. Moropoulou, A., Zacharias, N., Delegou, E.T., Apostolopoulou, M., Palamara, E., Kolaiti, A.: OSL mortar dating to elucidate the construction history of the Tomb Chamber of the Holy Aedicule of the Holy Sepulchre in Jerusalem. J. Archaeol. Sci. Rep. **19**, 80–91 (2018)

16. 10th Annual Conference on "New studies in the archaeology of Jerusalem" [Jerusalem, 26–28/10/2016, joint collaboration of the Israel Antiquities Authority, Tel Aviv University and The Hebrew University of Jerusalem]. Invited presentation: "Sustainable rehabilitation of the Holy Sepulchre: Interdisciplinary scientific study and monitoring – Scientific integrated governance of the project", a collective work presented by Prof. A. Moropoulou

17. 2nd International Conference on Art and Archaeology [Jerusalem, 11–14/12/2016]. Invited presentation: "Interdisciplinary planning and scientific support in the field of rehabilitation of the Holy Aedicule of the Holy Sepulchre", a collective work presented by Prof. A. Moropoulou

18. University of Oxford School of Archaeology: Fifth Anniversary of the Mica and Ahmet Ertegun Graduate Scholarship Programme in the Humanities [Oxford, 23/10/2017]. Invited presentation: "The Rehabilitation of the Holy Aedicule of the Holy Sepulcher", a collective work presented by Prof. A. Moropoulou

19. Institut für Christkatholische Theologie, Universität Bern [Bern, 21/03/2011]: Invited lecture: "The rehabilitation of the Holy Aedicule of the Holy Sepulchre", a collective work presented by Prof. A. Moropoulou

20. Salvaguardare la Memoria per Immaginare il Future. Atte della III edizione delle Giornate di Archeologia e Storia del vicino e medio Oriente [Milano, Italy, 05 Maggio 2017]; Lectio Magistralis: "Il restauro della Tomba di Cristo: Interventi di Conservazione, Rinforzo e Riparazione per il Restauro dell' Edicola del Santo Sepolcro a Gerusaleme", Prof. A. Moropoulou. Fondazione Terra Santa – Milano, pp. 101–118 (2018)

The Project of the Rehabilitation of Holy Sepulchre's Holy Aedicule as a Pilot Multispectral, Multidimensional, Novel Approach Through Transdisciplinarity and Cooperation in the Protection of Monuments

The Holy Sepulchre as a Religious Building

Stefanos Athanasiou[✉]

University of Bern, Bern, Switzerland
stefanos.athanasiou@theol.unibe.ch

Abstract. The Church of the Holy Sepulchre was shaken by a strong earthquake in 1927, which seriously damaged the Aedicule inside. The three denominations which were responsible for the Aedicule – the Greek Orthodox Patriarchate of Jerusalem, the Roman Catholic (Franciscan) Custodian and the Armenian apostolic Church – could only agree at that time on the most necessary work on the Holy Sepulchre to be done; as a result the Aedicule was in danger of collapse. In 1947 the former British mandate government was forced to support the Aedicule with steel pillars. The interior and exterior of the Aedicule was also damaged by candle soot, which affected the entire construction of the building. Particularly, older frescoes and writings adoring the Aedicule were covered in soot. Much later, in 2016/2017 the three denominations agreed on a restoration and it was commissioned to the renowned engineer Prof. Dr. Antonia Moropoulou and her team from NTU Athens. From the context of these disputes it becomes clear that the building of the Holy Sepulchre is not easy to be separated form its religious significance. Specifically, it has a special geopolitical significance in the whole history of its existence due to its religious dynamics. The paper will show the historical development of the Holy Sepulchre Church building through important sources in the first and second millennium. Furthermore, it will also be pointed the inter-religious significance of the Church. In particular, the ceremony of the Holy Fire on Holy Saturday will be mentioned, which was once celebrated not only as a Christian event, but also by Muslims on 9th and 10th centuries. This ceremony also played an important symbolic role in the separation between the East and West Christianity. The entire building of the Church of the Holy Sepulchre can be understood properly only if its religious significance in time can be recognized. The paper will focus for this reason on the historical-religious significance of the building. This church is an example of how religion can become the protector or the destroyer of Cultural Heritage.

Keywords: Christianity · Religion · Culture · Dialog · Jerusalem · Holy Sepulchre

Lecture in Systematic Theology at the Universities of Bern and Fribourg, the Theological College of Chur in Switzerland and on the Dormitio Theological study year Programm in Jerusalem.

A. Moropoulou et al. (Eds.): TMM_CH 2018, CCIS 961, pp. 29–43, 2019.
https://doi.org/10.1007/978-3-030-12957-6_2

1 Introduction

Simon Sebag Montefiore in the introduction of his biography about Jerusalem mentions, that "the history of Jerusalem is the history of the world" [1]. Even if this statement may seem exaggerated, it is to some extent justified that Jerusalem has profoundly influenced world history from the spiritual dynamic it has acquired throughout the human history, and that happens because Jerusalem seems to be an intersection between heaven and earth, especially for the monotheistic religions. Here, monotheism revealed itself, making Jerusalem a special place for the three greatest world religions. For the Jews, Jerusalem, is the city of their ancestors, revealed to them by God Himself. For Christians it is the place where Jesus Christ, the divine Logos of God, suffered, was crucified, died, rose and went to heaven. For the Muslims the place where Mohammed has experienced his ascension. The sacrality of this place goes so far that in the tradition and theology of the monotheistic Religion not only the earthly Jerusalem exists, but also heavenly Jerusalem, as the "transfigured" image of this city, representing the eternal hereafter. As a result, Jerusalem experiences an ecumenical-cosmic dimension that turns this city into something beyond any other city. In that sense, the history of Jerusalem is actually the history of the world. The heavenly Jerusalem becomes the city of eternity and it is still realized wherever the divine communion finds its expression. Ultimately, it is the earthly Jerusalem that will bring the heavenly Jerusalem - the Eschata, and that happens because this earthly Jerusalem is the prelude of the Last Judgment [2].

In Christianity, and especially in the Orthodox tradition, the image of the new/heavenly Jerusalem is the same Divine Liturgy, which is not only a symbolic act of the historical events of Jesus' life, but invites the faithful to participate truly in the same historical events. The faithful experience in the Divine Liturgy: the descent, the actions, the passion, the death, the resurrection, the ascension of Jesus Christ and the judgment are considered as real experiences for the believers. Thus it is said in a prayer before the transmutation of the gifts in the Chrysostom liturgy: "Remembering, therefore, this command of the Saviour, and all that came to pass for our sake, the cross, the tomb, the resurrection on the third day, the ascension into heaven, the enthronement at the right hand of the Father, and the second, glorious coming" [3].

The liturgy becomes a timeless place, even the heavenly Jerusalem, where the entire history is experienced in its fulfilment.

Moreover, in Orthodox services such as at Christmas, Epiphany, Good Friday, etc. are paradoxically always emphasized in the church chants that "today" Jesus is born or baptized, crucified or risen. Thus, every Divine Liturgy and every Orthodox liturgical service becomes a "Topos" in which anyone can themselves lived and experienced Jerusalem. Every liturgical Community in time and space witnesses the events of Jerusalem, and every church thereby becomes Jerusalem in itself. In every liturgy believers leaved the time in which they live and become witnesses of the events in Jerusalem. And as such witnesses, they are called, like the disciples, but in their respective time to confess and testify Jesus Christ. Although in a mystagogical way Jerusalem is experienced and lived in every eucharistic service. It is a fact that the earthly and historical Jerusalem has always been, and still is, a special place as "proto-topos" of the events of Jesus Christ. For this reason, it is not surprising that the diocese of Jerusalem became a bishop's place of honour in the geographical place of Palestine as a metropolis at the first Ecumenical Council of Nicaea in 325 AD and was even appointed as a patriarchate seat in the fourth ecumenical council of Chalcedon in 451 AD, even if was politically unimportant in that time [4]. However it had a very important religious significance, and that should be the reason that this city was protected by many political powers, as we will see. Especially the Church of the Holy Sepulchre with its Aedicule, where the tomb of Christ is covered by Rotunda since 4th Century, paradoxically represents the centre of Christian hope. The tomb as a symbol of the death, becomes a symbol of life and an expression of human hope. This represents the core of Orthodox theology of hope. The empty tomb becomes a symbol of full life.

Although each altar represents the tomb of Christ and each church at some point a depiction of the Church of the Holy Sepulchre, the authentic place has an unbelievable importance, and that is because the tomb of Christ is a historical testimony of the theological legitimation for the liturgical practice as it was described above. On the other hand every liturgy is a place where cultural heritage is preserved and that happens because the liturgy is the "Icon" of the Holy Sepulchre.

Even though the Church of the Holy Sepulchre is a place of ecumenical hope and life, human history has shown that such places became "places of death", if political forces try to use such places for ideological reasons and to change cultural Heritage for own benefit. However, this place is so ecumenical-global-cosmological, that its meaning cannot and must not be limited. In the following pages, the theological and ideological significance of this place shall be pointed out on a basis of selected historical testimonials and structural data. I believe, that the entire construction of the Aedicule and the Church of the Holy Sepulchre is probably not of a constructional peculiarity only. Without its theological or religious significance, the building would hardly have received any special interest from the polytechnic side. What makes this building so important and unique is its theological significance for billions of Christians and religious people.

2 In the Beginning was Death - The Grave

The great question concerning the Aedicule as burial place of Jesus Christ employs many scientists and researchers. Constantly, some of them try to prove that the tomb of Jesus Christ is not identified with the Aedicule.

We only have little topographic evidence in the New Testament regarding the tomb of Jesus Chris. We know that Jesus was led away from the Praetorium, the city centre of Jerusalem at that time, and out of the city [5]. The Gospel of John mentions that the place of the crucifixion was "near the city" [6]. The Jerusalem expert Max Küchler assumes that the crucifixion itself must have taken place in a busy road, which is for Küchler recognizable from the trilingual cross inscriptions (Latin, Greek, Hebrew). The multilingualism of the cross inscription "King of the Jews" proves to Küchler that the crucifixion took place on a street where Jews, Greeks and Romans passed by [7]. Küchler also relies his assumption on the fact that in Mk 15, 29 there is a mention of "passing persons" or in Jn 19:20 it is pointed out that "many Jews" attended the event [8]. For this reason, Küchler is sure that the crucifixion itself took please near a city gate or on a street leading to it [8]. It is sensible that if Jesus actually caused a agitation in the Jewish population of that time, Pontius Pilate would like to make an example of Jesus to all those who longed or even organized a political revolution against the Romans. Pilate had not probably realized that the kingdom that Jesus proclaimed was not supposed to be a politically liberated Palestine, but it was a kingdom out of this world [9]. Therefore Jesus propagated an ontological and not a political liberation, as Barabas for example tried to do [10]. In addition to the fact that Jesus was crucified on a street, the Gospel of Mark tells us, "… and they carried him to the site of Golgotha, which is the place of the skull" [11]. Apart from the etymological meaning of the word, this must have been an uplifting hill, as evidenced by the Gospel of Mark, which tells us about the possibility to watch the event "from afar" [12]. In the addition the pilgrim's report of the pilgrim from Bordeaux, in 334 we find the mention about a small mountain called Golgotha (Monticulus Golgotha) [13].

According to the Gospel of John, near the crucifixion place there was a garden with the private tomb, that Joseph of Arimatea had build actually for himself [14]. It may astonish today's pilgrims that the tomb of Jesus has been so close to the Golgotha hill. Some scholars have given an allegorical interpretation of the entire garden as it is portrayed in John Gospels and equated it allegorically with the Garden of Eden [15]. The fact is, that the archaeologists find tombs around the Old City of Jerusalem and that have proved that gardens around Jerusalem were popular places for tombs in the time of Jesus [16]. The fact that more tombs have been found around the tomb of Jesus proves, first and foremost, that this is a place that at the time of Jesus was outside the city, as it was unacceptable upon to build tombs inside the city walls. There is no doubt that Jesus' grave laid outside the city walls at that time. After the extension of the city wall in its third construction phase by Herodes in the years 41–44 AD. The entire Golgotha region has moved into the city centre of Jerusalem. The fact that Melito of Sardis (died 180 AD), in his Passover Homily, accuses the Jews three times of having crucified Christ "in the midst of Jerusalem" [17], proves that at the time of Melito the tomb of Jesus was already in the city centre of Jerusalem. Melito maybe was unaware of the fact that the Golgotha region exist outside of the city before 44 AD. The fact that Melito declares the burial ground in the middle of the city suggests that it was well known to Christians that the tomb of Jesus was within the city and well known as a place of pilgrimage.

Somebody can recognize here very well how cultural or religious heritage can be carried on by tradition or consciousness of the faithful - church.

However, the fact is, that it is not clear at all from the New Testament sources and geographical information's that we can found there that to tomb of Christ is the Aedicule of Jerusalem. The only and sufficient fact that it is the actual gravesite of Jesus Christ is the local tradition, which was well-preserved until the 4th century and which has been established as we could see in the Paschal Homily of the Meliton. The fact that Eusebius of Caesarea, the imperial historian of Constantine, reports that the architects of Constantine's Basilica searched the tomb of Jesus in the Forum of Emperor Hadrian, shows that this place has been remembered by the surviving Christians in Jerusalem. Eusebius mentions in his church history a list of bishops of Jerusalem from 35–135, that shows us, that in Jerusalem was still a living Christina community.

This suggests that the destruction of Jerusalem in 70 AD could not completely expel Christians from the region and preserve an existing local tradition concerning the tomb of Jesus Christ. Even though Eusebius tells us that "the grotto of the Redeemer had long been hidden in the darkness, since the wicked and the rejected deprecated it altogether in human beings," the place was discovered through "divine inspiration" [18]. Although this statement of Eusebius suggests that the Christians did not know the exact burial place of Jesus Christ in at the time of Constantine, Max Küchner suspects that the "old local tradition of the tomb of Jesus was present, but it was provided with the note of oblivion, to tell a dramatic discovery story. The emphasis of being forgotten is then almost a proof that the site was still well known. The ancient sanctuary must be forgotten so that it can be divinely legitimized and more gloriously restored in the present" [19]. The divine legitimation is for Küchner the important step that has led to the edification of the first Church of the Holy Sepulchre and the Aedicule.

With regard to the first Church of the Holy Sepulchre, known as the Constantinian Church of the Holy Sepulchre, we get from Eusebius lots of information's about the construction phase and structure of the Church. In the Vita Constantini Eusebius emphasizes:

"First of all, then, he adorned the sacred cave itself, as the chief part of the whole work, and the hallowed monument at which the angel radiant with light had once declared to all that regeneration which was first manifested in the Saviour's person. This monument, therefore, first of all, as the chief part of the whole, the emperor's zealous magnificence beautified with rare columns, profusely enriched with the most splendid decorations of every kind. The next object of his attention was a space of ground of great extent, and open to the pure air of heaven. This he adorned with a pavement of finely polished stone, and enclosed it on three sides with porticos of great length. For at the side opposite to the cave, which was the eastern side, the church itself was erected; a noble work rising to a vast height, and of great extent both in length and breadth. The interior of this structure was floored with marble slabs of various colours; while the external surface of the walls, which shone with polished stones exactly fitted together, exhibited a degree of splendour in no respect inferior to that of marble. With regard to the roof, it was covered on the outside with lead, as a protection against the rains of winter. But the inner part of the roof, which was finished with sculptured panel work, extended in a series of connected compartments, like a vast sea, over the whole church; and, being overlaid throughout with the purest gold, caused the entire building to glitter as it were with rays of light. Besides this were two porticos on each side, with upper and lower ranges of pillars, corresponding in length with the church itself; and these also had their roofs ornamented with gold. Of these porticos, those which were exterior to the church were supported by columns of great size, while those within these rested on piles of stone beautifully adorned on the surface. Three gates, placed exactly east, were intended to receive the multitudes who entered the church. Opposite these gates the crowning part of the whole was the hemisphere, which rose to the very summit of the church. This was encircled by twelve columns (according to the number of the apostles of our Saviour), having their capitals embellished with silver bowls of great size, which the emperor himself presented as a splendid offering to his God. In the next place he enclosed the atrium which occupied the space leading to the entrances in front of the church. This comprehended, first the court, then the porticos on each side, and lastly the gates of the court. After these, in the midst of the open market-place, the general entrance-gates, which were of exquisite workmanship, afforded to passers-by on the outside a view of the interior which could not fail to inspire astonishment. This temple, then, the emperor erected as a conspicuous monument of the Savior's resurrection, and embellished it throughout on an imperial scale of magnificence. He further enriched it with numberless offerings of inexpressible beauty and various materials,–gold, silver, and precious stones, the skillful and elaborate arrangement of which, in regard to their magnitude, number, and variety, we have not leisure at present to describe particularly" [20].

As we know from the sources, the construction work of the Church of the Holy Sepulchre began around the spring of 326 AD and was completed 10 years later. The inauguration of the church took place at 13–20 September 336 AD. In Jerusalem itself Bishop Makarios had the direction of the building. The name of the architect is also preserved - his name was Zenobios. Next to him, a third man, the Constantopolitan Presbyter Eustathios, is mentioned as the main construction assistant of the building [21]. At the inauguration of the Holy Sepulchre, the synod meeting in Tyrus was invited by the Emperor. Bishops were also present from Macedonia, Pannonia, Thrace,

(a)

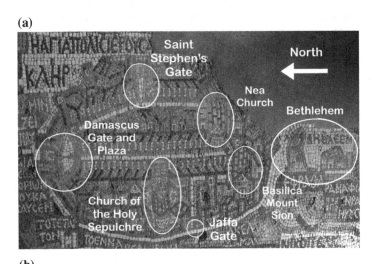

(b)

Fig. 1. (a) Mdaba Mosaic Map: Jerusalem. (b) Mdaba Mosaic Map: Holy Sepulchre

Mysia, Bithynia, Cilicia, Syria, Mesopotamia, Persia, Egypt and Libya [22]. The inauguration of the Church of the Holy Sepulchre undoubtedly symbolized the new position of Christianity in the Empire. Even Eusebius inform us about the Church of the Holy Sepulchre, he tells us nothing about the rotunda over the tomb of Jesus Christ. This probably suggests that the rotunda wasn't still built. However, it is certain that in the time of Cyril of Jerusalem (died 386) the rotunda existed over the tomb because he mention that the catechism of the newly baptized took place there [23].

However, the architecture of the Constantine basilica shows that it was structured in three parts. The main ship – the Martyrium, followed by an atrium in which the Golgotha Hill existed. Everything was aligned in the direction of the Resurrection Rotunda, where the grave of the Lord was located.

The Constantinian Church of the Holy Sepulchre is also recorded in the Madaba Mosaic map (6th c.) (Fig. 1) further to the Eusebian explanation of the Church of the Holy Sepulchre. Additionally, there are miniature models of the Aedicule on pilgrim bottle of Monza and Bobbio (Fig. 2) [24]. Also on the reliquary in the Capella Sancta Sanctorum in San Giovanni in Lateran (6th century), it is possible to guess exactly how the tomb must have looked in the Constantinian period (Fig. 3) [8]. Characteristic of this is, for example, that in this portrait the rotunda is recorded above the Aedicule. The reverence and the desire of the faithful to possess images of the Church of the Holy Sepulchre and to bring for example them to other believers who was far away form Jerusalem, has contributed in a certain way the Constantine Basilika and its appearance to the future.

Fig. 2. Monza and Bobbio bottle (6th Century)

Fig. 3. Reliquary in the Capella Sancta Sanctorum in San Giovanni in Lateran (6th century).
Draft: Max Küchler

3   The Church of the Holy Sepulchre as a Plaything of the Religious Communities

At 614 AD, we enter the second period of the Church of the Holy Sepulchre and its history. The period is characteristic of the gradual decline of Christians and specifically, Byzantine hegemony in the Middle East. Karl Schmaltz mentions this time as characteristic: "At the beginning of this period there are still the great names of Modesto, Sophronios, Johannes Damascenos, but then it becomes quieter and quieter. At the Holy Sepulchre, the fire magic of Easter Sunday has become the main festival of the year, and the crucifix relic the actual idol of this temple" [25]. According to Schmaltz, Jerusalem has slowly declined from a living theological community, which still influenced liturgical and theological in the time of Cyril of Jerusalem the entire Empire, to a quiet place of wonders. Jerusalem became a place of mysticism and wonders and that was the reason for Schmaltz, that Jerusalem continued to arouse the interest of the faithful all over the Empire. The first damage to the Constantine Church of the Holy Sepulchre on 4 May 614 in the course of the conquest of Jerusalem by the Persian King Chosroes II, devastated but did not destroy the Constantinian church. There followed a time when Jerusalem changed hands between Christians and Muslims. For example, in 630 Emperor Heraclius was able to return the Holy Cross on the hill of Golgotha, which had previously been brought to Ktesiphon to save it from the Muslims. Since the Islamic rulers in the Middle East also pursue some important trade relationship to Byzantium, they gave the Christians safe access to the pilgrimage sites of the Holy Land. Joseph, an Armenian hermit who visited the Holy Land in 659–663, stated:

"… and you will find here trustworthy information derived from an eye-witness. The sepulchre of Jesus, the Giver of Life, hewn about of the rock, is a fathom and a half from the middle oft he dome oft he life given tomb. And in the church, which is 100 cubits in height and 100 cubits in breadth all round, there are on each side 12 columns above and 12 below the gallery. And in the gallery are the lance, sponge and cup of Christ, laid up in gold. In the principal church, which is called the martyrs chapel, which is the church oft he Finding oft he cross, 20 cubits form the Resurrection, there are 65 columns alone the length above and below. The holy church of Golgotha, which is called the tomb of Adam is 10 paces from the Resurrection. And above is a sacramental table at the place, where Christ was crucified on the rock" [26].

Schmaltz comments this report and comes to the conclusion that the Armenian numbers should be treated with caution. What is clear from the report, is the fact that in the 7th century on the Golgotha hill already a two-floors building had existed. However, as far as the Constantinian Aedicule is concerned, it is likely that it survived until the complete destruction of the Church of the Holy Sepulchre in 1009 [27]. In 1009, the Fatimid caliph of Egypt Al Hakim ordered Amr Allah Yaruk the Governor of Ramallah "to destroy the Church of the Resurrection. Moreover, to remove the (Christian) symbols, and destroy all traces and the memory of them" [28]. In this point, certainly, the cross that the Constantinian tomb had still adorned disappeared.

The question why Al Hakim gave the order to destroy the Church of the Holy Sepulchre cannot be completely and surely answered by the sources. Arabic, Greek, and Western sources indicate very different reasons for the destruction of the Church of the Holy Sepulchre. The one side tell us about intrigues between Christians in Jerusalem and the other side about an action directed by European Jews [29]. The Syrian historian Ibn al-Qalanisi (1070–1160) explains in his scripture "History of Damascus" his view why the Church of the Holy Sepulchre was destroyed. According to Al-Qalanisi, the total destruction was related to the fact that more and more Muslims heard of the fire miracle and visited Jerusalem together with the Christians pilgrims, on Holy Saturday to take part to the ceremony of the holy fire. The destruction happened when Muslims doubted the Christian fire miracle [30], claiming that the Christians lit the fire with trick and that was a doubt against the metrical and the power of the Christian God [31]. Sources, such as that of the Persian Al Biruni, also underline that before the destruction of the Church of the Holy Sepulchre even Islamic leaders of the city of Jerusalem, the Imam, the Emir and the Muezzin participated in the miracle of fire and even brought their own oil lamps to receive it [32]. The miracle of fire it self played a very important role in interreligious Jerusalem and for all Christendom. Thus Pope Urban II referred to the miracle of light in 1095 in his speech at the Synod of Clermont and emphasized, among other things: "How precious is the much desired, incomparable place of the Lord's grave, even if God did not work the annual miracle there. On the days of his Passion all the lights in the tomb and around the church, which had been extinguished, are lit again by divine command. Whose heart is so stony, brethren, that it would not be touched by such a great miracle? Believe me, man must be animal and unfeeling, whose heart cannot be directed to faith by such divine proof of grace" [33].

The facts show us that the faith of the Holy Light strongly increased the significance of the Church of the Holy Sepulchre not only among the Christians but also among the Muslims. The Holy fire and the site of the tomb of Jesus Christ had such a dynamic that only a few years later, after its complete destruction, it had been rebuilt step by step. Already in 1047, Nasir describes the new rotunda of the Church of the Holy Sepulchre [34]. With the historical information of that period, Martin Biddle tries to provide a reconstruction of the second Aedicule: "Like the Constantinian previous building, the Aedicule of the 11th century included two main elements: a rounded western structure enclosing the burial chamber, and a narrower rectangular eastern Department that forms the entrance to the tomb. Four elements, which are at the same time part of the present Aedicule of 1809–10 and the Aedicule of 1555, appear for the first time in connection with this medial monument" [8].

When Al-Hakim destroyed the tomb, the tomb was almost completely destroyed. For this reason, it was now possible to put a vaulted roof on top of the grave. Here one has paid attention to install an opening in the new roof. This had probably both practical and a mystagogical-theological reason. On the one hand, it was possible to get the candle smoke out of the room through this ventilation, on the other hand now there still existed a new opening for the passage of the Holy fire [35]. Since the dome of the Church of the Holy Sepulchre was now open at the top, a second dome had to be installed on the Aedicule itself, which from now on has left its mark on the image of the Aule of the Church of the Holy Sepulchre until today. When the Crusaders conquered Jerusalem in 1099, they probably did not change much of the Aedicule, except from the fact that they put a Latin Christ figure [36] on the pedestal of the Aedicule, as the pilgrim Daniel inform us. Certainly, the exaltation of a Christ figure was a sign that now the grave was in the hands of the western Christians, that even as the Eastern Orthodoxy had recognized in the 7th Ecumenical Council of Nicaea in 787 the icons, where as a cultural feature rather the sanctuaries have prevailed, which still was frowned upon the East symbols of the pagan times.

However, since the dome of the rotunda, as mentioned, was open, the Christ figure had to remove relatively quickly, as it could not withstand the weather condition, to a gilded cross [37]. The cross on the Aedicule was thus a permanent sign that belonged to the image of the Aedicule from the beginning. Both, the Constantinian and the Byzantine Aedicule (also in the crusader times) were decorated by a cross [38]. When Saladin conquered Jerusalem on 2 October 1187, the golden cross became victim of plundering. This is particularly evident from the fact that the drawings of the Aedicule from 1187 is depicted without a cross, except the time of grand land, were a tradition until the end of 19 century, allowed the Christians to put a Cross and Hexaperia (six wings of the angels) in the top of the Aedicule (Fig. 4). Until 1555 it seems that restoration work hasn't been done on the Aedicule, probably because this was not allowed [39]. Only Boniface of Ragusa was able in 1555 with the help of the western world to undertake a restoration of the Aedicule. As early as 1728, a second short interior restoration had to be taken by Father Elzear Horn, which was probably confined to the fixing of some marble slabs. In 1808 the church of the Holy Sepulchre was damaged by a strong fire. Even if, for example, Martin Biddle does not say anything about the fire in his book about the rotunda, the Augsburg Odinari Postzeitung reports on 30 March 1809 about this happening:

"… the fire broke out in Jerusalem on October 12, 1808, as follows: The holy grave, into which the good, pious man Joseph of Arimathia had once placed the body of Jesus according to the Bible, is today covered from outside and inside with marble, 9 feet long, and 12 feet high. Above this, a small dome floats resting on 12 precious marble columns. Above the sacred tomb, also known as the Chapel of the Resurrection, is the famous church of the Holy Sepulchre. Built in the style of the rotunda in Rome, the pantheon of the pagan Romans, it has a high dome of cedar, from the mountains of Lebanon, covered with sheet metal. The Turks never want to declare that this dome would be built of stone, so that they would not surpass the Mohammedan Mosque in Jerusalem, which now stands on the place where ever the solomonic Temple has been in past. Inside the church are very beautiful galleries on the walls, resting on magnificent marble columns, around. On the eastern side of the Holy Sepulchre is the choir of the Greeks, on the southern side are seven floors everyone above each other. There are rooms, where the Armenians, the Catholics and the Pilgrims worship. Some of these rooms also serve as apartments, or for storing wallpaper, lamps, and other holy ornaments. The fire broke out at 11 o'clock in the night on the 12 October, first in the rooms of the Armenians, no one know in what way. Later after, the large wooden dome, the choir of the Greeks, and finally the rooms of the Catholics, with all the church utensils kept there, burned up. After five hours, the burning dome plunged down into the interior of the church, shattering the central Aedicule dome above the holy tomb, with the pillars upon which it rested. The holy grave itself stood for several hours in a glow that melted the metal (…) everyone believed that the interior of the holy grave must have a similar fate. But this alone remained undamaged, which even the Turks regarded as a miracle" [40].

Fig. 4. Carte Postale. Union Postale Nr. 16202 (End of 19th Century)

Fig. 5. Augsburgische Ordinari Postzeitung (30.3.1809)

The description from the Augsburg Ordinari Postzeitung (Fig. 5) report shows that the wooden dome of the Church of the Holy Sepulchre was made of cedar wood from Lebanon. It is obvious that Lebanese cedar has been used, as according to Psalm 104, 16 (103,16) these are the trees that the Lord himself planted. In addition, the newspaper reports that the fire has broken out of the Armenian apartments, but that the grandstand is unknown.

After this big fire, the Aedicule was restored from the ground up, by the Greek Orthodox Patriarchate of Jerusalem, through a decree (Firman) of Sultan Mahumud II in 1809, commissioned the restoration to the Greek Orthodox Patriarchate of Jerusalem. The Patriarchate commissioned, the well known architect from Constantinople, Nikolaos Komnenos from Mytilene with the order. Komnenos had previously restored the church Zoodochos Pege in Balukli and the church of the Panagia of Pera in Constantinople. Komnenos had as epithet the word Kalfa, that means masterbuilder in osmatic turkish [41]. Komnenos finished the work on the Aedicule and the Church of the Holy Sepulchre on September 13, 1810 (he travelled from Constantinople to Jerusalem on May 3, 1809). Although Komninos renewed the entire Church of the Holy Sepulchre, he left open the dome of the Church of the Holy Sepulchre, as it was before. As a result, the rainwater on the Aedicule read a lot of the iron cramps that held together the new Aedicule in their Istanbul Baroque style. Moreover, the additional weight that the Aedicule had gained through liturgical and ecclesiastical devices in the period of the great lent, as we already mention, is responsible for the vaulting of the exterior walls of the Aedicule. The tradition of fasting jewellery that the Aedicule raised several thousand kilos more weight and thus promoted the vault of the outer wall was abolished when in 1927 a major earthquake damaged the rotunda. The Aedicule had to be supported 1947 with steel pillars, so it does not collapse. (one of the last task of the British Mandate Government).

Fig. 6. The different confessions in the Holy Sepulchre

In this context, it was necessity to restore the Aedicule to keep it not only for the next generations but to allowed to the different confessions, who have rights in the Holy Sepulchre, to use it for liturgical necessities (Fig. 6). The restoration by the team of Prof. Dr. med. Moropoulou and the Athens Technical University in 2016, made possible by the historic common agreement of the three denominations, the Greek Orthodox Patriarchate of Jerusalem, the Franciscan Custody and the Armenian Patriarchate in Jerusalem, responsible for the Church of the Holy Sepulchre and the Aedicule. Particularly praise-worthy here is the fact that the Aedicule decorates a cross again as it did in the Constantinian and Byzantine Aedicule. Time demands for signs, and today's cross that adorns this Aedicule is a sign of its past, its present but also its future. The cross on the Aedicule expresses the freedom of time and give the hope of a peaceful future. And the glass window inside the Aedicule gives to the pilgrim the possibility of to see and to recognise the cave of the tomb of Christ. It is therefore a obligation to protect and to save such places, not only for history, but as places that can give hope and joy to millions of people. In every liturgical service the tomb of Christ became the image of the new Jerusalem and the prototype of every liturgy in the cosmos giving the intention as the inscription over the tombs confesses, to the hole world: "Η πηγή της ημων αναστάσεως".

4 Conclusion

We have seen how historically the Aedicule and the Church of the Holy Sepulchre were changed. But how the Topos of the cave, and the tomb of Jesus Christ was protected by memory and tradition of church. We have see also how the theological and especially the religion meaning of the place became the reason that this place was restore and rebuilt in history many times but also the reason to destroy this place. However, in the memory of the Church the Aedicule and the whole Holy Sepulchre are still present in every liturgy as we said. So the holy liturgy is for the Orthodox Church and Theology the protector not only of cultural heritage but also for historical and theological heritage. But it is also important for the Orthodox Theology, to protect such places like the Holy Sepulchre, because this are places were God has shown to the humanity in the holy economia/in history his grace. Places of Pilgrimage are places of hope, because there are places of grace and places where people can recognise the hope. In this way the old/new cross [42] on the Aedicule and the new glass window in the Aedicule are symbols who shows the researcher of the future, that in our time things was possible that in other times was not. It is a responsibility of every generation to protected the cultural heritage for the future generation and a responsibility of church to protected the theological heritage and the historical places where such heritage were created.

References

1. Montefiore, S.: Jerusalem. Die Biographie, p. 13. Frankfurt, Fischer Taschenbuch (2011)
2. Hoffmann, H.: Das Gesetz in der frühjüdischen Apokalyptik. Vandenhoeck und Ruprecht, Göttingen (1999)

3. Divine Liturgy of Saint John Chrysostom, p. 10, 01 July 2018. http://stpaulsirvine.org/music/wp-content/uploads/2015/06/choir-book-divine-liturgy-Jun-2015.pdf

4. Papadopoulou, Ch.: Ιστορία της Ἐκκλησιας Ιεροσολύμων, Athens, pp. 105–107 (2010)

5. Mk 15,20. Mt 27,32

6. Jn 19,17

7. Küchler, M.: Jerusalem, pp. 416–417. Ein Handbuch und Studienreiseführer zur Heiligen Stadt. Vandenhoeck und Ruprecht, Göttingen (2007)

8. ibid

9. Jn 18,36

10. Mt 27, 15-26. Mk 15, 6-15. Lk 23, 18. Jn 18, 39–40

11. Lk 23.33. See also Mt 27, 33 and Jn 19.17

12. Mk 15,40

13. Geyer, P.: Itinera hierosolymitana saecvil IIII–VIII. Corpus Scriptorum Ecclesiasticorum Latinorum, vol. 39, pp. 1–33. Tempsky, Vienna (1898)

14. Jn 19,41

15. Schnackenburg, R.: Das Johannesevangelium, HThKNT, Freiburg, vol. VI, p. 351 (1965)

16. Küchler, M.: Jerusalem. Ein Handbuch und Studienreiseführer zur Heiligen Stadt, p. 416

17. Melition von Sardis, Paschahomelie 94

18. Eusebius von Ceasarea, Vita Constantini, 3, pp. 26–28

19. Küchler, M.: Jerusalem. Ein Handbuch und Studienreiseführer zur Heiligen Stadt, p. 419

20. Eusebius von Ceasarea, Vita Constantini, 3, pp. 33–40

21. Theophanes, Prosper Aquit. chron., Migne PSL Vol 51, p. 576

22. Eusebius von Ceasarea, Vita Constantini, 3, pp. 25–50

23. Cyril from Jerusalem, Catecheses 18,33

24. Küchler, M.: Jerusalem. Ein Handbuch und Studienreiseführer zur Heiligen Stadt, pp. 436–437

25. Schmaltz, K.: Die Grabeskirche in Jerusalem, pp. 68–69. Heitz und Mündel, Strassburg (1918)

26. Brooks, W.: An Armenian visitor to Jerusalem in the VII century. English historical review II, pp. 93–94 (1896)

27. Biddle, M.: Das Grab Christi, p. 87

28. Cheikho, L., Carra de Vaux, B. (eds.): Annales Yahia Ibn Said Antiochensis, CSCO 51. Yahya History, Beirut & Paris, pp. 491–492

29. Küchner, M.: Jerusalem, Ein Handbuch und Studienreiseführer zur Heiligen Stadt, p. 448

30. The fire miracle of Jerusalem must be dated since the beginnings of the Church of the Holy Sepulchre, and that because already Egeria mention it on 385 AD. (Bishop Auxentio. (1993). The Paschal Fire in Jerusalem, (Saint John Chrysostom Press) Berkeley California p. 23.)

31. Peters, F.E.: Jerusalem: The holy city in the eyes of chronicles, Princeton, p. 258 (1985)

32. Skarlakidis, Ch.: Heiliges Licht. Das Wunder der Herabkunft des Auferstehungslichtes a am Grab Christi, Elea, Athens, pp. 58–63 (2017)

33. Krey, A.C.: The First Crusade, The accounts of Eyewitnesses and Participants, Princeton, p. 34. "Quam preciosus Sepulture Domini locus: locus concupiscibilis, locus incomparabilis. Neque sinquidem ibi Deus adhuc annuum pratermisit facere miraculum cum in diebus passionis sue extinctis ominus et in sepulchro in elcclesia circum circa luminibus, iubare divino lampades extimcte reaccenduntur. Cuius pectus silicinum fratres, tantum miraculum non emolliat? Creditemihi bestialis homo et insulsi capitis est, cuius cor viratus divina tam praesens, ad fidem non evebrat" (Baldric, Hisoria Ierosolimitana, in: Gesta Die per Francos, Hannover (1611), p. 87 (1921)

34. Biddel, M.: Das Grab Christi, p. 104

35. Biddel, M.: Das Grab Christi, p. 109

36. Biddel, M.: Das Grab Christi, p. 110
37. Biddel, M.: Das Grab Christi, p. 115
38. Anonymous ink drawing of the Holy Sepulchre, Biblioteca Vaticana, Cod Urbinate n 1362
39. Da Treviso (Publ.) Fonifazio Stefani, Liber de perenni cultu Terrae Sanctae et de fructuosa eius peregrinatione, Venice, p. 188 (1875)
40. Augsburgerische Ordinari Postzeitung, Nr. 76, pp. 1–2, Thursday 30 March 1809
41. Vincent, L., Abel, F.: Jerusalem. Recherches de topographie, d'archeologie et d'histoire, ii, Jerusalem Nouvelle, Paris, p. 298, Nr. 1 (1914)
42. The new cross on the top of the Aedicule can also be a symbol of better understanding among the confessions in the Holy Sepulchre. The ideologizing of Symbols in such places leaves room for forces who would like to harm this place. This cross can also be and is a sign of good cooperation as it was with the cross above the dome of the Holy Sepulchre 1996, between the Greek Orthodox and the catholic Church. (Kühnel G., Ein neues Kreuz für die Grabeskirche – Kunstgeschichtlich Überlegungen, in: Hammers, M. & Nagel J. (1999), Geist und Hände Werk. Metallarbeiten im kirchlichen Raum, Coleman (Ed.), Köln pp. 12–19)

Preliminary Assessment of the Structural Response of the Holy Tomb of Christ Under Static and Seismic Loading

Constantine C. Spyrakos[1], Charilaos A. Maniatakis[1(✉)] [ID],
and Antonia Moropoulou[2]

[1] School of Civil Engineering Laboratory for Earthquake Engineering,
National Technical University of Athens, Polytechnic Campus, 9 Heroon
Polytechniou Street, 15780 Zografos, Athens, Greece
cspyrakos@gmail.com, chamaniatakis@gmail.com
[2] School of Chemical Engineering, Section of Materials Science
and Engineering, National Technical University of Athens, Polytechnic Campus,
15780 Zografos, Athens, Greece
amoropul@central.ntua.gr

Abstract. The complex structure that houses the Holy Tomb of Christ, the so called Holy Aedicule, in the Most Holy Church of the Resurrection in Jerusalem has been imposed to considerable damage and structural reformations during its long history. Before the most recent rehabilitation works performed between 2016 and 2017 by the interdisciplinary team of the National Technical University of Athens, the Holy Aedicule has been reconstructed in 1810; however, no later than the first half of the 20th century, a supporting structure was placed at the monument in order to prevent it from further damage. Extensive structural and non-structural damages in recent years, have led to the urgent need for rehabilitation measures. This paper provides a first stage evaluation of the initial condition of the Holy Aedicule under static and seismic loading. This assessment is a part of the series of studies that preceded the rehabilitation works that were performed by the interdisciplinary team of the National Technical University of Athens, completed in March 2017. Based on this initial evaluation a retrofit scheme was applied in order to eliminate the weaknesses of the bearing structure.

Keywords: Holy Aedicule · Holy Tomb of Christ · Finite element method · Structural assessment

1 Introduction

Unlike the design of new constructions where a clear regulatory framework is available, the rehabilitation of historic buildings and monuments is governed by a number of directives, such as The Athens Charter [1], the Venice Charter [2] and the declaration of Amsterdam [3], which only provide the general principles and the philosophy of restoration without, however, going into details on the adequacy of all the available restoration techniques. This fact should not be a surprise, as it was the case until

© Springer Nature Switzerland AG 2019
A. Moropoulou et al. (Eds.): TMM_CH 2018, CCIS 961, pp. 44–57, 2019.
https://doi.org/10.1007/978-3-030-12957-6_3

recently, even for conventional existing constructions. For example, at a European level, only in 2004 a common policy on the retrofit of conventional buildings was adopted with the implementation of the Part 3 of the Eurocode 8 [4].

Because of this ambiguity and the lack of regulatory framework, inappropriate rehabilitation methods have been repeatedly applied to monuments in the past. An example of an inappropriate intervention that has been extensively applied in many cases of monumental structures in Europe and Asia is the replacement of old stones with new ones and the re-pointing with cement-based mortar. Even though the negative effects of the use of cement as a retrofit technique for masonry structures are well addressed in the literature, e.g., [5–7], the method is still applied at many heritage sites. In order to avoid choosing the wrong methods of intervention, the development of codified provisions is needed in global scale that should address different aspects of the rehabilitation works. However, such an effort presents significant difficulties since the behavior of every monument might be unique because of its structural evolution, the state of damage and the risks that threaten its integrity. For this reason a first thorough "structure by structure" understanding is needed [8] in order to determine the causes of vulnerability. In earthquake prone countries, seismic loading is usually the main cause of damage for monuments [9]; however, other causes of damage cannot be excluded, including the climate change or the use of inadequate restoration materials, as reported above.

International experience in the field of restoration of monuments has shown that successful rehabilitation requires the collaboration of specialists from different fields of science. During the last decade the growing concern regarding the seismic vulnerability of monuments lead to the establishment of several national provisions for the seismic strengthening of monuments, e.g., [10]. However, these provisions mainly refer to structural issues regarding intervention strategies, such as the effectiveness of different modeling approaches and the applicability of different analysis methods.

From a structural point of view the assessment of historic structures and monuments under static and seismic loads is a task that presents significant challenges related to the selection of: (a) an appropriate modeling method; (b) an appropriate analysis method; (c) an acceptable performance of the retrofitted structure. It has been observed that specially regarding structures made of unreinforced masonry the development of local collapse mechanisms should be expected under earthquake loads, rather than a behavior of a continuum structure, especially when initial cracking is present. In that case, additionally to the usual global analysis that considers all structural parts inter-connected [11–14], limit analysis should also be performed usually followed by a kinematic approach [14, 15]. In case adjacent parts of the structure have been built during different construction phases the significance of pounding phenomena should also be taken into consideration with appropriate modeling and analysis [16]. Another issue of concern is that, because of limitations stemming from internationally accepted guidelines for the interventions to cultural heritage structures, very often there is a difficulty to fulfill the performance level, required for new constructions within a specific conventional life. This critical issue is discussed in detail in [17, 18].

Based on the available provisions regarding the restoration of cultural heritage structures, [e.g., 1–3] several principles should be adopted by the restorers-engineers including: (i) durability of the retrofit technique; (ii) interventions that should be distinguished from the original structure; (iii) 'minimal impact' of the restoration and its potential reversibility; (iv) respect for the authenticity; (v) physical-chemical compatibility of the new materials.

The satisfaction of these principles should be achieved by a cross-disciplinary approach that may include, depending on the monument, architects, structural engineers, surveying engineers, chemical engineers, art historians, archaeologists, experts in the preservation of materials and conservators [19]. Even though the contribution of every aforementioned discipline for the protection of cultural heritage is well documented, independently from each other, in the literature, e.g., [9], there are very limited case studies that highlight this interdisciplinary approach [20].

From this point of view, the rehabilitation of the Holy Aedicule in Jerusalem suggests an exemplary case of intervention on a monument that held broad cooperation of many experts in order to apply an optimal method of rehabilitation with the maximum documentation of interventions. The evolution in different scientific fields has been applied and the expertise of different disciplines was used to fulfill the basic principles of restoration.

The Holy Aedicule, that is the structure that houses the remaining parts of the original tomb of Christ, is located inside the Holy Church of the Resurrection in Jerusalem at the west side of the Katholikon of the Resurrection, under the great dome, and is the pilgrimage center of the entire Temple complex. The original tomb, i.e., the Holy Sepulcher, was a typical Jewish monument carved into the rock according the evangelical references. During the construction phase of Constantine the Great (326–335 AD), the hill of Golgotha was carved in a way that the portion of the Sacred Rock, which was the Holy Sepulcher, became independent from the rest rocky slope. The other parts of the hill were leveled to construct the first worship structures [21, 22].

The Holy Aedicule is the composite protective structure that has integrated, inside the preserved part of the Sacred Rock of Golgotha, the original tomb. It has been evolved to its current form over the centuries, after repeated damages and reconstructions. After the initial construction phase, between the most important restoration works carried out at the Holy Aedicule during its long life, it is worth mentioning the following [22, 23]:

(i)   the restoration works that followed the disaster from Persians in 614 AD known as the construction phase of Patriarch Modestos;

(ii) the restoration works performed in the 11th century during the reign of Constantine IX the Monomachus;

(iii) the restoration works of 1119 AD carried out by the sculptor Renghiera Renghieri;

(iv) the restoration works of 1808–1810 AD carried out by Nikolaos Komnenos, a Greek Architect native of Lesvos island;

(v)   the most recent rehabilitation works performed by the interdisciplinary team of the National Technical University of Athens, NTUA, between March 2016 and March 2017.

The present work provides a part of the preliminary assessment of the structural condition of the Holy Aedicule, completed by February 2016, which allowed the interdisciplinary team of NTUA to draw conclusions regarding its structural integrity and to propose a suitable retrofit scheme [24]. This assessment utilizes the results of all the diagnostic work carried out by that time, in order to assess the adequacy of the monument to static and seismic loads through elaborated finite element models and analysis [25, 26]. The present work includes the initial part of the assessment that was completed prior to the initiation of the restoration works.

2 Recent Structural History and Observed Damages Prior Interventions

As already mentioned, before the NTUA restoration, the most recent extensive rehabilitation works took place by Komnenos and his collaborators, dating back to 1810. These restoration works followed a devastating fire that provoked significant damage to the Holy Church of the Resurrection and the Holy Aedicule in 1808. According to the descriptions of that time the fire caused the collapse of the great Rotunda dome and severe damage to the Aedicule itself while the whole remaining part of the original Holy Tomb has been revealed. After this catastrophic event a re-construction of the Holy Aedicule took place at the form that is known from thereafter [27].

In 1927 AD a significant earthquake took place with a seismic magnitude $M_S = 6.25$ [28] that provoked damage related to a seismic intensity in the order of MMS VII at the cities of Jericho and Jerusalem [29]. Significant damage has been recorded at the city of Jerusalem that is located within the range of 25–30 km from the Dead Sea, including 25 fatalities, 38 injuries and collapse of 250 houses in the old city area. There is also evidence that the ground acceleration reached particularly high values [28]. Damage was reported at the stone dome of the Katholikon and its supporting drum, while evidence of significant damage was reported also for the Holy Sepulcher [30].

In 1947 a supporting structure made of iron was placed externally of the Holy Aedicule by the British authorities to prevent deformation of the external marble panels that have been observed to present severe out-of-plane deformations. The structure consists of vertical, horizontal and crosswise beams. Timber wedges were placed between the external marble-stone walls and the steel girders along the northern and southern side; however, further loosening of the external walls has been reported since the initial installation of the iron structure [31, 32].

During the series of inspections performed by the NTUA team in 2015 and 2016 it became evident that the major structural deficiencies of the Holy Aedicule were the following:

(a) *Deformation of the external marble panels*. The external marble stones were tilted relatively to the vertical with deformations that reached approximately 7 cm at a height of 80 cm from the base [33]. However, it should be mentioned that this deformation was not observed also to the internal structural body of the Aedicule.

(b) *Deformation of the iron girders*. The use of wedges and the progressive deformation of the outer marble panels have caused significant deformations also to the vertical elements of the metal supporting structure in both horizontal directions [34].

(c) *Loss of mortar*. An extensive loss of adhesion and detachment between the internal mortars and the external covers with marble-stones was observed [33].

(d) *Loosening of the tendons of the iron cage*. The in situ measurements of the tendons connecting the metal girders of the supporting structure along the north-south direction [35] revealed that all but one tendons were relaxed.

In addition to these structural weaknesses, a series of non-structural damage influenced the aesthetics of the structure, in the form of corrosion and chromatic lesions of the external and internal surfaces that may be attributed to humidity. Another challenge in evaluating the monument was to document its possible behavior in the occurrence of a large earthquake.

3 Structural Parts and Modeling Assumptions

The current form of the Holy Aedicule, as already mentioned, has emerged following the most recent restorations carried out after the devastating fire of 1808 [22, 23]. The bearing body of the Holy Aedicule consists of layers of different materials that have been constructed in different historical phases to form structures that surround the Sacred Rock of Golgotha.

The Holy Sepulchre has maximum external dimensions of 7.85 m and 5.20 m in the east-west and north-south direction, respectively. It has a single entrance at the east façade. Internally, it consists of two chambers as shown in Fig. 1: (a) the hall or Chapel of the Angel with internal dimensions 2.9 m × 3.4 m, and (b) the main burial chamber with internal dimensions of 2.0 m × 1.9 m. Inside the Chapel's bearing structure there are two stairways that lead to the roof of the Holy Aedicule at a height of 5.6 m. The total thickness of the walls of the Holy Aedicule ranges from 0.9 m to 1.7 m. The main bearing structure was initially analyzed in its state prior interventions. Rich decoration of marble panels and columns has been placed in the internal and external surfaces.

On the west side of the roof and above the Burial Chamber an outer dome is raised as a distinct structure that is not visible from the inside with an upper level at approximately 12.10 m as shown in Fig. 1. These layers, as has emerged from the relative diagnostic studies [37] include, in addition to the part of the Sacred Rock, parts of masonry of the Crusaders' period mortar belonging to different historical periods and marble panels and columns.

All the data and information provided by the diagnostic work that was carried out by the research team of the NTUA were incorporated in the development of the finite element models. The results of the diagnostic works included: (i) detailed architectural and geometric documentation; (ii) evaluation of the mechanical and chemical properties of the materials through in situ and laboratory measurements and tests; (iii) determination of their boundaries within the bearing body of the structure; and (iv) measurement of modal characteristics of the structure in the field.

The dominant eigenfrequencies of the finite element model regarding the main bearing structure present acceptable accuracy in comparison with the measured frequencies; thus, this first simulation stage can be considered as reliable. However, the reliability of the models was improved with the progress of restoration work during which the removal of the outer layers allowed an enhanced knowledge of the boundaries of the materials and an enrichment of samples for the measurement of the

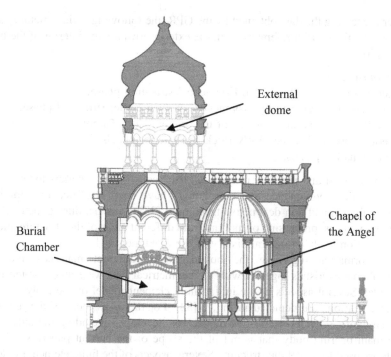

External
dome

Chapel of
the Angel

Burial
Chamber

Fig. 1. West-East elevation of the Holy Aedicule (based on [22, 36]).

mechanical and chemical characteristics. With the progress of the works further analysis was conducted using the updated data.

The thorough diagnostic studies, already conducted by the research team of the NTUA, allowed to explore the chemical composition and mechanical properties of the different materials and their boundaries within the bearing body of the Holy Aedicule [27, 37], to document the external and internal geometry of the surfaces [38] and to measure the dynamic characteristics of the structure [35].

For the needs of the geometrical documentation, a high resolution three-dimensional model was developed with the combination of terrestrial laser scanning at the outside and inside of the monument and an automated image based method. The usefulness of the model was threefold [38]: (a) allowed the performance of accurate geodetic measurements for the production of conventional 2D products, such as sections and facades; (b) provided visualizations; (c) supported decision making and documentation of the restoration works. The determination of geometry was of critical importance for the development of the finite element models.

The internal structural layers of the Holy Aedicule have been identified prior to the interventions by scanning the internal and external surfaces of the monument with a ground penetrating radar (GPR) [37]. The GPR technique is a non-destructive method that detects the stratification within the body of a material or a soil that is not visible by transmitting electromagnetic pulses and recording the obtained signal after reflection and superposition of the pulses on the boundaries of different materials [39].

After processing the data obtained by the GPR, the following main structural layers have been identified, with reference from the exterior towards the interior of the burial chamber [40]:

(a) exterior panels;
(b) filler mortar placed during the Komnenos' structural phase;
(c) main masonry dating to the Komnenos' or even earlier structural phase;
(d) Holy Rock, that is the remnants of the original Holy Tomb;
(e) interior masonry between Holy Rock and interior panels;
(f) interior marble panels.

The mechanical and physical properties of the constituent materials together with their composition, provenance and morphology have been identified by applying a series of analytical and non-destructive techniques [38]. Later on, during the evolution of restoration works, processing of mortars from different areas of the structure allowed the identification of the structural history of the monument [41].

The information regarding the boundaries of different materials and their mechanical characteristics together with the geometrical data were incorporated in the FEM model. Several aspects of the initial finite element model of the Holy Aedicule before interventions are depicted in Fig. 2 [34]. For the assessment of the retrofitted bearing structure, an elaborated FEM model was developed including the outer dome and the main bearing body that is out of the scope of the current paper. Solid finite elements are used to model the structure. Several aspects of the finite element model are depicted in Fig. 2 for the Holy Aedicule at its initial state [34]. A finite element model was also developed for the metal supporting structure that consists of linear elements as shown in Fig. 2c.

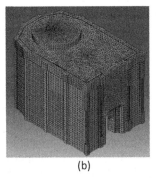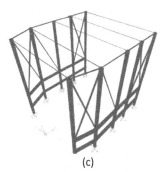

(a)   (b)   (c)

Fig. 2. Initial modeling phase for the assessment of the Holy Aedicule before interventions: (a) finite element mesh for the dome; (b) finite element mesh for the main bearing body; (c) finite element model of the peripheral metal supporting structure.

4 Static and Dynamic Loads

The static loads on the bearing structure are imposed from the self weight of various different materials and are calculated from the geometrical dimensions of the components and the specific weight of each material. In addition to the static loads from its constitutive materials, the bearing structure of the Holy Sepulcher was assessed for the seismic forces that may threaten its structural integrity.

The seismic forces were based on the historical seismicity of Jerusalem, following the current provisions of Eurocode 8 and relevant bibliography [42]. According to the seismic hazard map of Israel a peak ground acceleration of $a_{gR} = 0.13$ g accounts for rock conditions at Jerusalem [43, 44]. This peak ground acceleration refers to a 10% probability of exceedance in 50 years. In order to calculate the acceleration spectrum, the provisions of the current Eurocode 8, EC-8, [42] are applied for Type 1 design earthquake spectrum with a reference peak ground acceleration, $a_{gR} = 0.13$ g, ground type A, and the maximum value for the importance factor, $\gamma_I = 1.40$. The elastic horizontal spectrum is depicted in Fig. 3.

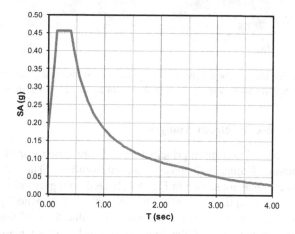

Fig. 3. Design acceleration spectrum according to the Israeli standard [44] and Eurocode 8 relationships [42] for $a_{gR} = 0.13$ g.

5 Analysis and Indicative Results

For the assessment of the Holy Aedicule before and after the application of the retrofit scheme, the following types of analysis were performed:

1. Analysis for static loads.
2. Modal analysis, MA, to determine the eigenmodes and eigenvectors.
3. Modal response spectrum analysis, MRSA, applying the design spectrum.

The calculations were based on the consideration that the holy Aedicule is monolithically supported by a homogeneous elastic half-space. In the following some indicative results are presented.

5.1 Modal Analysis

The first significant modes for the Holy Aedicule, at each initial state prior to the interventions, along the horizontal direction z-z (North-South) and x-x (East-West) are depicted in Fig. 4(a) and (b), respectively. The eigenperiods resulting from modal analysis agree to an acceptable level with the eigenperiods that have been measured in-situ applying modal testing [35].

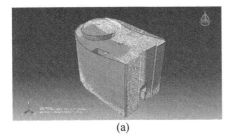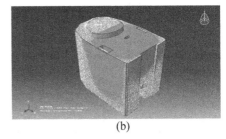

(a)   (b)

Fig. 4. Dominant eigenmodes for the main bearing structure in its initial state: (a) first dominant mode along z-z direction (north-south); (b) first dominant mode along x-x direction (east-west).

5.2 Modal Response Spectrum Analysis

In Fig. 5 the contours of maximum principal stresses are shown for the main bearing body of the structure (Komnenos' phase) prior interventions. The areas of principal stress strength exceedance are shown in Fig. 5 indicated with gray color and red circle. Based on the initial assumptions regarding the strength of Komnenos' phase masonry the analysis yields maximum stresses at the internal vaults, the wall that separates the two chambers and at the area of the internal stairwells where masonry thickness reduces. It became evident that a proper retrofit scheme was needed to provide adequate strength to the structure.

Modal response spectrum analysis was performed for the metal supporting structure applying the Eurocode 8 provisions and considering seismic hazard in the area as discussed above. In Fig. 6 the design results are shown following the provisions of Eurocode 3 [42].

Highlighting with red in Fig. 6 suggests exceedance of resistance. The results show a widespread deficiency in all columns at lower levels (wedging positions) and in the horizontal elements which are placed to connect the columns at their top. All deficiencies detected regard the interaction of axial load and biaxial bending. Based on these results the removal of the metal supporting structure was suggested.

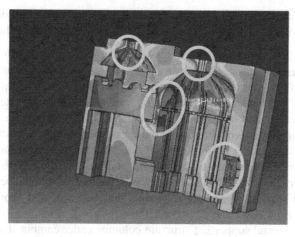

Fig. 5. Maximum principal stresses under seismic loading for the main bearing body of the Holy Aedicule (Komnenos' phase) at the initial state prior restoration. (Color figure online)

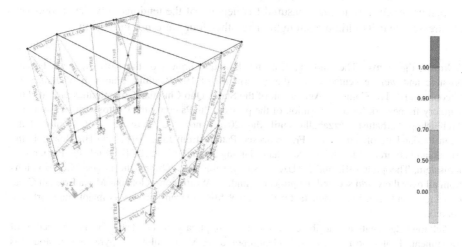

Fig. 6. Resistance assessment for the envelope of seismic combination. (Color figure online)

6 Conclusions

This paper provides the most important findings regarding the preliminary assessment of the structural integrity of the Holy Aedicule of the Holy Sepulcher in the Most Holy Church of the Resurrection in Jerusalem to static and seismic actions. The structural integrity of the Holy Aedicule is assessed by means of numerical models developed using the finite element method.

Based on the analytical work presented herein the bearing structure at its state prior the interventions was expected to present a number of failures under seismic loading the extent of which depends to a large extent on the strength of the individual structural materials. The failures in the Holy Aedicule can be summarized as follows:

a. Localized failure under tension of masonry that surrounds the Holy Rock. The failure is extended; the results however were based on assumptions regarding the tensile strength of mortar for which there was initially a lack of experimental data regarding its mechanical properties. The measurement of this tensile strength through refinement and the exact representation of the geometry allowed for a more accurate estimation of probable failure during the progress of the restoration works.
b. Localized failures under tension regarding the mortar of Komnenos' construction phase. They referred mainly to the internal stairwells at the Chapel of the Angel where masonry thickness drastically decreases, the base of internal bearing columns at the Burial Chamber, the internal vaults and the wall that separates the two chambers of Holy Aedicule.
c. Failures of the metal supporting structure columns under combined axial force and bi-directional bending under earthquake loads. Along the East-West direction the placement of the metal supporting structure and the transfer of forces only through friction were not expected to prevent the formation of mechanism for the marble panels. According to the measured frequencies of the tendons only the second was stressed near its ultimate strength. The other four were not in tension.

Acknowledgements. The study and the rehabilitation project of the Holy Aedicule became possible and were executed under the governance of His Beatitude Patriarch of Jerusalem, Theophilos III. The Common Agreement of the Status Quo Christian Communities provided the statutory framework for the execution of the project; His Paternity the Custos of the Holy Land, Archbishop Pierbattista Pizzaballa (until May 2016 – now the Apostolic Administrator of the Latin Patriarchate of Jerusalem), Fr. Francesco Patton (from June 2016), and His Beatitude the Armenian Patriarch of Jerusalem, Nourhan Manougian, authorized His Beatitude the Patriarch of Jerusalem, Theophilos III, and NTUA to perform this research and the project. Contributions from all over the world secured the project's funding. Worth noting Ioanna- Maria Ertegun Great Benefactor and Jack Shear Benefactor through WMF, Aegean Airlines as major transportation donor et al.

Acknowledgements are attributed to the interdisciplinary NTUA team for the Protection of Monuments, Professors Em. Korres, A. Georgopoulos, A. Moropoulou, C. Spyrakos, Ch. Mouzakis and specifically A. Moropoulou, as Chief Scientific Supervisor, of the rehabilitation project.

References

1. The Athens Charter for the Restoration of Historic Monuments. First International Congress of Architects and Technicians of Historic Monuments, Athens (1931)
2. International Council on Monuments and Sites – ICOMOS: International Charter for the Conservation and Restoration of Monuments and Sites (The Venice Charter 1964). Second International Congress of Architects and Technicians of Historic Monuments, Venice (1964)
3. Congress on the European Architectural Heritage: The Declaration of Amsterdam, Amsterdam (1975)

4. Comité Européen de Normalisation (CEN): EN 1998-3:2004. Eurocode 8: Design of structures for earthquake resistance - Part 3: Assessment and retrofitting of buildings. CEN, Brussels, Belgium (2004)
5. Dimes, F.G., Ashurst, J.: Conservation of Building and Decorative Stone. Routledge, London (2007)
6. Přikryl, R., Smith, B.J. (eds.): Building Stone Decay: From Diagnosis to Conservation. Geological Society of London, London (2007)
7. Quist, W.J.: Replacement of natural stone in conservation of historic buildings, Evaluation of replacement of natural stone at the church of Our Lady in Breda. Heron **54**(4), 251–278 (2009)
8. Oliveira, C.S.: Seismic vulnerability of historical constructions: a contribution. Bull. Earthq. Eng. **1**(1), 37–82 (2003)
9. Lagomarsino, S., Cattari, S.: PERPETUATE guidelines for seismic performance-based assessment of cultural heritage masonry structures. Bull. Earthq. Eng. **13**(1), 13–47 (2015)
10. Italian Building Code, Guidelines: Assessment and mitigation of seismic risk of cultural heritage with reference to the 2008 Italian Building Code (Linee Guida per la valutazione del rischio sismico del patrimonio culturale allineate alle nuove norme tecniche per le costruzioni), G.U. No. 47 (2011). (in Italian)
11. Spyrakos, C.C., Kiriakopoulos, P.D., Smyrou, E.: Seismic strengthening of the historic church of Sts Helen and Constantine in Piraeus. In: Proceedings of the 3rd International Conference on Techniques Methods in Structural Dynamics and Earthquake Engineering COMPDYN 2011, pp. 2401–2413. Institute of Structural Analysis and Antiseismic Research, National Technical University of Athens, Greece (2011)
12. Spyrakos, C.C., Touliatos, P., Patsilivas, D., Pelekis, G., Xampesis, A., Maniatakis, C.A.: Seismic analysis and retrofit of a historic masonry building. In: Syngellakis, S. (ed.) Retrofitting of Heritage Structures - Design and Evaluation of Strengthening Techniques, pp. 65–74. Wessex Institute of Technology Press, Southampton (2013)
13. Spyrakos, C.C., Francioso, A., Kiriakopoulos, P.D., Papoutsellis, S.: Seismic evaluation of the historic church of St. Nicholas in Piraeus before and after interventions. In: Proceedings of 4th International Conference on Techniques Methods in Structural Dynamics and Earthquake Engineering COMPDYN 2013, pp. 3015–3029. Institute of Structural Analysis and Antiseismic Research, National Technical University of Athens, Greece (2013)
14. Spyrakos, C.C., Maniatakis, C.A., Kiriakopoulos, P., Francioso, A., Taflampas, I.M.: Performance of a post-Byzantine triple-domed basilica under near and far fault seismic loads: analysis and intervention. In: Asteris, P.G., Plevris, V. (eds.) Handbook of Research on Seismic Assessment and Rehabilitation of Historic Structures, vol. II, pp. 831–867. IGI Global Editions, Hershey (2015)
15. Spyrakos, C.C., Pugi, F., Maniatakis, C.A., Francioso, A.: Evaluation of the dynamic response for a historic Byzantine crossed-dome church through block joint and kinematic analysis. In: Proceedings of the 5th International Conference on Techniques Methods in Structural Dynamics and Earthquake Engineering COMPDYN 2015, pp. 2354–2364. Institute of Structural Analysis and Antiseismic Research, National Technical University of Athens, Greece (2015)
16. Maniatakis, C.A., Spyrakos, C.C., Kiriakopoulos, P.D., Tsellos, K.P.: Seismic response of a historic church considering pounding phenomena. Bull. Earthq. Eng. **16**(7), 2913–2941 (2018)

17. Spyrakos, C.C.: Seismic risk of historic structures and monuments: a need for a unified policy. In: Proceedings of the 5th International Conference on Techniques Methods in Structural Dynamics and Earthquake Engineering COMPDYN 2015, pp. 2423–2439. Institute of Structural Analysis and Antiseismic Research, National Technical University of Athens, Greece (2015)

18. Spyrakos, C.C.: Bridging performance based seismic design with restricted interventions on cultural heritage structures. Eng. Struct. **160**, 34–43 (2018)

19. Carbonara, G.: An Italian contribution to architectural restoration. Front. Archit. Res. **1**(1), 2–9 (2012)

20. Spyrakos, C.C., Maniatakis, C.A.: Seismic protection of monuments and historic structures - the SEISMO research project. In: Proceedings of the VII European Congress on Computational Methods in Applied Sciences and Engineering ECCOMAS 2016, pp. 5382–5395. Institute of Structural Analysis and Antiseismic Research, National Technical University of Athens, Greece (2016)

21. Varvounis, M.G.: The Holy Sepulcher and the Temple of Resurrection (Ο Πανάγιος Τάφος και ο Ναός της Αναστάσεως). Helandion Editions, Athens (2009). (in Greek)

22. Lavvas, G.: The Most Holy Church of the Resurrection in Jerusalem (Ο Πανίερος Ναός της Αναστάσεως στα Ιεροσόλυμα). Academy of Athens editions, Athens (2009). (in Greek)

23. Mitropoulos, Th.: The Holy Sepulcher of the Most Holy Church of the Resurrection - Brief presentation of history and structural reformations from the ancient times until the fire of the Church of the Resurrection in 1808 (Ο Πανάγιος Τάφος του Πανίερου Ναού της Αναστάσεως - Σύντομη ιστορική και κτηριακή αναδρομή από τους αρχαιοτάτους χρόνους μέχρι την πυρκαϊά του Ναού της Αναστάσεως το 1808) (2015). (in Greek)

24. Moropoulou, A., et al.: Faithful rehabilitation. Civ. Eng. ASCE **87**(10), 54–61 & 78 (2017)

25. Spyrakos, C.C.: Finite Element Modeling in Engineering Practice. Algor Publishing Division, Pittsburg (1995)

26. Spyrakos, C.C., Raftoyannis, J.: Linear and Nonlinear Finite Element Analysis in Engineering Practice. Algor Publishing Division, Pittsburg (1997)

27. Moropoulou, A., et al.: Five-Dimensional (5D) modelling of the Holy Aedicule of the Church of the Holy Sepulchre through an innovative and interdisciplinary approach. Mixed Reality and Gamification for Cultural Heritage, pp. 247–270. Springer, Cham (2017). https://doi.org/10.1007/978-3-319-49607-8_9

28. Rutenberg, A., Levy, R.: Some Comments on Seismicity in Israel, Performance Assessment & Damage to Historic Monuments in Jerusalem. Technion-Israel Institute of Technology, PROHITECH – WP2, DAMAGE ASSESSMENT (2004)

29. Ben-Menahem, A.: Four thousand years of seismicity along the Dead Sea rift. J. Geophys. Res. **96**(812), 20195–20216 (1991)

30. Zohar, M., Rubin, R., Salamon, A.: Earthquake damage and repair: new evidence from Jerusalem on the 1927 Jericho earthquake. Seismol. Res. Lett. **85**(4), 912–922 (2014)

31. Cooper, M.A.R., Robson, S., Littleworh, R.M.: The tomb of Christ, Jerusalem; analytical photogrammetry and 3D computer modelling for archaeology and restoration. Int. Arch. Photogram. Remote Sens. **29**, 778 (1993)

32. Biddle, M., Cooper, M.A.R., Robson, S.: The Tomb of Christ, Jerusalem: a photogrammetric survey: a report of the work undertaken under the aegis of the Gresham Jerusalem Project. Photogram. Rec. **14**(79), 25–43 (1992)

33. Korres, M.: Church of the Holy Sepulchre: Form, Structure, Core Damages, Way of Repair - Interim Report (2016)

34. Spyrakos, C.C., Maniatakis C.A.: Assessment of Current Condition under Static and Seismic Loading and Proposal for Intervention for the Holy Aedicule of the Holy Sepulchre in the Most Holy Church of the Resurrection in Jerusalem (2016)

35. Mouzakis, H.P.: Dynamic Characteristic of The Holy Edicule of Christ's Tomb – Final Report (2016)
36. Lavvas, G., Balodimos, D.: Most Holy Temple of Resurrection, Topographical - Photogrammetric Documentation (Πανίερος Ναός Αναστάσεως, Τοπογραφική-Φωτογραμμετρική Τεκμηρίωση). Greek Orthodox Patriarchate of Jerusalem, Jerusalem (2003). (in Greek)
37. Lampropoulos, K.C., Moropoulou, A., Korres, M.: Ground penetrating radar prospection of the construction phases of the Holy Aedicula of the Holy Sepulchre in correlation with architectural analysis. Constr. Build. Mater. **155**, 307–322 (2017)
38. Georgopoulos, A., et al.: Merging geometric documentation with materials characterization and analysis of the history of the Holy Aedicule in the church of the Holy Sepulchre in Jerusalem. Int. Arch. Photogram. Remote Sens. Spat. Inform. Sci. **42**, 487–494 (2017)
39. Daniels, D.J.: Ground Penetrating Radar, 2nd edn., Radar, Sonar, Navigation and Avionics Series 15. Institute of Electrical Engineers, London (2004)
40. Agrafiotis, P., Lampropoulos, K., Georgopoulos, A., Moropoulou, A.: 3D modelling the invisible using ground penetrating radar. Int. Arch. Photogram. Remote Sens. Spat. Inf. Sci. **42**, 33 (2017)
41. Moropoulou, A., Zacharias, N., Delegou, E.T., Apostolopoulou, M., Palamara, E., Kolaiti, A.: OSL mortar dating to elucidate the construction history of the Tomb Chamber of the Holy Aedicule of the Holy Sepulchre in Jerusalem. J. Archaeol. Sci. Rep. **19**, 80–91 (2018)
42. Comité Européen de Normalisation, CEN: Eurocode 8 (EC8-1), 2004: Design of structures for earthquake resistance Part 1: General rules, seismic actions and rules for buildings. CEN, Brussels (2004)
43. Shapira, A.: An updated peak accelerations map for the Israeli code SI 413: Explanatory notes. GII Report 592/230/02, Geophysical Institute of Israel, Lod (2002)
44. SI 413, Design provisions for earthquake resistance of structures, amended December 1998, and May 2004. Standards Institution of Israel, Tel Aviv, Hebrew (1995)

Corrosion Protection Study of Metallic Structural Elements for the Holy Aedicule in Jerusalem

Eleni Rakanta, Eleni Daflou$^{(\boxtimes)}$, Angeliki Zacharopoulou,
George Batis, and Antonia Moropoulou

School of Chemical Engineering, Laboratory of Materials Science and
Engineering, National Technical University of Athens (NTUA), Athens, Greece
eldaflou@central.ntua.gr

Abstract. This paper concerns the corrosion evaluation and protection study of metallic structural elements embedded in mortars and stone. The corrosion protection study of the metallic structural elements was evaluated at the Holy Aedicule monuments in Jerusalem. Chapel Aedicule, which contains the Holy Sepulchre itself, located in the center of Rotunda. The Aedicule has two rooms, the first holding the Angel's Stone, which is believed to be a fragment of the large stone that sealed the tomb; the second is the tomb itself. In order to assess the environment and the corrosion rate of the metal components, visual inspection, corrosion potential of embedded metals (Open Circuit Potential) and the electrical conductivity of mortars and stones were measured. The afore-mentioned measurements resulted that the environment of the metal elements has low to moderate corrosive capacity. However, even if the corrosion environment is moderate, always in the case of monuments when metallic elements involved should be protected.

Keywords: Holy Aedicule · Corrosion protection · Organic coatings · Corrosion inhibitors

1 Introduction

One of the most important historical and holiest sites of the Christianity is the Church of the Holy Sepulchre (Church of the Resurrection) in Jerusalem. The Holy Aedicule's of the Holy Sepulchre structure envelopes what is believed to be the Tomb of Christ, the place where he was buried and the site of His resurrection. The Holy Aedicule evolved and transformed throughout the centuries, changing in form, size and shape as a result of various construction phases, damages and destructions, reconstructions, and protection interventions [1, 2]. From a typical hewn rock tomb of the 1st century CE, the Tomb of Christ transformed into a ciborium type structure (Aedicule) enveloping the rock tomb during the reign of Constantine the Great in the 4th century. Throughout the centuries, the Holy Aedicule sustained many damages (deliberate destructions, destructive earthquakes, fires) and consequent restorations and reconstructions, while it expanded to the east when the Crusaders added an ante-chamber, the Chapel of the

© Springer Nature Switzerland AG 2019
A. Moropoulou et al. (Eds.): TMM_CH 2018, CCIS 961, pp. 58–68, 2019.
https://doi.org/10.1007/978-3-030-12957-6_4

Angel, in the 12th century. Its polygonal form up to then most possibly changed at this period giving the Aedicule its characteristic since then horse-shoe shape [3, 4].

The latest reconstruction was implemented in 1809–1810 by the Architect "Kalfas" Komnenos, after a great fire broke out in the Rotunda area, severely damaging the external facades of the Aedicule. Serious deformations and displacements were observed on the Holy Aedicule structure already by the early 20th century, especially after the destructive earthquake of 1927. The observed buckling of the marble slabs and the intense deviation from verticality led to the installation of an iron grid from the British Mandate in order to avoid collapse of the structure [5–8].

In 2015, after invitation of his Beatitude, Patriarch of Jerusalem Theophilos III, the National Technical University of Athens (NTUA) Interdisciplinary Team for the Protection of Monuments, performed an "Integrated Diagnostic Research Project and Strategic Planning for Materials, Interventions Conservation and Rehabilitation of the Holy Aedicule of the Church of the Holy Sepulchre in Jerusalem" in order to examine the observed damages and deformations, explain their cause, and propose appropriate measures to restore the structure. Based on the results of this study and the rehabilitation interventions proposed, a rehabilitation project was initiated, supervised by NTUA, after the Common Agreement of the three Christian Communities responsible for the Holy Sepulchre. The rehabilitation project was successfully implemented and completed in March 2017 [9, 10].

This paper concerns the corrosion evaluation and protection study of metallic structural elements embedded in mortars and stone. The corrosion protection study of the metallic structural elements was evaluated at the Holy Aedicule of the Holy Sepulchre in Jerusalem.

The metallic elements which protected divided into three categories:

- Metallic elements exposed to the atmospheric environment. These are the metallic tie for the earthquake protection existed (Fig. 1) and the metallic supports of the roof of the Coptic Chapel (Fig. 2).
- Steel elements which were covered with stones and mortars, but revealed during the rehabilitation such as panels that support the Holy Aedicule of the Holy Sepulchre (Fig. 6).
- Steel elements that were embedded into the mortar (Fig. 5).

For the aforementioned categories a different method of protection was applied. Για κάθε μία κατηγορία ακολουθήθηκε διαφορετική μέθοδος προστασίας.

2 Evaluation Methods

In order to evaluate the induced corrosion process the following electrochemical techniques were performed.

(a) Measurements of Half-cell potential or Corrosion Potential E_{corr}. The half-cell potential of steel in the environment was measured versus a saturated calomel electrode (SCE) using a high impedance voltmeter (10 MΩ). The implementation of the method and the interpretation of the half cell potential values were performed according to *ASTM C876* "standard Test Method for Half-Cell Potential of Reinforcing Steel in Concrete".

(b) Electrical resistivity mortar measurements. These measurements were made with four – probe electrode (Wenner method) with resistivity meter of Proceq Testing Instruments. The correlation between resistivity values and corrosion risk were approached according to empirical data referring to depassivated steel [11].

(c) The protection of the iron elements in the mortars was done with corrosion inhibitors. In order to determine the corrosion inhibitor effectiveness in the aerated mortar [1, 6], a salt chamber test was carried out. A 50 cm long concrete rebar was placed on a rectangular mortar specimen (80 × 80 × 30 cm). VCI corrosion inhibitor emitters were placed on both of its sides (Fig. 3) and placed in a salting chamber for 30 days in order to examine the results in accelerated corrosion conditions. The corrosion behavior was assessed by measuring the corrosion current density during the 30 days interval by using the Gecor 8 corrosion rate meter of NTD James instruments INC.

3 Results and Discussion

3.1 Half Cell Potential and Electrical Resistivity Measurements

Since corrosion is an electrochemical phenomenon, the half cell corrosion potential of metal elements and the electrical resistivity of the mortar will have a bearing on the corrosion rate of metal elements as the electric current in the form of a flow of ions must pass from the anode to the cathodes for corrosion to occur. Half-cell potential measurements of steel elements is the most typical procedure to the routine inspection of structural metal elements embedded into the mortar regarding the corrosion risk and gives an indication of corrosion trend of the metal. Its use and interpretation are described in the ASTM C876 standard Test Method for Half-Cell Potential of Reinforcing Steel in Concrete. Potential readings, however, are not sufficient as criterion, since they are affected by number of factors, which include polarization by limited diffusion of oxygen, mortar porosity and the presence of highly resistive layer. According to ASTM C 876, potentials more negative than −350 mV, with respect to SCE, indicate greater than 90% probability of active metal corrosion. Values less negative than −200 mV SCE indicate a probability of corrosion below 10%, while those falling between −200 and −350 mV SCE indicate uncertainty of corrosion [11, 12]. The half cell potential measurements and the specific locations that measured are shown in Table 1 and Figs. 1, 2 respectively.

The electrical resistivity is an indication of the amount of moisture in the pores and the size and tortuosity of the pore system of mortar. Resistivity is strongly affected by mortar quality. The salt level does not affect resistivity directly as there are plenty of ions dissolved in the pore water. However, salts such as chlorides in mortar can be hydroscopic and trend the mortar to retain water resulting in a reduction of mortar resistivity. The electrical resistivity of mortars was measured with Wenner method using four probe electrodes. In Wenner method current is applied between the two outer probes and the potential difference measured across the two inner probes. This approach eliminates any effects due to surface contact resistance. For mortar the resistivity ρ is given by the following equation:

$$\rho = 2\pi\alpha V/I$$

Where, α is the electrode spacing, I is the applied current across the outer probes and the V is the potential measured across the inner probes. The electrical resistivity measurements of mortars of Holy Aedicule and Coptic Chapel are presented in Table 2.

Table 1. Half cell potential measurements of metallic elements

	Location	Half Cell Potential vs SCE (mV)
Coptic Chapel	1	−153
	2	−161
	3	−382
	4	−290
	5	−185
	6	−281
	7	−122
	8	−126
Metallic tie of Holy Aedicule - Church of the Holy Sepulchre	1	−83
	2	−94
	3	−130
	4	−132
	5	−120
	6	−105

Fig. 1. Locations of Half cell potential measurements in metallic tie of Holy Aedicule - Church of the Holy Sepulchre.

Fig. 2. Locations of Half cell potential measurements in Metal elements of Coptic Chapel.

Table 2. Electrical Resistivity measurements of mortars

Electrical resistivity of mortar (KΩ.cm)													
Coptic Chapel								Holy Aedicule - Church of the Holy Sepulchre					
90	95	53	85	64	>99	>99	86	90	>99	95	>99	>99	>99

The Coptic Chapel, however, which is a small, partially enclosed space framed in iron lattices, was relocated from the west side of the aedicule to another section within the Rotunda. A series of iron rods that had been used in the construction of the aedicule had become partially oxidized, and in some locations had even completely disintegrated. According to ASTM C876, the half cell potential values of the examined metal elements suggest that the corrosion risk is moderate to high. In addition the mortars in which metallic elements are embedded are moderate to high in electrical conductivity therefore its corrosive effect is not negligible. In addition, the physicochemical characterization of mortars that carried out from Prof. Moropoulou, A. scientific research team showed high percentages of total soluble salts, with values exceeding 3% in all cases, and reaching up to 14.58%. Chloride salts were detected in all samples through spot tests [1].

The corrosion current density versus time is given in Fig. 4. The decrease of the corrosion current density is typical and similar to that in various concrete or cement mortar experiments. The reduction of the corrosion current density was considered adequate to assume that the corrosion inhibition was successfully functioned.

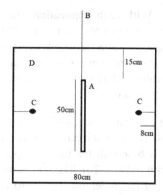

Fig. 3. Schematic illustration of the experiment. Where A: Rebar, B: Cable, C: Corrosion inhibitor emitter and D: Mortar

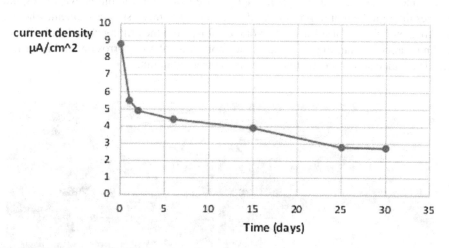

Fig. 4. Corrosion current density of reinforced mortar specimens treated with VCI emitters versus time.

3.2   Interventions to Protect Metal Elements in the Masonry

In order to prevent the corrosion mechanism process corrosion inhibitors and resin coatings was applied. In addition, a number of structural new elements were constructed by titanium bars.

The use of corrosion inhibitors for metal elements embedded within the masonry and in some cases in the ground, the following techniques were implemented according to the specific case.

Case 1: In the case where the reinforcements were exposed to the open air during the restorations, liquid corrosion inhibitor was applied at the surface, in order to prevent the corrosion continuity due to atmospheric environment [12–17]. The spray application was implemented through three distinct applications and each layer implemented at least 6 h after the prior one. With each application, the minimum consumption of sprayed corrosion inhibitor was 0.333 L/m² of mortar. Subsequently, corrosion inhibitor surface treated steel elements were covered with lime-kaolin repaired mortar. Figure 6 shows the application of the liquid corrosion inhibitor to the surface of the metal element prior to application of the repaired mortar. The new reinforcements shown in Fig. 6 are made from titanium.

Case 2: If the metallic element was in a vertical position in relation to the masonry surface or in a parallel position the uses of corrosion inhibitors in the form of cartridges or emitters were preferred. This method is based on the use of vapor phase corrosion inhibitors (VCI = Vapor Corrosion Inhibitors) [18–20]. The corrosion inhibitor is placed within a cylinder with 1 cm diameter and 10 cm length. The corrosion inhibitor is enclosed in a porous material in order to allow sublimation. One cartridge usually offers protection at one meter radius. The application was done through drilling a hole into the mortar and inserting the VCI. If the metal element was at a depth of one meter the drilling stopped at 55 cm and a cartridge was inserted. Vapor phase Corrosion Inhibitors evaporates and diffuses through the mortar pores towards the vicinity of metal surface element, where it absorbed onto the metal surface. The void resulting from the drilling closed with a lime-kaolin repaired mortar. Figure 5 shows the application of VCI corrosion inhibitor emitters. The lattice shown in this figure was made from titanium.

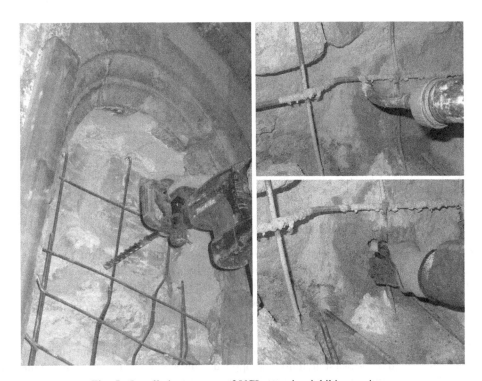

Fig. 5. Installation process of VCI corrosion inhibitor emitters.

Case 3: If the metallic element partially was embedded into the mortars and stones and partially projects into the open atmosphere, the protection of the embedded section into the mortars and stones protected through the use of VCIs, as aforementioned described. The section exposed directly to the environment protected through the use of transparent organic coatings (varnishes). In this case it was assessed whether the coatings should be reversible (easily removed) or permanent.

For the case of reversible coatings (varnishes), the specialized wax varnishes were suggested which can be removed with warm water. However, this material is not preferable for use in warm climates, such as Israel's. A material suggested for use in warm climates was the one-compound polyurethane system, where the varnish was in one bottle. These varnishes "dry" through the moisture present in the atmosphere. The first layer application was applied on steel with oxides which may be porous, as long as greases or oils and loose oxides have been removed with a brush. Its thickness was up to 20 μm. The second application layer was applied onto the first one. Its thickness did not exceed 40 μm. The application was done either with a brush or via a high pressure spray gun. These coatings have been used extensively in Greece presenting excellent results, even in coastal environments [21–23].

Fig. 6. Implementation of liquid corrosion inhibitor to the metallic elements that that have been covered with repaired lime-kaolin mortar.

If reversibility was not deemed mandatory, epoxy varnishes were applied at rebars exposed in atmosphere. Specialized materials were used such as priming in combination with the two compound epoxy varnishes. However in this case it must again be noted that their removal is extremely difficult [24–26]. This system was composed of:

- One application layer of an epoxy priming varnish.
- Two application layers of epoxy varrnish.
- One application layer of polyurethane varnish.

The three first layers were applied in order to ensure protection against corrosion and the fourth one were applied in order to offer protection from ultraviolet radiation.

4   Conclusions

The protection of the metal parts outside the masonry was carried out using organic coatings. An anti-corrosion primer was used in one layer and anti-corrosion varnishes in two layers.

The protection of the metallic elements within the masonry was carried out with vapor phase corrosion inhibitors in the form of suitable emitters (cartridges). The emitters (cartridges) were placed at appropriate locations within the masonry to achieve complete protection against corrosion.

The metal elements, one part of which was inside the masonry and the other accept a complex method of protection. The part in the masonry was protected with vapor phase corrosion inhibitors. The non-masonry section was protected with organic coatings.

Acknowledgements. The study and the rehabilitation project of the Holy Aedicule became possible and were executed under the governance of His Beatitude Patriarch of Jerusalem, Theophilos III. The Common Agreement of the Status Quo Christian Communities provided the statutory framework for the execution of the project; His Paternity the Custos of the Holy Land, Archbishop Pierbattista Pizzaballa (until May 2016 – now the Apostolic Administrator of the Latin Patriarchate of Jerusalem), Fr. Francesco Patton (from June 2016), and His Beatitude the Armenian Patriarch of Jerusalem, Nourhan Manougian, authorized His Beatitude the Patriarch of Jerusalem, Theophilos III, and NTUA to perform this research and the project.

Contributions from all over the world secured the project's funding. Worth noting Mica Ertegun's and Jack Shear's donations through WMF, Aegean Airlines et al.

The interdisciplinary NTUA team for the Protection of Monuments, Em. Korres, A. Georgopoulos, A. Moropoulou, C. Spyrakos, Ch. Mouzakis, were responsible for the rehabilitation project and A. Moropoulou, as Chief Scientific Supervisor, was responsible for its scientific supervision.

References

1. Apostolopoulou, M., et al.: Study of the historical mortars of the Holy Aedicule as a basis for the design, application and assessment of repair mortars: a multispectral approach applied on the Holy Aedicule. Constr. Build. Mater. **181**, 618–637 (2018). COST Action TU1404, Early Age Cracking and Serviceability in Cement-based Materials and Structures
2. Cameron, A., Hall, S.: Eusebius' Life of Constantine. Clarendon Press, Oxford (1999)
3. Pringle, D.: The Churches of the Crusader Kingdom of Jerusalem. A Corpus. The City of Jerusalem, Vol III. Cambridge University Press, Cambridge (2010)
4. Biddle, M.: The Tomb of Christ. Sutton, Gloucestershire (1999)

5. Mitropoulos, Th.: The Church of Holy Sepulchre – The Work of Kalfas Komnenos. European Centre of Byzantine and Post-Byzantine Monuments (2009)
6. Moropoulou, A., Zacharias, N., Delegou, E.T., Apostolopoulou, M., Palamara, E., Kolaiti, A.: OSL mortar dating to elucidate the construction history of the Tomb Chamber of the Holy Aedicule of the Holy Sepulchre in Jerusalem. J. Archaeol. Sci. Rep. **19**, 80–91 (2018)
7. Lampropoulos, K.C., Moropoulou, A., Korres, M.: Ground penetrating radar prospection of the construction phases of the Holy Aedicula of the Holy Sepulchre in correlation with architectural analysis. Constr. Build. Mater. **155**, 307–322 (2017)
8. Moropoulou, A.: Materials & Conservation, Reinforcement and Rehabilitation Interventions in the Holy Edicule of the Holy Sepulchre (National Technical University of Athens, Athens (2016)
9. Labropoulos, K., Moropoulou, A.: Ground penetrating radar investigation of the bell tower of the church of the Holy Sepulchre. Constr. Build. Mater. **47**, 689–700 (2013)
10. Moropoulou, A., et al.: Five-dimensional (5D) modelling of the holy aedicule of the church of the holy sepulchre through an innovative and interdisciplinary approach. In: Ioannides, M., Magnenat-Thalmann, N., Papagiannakis, G. (eds.) Mixed Reality and Gamification for Cultural Heritage, pp. 247–270. Springer, Cham (2017). https://doi.org/10.1007/978-3-319-49607-8_9
11. Broomfield, J.P.: Corrosion of Steel in Concrete Understanding, Investigation and Repair, pp. 63–65. E & FN SPON, London (1997)
12. Batis, G., Routoulas, A., Rakanta, E.: Effects of migrating inhibitors on corrosion of reinforcing steel covered with repair mortar. Cem. Concr. Compos. **25**, 109–115 (2003)
13. Batis, G., Rakanta, E.: Corrosion of steel reinforcement due to atmospheric pollution. Cem. Concr. Compos. **27**, 269–275 (2005)
14. Rakanta, E., Daflou, E., Batis, G.: Evaluation of corrosion problems in a closed air-conditioning system: a case study. Desalination **213**, 9–17 (2007)
15. Zafeiropoulou, T., Rakanta, E., Batis, G.: Performance evaluation of organic coatings against corrosion in reinforced cement mortars. Prog. Org. Coat. **72**(1–2), 175–180 (2011)
16. Zafeiropoulou, Th., Rakanta, E., Batis, G.: Industrial coatings for high performance application: physico- chemical characteristics and anti-corrosive behavior. In: Brick and mortar research, Chapter 9, pp. 245–258 (2012)
17. Rakanta, E., Zafeiropoulou, Th, Batis, G.: Corrosion protection of steel with DMEA-based organic inhibitor. Constr. Build. Mater. **44**, 507–513 (2013)
18. Vyrides, I., Rakanta, E., Zafeiropoulou, Th, Batis, G.: Efficiency of amino alcohols as corrosion inhibitors in reinforced concrete. Open J. Civ. Eng. **3**(2A), 8 (2013)
19. Batis, G., Rakanta, E., Daflou, E.: Corrosion protection of steel reinforcement with DMEA in the presence of chloride ions. In: European Corrosion Conference Long Term Prediction & Modeling of Corrosion, EUROCORR 2004, Nice, France (2004)
20. Daflou, E., Rakanta, E., Batis, G.: Corrosion protection methods of structural steel against atmospheric corrosion. In: 16th International Corrosion Congress, Beijing, China (2005)
21. Rakanta, E., Daflou, E., Batis, G.: Evaluation of organic corrosion inhibitor effectiveness in to concrete. In: Konsta-Gdoutos, M.S. (ed.) Measuring, Monitoring, and Modelling Concrete Properties, pp. 605–612. Springer, Alexandroupoli (2006). https://doi.org/10.1007/978-1-4020-5104-3_73
22. Martínez, I., Andrade, C., Rebolledo, N., Luo, L., De Schutter, G.: Corrosion-inhibitor efficiency control: comparison by means of different portable corrosion rate meters. Corros. J. Sci. Eng. **66**(2), 026001–026012 (2010)

23. Moropoulou, G., et al.: Criteria and methodology for diagnosis of corrosion of steel reinforcements in restored monuments. In: Konsta-Gdoutos, M.S. (ed.) Measuring, Monitoring, and Modelling Concrete Properties, pp. 563–574. Springer, Alexandroupoli (2006). https://doi.org/10.1007/978-1-4020-5104-3_68

24. Zafeiropoulou, Th., Rakanta, E., Batis, G.: Novel coatings: physicochemical characteristics and corrosion evaluation. In: 12th International Conference on Recent Advances in Concrete Technology and Sustainability Issues, Prague, Czech Republic, 24 pages (2012)

25. Zafeiropoulou, Th., Rakanta, E., Batis, G.: Reinforced concrete corrosion control with the usage of nano coatings: comparison with traditional and high performance application systems. In: UKIERI Concrete Congress Innovations in Concrete Construction, Punjab, India (2013)

26. Andrade, C.: Modelling the concrete-real environment interaction to predict service life. Struct. Concr. **16**(2), 159–160 (2015)

Innovative Methodology for Personalized 3D Representation and Big Data Management in Cultural Heritage

Emmanouil Alexakis[1]([✉]), Evgenia Kapassa[2], Marios Touloupou[2],
Dimosthenis Kyriazis[2], Andreas Georgopoulos[3],
and Antonia Moropoulou[1]

[1] School of Chemical Engineering,
National Technical University of Athens, Athens, Greece
alexman@central.ntua.gr
[2] Department of Digital Systems, University of Piraeus, Piraeus, Greece
[3] School of Rural and Surveying Engineering,
National Technical University of Athens, Athens, Greece

Abstract. Three-dimensional (3D) visualization is an effective way not only to study but also to disseminate the built environment particularly the one of the cultural heritage (CH). The extended growth of the technical means have enabled standardized methods for (semi) automatic 3D model production motivating more and more researchers from various backgrounds (e.g. material scientists) to integrate the aspect of 3D modeling within their work despite the certain limitations concerning the amount and types of the data that have to be managed. In order to facilitate the needs of data management, a scalable, interactive, web-based platform is proposed within this work. The case study over which the proposed methodology has been implemented is the recently rehabilitated Holy Aedicule of the Holy Sepulchre in Jerusalem due to the vast amount of heterogenous data produced for the needs of monument's documentation before and during rehabilitation. To address this issue, the present study will provide the missing layer that enables the semantic data labeling, through data processing and filtering, and their spatial representation - visualization. Additionally, a straightforward methodology for affordable but of high quality and precision modeling of cultural heritage data and its exploitation towards the delivery of personalized content is achieved. As a result, the derivative georeferenced high-resolution 3D models are transformed resulting in a multi-resolution model management which under the appropriate configuration of the JavaScript components on an interactive web platform.

Keywords: Cultural heritage · 3D visualization ·
Interactive web-based platform · Data interoperability

A. Moropoulou et al. (Eds.): TMM_CH 2018, CCIS 961, pp. 69–77, 2019.
https://doi.org/10.1007/978-3-030-12957-6_5

1 Introduction

Surrounded by the remains of previous generations, as remainders of the glorious times, events and cultures, cultural heritage (CH) could be considered as a silent witness of the past. Despite their significance, most of those remains do not reflect the magnificence of the objects in their time. Nowadays, digital technologies provide tools to recreate the original shape and appearance of cultural heritage objects enabling the virtual journey through the past. The extended growth of the technical means has enabled three-dimensional (3D) representation as an effective mean for studying and disseminating the built cultural heritage environment [1]. There have been several projects comprehending interactive visualization of cultural heritage assets including from 3D reconstruction of damaged artworks, artifacts to the ones of whole buildings and monuments. Though, ancient history comes to life through virtual walks in Ancient Olympia [2] and the valley of pyramids in Giza [3]. Additionally, Hellenic Cosmos Theater in Athens, hosts a Virtual Reality Programme where visitors can make the tour at the peninsula and the city of Miletus, back in time as it was 2000 years ago.

Appearance and shape visualization of the cultural heritage assets alone cannot satisfy the needs of a holistic digital documentation without the contribution of data coming from various scientific areas e.g. Material Sciences. Thus, 3D modelling and data integration is demanded but there are certain limitations concerning not only the amount but also the diversification of the data that has to be eventually effectively managed [4, 5]. To overcome these restraints, there is the need of a scalable platform to manage this cultural heritage data in three levels: (a) raw data collection and access (acquisition, storage and retrieval), (b) data analytics (correlation, contextualization, knowledge extraction and user-specific annotations), (c) data representation (interpretation and visualization) and data sharing and reusability (interoperability and metadata enriching). For a successful implementation, the interoperability of the data is a prerequisite as it allows the interlinking and correlation among them only if they have been annotated in a certain manner.

The approach suggested in the present work, has been implemented in the case study of the recently rehabilitated Holy Aedicule of the Holy Sepulchre in Jerusalem. The Holy Aedicule of the Holy Sepulchre, after the restoration works took place to amend the destruction caused by the catastrophic fire in the Church of Resurrection in 1808, confronted with deformation issues upon its facades. A diagnostic study took place in order to identify and reveal the reasons behind the deformation phenomena. This study not only illuminated the factors and the processes contributed to the current monument's preservation state but also came up with certain suggestions – proposals, delivered in a rehabilitation works study that eventually was indeed implemented. The phases of the rehabilitation works follow a direct timeline beginning from (a) the dismantling and documentation of the facades building stones, (b) the revelation of the internal historic masonry and the removal of the disintegrated filling and joint mortars, (c) the repointing of the joints, (d) the grouting of the panels for the homogenization of the revealed construction chases and the protection and preservation of the Holy Rock (e) The titanium reinforcement of the structure, (f) reassembly of the stone panels (g) Rehabilitation of the Top Zone, Baluster and Terrace (h) Rehabilitation of the

Onion Dome (i) conservation and rehabilitation interventions inside the Holy Aedicule and (j) the removal of the outer facade metal frame supporting the structure. All of the aforementioned rehabilitation phases have been thoroughly documented and monitored with both Non-Destructive Testing and Analytical Techniques approaches before and during of the rehabilitation works, respectively [6–8].

The vast amount of diverse types of data produced before and during of every phase of the rehabilitation works includes: History and architecture of the building, structural and seismic assessment, materials and geometry. There are already several published studies [7–9], that concern the documentation of the monument within which, the range, the scale and the heterogeneity of the data availability is highlighted. In order to serve the need of an effective data management for scientific decision making, knowledge acquisition and information sharing to the several types of end-users, summarized as the objective of the current work, it is realized through the development of an innovative web-based platform. The utilization of data and more specifically their integration for achieving a multilayered, spatial management of information enhanced in a visual manner, is the innovation lying within the current work. Additional to the ability to store, retrieve and query data (and metadata) of various thematic areas like data types (Historic, Materials, etc.) and evolution in time through the rehabilitation phases (Dismantling the stone panels, titanium reinforcement of the structure, etc.), the platform is ready for cross-discipline data analytics for diagnostic and/or monitoring purposes (i.e. identifying automatically threating events and triggering of alerts).

2   Web Based Platform Architecture

The advantages outlined in the previous section guided the development of a web-based platform for the storage and retrieval of the documentation data collected before, during and after the end of the rehabilitation works of the Holy Aedicule of the Holy Sepulchre. In order to properly manage and integrate the various 3D models within the platform some semantic structuring of the models themselves is required. This is not only for representing the differences in terms of materials, but also for allowing an efficient visualization and structuring the data in an appropriate and accurate way [10]. The baseline software for web interface platform is a 3D Heritage Online Presenter (3DHOP). It is an open-source software package and it has been designed to allow the creation of interactive multimedia web presentations based on 3D digital models in a simple manner. 3DHOP target audience goes from the museum curators with some IT experience to the experienced Web designers who want to embed 3D contents in their creations, from students in the CH field to small companies developing web applications for museum and CH institutions. 3DHOP enables the creation of interactive visualization of 3D models on a common web page, simply by including JavaScript functions in a standard HTML page (by integrating JavaScript files and start up functions [11], and adding specific HTML elements), that can be configured in a declarative fashion [12].

However, to be able to integrate dynamic processes within the modeling, more advanced behavior and integration in the web page is necessary. In order to support the aforementioned challenges and being able to describe more complex configuration of each 3D scene, customized JavaScript functions are included in the web platform, in a way to allow interactively adding georeferenced data and enhance them with additional metadata [13]. Metadata derive meaningful knowledge, such as "hotspot's" coordinates, date, temperature, description etc. The "hotspot" along with its metadata are stored in a database and they are accessible for digitized visualization at the 3D scene. The provided database serves the users via functions for storing, searching, and retrieving "hotspots" information, annotating them with additional representation (i.e. metadata). The role of the hotspots' metadata is of paramount significance in the Web Based Platform functionalities, rendering it an information-driven platform. The information of all stored objects allows the efficient search, correlation between inputs from several end-users, versioning, updates etc. The platform architecture constisted of three-tier layers; (a) Presentation (b) Business Logic and (c) Data Access, is illustrated in Fig. 1.

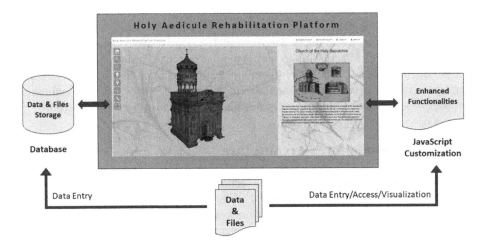

Fig. 1. Web-based integration platform architecture – three-tier architecture layers: presentation, business logic and data access.

3   Results and Discussion

3.1   JavaScript Customization

The customization of the JavaScript functions is aligned to the dynamic changes of the users' needs. In particular, the default features-tools shortly explained in Fig. 2 have been enhanced with additional functionalities, as illustrated in Fig. 3, that concern the:

(a) Visualization of the available data and metadata
(b) Data filtering for efficient data management and provision of personalized content. The filtering is based in many different model levels. Specifically:

- Access data based on rehabilitation phase: (i) initial phase (ii) final phase
- Access data based on 3D model scene: (a) inner model (b) outer model

- Access data based on different layers: (i) Hotspot Types (e.g. Non-Destructive Testing data, Geometric data, etc.), (ii) Data Type (e.g. presentation or/and research purposes)

(c) Creation of new data hotspots in georeferenced coordinates able to associate with corresponding metadata and files.

(d) User profile set up; Each user have their own profile with personalized information allowing the gathering of user activity data and the generation of statistical models that can be used to personalize the visualized content for individual users.

(e) Registration of new platform users (Every user has different accessibility rights).

(f) Add new customized phases.

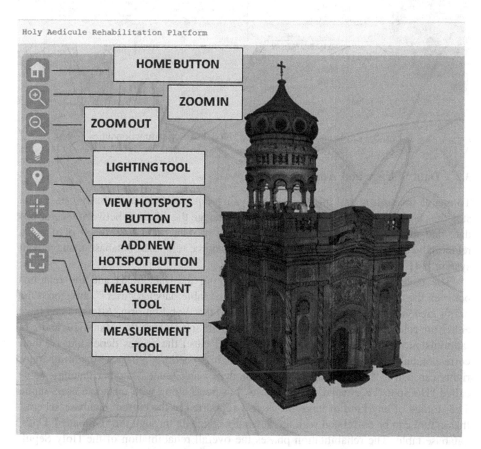

Fig. 2. Explanation of default GUI functionalities

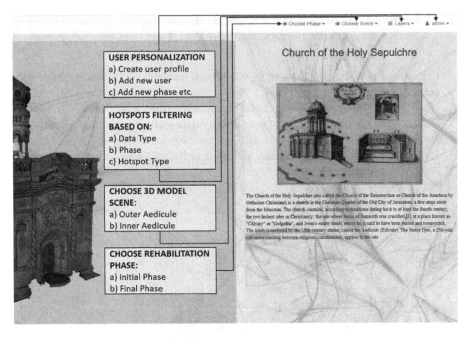

Fig. 3. Explanation of customized GUI JavaScript functionalities

3.2 Data Storage and Analytics

To enable the storing and information analyzing capacity for the platform user, a data store has been implemented. To appropriately manage the relations between the various data and the scene (comprising of geo-referenced data points) the need of using a relational database emerges; MySQL, an open-source relational database management system [14].

The tables created inside the database among the data fields and the relations between them, are illustrated in Fig. 4. In the first table, labeled "users", are stored the user's credentials in order to give them access to the Web Platform. All users can have either the role of administrator or the one of single user where they can view and edit (administrator case) or simply view (single user case) the scenes depending on their corresponding accessibility rights. After the login procedure, a user with administrator rights can add a "hotspot" upon the 3D model of any scene. Another database table called "HotSpots" is created so as to store "hotspots" along with metadata users input. The storage of any kind of large file is also supported in the current database schema, that in turn can be correlated with the hotspot's metadata through the "Uploaded_Data" database table. The rehabilitation phases the overall rehabilitation of the Holy Sepulchre in Jerusalem. Those two tables, "Initial Phase" and "Final_Phase", keeps record of the dates that the two phases started but also ended. Furthermore, table "Phases_Period" keeps track of different sub-phases of the rehabilitation phases, for further analysis and correlation with the corresponding hotspot's metadata.

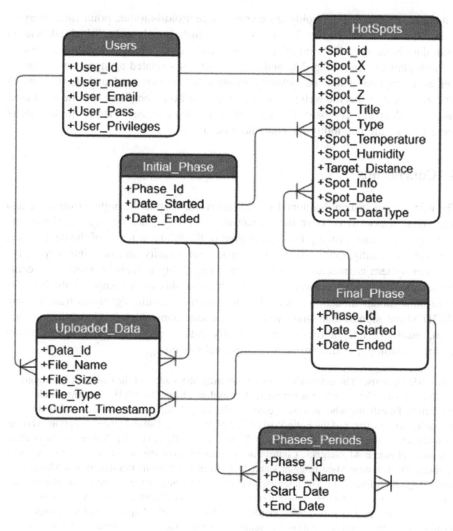

Fig. 4. Database ER model

The personalized content delivery is achieved based on analytics aiming at identifying users' behaviors so as to propose specific and user-tailored content to the correspondent user. For example, users that browse data of a specific discipline - such as materials - are provided with up-to-date data related to this discipline. To facilitate the latter, clustering algorithms have been utilized. Clustering is a Machine Learning technique that involves the grouping of datasets. Given a set of datasets, we can classify each data point into a specific group. More precisely, we use the Clustering algorithm K-Means for the analysis of our data, one unsupervised learning algorithm that solve the well-known clustering problem. The k-means clustering algorithm attempts to split a given anonymous data set into a fixed number (k) of clusters. Initially k number of so-called centroids are chosen. A centroid is a data point (imaginary or real) at the center of a cluster. We use K-means clustering algorithm which takes as an input data based on what kind of information the end-user previously shown through the web platform, browsing data and history, activities related to ingestion of data as well as description (annotated metadata) of the data being ingested, and also how much time was spent in specific web pages (i.e. which rehabilitation phase was the most used one). The algorithm compares this kind of data with equivalent data from other users and estimates the possibility for equivalent behavior.

4 Conclusions

The information system developed within this work, despite the high performance and scalability demonstrates over the management of Big Data (storage, retrieval and exchange), it is also proving the high degree of flexibility in terms of the deployment platform and configuration. The improved system security and data integrity of the three-tier system architecture reassures better reusability especially when it concerns the personalized representation of data. To this end, this work comprises the base for multi-dimensional modeling where the creation of semantic signatures transforming 2D/3D visual signals-data into symbolic representations (high-level entities - metadata), enable the integration of every available data over a certain time frame, within the same ontology, leading to an n-dimensional modelling.

Acknowledgments. The authors wish to acknowledge the leaders of the Christian communities: His Beatitude the Greek Orthodox Patriarch of Jerusalem Theophilos III; His Paternity Archbishop Pierbattista Pizzaballa, who was the Custos of the Holy Land until May 2016 and is now the Apostolic Administrator of the Latin Patriarchate of Jerusalem; Father Francesco Patton, who has been Custos of the Holy Land since June 2016; and His Beatitude the Armenian Patriarch of Jerusalem Nourhan Manougian. Contributions from all over the world secured the project's funding. Worth noting Mica Ertegun's and Jack Shear's donations through World Monuments Fund, Aegean Airlines et al. The interdisciplinary NTUA team for the Protection of Monuments, Em. Korres, A. Georgopoulos, A. Moropoulou, C. Spyrakos, Ch. Mouzakis, were responsible for the rehabilitation project and A. Moropoulou, as Chief Scientific Supervisor of the project, was responsible for its scientific supervision. Acknowledgements are also attributed to the members of the Holy Sepulchre Common Technical Bureau: Th. Mitropoulos, Ph.D., the director of the Common Technical Bureau and construction site manager; O. Hamdan; C. Benelli; and I. Badalian.

References

1. Quintero, M.S., Georgopoulos, A., Stylianidis, E., Lerma García, J.L., Remondino, F.: CIPA's mission: digitally documenting cultural heritage. APT Bull.: J. Preserv. Technol. **48**(4), 51–54 (2017)
2. Gaitatzes, A., Christopoulos, D., Papaioannou, G.: Virtual reality systems and applications: the ancient olympic games. In: Bozanis, Panayiotis, Houstis, E.N. (eds.) PCI 2005. LNCS, vol. 3746, pp. 155–165. Springer, Heidelberg (2005). https://doi.org/10.1007/11573036_15

3. Der Manuelian, P.: Giza 3D: digital archaeology and scholarly access to the Giza Pyramids - the Giza Project at Harvard University. In: Proceedings of Digital Heritage 2013 (Digital Heritage International Congress), pp. 727–734. Institute of Electrical and Electronics Engineers, Marseille (2013)

4. Hassani, F.: Documentation of cultural heritage techniques, potentials and constraints. In: Yen, Y.-N., Weng, K.-H., (eds.) The International Archives of the Photogrammetry, Remote Sensing and Spatial Information Sciences, 25th International CIPA Symposium 2015, vol. XL-5/W7, Tapei, Taiwan (2015)

5. Van Balen, K.: Challenges that preventive conservation poses to the cultural heritage documentation field. In: The International Archives of the Photogrammetry, Remote Sensing and Spatial Information Sciences, 26th International CIPA Symposium 2017, vol. XLII-2/W5, Ottawa, Canada (2017)

6. Alexakis, E., Delegou, E.T., Lampropoulos, K.C., Apostolopoulou, M., Ntoutsi, I., Moropoulou, A.: NDT as a monitoring tool of the works progress and the assessment of materials and rehabilitation interventions at the Holy Aedicule of the Holy Sepulchre. Constr. Build. Mater. **189**(20), 512–526 (2018)

7. Moropoulou, A., et al.: Five-dimensional (5D) modelling of the Holy Aedicule of the Church of the Holy Sepulchre through an innovative and interdisciplinary approach. Mixed Reality and Gamification for Cultural Heritage, pp. 247–270. Springer, Cham (2017). https://doi.org/10.1007/978-3-319-49607-8_9

8. Georgopoulos, A., et al.: Merging geometric documentation with materials characterization and analysis of the history of the Holy Aedicule in the Church of the Holy Sepulchre in Jerusalem. In: The International Archives of the Photogrammetry, Remote Sensing and Spatial Information Sciences, Geomatics and Restoration - Conservation of Cultural Heritage in the Digital Era, Florence, Italy, vol. XLII-5/W1, pp. 487–494 (2017)

9. Apostolopoulou, M., et al.: Study of the historical mortars of the Holy Aedicule as a basis for the design, application and assessment of repair mortars: a multispectral approach applied on the Holy Aedicule. Constr. Build. Mater. **181**, 618–637 (2018)

10. Apollonio, F.I., et al.: A 3D-centered information system for the documentation of a complex restoration intervention. J. Cult. Heritage **29**, 88–99 (2018)

11. Code School (2016). https://www.javascript.com

12. Potenziani, M., Callieri, M., Dellepiane, M., Corsini, M., Ponchio, F., Scopigno, R.: 3DHOP: 3D heritage online presenter. Comput. Graph. **52**, 129–141 (2015)

13. WhatIs (2017). http://whatis.techtarget.com/definition/metadata

14. MySQL (2017). https://www.mysql.com

Governance and Management of the Holy Edicule Rehabilitation Project

Antonia Moropoulou[1] and Nikolaos Moropoulos[2(✉)]

[1] School of Chemical Engineering, NTUA, Athens, Greece
[2] Athens, Greece
nikolaos.moropoulos@gmail.com

Abstract. This paper presents the key aspects of Governance and Management of the project that rehabilitated, reinforced and conserved the Holy Edicule in the Holy Church of the Resurrection in Jerusalem. The overall approach was based on the continuous communication and collaboration with the three Christian communities who share the principal responsibility for the Church of the Holy Sepulchre. This was combined with full transparency on all aspects of the project and intense publicity and external communication so that the progress of the work would be shared and publicized to the media of the world at large. The Church of the Holy Sepulchre and the Holy Edicule have been monuments of ecumenical love and devotion throughout the centuries and we wanted to accentuate this by making the project available to the world. The coordination of the scientific and the managerial team was founded on frequent meetings where all key people would participate and contribute to the resolution of the issues and sound decision making. The quality of the work and the decision making was made possible by the analysis and storage of all emergent data with the full deployment of scientific equipment and digital technologies. The high level of uncertainty in the first four months of the project made it necessary to adopt an agile approach to decision making and management. The stakeholders and the project teams should be ready and able to respond quickly to emergent data about the monument and its features, by making the necessary adjustments to the project plan, schedule and budget, and ensuring that we would put the knowledge gained to good use. The project was successfully completed on time and with a small increase of total expenditure compared to the original budget.

Keywords: Governance · Management · Project · Holy · Edicule · Rehabilitation

1 Introduction

"Wednesday, March 22, 2017, was an historic day in the long history of the Basilica of the Holy Sepulchre in Jerusalem. That morning, an ecumenical celebration marked the end of the restoration work on the Edicule that encloses the remains of the tomb of the Risen Jesus." The Holy Land Review [1]

The Church of the Holy Resurrection in Jerusalem was originally constructed in the fourth century over a tomb that is universally accepted as the burial place of Christ.

© Springer Nature Switzerland AG 2019
A. Moropoulou et al. (Eds.): TMM_CH 2018, CCIS 961, pp. 78–102, 2019.
https://doi.org/10.1007/978-3-030-12957-6_6

The rock-cut tomb is enclosed in an Edicule (small edifice), which lies inside the Rotunda of the church. The current Edicule is the fourth structure to have covered the tomb since the construction of the church in the fourth century. The present Edicule was built by a Greek architect, Nikolaos Komnenos (1770–1821) in 1808–10, following a fire. Under his supervision, a local workforce carried out extensive repairs of the damaged church and rebuilt the Edicule from its foundations. The structure, like previous ones, encloses two spaces: The Tomb Chamber, to the west, to which access is gained through the Chapel of the Angel, to the east. The Tomb Chamber is surmounted by a cupola. Komnenos' work extended to the interior of the Chapel of the Angel and the vaulting over the Tomb Chamber, but it did not include the marble cladding inside the Tomb Chamber, which had survived the fire intact.

In 2015, the extensive deformations of the monument and the steel cage put around it by the British in 1947 to support it, led the Israeli police to shut it down as it was considered to be unsafe for the pilgrims. Following this incident, the Greek Orthodox Patriarch Theophilos III, invited the National Technical University of Athens to perform a study to document the Edicule, diagnose the conditions affecting its preservation, and propose an appropriate course of action for its future. A relevant framework agreement was signed, and the study was completed in January 2016.

The project's scope was defined as the implementation of the 2016 NTUA study, and it became possible following the common agreement between the three principal churches which act as custodians of the Church of the Holy Sepulchre. This is a historic agreement as it is the first time for more than 200 years that the three custodians have agreed to perform significant interventions on the monument.

After the necessary preparations, the project started in June 2016 and ended successfully in March 2017.

The second section of the paper (after the introduction) highlights project governance, the third section presents the scientific and managerial roles, whereas the fourth section focuses on the construction site team and its structure.

The fifth section presents some of the project's constraints, and the sixth outlines the key attributes of the management approach. The seventh section deals with structuring and scheduling of the work, and the eighth on key project processes.

The penultimate section presents key aspects of the project's publicity and external communications, followed by the results. The conclusions wrap up the paper.

2 Project Governance

The common agreement of the three principal Churches acting as custodians of the Church of the Holy Sepulchre provided the framework for the governance of the project, outlining the formation and operation of the Project Owners' Committee and the Steering Committee.

2.1 The Stakeholders

The three principal stakeholders of the project were the three Christian Churches which act as custodians of the Church of the Holy Sepulchre: the Greek Orthodox Patriarchate of Jerusalem, the Franciscan Order, and the Armenian Patriarchate of Jerusalem. The three churches will be called for brevity "the status quo communities", in reference to the first official declaration freezing the rights of worship and possession of the religious denominations within the church of the Holy Sepulchre, which was issued in 1852 by Sultan Abdul Mejid, in a decree known as the Status Quo.

They formed the Project Owners' Committee, the highest authority of the project, which is presented in a section below.

Continuous communication with the status quo communities and quick resolution of emerging issues was a critical success factor for the project. The project had a tight deadline and we could not afford any delays in the decision-making process. We tackled this challenge successfully on most occasions, but failed in some, and the relevant issue dragged on. Regardless of that, we were able to contain the delays and there was no overall negative impact on the project.

World Monuments Fund (WMF), a private, international, not-for-profit organization, supported the restoration project. Financial support through WMF, thanks to two individual donors, made the organization a key stakeholder, represented in the project Steering Committee. Funds were released to the project in installments, based on a timeline, and contingent on the progress of the project. Progress was satisfactory throughout, and all installments were released on time. While the project was ongoing, WMF was able to secure additional funds and increase its initial contribution to the project. In addition to lead financial support, WMF provided an additional layer of technical oversight for the project.

National Geographic Society, a global non-profit organization, was the project stakeholder engaged in the project's external communication and publicity.

2.2   The Common Agreement

In March 2016 all three principal custodians of the Church of the Holy Sepulchre: The Greek Orthodox Patriarchate of Jerusalem, the Franciscan Order in the Holy Land, and the Armenian Patriarchate in Jerusalem approved of the project as per their "**Common Agreement**" dated 22nd March 2016 [2].

The common agreement was signed by the at the time heads of the three "status quo communities":

- the Patriarch of Jerusalem, His Beatitude, Theophilos III,
- the Custos of the Holy Land, His Paternity, Most. Rev. Fr Pierbattista Pizzaballa, and
- the Armenian Patriarch in Jerusalem, His Beatitude, Archbishop Nourhan Manougian

On the 24th June 2016, Fr. Pierbattista Pizzaballa, was appointed titular Archbishop of Verbe, and Apostolic Administrator of Jerusalem. He was replaced by Fr. Francesco Patton.

We will quote the common agreement in various sections of this paper as it provided the basic framework for the governance of the project.

2.3 Project Owners' Committee

The Project's highest authority was the Project Owners' Committee (POC). According to the "Common Agreement":

> *"2.1 The meeting of the Heads of the three major Communities performing as "project owners' committee" (POC) will undertake the responsibility for all strategic decision making."*

The heads of the three "status quo communities" were the POC members:

- the Patriarch of Jerusalem, His Beatitude, Theophilos III,
- the Custos of the Holy Land, His Paternity, Most. Rev. Fr Francesco Patton, and
- the Armenian Patriarch in Jerusalem, His Beatitude, Archbishop Nourhan Manougian

The chairman of the committee, His Beatitude Theophilos III, Patriarch of Jerusalem, was also the chairman of the project steering committee.

The Project Owners' Committee had complete authority over the project and its decisions were irrevocable. The committee convened at the start of the project, its middle, and at the end of the project.

Following each Steering Committee meeting, His Beatitude briefed the project owners' committee on the proceedings and the decisions. All Steering Committee decisions were sanctioned by the project owners' committee.

There have been regular and ad hoc POC meetings.

The regular meetings took place immediately after the Steering Committee meetings, as on the 20th July 2016.

Ad hoc meetings took place to address and resolve issues. Examples are the meetings of the 26th May 2016 (project working hours and community religious services) and 18th October 2016 (three-day closure of the Holy Edicule - see also the "Project Constraints" section.)

2.4 Steering Committee

According to the "Common Agreement":

> *"2.7 The (POC) project owners' Committee authorizes the Steering Committee (SC) to cope with the current problems of integrated project governance with the participation of the CSS (Chief Scientific Supervisor), the CSM (Construction Site Manager) and the PM (Project Manager). The Patriarch of Jerusalem or His Deputy is chairing the SC with the obligation to inform the project owners Committee."*

The Steering Committee (SC) of the Project had the highest governing authority. It convened regularly to review progress, identify and address project issues and make decisions.

The Committee was chaired by His Beatitude Theophilos III, Patriarch of the Holy City of Jerusalem and All Palestine. The SC had the following seven members in total:

- His Beatitude, Patriarch of Jerusalem, Theophilos III, Chairman
- Archbishop Aristarchos of Constantina, Deputy Chairman
- Archbishop Isidoros of Hierapolis, substitute Deputy Chairman

- Prof. A. Moropoulou, Chief Supervising Scientist, Member
- Dr. Th. Mitropoulos, Construction Site Manager, Member
- Prof. Harris Mouzakis, Deputy Construction Site Manager, Member
- Nikolaos Moropoulos, Project Manager, Member

Depending on the agenda of the SC meeting, the Chairman invited representatives of the religious communities, donor representatives, Government officials, as well as project team members to participate in the relevant proceedings.

The first meeting of the Steering Committee was also the Project's Kick Off meeting and it took place on the 20th May 2016.

The five SC meetings that followed had two parts: in the first part Professor Antonia Moropoulou, the chief scientific supervisor of the project, presented the scientific report of the relevant period; in the second part Mr. Nikolaos Moropoulos, the project manager, presented the relevant progress report.

There have been six Steering Committee meetings in the period from May 2016 to March 2017 as follows:

- Final Meeting – 22 March 2017
- Fifth Meeting – 21 February 2017
- Fourth Meeting – 16 December 2016
- Third Meeting – 7 October 2016
- Second Meeting – 21 July 2016
- First Meeting – 20 May 2016

2.5 World Monuments Fund (WMF)

In addition to the three Status Quo communities, World Monuments Fund was the main donor to the project and a key project stakeholder. An agreement to that effect was signed between the Greek Orthodox Patriarchate of Jerusalem and WMF in July 2016.

Support from WMF was made possible thanks to a contribution from Mica Ertegun, a longtime donor and member of the Board of Trustees of the organization. The project received USD 1,100,000 from Mica Ertegun through WMF, a contribution which formed the basis of the July 2016 agreement. Additional funding of USD 150,000 was provided by Jack Shear, another longtime donor and Trustee, in December 2016.

Yiannis Avramides, Program Manager, was responsible for reviewing all progress reports and liaising with the Project Manager on behalf of WMF.

In addition to WMF, copies of all Regular Progress Reports were sent by the Project Manager to Mrs. Linda Wachner of New York and to Fr. Alex Karloutsos, a Protopresbyter of the Ecumenical Patriarchate and a member of the staff of the Greek Orthodox Archdiocese of America with responsibility for Public Affairs.

According to the agreement between the Greek Orthodox Patriarchate of Jerusalem and WMF, disbursement of funds was contingent upon timely submission of reports, according to a mutually agreed-upon schedule, and review and approval of project reports by WMF staff.

In the course of the project all reports were issued on time, and were approved by WMF staff, and as a result the disbursement of funds throughout the project proceeded according to schedule.

In addition, WMF representatives were invited and participated in all Steering Committee meetings, with the exception of the first meeting on May 20, 2016, which took place prior to the signing of the agreement.

Mrs. Ertegun, Mr. Shear, Mrs. Wachner, Fr. Karloutsos, Mrs. Bonnie Burnham (President Emerita of World Monuments Fund), and Mr. Avramides all attended the project closure and completion ceremonies on March 22, 2017 in Jerusalem.

3 Scientific and Management Roles

The project engaged scientists, conservators, restorers and masons. Coordinating the teams was a major challenge. It was successfully undertaken by the Chief Scientific Supervisor and the Project Manager. The scientific team prepared, analyzed and made available all the necessary data provided by the scientific equipment in place and the construction site work as it progressed. The key scientific and management project officials were then able to make relevant decisions. The decisions were made on a timely fashion and no delays were observed in this respect.

3.1 Scientific Roles

The scientific supervision of the project was one of the key success factors. According to the Common Agreement:

> "2.5 The Scientific Supervision will be performed by the interdisciplinary NTUA Study Team, headed by Professor A. Moropoulou (CSS). She has the overall responsibility for the scientific monitoring of the work and is the director of the interdisciplinary scientific monitoring laboratory which will be set up in the construction site. In collaboration with the interdisciplinary NTUA scientific team, the Project Manager (PM) and the CSM she will monitor and control the work."

The key roles in the project from a scientific perspective were the following:

- Chief Supervising Scientist (CSci), NTUA Professor A. Moropoulou. She had the overall responsibility for the scientific monitoring of the work and was the director of the interdisciplinary scientific monitoring laboratory which was set up in the construction site. In collaboration with the interdisciplinary NTUA scientific team, the Project Manager (PM) and the CSM and his deputy she monitored and controlled the work, to ensure it was progressing and completed according to the scientific specifications set by the design study, contributed to the regular progress report and recommended, when necessary, adjustments to the design guidelines and directives, the project's schedule and the budget. When necessary she escalated the issues and/or risks to the Project Manager. She was responsible to communicate the scientific project progress to the International Community, to disseminate innovation aspects, and promote on-site training and education. The role of K. Labropoulos, E. Delegou, M. Apostolopoulou, Emm. Alexakis from the Scientific team CSS Office to the technical editing of the scientific reports, as well as, the executive role of A. Lampropoulou at the dissemination and education plan and documentation were highly appreciated.

- CSS was acting as well as director of materials, repair, reinforcement and conservation interventions (DMC). Professor A. Moropoulou. She had the overall responsibility for the materials and conservation interventions, as well as the measurement of the impact and assessment of the work done. She addressed relevant issues and risks.
- Director of rehabilitation (DRH). Professor Emm. Korres. He had the overall responsibility for the rehabilitation work and addressed relevant issues and risks. When necessary, he escalated the issues and/or risks to the Chief Supervising Scientist.
- Director of structural assessment (DSA). Professor C. Spyrakos. He had the overall responsibility for the reinforcement work and addressed relevant issues and risks. When necessary, he escalated the issues and/or risks to the Chief Supervising Scientist.
- Director of geometric documentation (DGD). Professor Georgopoulos. He had the overall responsibility for the geometric documentation of the work done and addressed relevant issues and risks. When necessary, he escalated the issues and/or risks to the Chief Supervising Scientist.

The National Technical University of Athens deployed a large team of scientists to work under the leadership of the five directors. The complete is shown in Appendix 1.

3.2 Management Roles

This section presents the project's management roles. According to the Common Agreement:

> "2.6 The project management will implement the project charter, report on the work progress according to the schedule and budget and coordinate the construction and the scientific supervision teams in order to complete the work successfully and on time and to manage risks on regular basis."

The key roles in the project from a project management perspective were the following:

- Construction Site Manager (CSM), Dr. Th. Mitropoulos, Chief Engineer of the Holy Sepulchre Common Technical Bureau. The CTB (Common Technical Bureau of the Church of the Holy Sepulchre), staffed by three Architects by the three Communities, was responsible for overseeing the execution of the project so that it followed the scientific studies and directives set by the National Technical University of Athens.
- The CSM had the overall responsibility for the construction site's operation within the health and safety directives set forward by the relevant authorities. He has been working closely with the especially appointed "Safety Advisor", in implementing all local regulations and guidelines, while adhering to the laws of the country of Israel. He was supported in his work by the Deputy CSM.
- Deputy Construction Site Manager (dCSM), Professor H. Mouzakis. He had the overall responsibility for the implementation of the project's engineering design and the correct use of all the facilities and equipment in the construction site. He also

assumed the role of the Superintendent of all construction work. He worked closely with the authorized construction site manager and the team leader of the conservation team to whom he delegated tasks and responsibilities as the needs of the project dictated.

- Project Manager (PM), Mr. Nikolaos Moropoulos. He had the responsibility to coordinate the construction and the scientific supervision teams to successfully complete the work. The project manager, working with the Chief Supervising Scientist, the Construction Site Manager, his deputy and the other project officers maintained the project schedule, cash flow and budget and prepared the regular progress reports. He coordinated the teams in identifying and addressing the project issues and risks on a regular basis and ensured that the project standards and procedures were adhered to.

3.3 Project Management Office

The Project Manager was supported in his work by the Greek Orthodox Patriarchate's (GOP) Secretariat which maintained the project archive and processed all documents in the purchasing cycle. This support was invaluable, especially if one considers that the project workload was carried by the Secretariat in addition to its regular work load, which is considerable.

The Secretariat organized all Steering Committee and Financial Committee meetings, the transfers of team members to/from the Tel Aviv airport, the relevant hotel reservations, as well as the Monument's Inauguration events of March 2017.

The indefatigable Archbishop Aristarchos of Constantina, Elder Secretary-General of the Greek Orthodox Patriarchate, led and continuously supported the effort of the Secretariat team.

Equally valuable was the support of the Greek Orthodox Patriarchate's Financial Committee who was managing the donations to the project.

In addition to the above resources of the Greek Orthodox Patriarchate, a major contribution to the Project Management Office was made by the Athens Greece based NTUA team of the Chief Scientific Supervisor (CSS). The team compiled all technical specifications for the equipment and materials that were to be acquired in cooperation – as needed - with the Construction Site Management Team and conducted the necessary market research to identify reliable suppliers who would then be invited by the GOP Secretariat to submit a quotation in the context of the purchasing cycle (see also the section on supplier management).

CSS's team also handled the planning of the project team's travel to/from Jerusalem. They compiled and maintained a weekly schedule and liaised with the GOP Secretariat and Aegean Airlines, the air transportation sponsor of the project in order to make the air travel bookings and issue the relevant tickets. The efforts of the CSS relevant administrator Mrs. Georgia Skoulaki and her deputy Ms. Katerina Kolaiti are greatly appreciated.

3.4 Common Technical Bureau

"2.4 The CTB (Common Technical Bureau of the Church of the Holy Sepulcher), staffed by three Architects by the three Communities, will be responsible for the correct execution of the project according to the scientific studies and directives realized by the National Technical University of Athens. The representative of the Common Technical Bureau of the Church of the Holy Sepulcher (Dr. Theodosios Mitropoulos), as Construction Site Manager (CSM), will be responsible for the construction site's operation within the directives set forward by the relevant authorities." (The Common Agreement)

The members of the Common Technical Bureau who have been involved and made contribution to the project are:

- Dr. Theodosios Mitropoulos, Architect, representing the Greek Orthodox Patriarchate of Jerusalem
- Mr. Osama Hamdan, Architect, representing the Custody
- Ms. Carla Benelli, Art Historian, representing the Custody, and
- Irene Badalian, Architect, representing the Armenian Patriarchate

The Chief Scientific Supervisor, Prof. A. Moropoulou, met regularly with the Common Technical Bureau and consulted with them regarding the materials used and the relevant interventions in the Edicule. She also cooperated with them in addressing the problem that emerged regarding "Pavement preservation and rehabilitation" (see section on "Additional Problems".)

The Governance and Management structure of the project is shown in Fig. 1.

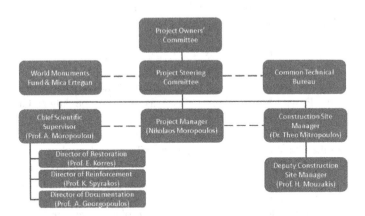

Fig. 1. Project governance and organisation

4   The Construction Site Team

The construction site team comprised the restorers and masons' team, the conservators' team, and the safety advisor of the project. The team was fully deployed in the second week of June 2016. As the project progressed, some restructuring and strengthening was necessary.

4.1 Restorers and Masons Team

V. Zafeiris, Civil Engineer, Team Leader, Authorised Construction Site Manager
G. Anastasiadis, Senior Marble Mason
G. Palamaris, Marble Mason
 I. Andritsopoulos, restorer
C. Theodorakis, restorer
A. Karydis, restorer
P. Chaloftis, restorer

Team Restructuring (26 November 2016)

There have been some occasions where the construction site team had to be restructured to address critical project requirements. An example is the restructuring that was implemented at the end of November 2016 to meet the deadlines of a critical path activity, the reinstalling of the external stones. This was a show-stopper activity. Missing the deadline for its completion would mean that the project would miss the 22nd March 2017 project completion deadline.

As stated in the Construction Site Management Meeting Minutes of the 26th November 2016:

> *"Effective immediately, Mr. G. Palamaris and Mr. Th. Carydis will work as a team with Mr. G. Anastassiadis. As soon as the work on the staircases by the entrance of the Edicule is complete and the titanium mesh is placed in its vertical position, the team will be enhanced by the addition of Mr. Ch. Theodorakis and Mr. P. Chaloftis. This restructuring will remain in effect until the 7th February or thereafter."*

This change was necessary to support the 'Preparation for reinstalling the external stones' task. As stated in the same meeting's minutes:

> *"All the stones of zones Delta, Epsilon and Zeta will be placed by the relevant panel, and then they will be positioned to their final panel position, marked accordingly, and prepared for anchoring as needed. Once this "mock" reinstalling is done, each stone will be returned to the area by the panel. This task will be done from the 1st to the 10th December. The team of the task comprises G. Anastassiadis, G. Palamaris, and Th. Carydis. Ch. Theodorakis and P. Chaloftis will join the team once they finish their other assignments."*

4.2 Conservators Team

Th. Mavridis, M.Sc., Conservator, Greek Ministry of Culture, Team Leader
M. Troullinos, Senior Conservator (joined in early January 2017)
K. Karathanou, M.Sc. Archaeol.- Cons., Greek Ministry of Culture
Am. Troullinou, Conservator (joined in early January 2017)
Ar. Troullinou, Conservator (joined in early January 2017)

Team Strengthening

In early January 2017 the conservation workload exceeded the significantly capacity of the conservators' team. To address the workload peak, three experienced conservators (M. Troullinos, Am. Troullinou, Ar. Troullinou) joined the team. This resulted in the successful and timely completion of the conservation tasks prior to the 22nd March deadline.

4.3 Safety Advisor

In early June 2016, the project engaged Mr. Uri Agame as safety advisor. The first working permit for the project was issued by Mr. Agame on the 14th June 2016.

Mr. Agame did an excellent job and ensured that the safety of the construction site was properly maintained and provided for throughout the project. There have been no safety incidents.

The construction site team structure is shown in Fig. 2.

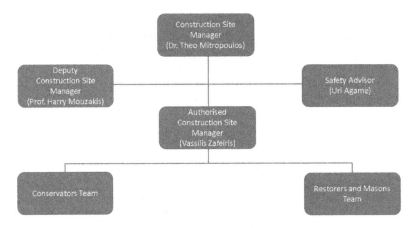

Fig. 2. Construction site team

5   Project Constraints

"2.3 b. The works, which will be completed in approximately eight months to one year, will not prevent the religious services in the Holy Sepulcher or, more specifically, in the Edicule, nor prevent the access of pilgrims into these places." (The Common Agreement)

The project was implemented under a set of constraints, the most important of which was the completion deadline. The Holy Edicule had to be delivered to the status quo communities on the 22nd March 2017 so that the Easter ceremonies would go ahead as always. The deadline was met. The restored and conserved Edicule was delivered to the status quo communities on the 22nd March 2017, ten months after the start of the project.

Another important constraint was the one stated in the common agreement. The project should be implemented in a way that the religious services would proceed as normal and the pilgrims would be able to access and visit the Holy Edicule.

The religious services constraint was met but there were some difficulties, especially at the beginning, due to inadequate communication. As mentioned in the Construction Site Meeting Minutes (27th May 2016), the following incident took place on the 26th May 2016:

"Apparently, members of the Franciscan Order in the Church of the Holy Sepulchre were not aware of the work schedule that has been approved by the Steering Committee of the 20th May 2016, and repeatedly stopped the work of moving the construction site equipment in the Church designated areas."

His Beatitude Theophilos III, Patriarch of Jerusalem, was notified and he instructed Archbishop Isidoros, the Greek Orthodox Superior of the Holy Sepulchre Church, to organize a meeting with the status quo communities in the Church of the Holy Sepulchre, to explain to them the work schedule and ask for their cooperation.

The pilgrims' access constraint was also met. Work on the Edicule took place during the night, from 7pm to 6am every day. The only period the Edicule was closed to the public was from 1800 h of the 26th October 2016 to 0600 h of the 29th October 2016. This was necessary to make necessary interventions in the Holy Tomb and the Holy Rock.

To ensure the full support of the status quo communities, the Chief Scientific Supervisor, Prof. Moropoulou, met with their leaders in Jerusalem on Tuesday 18th October 2016, in the presence of the Chief Secretary of the Greek Orthodox Patriarchate of Jerusalem, Archbishop Aristarchus of Constantia.

Another set of constraints is relevant to space. To operate a construction site, we needed storage space for the equipment and materials. The space available inside the Church of the Holy Sepulchre was not big enough, so most of the stones that were used in the restoration had to be left in the piazza outside the Church. Moving these heavy stones in an area without the relevant infrastructure was time consuming and required to hire laborers to do the job, as the construction site team was not set up for this type of task. In addition, extra care had to be taken to avoid accidents.

The location of the Church posed another set of constraints to the project. Equipment and materials would be delivered to Jaffa Gate (see Fig. 3) and would have to be transported to the Church of the Holy sepulcher using small vehicles that can negotiate the bumpy narrow passages of Jerusalem's Old Town.

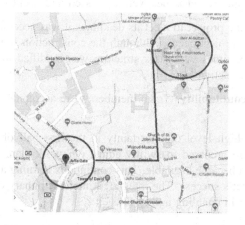

Fig. 3. From Jaffa Gate to the Holy Sepulchre

Another critical constraint was the existence of the Copts' chapel in the western side of the monument. To proceed with the project work, the chapel had to be repositioned. This entailed communication of His Beatitude Theophilos III, Patriarch of Jerusalem with the Patriarch of the Copts to explain the necessity of the move and ensure that the tentative site where the chapel would be removed, on the southern side of the monument, would function equally well as the original. A tentative structure was put in place and the chapel was successfully repositioned on the 18th July 2016.

6 Management Approach

The approach we followed in project management had the following attributes:

Transparency. All project information was available to all stakeholders continuously during the project. Key scientific, managerial, and financial data were presented and shared during the steering committee meetings. All the relevant project documents were released on time to the status quo communities.

Timeliness. All members of the joint scientific and management team had an acute sense of urgency and strived to perform their tasks in a timely fashion. Estimating was particularly difficult especially in the first half of the project, where we had no hard data of the team's performance on which we could base our estimates. To counter this uncertainty, we inserted time buffers in the project schedule and monitored planned versus actual on a continuous basis. More often than not, critical factors for the completion of tasks were related to the availability of the required materials and equipment.

Effectiveness. The teams were given all the required resources in order to get the job done. They had to ensure good enough execution of their tasks in spite of the inherent uncertainties. This implies that the team members had to exercise their initiative and adjust on the spot, if they assessed that this was the right thing to do. The teams were empowered to do so.

Participation. The key team members participated in the construction site management meeting where all project issues were discussed and decided upon.

Collaboration. For the project to meet its deadlines it was necessary for the teams to collaborate as per the project needs. Although the construction site team was properly structured, there were times when this structure was modified in order to allow for tackling key tasks and activities.

Ownership and Accountability. Team members were accountable for the tasks they owned.

Agility. Due to the high level of uncertainty in the first half of the project, it was essential that we were prepared to swiftly adjust our work plan, schedule and effort estimates as new data became available. This continuous learning and adjustment were essential to the project's success. The integrated interdisciplinary decision making we deployed was critical in enabling agility [7].

7 Structuring and Scheduling the Work

Determining the elements and components of the monument and the relevant interventions, was the basis of the project's work breakdown structure. Scheduling of the work was iterative and continuously considered the new data that emerged as work progressed.

7.1 Work Breakdown Structure

The project work breakdown structure was primarily defined based as a set of interventions on elements and components of the monument defined using the plan shown on Fig. 4 and the façade shown on Fig. 5.

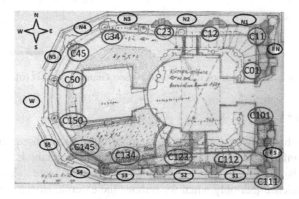

Fig. 4. Holy Edicule plan. Based on Prof. Korres's drawing [8]. (Color figure online)

In terms of vertical elements, the monument comprises 13 vertical panels (bays) (marked in black ellipses), 2 staircases and 14 columns (marked in red ellipses).

In terms of horizontal layers, the monument comprises 15 layers, denoted by the Greek letters from Alpha to Omicron, as shown on the left of Fig. 5. These layers were grouped into three zones, starting from the ground up.

The following components of the monument were also included in the project work breakdown structure:

- Additional lateral support
- Retaining structure to support the western side
- The British Steel Cage
- Holy Tomb Chamber
- Myrrh bearers wall-painting in the Holy Tomb Chamber
- Onion Dome
- Chapel of the Angel

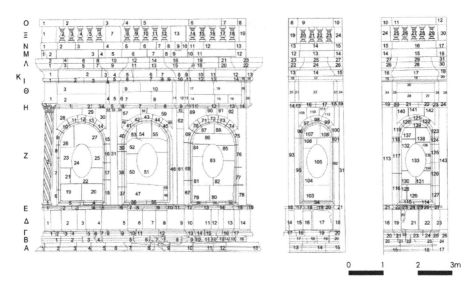

Fig. 5. Facade of the north side. See Giannakopoulos [6]

7.2   Scheduling

A preliminary schedule was drafted at the start of the project based on the technical studies and the information supplied by the non-destructive methods.

However, this was tentative, and all members of the team were aware of this, as we knew from experience that only when we remove the external stones we would have a clear picture of the tasks in hand.

The first major additional activity that was added to the project was the installation of additional lateral support, as the existing steel grid was severely deformed and might not hold the monument once we removed the external stones.

Following that, when we removed the stones from the panels N2 and N3 it became clear that after removing the loose mortar and cleaning the masonry, we would only need to repair the masonry.

A different picture emerged from panels N4, N5, W, S5 and S4, where the masonry was in such a bad state that it had to be completely rebuilt. The rebuilt in turn required that we had to design and install a retaining structure that would hold the upper part of the monument while the restorers rebuilt the masonry of the lower part.

The project schedule was therefore continuously adjusted, to account for the new findings and the relevant additional and/or modified tasks.

In the period from May 2016 to December 2016 the granularity of the schedule was medium. The activities were defined in such a way that they could be assigned to a team (e.g. restorers and masons) and it was the team that would then develop a detailed day by day schedule for implementing the activity.

The granularity changed to fine in December 2016, when we only had three months to complete the project and the level of uncertainty has been reduced substantially. Working with the scientific and the construction teams, we were able to prepare a detailed list of tasks, allocate team members to each, and estimate the time for each task's completion.

To properly update the schedule to include the daily progress of the work, a "Daily Activity Report" would be prepared by the Authorized Construction Site Manager, including the active tasks, the effort extended to each by each assigned team member, and any issues that were relevant. The Project manager would review the Daily Activity Report and update the project schedule to reflect the progress. All relevant issues were either addressed on the spot, if they were simple, or discussed and resolved in the Construction Site Meeting.

8 Project Processes

Supplier management was one of the project's critical processes as the timely supply of equipment and materials was linked to the project milestones. The regular construction site management meeting developed into a key project process for issue identification and resolution. Scientific and progress reporting was a standard feature of all steering committee meetings. Equally important were financial reports and meetings of the project's financial committee.

8.1 Supplier Management

There were multiple issues with the suppliers of the project. The first issue had to do with a monopoly.

In the transport of materials and equipment from Jaffa Gate to the Church of the Holy Sepulchre, we did not have a choice, as there is only one operator in Jerusalem's Old Town. It was necessary to spend time to explain to the supplier the procedure to be followed every time we used them. They key issue is that upon completion of the transport, we had to receive from the supplier a signed statement of the work effort and transport cost, and approve it on the spot, or reject it, asking the mover to make changes. This was new to the mover, who until then only issued an invoice without a document that stated the services received and the relevant quantification.

Eventually the mover agreed to follow the procedure, but 'negotiating' the numbers on the 'services received' document was always an issue. The help and support of Archbishop Isidoros, the Greek Orthodox Superior of the Holy Sepulchre Church, in achieving this was decisive.

The second issue had to do with the timing of the works of the major local supplier of construction services. We had trouble in communicating to the supplier the deadlines of the project. A typical example was the installation of additional lateral support. We experienced a deadline creep and had to summon the representative of the supplier to a meeting with His Beatitude, Theophilos III, Patriarch of Jerusalem. The issue was eventually resolved, but by then we had suffered a delay. Fortunately, this occurred early in the project and we were able to catch up and absorb the delay.

Another problem we experienced with the major supplier was the quality of the invoices. Some of them were obscure and we could not easily match them to our purchase orders and the relevant deliveries. As a result, payment was withheld and the Project Manager, the Construction Site Manager, his Deputy and the Authorized Construction Site Manager had to spend considerable time cross checking and proofing the invoices. Eventually we held a meeting in early January 2017 to clear the backlog of pending invoices and ensure that all the documents were in order.

Another type of issue we experienced with local suppliers was that the materials and/or equipment we required were not available and had to be ordered or we had to order another item. In addition to the inconvenience, this complication resulted in loss of time and potentially additional cost.

The following excerpt from the minutes of a construction site meeting is indicative.

"Assuming that DS Construction deliver the steel beams and the Bedec and Hilti tools and equipment are also delivered by the 18th August, we will be in a position to start the rebuilding work on the 21st August." (Minutes of the Construction Site Management Meeting of the 10th August 2016.)

The timely delivery of materials purchased from suppliers abroad and shipped to Jerusalem was also an issue. The following excerpt from the minutes of a construction site meeting is indicative.

"The crucial milestone for the continuation of the project is the delivery of the titanium bars and anchors by the 15th November 2016, so that the anchoring of the external columns can begin on the 16th November 2016. This is on the project critical path." (Minutes of the Construction Site Management Meeting of the 30th October 2016.)

8.2   Construction Site Management Meeting

A key project process has been the Construction Site Management meeting. The meeting took place on average once every two weeks and resolved all pending issues or referred them to the scientific committee and/or the Chairman of the Steering Committee, as appropriate. A total of 28 meetings took place during the project.

The weekly meeting had the following standard agenda:

- Review of progress
- Review of issues
- Any other business
- Next meeting

The Construction Site Management meeting participants were the following:

- Prof. Antonia Moropoulou, Chief Supervising Scientist
- Dr. Th. Mitropoulos, Construction Site Manager
- Prof. Harry Mouzakis, Deputy Construction Site Manager
- Mr. Vassileios Zafeiris, Authorized Construction Site Manager
- Mr. Theodore Mavridis, Conservators Team Leader
- Mr. George Anastassiadis, Senior Stonemason
- Mr. Michail Troullinos, Senior Conservator (since January 2017)
- Mr. Nikolaos Moropoulos, Project Manager

The Construction Site Meeting was the key coordination and communication process in the project. It enabled the scientific and the construction site team to create a common ground, identify and effectively address the project's issues. It also functioned in a way that facilitated decision making and avoided delays due to internal miscommunication and misunderstandings.

The smooth and effective functioning of the Construction Site Meeting did not happen overnight. The first couple of months were trying and challenging, as each participant in the meeting came to it with different experiences and daily practice routines.

It was the sharing of a common vision and aspiration, to deliver the project on time and within the required quality, that brought all these different parties together and helped forge a common ground.

The leadership provided by His Beatitude, Theophilos III, Patriarch of Jerusalem, focused our minds on the task in hand and away from separate and divisive agendas.

To maximize the quality of the process, we invited to the meeting the senior stone mason, Mr. Anastassiadis, who was a key member of the team and made a significant contribution to each meeting. We did the same with the senior conservator Mr. Troullinos, when he joined the team in January 2017.

8.3   Regular Progress Reporting

Regular Progress Reports contained the following information:

I. Project Status

- Key achievements of the reporting period
- Improvement opportunities
- Status of tasks due
- Summary of project expenditures

II. Risks and Issues

- Issues
- Risks

III. Area Reports

- Report by the Chief Supervising Scientist
- Report by the Directors of Rehabilitation, Reinforcement, Materials and Conservation Interventions, and Geometric Documentation
- Report by the Construction Site Manager

IV. Decisions

Decision recommendations to be considered by the Steering Committee.

Regular progress reports were prepared and distributed on the following dates:

- 20 May 2016
- 15 July 2016
- 6 October 2016
- 15 December 2016
- 21 February 2017
- 19 March 2017 – Closure Report: Financing and Expenditures

All regular progress reports have been presented during the relevant Steering Committee Meeting (except the last one, because there has not been a relevant meeting) and distributed to the three status quo communities and the WMF.

8.4 Scientific Reporting

A total of four scientific reports have been compiled and delivered to the Steering Committee and the Project Owners' Committee as follows, starting with the second steering committee meeting and then continuing for each steering committee meeting until the completion of the project [3].

- 20th July 2016
- 6th October 2016
- 15th December 2016
- 22nd February 2017

The directors of the scientific team made a presentation of the project results in the context of the project closing meetings of the 22nd March 2017 [4].

In addition to the published reports above, the interdisciplinary team has produced numerous interim reports as the project progressed. An example is the technical report of the 8th of July 2016, regarding the findings following the removal of the first stone slabs. The report was compiled by Prof. M. Korres, Director of Rehabilitation and H. Mouzakis, Deputy Construction Site Manager, and was discussed at a meeting of the scientific team in the presence of the project manager. Decisions were made based on the report's findings.

8.5 Project Funding and Cash Flow

The project was funded by private donors and the status quo communities. At the end of the project the funds were more than adequate for the project needs. Total funding amounted to approximately 3.7 million Euros.

However, the timing of the funds' availability was a parameter that had to be carefully and closely monitored and managed. In its lifecycle, the project was always liquid. There have been a few occasions where cash was tight, but scheduled installments of already secured donations and/or new donations relieved the pressure.

At the beginning of the project the secured funds where approximately one third of the budgeted expenditure. At the end of the first three months of the project the secured funds were approximately half of the project expenditure. The solid progress of the project and the successful intervention in the Holy Tomb and the Holy Rock in October 2016 generated a new momentum in the funding of the project. By January 2017, the secured funds were approaching the full project expenditure.

The Project Manager prepared and maintained the project's cash flow, incorporating the actuals and projecting future inflows and outflows. The cash flow was a topic in all Steering Committee meetings.

8.6   Project Expenditure

Project expenditure was kept under continuous monitoring and control and all relevant information was shared with the stakeholders on a regular base and included in the regular project reports. Ad hoc requests for relevant information were answered in less than one week's time.

Under normal circumstances, the purchasing cycle started with a request for quotation by the Secretariat of the Greek Orthodox Patriarchate. The quotation was reviewed and, if it was considered reasonable, it would be approved by the project manager and the Secretariat would proceed to place a relevant order. On occasions, clarifications and modifications were requested from the supplier.

Purchases from suppliers who enjoyed a position of monopoly, e.g. the Old Town transportation firm, and/or requiring small amounts would be ordered and acquired without following the complete procedure, provided that a project official, e.g. the construction site manager would authorize it. In case of time and materials purchases, the ordering official would secure a document signed by himself and the supplier, detailing the hours worked and other units relevant to the provided service.

All purchases were supported by the relevant invoice. Invoices received were reviewed, matched to the purchased product/service and either approved, or put on hold so that the supplier would provide additional information and/or make changes.

The full documentation of project purchases is today included in the project archive.

Total project expenditure slightly exceeded 3.5 million Euros, an increase of approximately 16% of the initial project budget.

8.7   Financial Reporting and Meetings

The finances of the project were regularly reviewed by a committee coordinated by the Project Manager, comprising representatives of all three status quo communities.

- Brother Dobromir Jasztal, representing the Custody
- Brother David Grenier, representing the Custody
- Brother Athanasius Macora, representing the Custody
- Father Samuel Aghoyan, representing the Armenian Patriarchate
- Father Hovnan Baghdasaryan, representing the Armenian Patriarchate
- Nikolaos Moropoulos, Project Manager

The Project Manager prepared the following financial reports on the respective dates.

- 10 June 2016 – Budget, Cash Flow, Materials, Equipment and Infrastructure Expenditure Memorandum
- 20 July 2016 – Financial Review (PowerPoint Presentation)
- 21 July 2016 – Funding Memorandum
- 30 August 2016 – Funding Memorandum
- 8 September 2016 – Funding Apportioning Approach (PowerPoint Presentation)
- 2 February 2017 – Project Expenditures (prepared following a 31st January 2017 request by the WMF)

The following meetings of the finance committee took place on the respective dates. The relevant finance report(s) were presented and discussed in the meetings.

- Fourth Meeting – 13 February 2017
- Third Meeting – 8 September 2016
- Second Meeting – 20 July 2016
- First Meeting – 20 June 2016

The meetings and/or reports were triggered either by regular reporting and/or specific queries by the communities or the WMF.

As an example, the meeting of the 8th September was triggered by the following query submitted by the Armenian Patriarchate's accounting department.

"The Accounting Department of the Armenians has the below inquiry:

1. The name of the donors and whether all the donations are for restoration of the Church or private to Our Patriarchate. If the donations are for the restoration, they need the total amount so that they calculate how much is the remaining amount until they pay their 1/3.
2. If any of the 3 communities have paid the agreed amount of 350,000 euros discussed in the latest (July) financial meeting. If they have to pay, they need bills of companies so that they transfer the amount."

(email sent by the Secretariat to the Project Manager, 24th August 2016).

The report of 2nd February 2017 was prepared following a request by the WMF Program Manager, Mr. Yiannis Avramides. The report was to detail the project expenditures so that Mrs. Mica Ertegun and the WMF officials would be able to conduct a project review.

The report was well received by the WMF and was subsequently shared with the status quo communities.

9   Publicity and External Communication

The importance of publicity and external communication was clear right from the start of the project. The Church of the Holy Sepulchre has been and continues to be a monument that belongs to humanity regardless of religion, nationality, race. The millions of visitors every year are the living proof of this. Its location in the heart of the Old City of Jerusalem, makes it a monument that unites people and projects the ability of human beings to coexist and share, rather than fight and split.

We experienced the aura of peace and love that cuts through barriers from the first day we were engaged in the project and we felt that it was our obligation to honor the legacy of the monument and make the project transparent to all, providing the necessary publicity and external communication.

We shared this view with the status quo communities and WMF and we received a positive response to the wide and comprehensive publicity and comprehensive communication approach.

This aspect of the project was significantly strengthened when National Geographic Society, a global non-profit organization joined us as our publicity strategic partner and stakeholder. Gary E. Knell, President and CEO, and Jean Case, Chairman of the Board of Trustees, gave their wholehearted support to the project and committed all the necessary resources so that our project would become the project of the world.

The first major contribution of National Geographic Society (NGS) to the project materialized in October 2016, when the interventions in the Holy Tomb and the Holy Rock took place. NG immortalized the relevant activities with photographs and videos that were publicized with minimal delay.

With their experience in the field and top quality of professionals, led by Fredrik Hiebert, PhD, Archaeologist-in-Residence, NG have given the project worldwide exposure. The articles of Kristin Romey, NG editor and writer, are indicative of the quality and the depth of this coverage.

The NG contribution did not end with the completion of the project. Under the leadership of Kathryn Keane, Vice President of Exhibitions, and in close cooperation with the National Technical University of Athens, the NG team put together a digital exhibition "The Tomb of Christ" in Washington DC. The exhibition is curated by Fredrik Hiebert and is on show until January 2019.

10   Results

The project was successfully completed on time and with a small increase of total expenditure compared to the original budget [5]. In the course of the project we identified some additional problems that were outside its scope but need to be tackled by the status quo communities as they may adversely impact the Holy Edicule and the Church of the Holy Sepulchre at large.

More specifically, studies were carried out by the National Technical University of Athens to prospect and document the underground structures, tunnels, canals, cisterns and cavities in the Rotunda and the Church.

These studies highlighted the need for:

- Foundation Interventions for the underpinning, reinforcement, water and humidity control of the Holy Edicule
- Control of the rising damp and installation of proper sewage and water drainage system
- Pavement preservation and rehabilitation

Additional studies put forward proposals for addressing these areas of concern. All the studies have been presented and submitted to the status quo communities which now have all the information needed to make the relevant decisions.

11   Conclusions

The leadership and continuous support provided to the team by His Beatitude Theophilos III, Patriarch of Jerusalem, and the other Status Quo Communities enabled the governance and management approach we followed. The successful delivery of the rehabilitated Holy Edicule is largely due the following features of the governance and management approach.

- Continuous communication and collaboration with the three Christian communities who share the principal responsibility for the Church of the Holy Sepulchre
- Full transparency on all aspects of the project
- Intense publicity and external communication so that the progress of the work would be shared and publicized to the media of the world at large
- Coordination of the scientific and the managerial team, founded on frequent meetings where all key people would participate and contribute to the resolution of the issues and sound decision making
- Analysis of all emergent data with the full deployment of scientific equipment and digital technologies
- Adoption of an agile approach to decision making and management, so that the stakeholders and the project teams would be ready and able to respond quickly to emergent data about the monument and its features, by making the necessary adjustments to the project plan, schedule and budget, and ensuring that we would put the knowledge gained to good use.

Acknowledgements. The study and the rehabilitation project of the Holy Edicule became possible and were executed under the governance of His Beatitude, the Patriarch of Jerusalem Theophilos III. The Common Agreement of the Status Quo Christian Communities provided the statutory framework for the execution of the project; His Paternity the Custos of the Holy Land, Archbishop Pierbattista Pizzaballa (until May 2016 – now the Apostolic Administrator of the Latin Patriarchate of Jerusalem), Fr. Francesco Patton (from June 2016), and His Beatitude the Armenian Patriarch of Jerusalem, Nourhan Manougian, authorized His Beatitude the Patriarch of Jerusalem, Theophilos III, and NTUA to perform the research study and implement the project.

The project's funding was secured by donations from all over the world. Worth noting due to their size and/or timing are the donations (through WMF) by Mica Ertegun and Jack Shear and Aegean Airlines who donated the air transportation tickets from Greece to Israel.

The interdisciplinary NTUA team for the Protection of Monuments, Emmanouil Korres, Andreas Georgopoulos, Antonia Moropoulou, Costas Spyrakos, and Charis Mouzakis, were responsible for the rehabilitation project.

Appendix 1: The NTUA Scientific Team

The National Technical University of Athens Interdisciplinary Team for the "Protection of Monuments" scientific responsible for the Project:

- Chief Scientific Supervisor of the Project with executive authority: Prof. A. Moropoulou
- Interdisciplinary Team: Prof. Emeritus Emm. Korres (member of the Academy of Athens), Prof. A. Georgopoulos, Prof. A. Moropoulou, Prof. C. Spyrakos, Ass. Prof. Ch. Mouzakis

- NTUA School of Civil Engineering: Prof. C. Spyrakos, Ass. Prof. Ch. Mouzakis, Prof. Emeritus P. Marinos, Assoc. Prof. M. Kavvadas, EDIP S. Asimakopoulos, EDIP Dr. L. Karapitta, Dr. Ch. Maniatakis, PhD Cand. L. Panoutsopoulou
- NTUA School of Architecture: Prof. Emeritus Emm. Korres, Architectural Engineer V. Chasapis
- NTUA School of Chemical Engineering: Prof. A. Moropoulou, Prof. Emeritus G. Batis, Assis. Prof. A. Bakolas, EDIP Dr. E. T. Delegou, EDIP Dr. M. Karoglou, EDIP Dr. K. C. Lampropoulos, EDIP Dr. P. Moundoulas†, Dr. N. Vesic (Father Ambrosius), PhD Cand. Emm. Alexakis, PhD Cand. M. Apostolopoulou, PhD Cand. D. Giannakopoulos, PhD Cand. V. Keramidas, PhD Cand. A. Kolaiti, PhD. Cand. M. Kroustallaki, PhD Cand. I. Ntoutsi, PhD Cand. E. Tsilimantou, Dr. A. Zacharopoulou, Chemical Engineer N. Galanaki, Chemical Engineer M. Kalofonou, Architectural Engineer Z. Karekou. Communication and administrative support, A. C. Lampropoulou. Managerial and administrative support, G. Skoulaki. Technical support, I. Mountrichas.
- School of Rural and Surveying Engineering: Prof. A. Georgopoulos, Prof. Ch. Ioannidis, Prof. G. Pantazis, Assoc. Prof. E. Lambrou, Ass. Prof. A. Doulamis, ETEP S. Soile, ETEP S. Tapinaki, ETEP R. Chliverou, PhD Cand. P. Agrafiotis, PhD Cand. E. Stathopoulou, L. Kotoula, F. Bourexis, A. Papadaki, N. Tsonakas, P. Nikolakakou, M. Skamantzari
- The Interdisciplinary NTUA team cooperated with other Schools, Laboratories and scientific collaborators: Prof. S. Kourkoulis and Dr. E. Passiou from NTUA School of Applied Mathematics and Physical Science, Sector of Mechanics, Dr. A. Menychtas, NTUA School of Electrical and Computer Engineering, Mech. Eng. M. Agapakis, A. Fragkiadoulakis, S. Theocharis and Chem. Eng. I. Agapakis
- NTUA Inter-Departmental Postgraduate Program "Protection of Monuments", Direction "Materials and Conservation Interventions" graduate students Emm. Alexakis, D. Giannakopoulos, A. Zargli, A. Kolaiti, E. Koukouras, M. Kroustallaki have conducted Master Thesis interconnected to the project
- The Interdisciplinary NTUA team cooperated with: University of Pireaus, University of Peloponnese, Agricultural University of Athens, Institute of Geology & Mineral Exploration (I.G.M.E.), Athens Water Supply and Sewerage Company (EYDAP S.A.). Specifically, Assis. Prof. D. Kyriazis (Electrical Engineer) from the University of Pireaus, Assoc. Prof. N. Zacharias from the University of Peloponnese, EDIP Dr. A. Tsagkarakis from the Agricultural University of Athens, Dr. G. Economou, Dr. Ch. Papatrechas from the Institute of Geology & Mineral Exploration (I.G.M.E.) and A. Aggelopoulos, E. Karampelas and D. Tamvakeras from EYDAP S.A., Dr. P. Sotiropoulos and S. Maroulakis from Terra Marine

References

1. The Restored Edicule, The Holy Land Review, 1 June 2017
2. Integrated Diagnostic Research Project and Strategic Planning for Materials, Interventions Conservation and Rehabilitation of the Holy Aedicula of the Church of the Holy Sepulchre in Jerusalem. Scientific Responsible: A. Moropoulou, National Technical University of Athens (2016)
3. Common Agreement of the three Communities, the historic guardians and servants of the Holy Places, for the Restoration of the Sacred Edicule in the Church of the Anastasis. Signed by: Theophilos III, Patriarch of Jerusalem; Pierbattista Pizzaballa, Custos of the Holy Land; Nourhan Manougian, Armenian Patriarch of Jerusalem, Jerusalem, 22 March 2016
4. Scientific Reports on the scientific supervision, monitoring and decision making for the project for the conservation, reinforcement and repair interventions for the rehabilitation of the Holy Edicule of the Holy Sepulchre in the All-Holy Church of Resurrection in Jerusalem: 1st Steering Committee Meeting: Specific implementation studies. Jerusalem Patriarchate, 20.05.2016, 2nd Steering Committee Meeting, 20.07.2016; 3rd Steering Committee Meeting, 06.10. 2016; 4th Steering Committee Meeting, 15.12.2016; 5th Steering Committee Meeting, 22.02.2017 presented by A. Moropoulou, chief scientific supervisor on behalf of the interdisciplinary team em. prof. Emm. Korres, prof. A. Moropoulou, prof. A. Georgopoulos, prof. K. Spryrakos, Ass. Prof. H. Mouzakis
5. Moropoulou, A., Korres, E., Georgopoulos, A., Spyrakos, C., Mouzakis, Ch.: Presentation upon completion of the Holy Sepulchre's Holy Edicule Rehabilitation, National Technical University of Athens, 2017, p. 12. ISBN: 978-618-82196-4-9
6. Moropoulou, A., et al.: Faithful rehabilitation. History, culture, religion, and engineering all intertwined in a recent project in Jerusalem to rehabilitate and strengthen the site believed to be the tomb of Jesus of Nazareth. J. Am. Soc. Civ. Eng. **78**, 54–61 (2017)
7. Giannakopoulos D.: Architect: architectural documentation in interactive relation to the integrated study of protection of the Holy Edicule of the Holy Sepulchre in Jerusalem. M.Sc. thesis, NTUA Interdisciplinary program of postgraduate studies in Protection of Monuments —Direction Materials and Conservation Interventions, Supervisors: Prof. M. Korres, Prof. A. Georgopoulos, Prof. Ch. Ioannidis, Prof. A. Moropoulou, Ass. Prof. H. Mouzakis, February 2017
8. Moropoulou, A., Farmakidi, C.M., Lampropoulos, K., Apostolopoulou, M.: Interdisciplinary planning and scientific support to rehabilitate and preserve the values of the Holy Edicule of the Holy Sepulchre in interrelation with social accessibility. Sociol. Anthropol. **6**(6), 534–546 (2018). https://doi.org/10.13189/sa.2018.060603

Digital Heritage

Branding Strategies for Cultural Landscape Promotion: Organizing Real and Virtual Place Networks

Konstantinos Moraitis$^{(\boxtimes)}$

National Technical University of Athens, 8A Hadjikosta Street,
11521 Athens, Greece
mor@arsisarc.gr

Abstract. Neoteric Western civilization refers to the historic past of the ancient Greek city of Sparta proclaiming it as an initial paradigm of up-standing, virtuous governance and political formation. It was the acknowledgment of the previous correlation between neoteric references and Hellenic classical antiquity that drove the School of Architecture of the National Technical University of Athens (NTUA), in collaboration with the University of Peloponnese, to undertake a research program concerning the development of a 'place branding' strategy for the promotion of the cultural landscape of the territory of contemporary municipality of Sparta.

The research proposal that will be presented insists on the creation of landscape visitors' networks, connecting important focal places of historic importance in the interior of the urban structure, as well as in the surroundings of the contemporary city, with exceptional interest for the zone of river Eurotas. However the most important concern of the proposal is associated not with the real landscape treatment, but rather with the interrelation of material landscape networks to virtual space narrative references, attempting to bring visitors in immediate contact with libraries, museums and galleries of the world, where the mythological and historic past of the real, contemporary landscape of Sparta is treasured up.

Keywords: Cultural landscape promotion · Real place networks ·
Virtual narratives

1 Introduction: Networks in Real Landscape or Virtual Narrative Space, in Reference to the Territory of Sparta

The present paper refers to a research program, organized by the School of Architecture of the National Technical University of Athens (NTUA), in collaboration with the University of Peloponnese,[1] aiming to develop a 'place branding' strategy for the promotion of the cultural landscape of the city Sparta and its surroundings. We ought to

---

[1] With the participation of Professor K. Moraitis of the School of Architecture NTUA and Associate Professor Ioanna Spiliopoulou of the Department of History, Archaeology and Cultural Resources Management of the School of Humanities and Cultural Studies of the University of the Peloponnese.

© Springer Nature Switzerland AG 2019
A. Moropoulou et al. (Eds.): TMM_CH 2018, CCIS 961, pp. 105–118, 2019.
https://doi.org/10.1007/978-3-030-12957-6_7

note in advance that the promotion of cultural landscape identity is not only correlated to practical needs of touristic and economic invigoration, concerning the places of reference. Furthermore it may offer the promise of more important, 'political' advantages, as those concerning the development of conscience of the local population and the preservation of its cultural heritage.

Presenting in detail the above research context, we have to explain that it intends to work with the creation of visitors' networks, connecting all the focal points of mythological and historic importance in the interior of the city of Sparta, as well as in the surrounding landscape, especially at the fluvial zone of river Eurotas. The formation of such networks may offer a better space feeling, an organizational 'grid', on which partial, material elements of perception or partial 'intangible' elements of a narrative sequence, may be projected and organized.

A crucial concept, underlying this research project, has to do with the realization that a reference to actual geographical networks could be regarded as a limited application, in comparison to the possibilities of contemporary informatics technology. Reference to historic memory is par excellence a 'virtual' procedure, attempting to recreate a lost reality, in mental, non-material terms or to re-materialize it, in pictorial depiction and monumental formations. Thus recreation of the imaginative mythical or historical past could be regarded as a narrative challenge for contemporary simulation technology, not only in fictional attempts but also as a didactic, educative procedure. The research in question, proposes the organization of a virtual network in immediate correlation to the real geographic networks of Sparta. Information retrieved through mobile phone application or information presented in parallel to the real perception through augmented reality techniques, could bring the visitors in immediate contact with libraries, museums and galleries of the world, where the mythological and historic past of the real contemporary landscape of Sparta is treasured up.

2   Locating the Immaterial: Tracing the Intangible, for Economic, Cultural and Political Reasons

In her famous book on 'mnemotechnics', *The Art of Memory*, Yates [1] clearly indicated that preservation of memory references may be correlated to places and place sequences. We could thus speak of 'landscapes of memory', being completely immaterial, creating the imaginary, mental scenography where memory references may be installed. Or, we may speak of 'real', material landscape formations, where tangible indications reveal memory references and intangible values.

In a broader sense, public architecture and even a large part of urban design interventions have to do with cultural values, presented and promoted not only in monuments, but also in buildings and city formations. Contemporary place branding may ask for analogous emblematic connotations of the urbanscape or cultural landscape, inside or outside cities, correlating this demand with the economic invigoration of places. However, material references in landscape may contribute, as already stated, to the presentation and promotion of 'profound' cultural and political values, being in many cases much more important than the economic and commercial 'surface' of the place usage. In our presentation we shall try to associate the previous intentions, concerning the promotion of intangible qualities, with material landscape interventions, for economic and commercial or for cultural and political reasons.

Furthermore, in our presentation we shall insist on the fact that the cultural land-scape presents, in the case of Sparta, a unique set of references of constitutional importance for the creation of the neoteric Western civilization and neoteric Western political ethics. It is this unique historic significance, the monadical historic validity of Sparta that underlies the place branding proposal presented, imposing to the research in question an academic didactic commitment much broader than the desire for the touristic and economic invigoration of the territory of Laconia.

3 Sparta Regarded as Cradle of Political and Military Western Virtue

In the 18th century English park of Stowe, in Buckinghamshire, in the interior of the 'Temple of Ancient Virtue', Lycurgus' statue commemorates the political influence of the ancient Spartan governor and legislator to the European and Western world. It is for the same reason that a bas relief depiction of Lycurgus' profile decorates the entrance hall of the House of Representatives, in Washington City, USA [3, 8]. *In both cases neoteric Western ethics have to refer to the historic past of the ancient Greek city, by accepting it as an initial paradigm of up-standing, virtuous governance and political formation, Lycurgus being the quasi-legendary creator of its constitution* (Fig. 1).

Fig. 1. On the left, 'The Temple of Ancient Virtue' – Stowe Park, Buckinghamshire by William Kent (circa 1734). In the interior of the Temple, Lycurgus' statue commemorates the ancient Spartan governor and legislator (Personal picture of the author). On the right, bas relief of Lycurgus' profile in the entrance hall of the House of Representatives, in Washington City, USA (Public Domain. In: https://en.wikipedia.org/wiki/Lycurgus_of_Sparta#/media/File:Lycurgus_bas-relief_in_the_U.S._House_of_Representatives_chamber.jpg, last accessed 2018/07/31).

It is not clear if Lycurgus (c.820 BC) was an actual historical figure. However he was referred to, by a number of ancient historians and philosophers as Herodotus, Xenophon, Plato, Polybius, Plutarch [5], and Epictetus, as the lawgiver who established the military-oriented reformation of Spartan society, in accordance with an oracle of the god Apollo at Delphi.[2] Lycurgus immortalized, through his reforms, the three principal Spartan virtues; equality among citizens, military fitness, and austerity [6], as described in his political proclamation 'Megali Rhetra'[3] and thus instituted the communalistic and militaristic identity of the ancient Spartan society.

In recent times the cultural and political interest for ancient Sparta persists. In literature and film making, even in comic-strip publications, the political decision of ancient Spartans to resist in Thermopylae to the Asiatic Persian invasion, is constantly coming in discussion. We refer to productions of great international publicity, as the epic description of the battle of Thermopylae, by Steven Pressfield in his historical fiction novel *Gates of Fire* (1998), or *300* an American epic war directed by Zack Snyder (2006) and based on the previously successfully published homonymous comic series *300*, by Frank Miller and Lynn Varley (1998).

In all previous cases, the ancient military example is not used as an exemplary reference to the contemporary need for sound political ethics in general. It is rather presented, in particular, as a historic predecessor of the Western need to 'answer', in a dynamic way, to the 'fundamentalist' expression of the contemporary oriental, Asiatic aggressiveness. A stochastic evaluation of the above attitude could critically insist on the fact that all those recent expressive examples are rather associated to an equally aggressive tendency, a post-colonial aggressive tendency of Western origin, directed from Western world towards Asiatic societies, and, moreover attempting to legalize Western geopolitical practices and absolve them from any possible political and historic blame. In any case Spartan reference is constantly used, for a number of centuries, as an omnipotent historic paradigm able to justify, through more or less acceptable arguments, Western political disposition. It is even used as a principle reference correlated with the formation of the Western body-image, with the formation of fitness and aesthetic principles that may regulate the sound behavior and the attractiveness of Western neoteric physique.[4]

It was under the reconnaissance of the previously mentioned importance of the ancient Spartan paradigm for the formation of the neoteric Western culture that the

---

[2] Apollo being, according to the ancient Greek mythology, the god of sun, truth and prophecy.

[3] 'Megali Rhetra – Μεγάλη Ρήτρα', literally 'Great Saying' or 'Great Proclamation': the oral description of the Spartan constitution, probably in accordance to the initial oracle of Delphi, which was believed to have contained the entire body of the constitutional principles in verse.

[4] This body reference was principally associated with the masculine ideal of the athletic or military virility; however it is rather well-known that ancient Spartan women were also accustomed to athletic activities, possessing among all other ancient Greek women the more attractive, athletic physical structure. This common interest for athletic exercise, for male and female Spartans equally, is emphatically presented in an early oil, on canvas painting by French impressionist painter Edgar Degas, under the title *Young Spartans Exercising*, also known as *Young Spartans* or *Young Spartan Girls Challenging Boys* (presented in Fig. 6). The work depicts two groups of male and female Spartan youth exercising and challenging each other in a state of astonishing gender equality, even judged in the frame of 'liberated' contemporary morality.

research presented was proposed. We deeply understood the 'amphisemic' significance of this paradigm and its possible correlation with contemporary Western political aggressiveness. *However we are also conscious of its positive connotations; concerning Spartans' dedication to their city-state, their commitment to the independence of their society and their sound appreciation for physical exercise and for the aesthetic pleasure that derives from the perfection of the human body.*

4 Neoclassicism Was Not Just an Expressive Style[5]

We may correlate all aforementioned positive references with the classical aura superimposed on Western cultural and political orientation for a period of five centuries, as early as in Florentine Quattrocento till the first biggest part of 19th century and, in many cases, even later. Classicism as a mimetic tendency, expressed in the decorative and visual arts, literature, theatre, music, and architecture, was oriented towards ancient Greece and ancient Rome, indicating the cultural prestige that those historic paradigms possessed for neoteric Europe and Western world. It was conveyed under the expressive surface of the formal and aesthetic resemblance to the ancient world; it was nevertheless founded on deeper references, associated to the political status of the two ancient political states.

For royal sovereigns as Louis IXV and even for outstanding clerical personalities as Italian cardinal Ippolito d' Este, ancient reference connoted the undeniable political power. It was for this reason that the French king was compared to the solar deity of the ancient god Apollo, as personified in the golden sculptural complex emerging from the central fountain of the Versailles gardens. It was for the same reason that cardinal d' Este was compared to omnipotent Hercules, a hero many times correlated to landscape and garden formations in all over Europe.[6] However the correlation of the Renaissance spirit in Medici's Florence, as well as the correlation of 17th century Netherlands with classical antiquity, possessed a more specific meaning. It referred to the political paradigm of the ancient democracies in comparison to the bourgeois oriented political tendencies, which firstly appeared in the Italian peninsula and then acquired a more mature expression in the Dutch protestant ground. In all previous cases landscape symbolism coincided with the need of political promotion, with the political 'branding' of prominent political personalities or political regimes.

However we have to arrive to the period of Enlightenment in order to perceive the most powerful association of neoteric civilization with classical 'formality' and classical expression, under the more precise description of 'neoclassicism'. In a conscious

---

[5] Compare the subtitle of this part of the text to the title of the article "Classicism is not a style" by the architect Demetri Porphyrios [7].

[6] Ippolito's d' Este gardens in the famous Villa d' Este referred to Hercules, as a metaphor of the cardinal's political ambition compared to the ancient demigod's power [8, pp. 84–88]. It was for the same reason that Het Loo gardens in Netherlands, created for Willem of Orange initially 'Stadtholder' of Holland and later king William III of England, were also dedicated to Hercules [8, p. 158]. We may also mention Hercules' gigantic statue at the vanishing point of the central compositional axis of Vaux-le-Vicomte gardens, in Sein-le-Marne, near Paris [8, pp. 36–39].

way neoclassicism was associated to the subversive desire of the bourgeois class, being already mature in theory as well as in practical terms for French revolution. It is worth noting that, during this period of change, every mode of expression, visual arts, architecture and landscape architecture, urban design as well, were all characterized by classicistic inspiration. Yet neoclassicism was not just an expressive style, was not just limited to formal exteriority. In a much more profound way neoclassicism rather represented a neoteric political state of thinking, an 'emblematic' exteriorization of new political conscience. It is described as 'emblème de la Raison' [9], 'emblem of Reason', of political bourgeois democratic Reason (Fig. 2).

Fig. 2. *Leonidas at Thermopylae* (1814) by Jacques-Louis David. The painter chose his subject in the aftermath of the French Revolution, as a model of "civic duty and self-sacrifice", but also as a contemplation of loss and death, presenting Leonidas quietly poised and heroically nude. (Public Domain. In: https://en.wikipedia.org/wiki/Leonidas_at_Thermopylae#/media/File:L%C3%A9onidas_aux_Thermopyles_(Jacques-Louis_David).PNG, last accessed 2018/07/31).

It was under this connotative atmosphere that neoclassical painter Jacques-Louis David depicted Leonidas of Sparta at a state of serene heroic nudity, staring to the newly born European democracy, through the surface of the painting. Neoteric Europeans ought to preserve the territorial and political integrity of their state, in an uncompromised way. It is here again, in David's depiction, that the political virtue of ancient Spartans is represented as a paradigm for the possible political ethics of the modern world; as an indication analogous to the one of the 'Temple of Ancient Virtue', in the 18th century English park of Stowe, in Buckinghamshire, as well as to the bas relief of Lycurgus' profile in the entrance hall of the House of Representatives, in Washington City, USA, both previously mentioned in our text.

5 Symbolic Economy and the Connotative Approach to Places and Landscape

We have already commented on the emblematic function of neoclassicism and more specifically of the references to ancient Sparta for neoteric and, why not for contemporary Western imagery. Furthermore we have described the material expression of the neoclassical attitude and the space centered presentation of Spartan mythological and historic narratives. We have to remark in addition that in contemporary 'place branding' strategies we usually propose analogous emblematic connotations of urbanscape or cultural landscape, inside or outside cities, correlating these proposals with the economic invigoration of places. We also present these proposals under the terms 'symbolic economy', having in mind that contemporary societies are largely motivated in their consuming procedures through the stimulation of their imagery. Thus 'symbolic economy', a term recently used to present branding procedures in general or place branding procedures in particular, indicates the possibility to over-qualify material commercial reality through semantic, 'symbolic' intervention.

However material objectivity is always, constantly correlated to semantic, symbolic values, in the interior of economic procedures or outside them. The denotation of objects or the 'objectified' reality is constantly correlated to connotative significations...*it is constantly 'eroded' though symbolic meaning and symbolic imagery*. Why? Because, generally speaking, symbol thinking does not only refer to the formation of images analogous to the cross or a national flag. It rather generally refers to two intrinsic mental activities, *abstraction and creation of mental structures. Thus we cannot manipulate material reality without immaterial, intangible abstractions and immaterial, intangible structures. Human intelligence is a mental function correlated to abstraction and creation of abstract mental structures. We cannot manipulate 'reality' and 'objectivity' without them.*

What we discuss in our essay, are methods of making perceivable those 'immaterial' conditions of social and cultural production. We tend to associate symbolic economy to 'Place Branding' economically centred procedures; however, material references in landscape may contribute to the presentation and promotion of 'profound' cultural and political values, being in many cases, as we already stated, much more important than the economic and commercial 'surface' of the place usage.

6 Formation of Visiting Networks in Real Landscape

Returning to the present-day Sparta in south Peloponnese, Greece, it is obvious that the historic memory of the ancient city largely surpasses, in importance, its nowadays built formation. Thus it was a design challenge for academic experts, organizing a promotional strategy for the city, to transcend its material boundaries; to extend its allure to the surrounding landscape, to the background of mount Taygetus and to the river-zone of Eurotas. Moreover it was even more important for the inhabitants of Sparta and for the future tourists and visitors, to be able to come to contact with the universe of the abundant immaterial references, provided by Western historiography and literature, or depicted in

Western pictorial arts. Thus real landscape networks were proposed in the surroundings of the city, in combination with virtual networks organizing a second, 'virtual landscape' of references, supporting real place visitors with audio-visual information.

Fig. 3. Contemporary Sparta and its correlation to Eurotas river zone (the river is depicted with light blue dotted line), and the important archeological sites: the Acropolis of ancient Sparta (AC), the remains of the roman bridge over Eurotas (1), the remains of a roman villa (2), Eurotas Altar (3), Astravacus memorial (4), ancient sanctuary (5), sanctuary of Artemis Orthia (6) and Psychiko Altar (7). (The maps were produced as part of the diploma thesis "Designing an educational and recreational park at the riverside of Eurotas" by A. Mitsia, in the School of Architecture of the National Technical Univ. of Athens (NTUA) – March 2017, responsible professors M. Markou and K. Moraitis).

Insisting on the material landscape visiting networks, we may refer to the Acropolis of ancient Sparta (Fig. 3–AC), located at the peripheral, peri-urban territory, northwest of the contemporary city. The site was firstly excavated by the pioneer research of the British School of Archaeology, as early as in 1910 and is associated to a number of significant monuments, unearthed in its immediate vicinity.

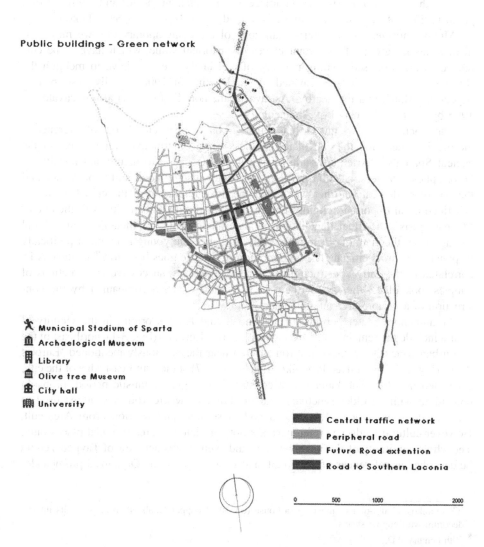

Public buildings - Green network

🏃 Municipal Stadium of Sparta
🏛 Archaelogical Museum
🏭 Library
🏢 Olive tree Museum
🏫 City hall
🏯 University

▮ Central traffic network
▮ Peripheral road
▮ Future Road extention
▮ Road to Southern Laconia

0   500   1000   2000

Fig. 4. Contemporary Sparta and its correlation to public buildings and open-air public spaces. (The maps were produced as part of the diploma thesis "Designing an educational and recreational park at the riverside of Eurotas" by A. Mitsia, in the School of Architecture of the National Technical Univ. of Athens (NTUA) – March 2017, responsible professors M. Markou and K. Moraitis.

Among the monuments of the Acropolis archeological site we have to indicate the temple of 'Athena Chalkioikos',[7] designed by the architect Vathyklis from Magnesia, alternatively called temple of 'Athena Poliouchos - Athena Guardian of the City'; also the ancient theater of Sparta at the south side of the Acropolis, a product of the early Imperial Period. Other important monuments of Acropolis site are a circular structure of unknown function, usually called the 'Circular Building', the remains of merchant stalls of the Roman Imperial period adjacent to the ancient theater and the relics of a grand basilica of the mid Byzantine Era, linked to the Basilica of Saint Nikon.[8]

Moving northeast of the peri-urban zone of the contemporary city we may enumerate another group of important places of historical value, which have to be correlated to the proposed visiting network of real landscape. We have to mention the 'Eurotas Altar' (Fig. 3-3), excavated by G. Dickins in 1906, usually described as 'Lycurgus Altar', and the nearby 'Astravacus memorial' (Fig. 3-4) also excavated in 1906 by the British School of Archaeology.

The sanctuary of 'Artemis Orthia' (Fig. 3-6) has to be a central node of reference of the previous network, the cult of the goddess being of principal significance for the ancient Spartans.[9] 'Artemis Orthia' was a deity associated to nature and fertility, 'a kurotrophos' goddess, a 'child nurturer', protector of childhood and adolescence and the sanctuary dedicated to her was rather the most important of the city. It could be described as an important religious center, associated with archaic rituals as the one of 'diamastigosis', the ritual flagellation of male adolescents,[10] indicating their cultural passage from the state of an immature boy to that of the young man and a politically responsible Spartan citizen. The consecration of the site goes back to 9th century A.D. correlated to the early construction of an altar. Later on, successive constructions of temples took place, the last of them in 3rd century P.D. accompanied by the construction of an impressive amphitheater.

Monuments of later periods, of Hellenistic and Roman origin, in the vicinity of Sparta include the remains of the roman bridge over Eurotas ((Fig. 3-1), northeast of the peri-urban zone, the remains of a roman villa near the previously mentioned 'Eurotas Altar' (Fig. 3-2), as well as 'Psychiko Altar' (Fig. 3-7) at the south-east edge of the city, excavated in 1962. The latter was a construction of the Hellinistic period, probably coinciding with an older sanctuary that of 'Phiveo', where young Spartans used to sacrifice dogs in favor of Enyalios, a god of soldiers and warriors from Ares cult. However cultural landscape references are not correlated to precise nodal places only. The whole continuity of Eurotas river zone and even the distant view of Taygetus could be described as references of important landscape significance. Organized promenades

---

[7] 'Chalkioikos, chalki-oikos', meaning 'a house (oikos) of copper (chalkos)' because of its interior decorum with copper sheets.

[8] 10th century AD.

[9] It was also excavated by the British School of Archaeology, between 1906 and 1910.

[10] The origin of 'diamastigosis' was analytically presented by the Greek traveler and geographer Pausanias (Pausanias 1918: III, 16, 9–11), in his famous work *Description of Greece* (Ελλάδος Περιήγησις, *Hellados Periegesis* in ancient Greek) [4]. Pausamias lived in the time of Roman emperors Hadrian, Antoninus Pius, and Marcus Aurelius (2nd century AD), and described ancient Greece from his first-hand observations.

Fig. 5. Exemplary partial mapping of Eurotas riverside zone. (The maps were produced as part of the diploma thesis "Designing an educational and recreational park at the riverside of Eurotas" by A. Mitsia, in the School of Architecture of the National Technical Univ. of Athens (NTUA) – March 2017, responsible professors M. Markou and K. Moraitis).

alongside the riverside zone, in correlation to the ancient monuments could be regarded as an integral part of the proposed 'real' landscape networks, bringing into mind the intangible, yet still alive memories of a remote noble historical past, glorified thereinafter by the intense admiration of the Western civilization (Fig. 5).

Furthermore organized promenades lengthwise Eurotas or through the nodal archeological sites outside the contemporary city, have to be connected with the network of public building and open spaces in the interior of the urban territory and principally with the archeological museum of Sparta, the library, the 'Olive Tree museum', the University of Peloponnese building complex, the city hall, the stadium and the principal squares of the city. In any case, the association of tangible, material, 'real' landscape with memory projection would help us to imagine an intangible, 'virtual' layer of information, of textual references or depictions, superimposed on immediate material reality.

7   Formation of Visiting Networks in Virtual Landscape and Their Coexistence with Real Sites of Sparta, in an 'Augmented Reality' Network

An important first organizational approach of the research team had to do with the mapping of the open public spaces, of the monuments and public buildings, as well as of commercial uses of contemporary Sparta. Then visiting itineraries were proposed on the 'real', material landscape of the city and its surroundings, extended from its urban texture till the zone of the river on its eastern part. In addition to those real landscape networks APPs for mobile phone or Vuzix AR3000 Augmented Reality Smart Glasses were also proposed, in an effort to connect intangible references with tangible, perceivable indications and offer to the user the simultaneous possibility to navigate in a virtual network, in a 'virtual landscape' much larger than the real one. Such a virtual landscape could be conceived as a multi-layered system composed of multiple partial networks of information. We have previously cited the information networks of historiography and literature concerning ancient Sparta, also the 'virtual' gallery with its pictorial references. We may describe in addition networks of images and information from other museums or historic landscapes, associated to the ancient city, to its mythological origin and its historic past.

All previous 'material' of possible references, presents a huge collection of objectified, 'materialized' associations of memory, already existing, however dispersed in an extended number of different places, all over Greece and even more all over the Western world. What the research program proposed by the School of Architecture of the National Technical University of Athens describes, is the effort to correlate all those dispersed references to a huge memory 'platform', a huge electronic virtual platform, to a total virtual museum, a total virtual gallery, a total virtual landscape of references correlated to the real visiting parts of the contemporary city of Sparta and its surroundings.

To this combination of virtual and real experience not only historic and archeological information transmitted to the visitor would be important. The public life of the contemporary city, its contemporary way of living in the same part of the world, where its ancient predecessor existed, under the same Hellenic sky, in the vicinity of the same river, with the presence of the same mountain at its landscape background, would be equally important to the organized historic and archeological guided tours, they could equally enrich the experience of the visitor (Fig. 4).

Fig. 6. *Young Spartans Exercising*, also known as *Young Spartans*, or as *Young Spartan Girls Challenging Boys*, by Edgar Degas. We could imagine 'visions' of ancient memory superimposed on the actual present activities, in the real locus of Sparta and its surrounding landscape. In Degas' depiction the girls are positioned to the left of the painting and the boys to the right. In between the two groups, in the background, a third group of participants appears watching them; they are fully dressed while the youth in the foreground stand naked or topless. Behind the onlookers, identified as Lycurgus and the mothers of the children, lies the city of Sparta, dominated by Mount Taygetus. Public Domain. In: https://en.wikipedia.org/wiki/Young_Spartans_Exercising#/media/File:Young_Spartans_National_Gallery_NG3860.jpg, last accessed 2018/07/31.)

8   Conclusion: Visions of Memory Superimposed on Real, Material Presence

We could conclude, by imagining happy groups of visitors walking by Eurotas banks in the next years, or we could imagine groups of young boys and girls, exercising by the river in the next years, under the distant presence of Taygetus, in an effort of revitalization of the ancient memory... and we could also imagine 'visions' of this ancient memory superimposed on the actual human reality on multiple virtual layers, illusionary visions imposed on our real experience, existing as vividly as the real locus of Sparta and its surrounding landscape.

Every time I visit Sparta I remember the Irish poet Patrick Kavanagh and his poetic imagery about a future time, "in a hundred years or so".[11] By that time the ghost of the poet "disheveled" would play "with little children, whose children have long since died". In a similar manner I fantasize a ghost of my own self, at the outskirts of the future Sparta, playing with ancient city adolescents "whose children have long since died"; playing with them on the riverside, on the blooming banks of Eurotas... Is it just a fantasy of mine or a possible virtual image of 'augmented reality', able to be produced by contemporary electronic technology?

References

1. Yates, A.F.: The art of memory. In: Frances Yates Selected Works, vol. III. Routledge & Kegan Paul, London (1966). http://www.alzhup.com/Reta/Docs/ArtOfMemory.pdf. Accessed 31 July 2018
2. Kavanagh, P.: The Complete Poems of Patrick Kavanagh. The Goldsmith Press Ltd., Kildare (1984)
3. Moraitis, K.: Το Ελληνικό Τοπίο – Θεωρητικό Σχεδίασμα με αφορμή το νησί της Σύρου (edited in Greek, English title: Hellenic Landscape – A theoretical outline in reference to Syros island). In: Tabula Gratulatoria in Favor of Prof. J. Stefanou, Syros Institute Edit, Syros, pp. 7–8 (2015)
4. Pausanias: Description of Greece. Translated by Jones, W.H.S.C., Ormerod, H.A. Harvard University Press, Cambridge Mass (1918)
5. Plutarch: Apophthegmata Laconica. Vol. III of the Loeb Classical Library edition. Harvard University Press, Cambridge Mass. (1931). http://penelope.uchicago.edu/Thayer/E/Roman/Texts/Plutarch/Moralia/Sayings_of_Spartans*/main.html. Accessed 31 July 2018
6. Forrest, W.G.: A History of Sparta 950-192, p. 50. B.C. Norton, New York (1963)
7. Porphyrios, D.: Classicism is not a style. http://jhenniferamundson.net/wp-content/uploads/2015/12/1983-Porphyrios.pdf. Accessed 31 July 2018
8. Moraitis, K.: Η Τέχνη του Τοπίου – Πολιτιστική επισκόπηση των νεωτερικών τοπιακών θεωρήσεων και διαμορφώσεων (edited in Greek, English title: The Art of Landscape – A cultural overview of the neoteric landscape theoretical and formative approaches). Kallipos NTUA Edit., Athens (2015). https://repository.kallipos.gr/bitstream/11419/2621/1/00_master_document.pdf. Accessed 31 July 2018
9. Starobinski, J.: 1789 - Les Emblèmes de la Raison. Ed. Flammarion, Paris (1973)

---

[11] We refer to Kavanagh's poem under the title "If Ever You Go To Dublin Town" (1984). The full 'stanza' of Kavanagh's poem, having nothing to do with ancient or contemporary Sparta but certainly correlated to memory recall and to its space references, is: "On Pembroke Road look out for me ghost, /Dishevelled with shoes untied, /Playing through the railings with little children / Whose children have long since died. /O he was a nice man, /Fol do the di do, /He was a nice man And I tell you" [2].

3D Survey of a Neoclassical Building Using a Handheld Laser Scanner as Basis for the Development of a BIM-Ready Model

Dimitrios-Ioannis Psaltakis[1], Katerina Kalentzi[1],
Athena-Panagiota Mariettaki[2], and Antonios Antonopoulos[3(✉)]

[1] LANDMARK, Athens, Greece
{support,info}@landmark.com.gr
[2] NTUA School of Architecture, Athens, Greece
athenamariettaki@gmail.com
[3] ALL3D, Athens, Greece
info@all3d.gr

Abstract. This paper explores the application of two widely-used digital technologies, Digital documentation using laser scanning and Building Information Modelling (BIM), in the case of neoclassical buildings. Laserscanning with a handheld scanner was used for the 3D documentation of a neoclassical building in Athens, including exterior and internal spaces. The resulting point cloud dataset constitutes the primary survey record of the neoclassical building in its current state (as-existing). Finally, a BIM-ready model of the existing structure was proposed as an alternative method for the production of coordinated 2D drawings and facilitating requirements of subsequent development of the project.

Keywords: Digital documentation · Handheld laser scanner · HBIM · 3D model

1 Introduction

In the last two decades Laser scanning has been used extensively as a method for 3D documentation of cultural heritage, ranging from large-scale landscapes, historic buildings, monuments, museum objects and artifacts. A growing range of LS (Laser scanning) software applications are currently available to an expanding base of users in the heritage sector, corresponding with significant volume of academic research and publications in the field. The use of this technology for the digitization of museum collections is becoming ever more widely practiced, as evidenced by the growing output of 3D digital models of cultural heritage objects in web-based 3D viewing and sharing platforms. The simplicity of the method makes Laser scanning an attractive solution for 3D recording of complex historic monuments. Building Information Modeling is a technology more widely used in the new-build construction industry. However, in the last five years, the concept of Heritage (or Historic) Building Information Modeling (HBIM) has become of hot topic for research, while in some

© Springer Nature Switzerland AG 2019
A. Moropoulou et al. (Eds.): TMM_CH 2018, CCIS 961, pp. 119–127, 2019.
https://doi.org/10.1007/978-3-030-12957-6_8

countries relevant legislation encourages or even enforces the uptake of BIM by the heritage sector (mainly concentrating on construction projects involving historic buildings). Greece has no such legislation at the moment. Unlike BIM for new-build, where the model develops along with the design and construction phases of a project, BIM for existing buildings (EBIM) and BIM for historic buildings (HBIM) generally requires a significant upfront investment of time and associated costs to create a fully developed as-existing BIM model, which represents not only the geometry but includes also material properties and other information about the various components of a building. The term 'BIM-ready' has been used to describe a 3D model created as an assembly of components in a BIM environment, which represents the geometry of the existing fabric, without incorporating any additional information. This paper looks at the application of portable LS for the survey of a significant example of neoclassical architecture in Greece and the development of a BIM-ready model of the building using Autodesk Revit.

2   Digital Documentation

"Digital documentation" is the digital surveying of an object as it is found (as build). There are a lot of digital documentation applications ranging from the "simple visualization" to the 3D models development, that are then used in planning, repair and reconstruction projects etc.

During the digital survey procedure there are a lot of points printed out on every single surface of an object, that describe the topology of every surface as a dense total of points with familiar XYZ coordinates. This sum of points is also known as «Pointcloud» . These points that «Pointcloud» is comprised from, also have in many cases one sort of information regarding the color of the point (RGB Value), a fact also very helpful during the visualization procedure, since this allows the 3D models true color creation.

The Pointcloud is the information carrier of all the available information for an object, and it can be used for a great variety of different purposes, such as the detailed production of 2D drawings (views, floor plans, sections, etc.), the creation of 3D models, the visualization, Building Information Modeling (BIM) etc. Digital documentation can be applied to various objects, ranging from tiny ones (millimeter), to huge areas of interest. The scale of the object, the level of Detail-LOD and the level of Accuracy (LOA), are the main elements to be considered when choosing the appropriate method.

Chart in Fig. 1 shows the methods used for 3D digital survey with respect to the size of the object and the LOD (detail level).

2.1   Laserscanning

Laser scanner is an instrument that sends a laser light beam to the surface of an object and receives reflection of this beam, calculating the distance from the object. A mirror on the scanner is rotating and captures, along vertical lines, whatever is on an almost spherical surface around it. Two calculation methods are used in order to measure the

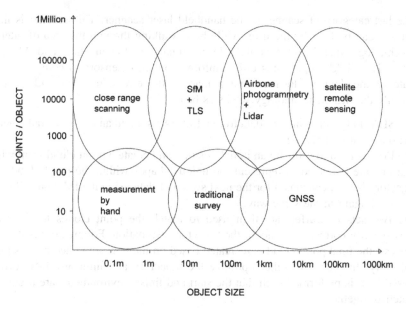

Fig. 1. 3D survey methods with respect to the size of the object and the LOD (detail level).

distance between the instrument and the object. The "Time of Flight" method, calculates the distance, via measuring the flight time from the route of a pulse to its return, and the "Phase shift" method, calculates the distance via the phase shift of sending and return.

Whatever system LaserScanners operates with, the user should be aware of the specific application capabilities and the limits of these measuring instruments. Thus, the quality of the results of a scanning depends on many different factors. For example, the degree of the reflection, affects directly the signal-to-noise ratio, depending on the laser wavelength used by the scanner. It is likely that significant problems could occur in laser scanning measurements, regarding areas of interest that are formed from various materials with big differences in the degree of reflectance.

The user must also be aware of the theory of errors and the errors propagation during the Scanning procedure. In this way, the planning of the measurements planning, the selection of the scanner related to the application, the use (or not) of a surveying control network, and finally the processing of the point cloud, should be done by experts. It is worth mentioning, that the new generation Scanners during the scanning, are doing photos and after correcting them, they produce an RGB color value at each point of the Pointcloud.

All the above regard the terrestrial scanners. The mobile scanners that are placed on a mobile base, (such as car, an airplane, a trolley) operate in a different way. Every single moment, the position of the mobile scanner is calculated with the use of GPS and other inertial systems, and subsequently, the coordinates of the points of the area of interest are being calculated. Apparently the precision of the final coordinates, cannot be as high as the precision that the terrestrial laser scanners could reach.

The last category of scanner is the handheld laser scanner. This scanner is light-weight and can be held by the user, while he is walking through the area of interest. The technology that is used is the 3D Simultaneous Localization and Mapping (SLAM). The SLAM algorithms utilize information from sensors to compute a best estimate of the device's location and a map of the environment around it. More specifically the SLAM technology works as below:

1. The SLAM algorithm utilizes data from a Lidar sensor and an industry grade inertial measurement unit (IMU).
2. The IMU is used to estimate an initial position and create a point cloud from which 'Surfels' are extracted to represent the unique shapes within the point cloud. The trajectory is then calculated for the next sweep of data using the IMU and 'Surfels' extracted again in the same way.
3. The two sets of Surfels are then used to match the point clouds together and subsequently correct and smooth the trajectory estimation. Following this iterative process, the final point cloud is recreated based on the new smoothed best estimate trajectory. In order to further optimize the trajectory and limit any IMU drift, a closed loop is performed such that the start and finish environments are accurately matched together.

The whole procedure is fully automatic and the user cannot control the quality of the end-results, but by following some guidelines, the best possible results can be achieved.

2.2   SfM Photogrammetry

Until recently, Photogrammetry was used mostly for large scale objects and it was not suitable for small ones; overall costs were very high due to the need for airplane or helicopter, expensive photographical instruments and expensive technical support (including data analysis and software).

The last few years, a number of technological and market developments have changed photogrammetry significantly. UAVs allow accurate photography at very low cost. Digital cameras have evolved to provide excellent accuracy at very low cost. Finally, the development of SfM (Structure from Motion) and associated logistical support (software and analytical methods) makes it possible to develop accurate 3-D representations of objects, which is a Pointcloud, similar to the one produced by Laser Scanning.

3   Building Information Modeling

3.1   BIM in the New-Build Construction Sector

British standard PAS 1192-2:2013 defines Building Information Modeling (BIM) as "the process of designing, constructing or operating a building or infrastructure asset using electronic object oriented information. Elsewhere, BIM is defined as "a digital representation of physical and functional characteristics of a facility. A BIM model is a shared knowledge resource for information about a facility forming a reliable basis for decisions during its life-cycle, defined as existing from earliest conception to demolition." Regardless of the formal definition, most experts agree that BIM describes a process, rather than a specific software or even a digital object (model). BIM refers to a framework for "sharing structured information" between project stakeholders. In technical terms, BIM can be described as object-based parametric modeling, a digital technology with origins in mechanical systems design. The modeling process involves the assembly of 'intelligent' parametric objects into a virtual representation of a building or facility. The use of parametric components, which "consist of geometric definitions and associated data and rules", are central to BIM philosophy. Various types of information, including materials, properties, cost, structural and environmental performance, are integrated in a structured way within the building model, which constitutes a digital information system. It is therefore evident that BIM differs from CAD and 3D modeling software, which are mostly limited to the digital representation of geometric information. A BIM model, empowered by the information attached to building components, operates like a virtual diagram of how the actual building is expected to perform.

3.2   BIM for Heritage - HBIM

BIM is worldwide applied in the new-build construction sector for a number of years while BIM for Heritage (HBIM) is a relatively new topic in academic research. HBIM as a shared system of building information, which includes both geometric and non-geometric information. Building components (walls, doors, windows, furnishings) are represented geometrically with the required level of detail both in 2D and 3D interconnected views. Non-geometric information is also represented in the model: materials, colour, etc. This is particularly important in the field of cultural heritage, where non-geometric information can include both tangible and intangible values. Each HBIM platform has a set of different types of information (attributes) preprogrammed in the software, but new properties can easily be added in accordance with specific project needs. All this information is structured and object-oriented, namely information is always linked to the building element it logically refers to.

At present information depending on archaeological sites and historic buildings is represented as a collection of drawings, reports, documents, various datasets and files of 2D or 3D CAD Software, provided by different scientists, each working with his own Datum, standards and tools. This information is dispersed in different places, databases and archives. Parts of the information for a single object after some years is lost or useless.

HBIM is a type of historic asset information model that:

- Represents the appearance of existing historic fabric using high-quality digital survey pointclouds.
- Offers a framework for collaborative working multi-disciplinary team.
- Incorporates all qualitative and quantitative information of a build, including physical and functional characteristics.
- Intergrades intangible characteristics of the build.

4 Scanning a Neoclassical Building

In this paper the scan to BIM procedure is represented in the case of a Neoclassical building in Plaka Athens. The construction of this three-storey building started in 18th or 19th century and the last renovation was done in 1985.

The scanning was performed using the handheld ZEB-REVO laser scanner from GeoSLAM. The whole building was scanned during one single scan mission, which lasted 24 min. The handheld laser scanner was very fast and simple to use; it was only needed to hold it and walk through all the rooms of the building. There was no need for any preparation or installation of targets.

The raw data were imported to the GeoSLAM Hub software, where the registration was done automatically in about 30 min. The resulted point-cloud of 42 million points is compatible with any point-cloud software (Fig. 2).

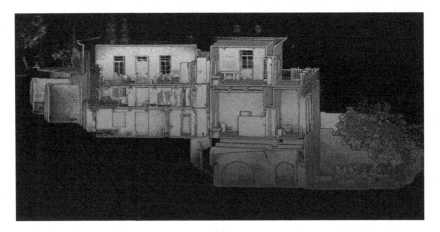

Fig. 2. Part of the pointcloud produced by the handheld laser scanner.

The unique problem in the scanning procedure was the upper part of the roof, what is common in laserscanning because the scanner can't see the roof from above.

5 Modeling the Neoclassical Building

The software used for the creation of the BIM model was Autodesk Revit. At the beginning of the process, the Pointcloud file is imported into Revit. Once imported, it appears on the drawing area as a set of data points in space covering all visible surfaces of the building and is used as a guide for the placement of the BIM objects - building parts in the correct position. This process is carried out by the engineer through hand stitching and will be briefly described below.

At first, vertical and horizontal cutting planes are drawn in order to create section views on the point cloud, which help facilitate its use and build the model with precision. The number and properties of these planes/views can be modified at any moment. The plans, sections and elevations of the model are based on these as well (Fig. 3).

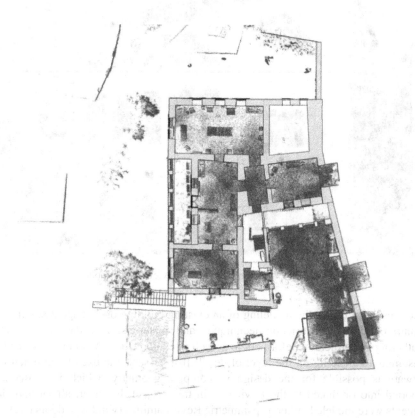

Fig. 3. Plan view of point cloud and model.

The model is generated as an assembly of BIM components (walls, windows etc.), the geometry of which is dictated by the survey data. They are placed in the correct position and configured so that the resulting shape coincides with the point cloud as closely as possible, depending on the required LOD. The model at this point can be considered as BIM-ready. Each one of the placed components can be associated with non-geometric information, such as material, cost, construction phase, thermal properties etc., thus creating a full-BIM. The model can be viewed and modified both in 2D (plans, sections, elevations) and in 3D (isometric, perspective) representations (Fig. 4).

The future user of the model can interact with the properties (geometric and non-geometric) of each element, add new parameters, enter additional information in the available fields, filter and schedule out any item according to their needs and their purpose. This is an essential feature of the BIM process and philosophy and a fundamental difference between a BIM and a 3D CAD model.

The difficulties encountered during the procedure associated to the fact that BIM software, such as Autodesk Revit, is developed mainly as a tool for the design of new

Fig. 4. Perspective photorealistic view of the model

structures. As a result, the modeling of an existing building or heritage asset, that may contain out-dated construction techniques and materials, irregular shapes due to weathering and structural deformation, as well as elaborate ornamentation, can be time-consuming, depending on the level of detail. If necessary, the use of additional CAD software is possible for the design of complex geometry, which is subsequently imported into or linked to the Revit file. In this project, however, all the individual elements were modeled either as parametric Revit Families (windows, doors etc.) or, in few cases, as generic 3D geometry (non-parametric In-Place elements.

6 Conclusions

- Although the terrestrial laser scanners can achieve better accuracy, have higher density in the point-cloud and can give a RGB value to the points, it was decided to scan the building with the handheld laser scanner ZEB-REVO. The decision was based on the users' requirement for level of detail and level of accuracy. The handheld laser scanner offered an unparalleled speed in complete scanning of all areas, even small rooms and staircases. Furthermore, the handheld scanner does not require to empty the building. In fact, scanning was performed with moving people inside the building.
- The accuracy and density of the scanning was adequate for the users' requirements to produce 2D plans and the BIM model of the building.
- The BIM model constitutes an accurate geometric representation as well as a digital source of information about a built asset.

- A common data environment is created that provides the framework used to support interdisciplinary collaboration, which is essential in every heritage project.
- BIM has proven to be efficient in terms of overall cost and timescale by allowing a higher level of communication between professionals.
- A potential disadvantage for BIM is that it has been developed mainly for the new-build sector with more regular geometry. As a result irregular geometry, which is common in Heritage buildings, is more time-consuming to model in Revit. Of course, this is not a disadvantage for most common applications.
- Overall, BIM software has proven very useful for most applications. Potential limitations to irregular geometries can be overcome by clear specification of the minimum level of detail and accuracy, sufficient for each application.

References

1. Buhrow, T.: Genauigkeitsuntersuchungen von Laserscannermessungenam Beispiel des I-SiTE 3D Laser Imaging Systems, Diplomarbeit, unveröffentlicht, Berlin (2002)
2. Historic England: 3D Laser Scanning for Heritage (second edition) (2011)
3. Kontogianni, G., Georgopoulos, A.: Exploiting Textured 3D Models for Developing Serious Games. ISPRS XL-5/W7, vol. 57, pp. 249–255 (2015)
4. Antonopoulou, S.: BIM for heritage: Parametric modeling for contemporary conservation practice. MSc Diss.(unpubl.), University of Edinburgh (2015)
5. Raymond, C.: Real-world comparisons between target-based and targetless Point-cloud registration in FARO Scene, Trimble Real Works and Autodesk Recap, unpubl. Dissertation, University of Southern Queensland (2015)
6. Doulamis, A., et al.: 5D modelling: an efficient approach for creating spatiotemporal predictive 3D maps of large-scale cultural resources. ISPRS Ann. Photogramm. Remote Sens. Spatial Inf., Sci (2015)
7. Historic England: Traversing The past: the total station theodolite. In: Archaeological Landscape Survey, 2nd edn. Historic England, Swindon (2016)
8. Osello, A., Rinaudo, F.: Cultural heritage management tools: the role of GIS and BIM. In: Stylianidis, E., Remondino, F. (eds.) 3D Recording Documentation and Management in Cultural Heritage. Whittle Publishing, Dunbeath (2016)
9. Remondino, F., Nocerino, E., Toschi, I., Menna, F.: A critical review of automated photogrammetric processing of large datasets. ISPRS XLII-2/W5, vol. 25, pp. 591–599 (2017)
10. Historic England: Photogrammetric Applications for Cultural Heritage (2017)
11. Historic England: BIM for Heritage: Developing a Historic Building Information Model. Historic England, Swindon (2017)
12. Antonopoulos, A., Antonopoulou, S.: 3D survey and BIM-ready modelling of a Greek Orthodox Church in Athens. In: IMEKO International Conference on Metrology for Archaeology and Cultural Heritage, Lecce, Italy, 23–25 October (2017)

Exploring the Possibilities of Immersive Reality Tools in Virtual Reconstruction of Monuments

Panagiotis Parthenios[1] and Theano Androulaki[1,2(✉)]

[1] School of Architecture, Technical University of Crete, Chania Crete, Greece
parthenios@arch.tuc.gr, tandroulaki@isc.tuc.gr
[2] Ephorate of Antiquities of Chania, Hellenic Ministry of Culture and Sports,
Chania Crete, Greece

Abstract. In the last decades digital technologies have been employed in the field of cultural heritage for various purposes. Immersive visualization, digital reconstruction of archaeological sites and findings and virtual reality applications are only a few potential tools available when studying the past. The aim of this paper is to present the digital reconstruction of an archaic column, a research conducted at the Digital Media Lab, Technical University of Crete, in coordination with the Hellenic Ministry of Culture and Sports. The 3D models of five very heavy parts of an archaic column were used for studying and virtually reconstructing the complete column. The column was part of an archaic temple unique in size and in type for the area of Chania, in West Crete, Greece. Structure from Motion technique was applied for the reproduction of high quality and accurate digital models of five sandstone drums. Specifically Agisoft Photoscan software was combined with fast, easy and low cost equipment. Furthermore our research team is currently investigating ways to utilize immersive reality for the reconstruction of the archaic column. The five 3D models that were produced with the SfM, are being uploaded as .obj files into Google Tilt Brush. Subsequently the user can experiment by moving, rotating and scaling the individual 3D parts in a 3D environment in real time, in Vive HTC, thus drastically simplifying the digital reconstruction process for similar projects. Finally, a hypothetical façade and a plan view of a similar archaic temple were transcribed in opaque sketches and imported in the immersive reality environment in order to serve as the canvas on which the 3D reconstruction of the column can take place in real scale.

Keywords: Cultural heritage · Structure from motion · Virtual reconstruction · Immersive reality

1 Introduction

This paper aims to present our research realized in Digital Media Lab, Technical University of Crete, supported by the Hellenic Ministry of Culture and Sports, via Ephorate of Antiquities of Chania. Using Structure from Motion Techniques, 3D models of five very heavy parts of an archaic column were produced. The five drums of the column were found in 1997, in a salvage excavation in the town of Chania, in West Crete, Greece. The archaeologists believe that they were parts of an archaic temple

© Springer Nature Switzerland AG 2019
A. Moropoulou et al. (Eds.): TMM_CH 2018, CCIS 961, pp. 128–140, 2019.
https://doi.org/10.1007/978-3-030-12957-6_9

unique in size and in type for the area. Nowadays the five drums are exposed, in random place in the yard of Archaeological Museum of Chania (Fig. 1) but soon they are going to be moved and reconstructed as column at the New Archaeological Museum. The items are extremely heavy so the experiment for their manipulation in a 3D environment, could be useful for the curators to find their original place. Agisoft Photoscan software produced the accurate 3D models of the items and the .obj files were imported in 3dsMax software where the drums were scaled and placed the one up to the other, until the complete column was created. Meanwhile the reconstruction of the column could be done in 3D environment, in Vive HTC by using the Google Tilt Brush in which the imported .obj files of the five drums could been moved, rotated or scaled.

Fig. 1. The five drums of the archaic column exposed in the yard of the Archaeological Museum of Chania, in West Crete, Greece.

2   3D Modeling in Cultural Heritage

Nowadays, new digital technologies for archaeological research are applied for recording, presenting and promoting famous or unknown archaeological sites, monuments and findings (Cosmas et al. 2001). The 3D visualization can be done with very expensive and specialized equipment such as laser scanners but also by photogrammetric methods i.e. using non metric digital cameras and 3d modeling and rendering software (El-Hakim 2002). The use of three-dimensional models of Cultural Heritage is useful in mapping archaeological sites, monuments and objects or for archival and scientific purposes. Digital models can be used in studies for conservation status, for damage recording or mapping the initial state of the monuments before the begging of restoration (Scopigno et al. 2011).

3   Virtual Reality and Cultural Heritage

Virtual Reality (VR) technologies are going to have a significant influence in life and will bring changes in our future work in archaeological field. Computer technologies in VR field, use software to generate realistic images, sounds and interactions that reproduce a real environment, and simulate a user's physical presence in this environment. Moreover, VR fulfils the creation of realistic and immersive simulation of a three-dimensional environment, using interactive software and hardware, and experienced or controlled by movement of the user's body or as an immersive, interactive experience generated by a computer. Furthermore in archaeological sciences, VR offers a fascinating opportunity to visit monuments in the past or places, which are not easily approachable, often from positions which are not possible in real life (Parthenios et al. 2016).

Using VR it is possible to explore artefacts in a computer-generated environment on a different reality, and to immerse oneself into the past or in a virtual museum without leaving the current real-life situation. For the VR experience, the user should only see the virtual world and he needs to wear a VR headset which fits around the head and over the eyes to visually separate themselves from the physical world. As with any other language, VR will be useful to transmit information over different media. In the precise area of museums: VR will help for the preservation of the heritage thanks to virtual replicas or reuniting disperse remains (reconstruction of objects and monuments linked to their original context) (Pujol 2004).

4   Existing Projects

4.1   Foundation of the Hellenic World

The Foundation of the Hellenic World (FHW), based in Greece, is a non-profit cultural heritage institution. The goal of the Foundation is to bring together archaeologists, historians, computer scientists, and artists in order to visualize their ideas and utilize the highest level of technology and resources for research and education within the context of Hellenic cultural heritage. To this purpose FHW has established two immersive VR systems. Some of the main projects undertaken by the VR team at FHW include: (i) the reconstruction and virtual journey through the ancient city of Miletus by the coast, (ii) A view of the Temple of Zeus at Olympia in virtual reality, (iii) The famous statue of Zeus at Olympia as seen through the doors of the temple of Asia Minor, (iv) the Temple of Zeus at Olympia, (v) a series of interactive educational environments that bring to life the magical world of Hellenic costume, and more. (Roussou 2001, 2014).

4.2   The Selimiye Mosque of Erdine, Turkey

The project of the Selimiye Mosque of Erdine, Turkey was carried out by the co-operation between BİMTAŞ, a company of the Greater Municipality of Instanbul, and the Photogrammetry & Laser Scanning Lab of the HafenCity University Hanburg, Germany. The virtual 3D model of the mosque is used for the demonstration of an immersive and interactive visualisation using the new VR system HTC Vive (Kersten 2017).

5 The Greek Doric Order

Ancient Greek architecture is best known from its temples which were divided in three defined orders: the Doric, the Ionic, and the Corinthian Order (Fig. 2). The Greek ancient temples were the most important and most extensive building type in ancient Greek architecture. Mainly the structures were built to house deity statues of ancient Greek religion and frequently the interior of the temples were used to store votive offerings.

Fig. 2. Sketch of the three ancient architect orders, the Doric, the Ionic, and the Corinthian Order. Giorgio Rocco, « Guida alla Lettura degli Ordini Architettonici Antichi, I.Il Dorico » , Tav.I (Rocco 1994)

The Doric Order (Fig. 3) was the first style of Classical Architecture, which is the sophisticated architectural styles of ancient Greece and Rome that set the standards for beauty, harmony, and strength for European architecture. It was the earliest of the three orders of Greek and later Roman of stone temple architecture, it became popular in Archaic period, roughly 750–480 BCE and replaced the previous style of basic, wood structures. It is most easily recognized by two basic features: the columns and the entablature. The purpose of the columns was to support the weight of the ceiling and in Doric order the column shaft is simple and tapered, meaning it is wider at the base than the top. The Doric columns, in their original Greek version, stood directly on the flat pavement of the temple without a base, called the stylobate. The top of the column has a wide flat section called the capital and its role was to support directly the weight of the ceiling. Capitals in Doric Order are smooth, without decoration and are flared, meaning the top is wider than the base. It was the earliest and in its essence the simplest of the orders, though still with complex details in the entablature above. With a height only four to eight times their diameter, the columns were the most flat of all the classical orders; their drums were divided in 20 stripes; and they were topped by a smooth capital that flared from the column to meet a square abacus at the intersection with the horizontal beam (architrave) that they carried (Lippolis et al. 2007).

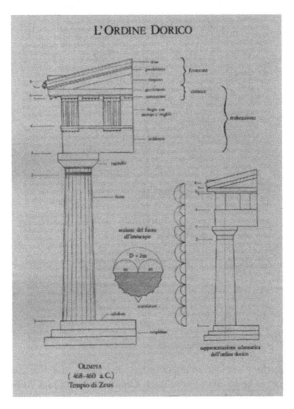

Fig. 3. Sketch of Doric Order, the earliest of the three orders of Greek and later Roman architecture. Giorgio Rocco, « Guida alla Lettura degli Ordini Architettonici Antichi, I.Il Dorico » , Tav.V (Rocco 1994)

6   Digital Reconstruction of an Archaic Column from Five Drums

In 1997 in a rescue excavation, in the centre of the city of Chania, at Michelidakis street and on the Raisakis-Benakis plot were revealed five striped drums of sandstone, placed in the ground in random place (Fig. 4) The five drums have similar dimensions, their height is about 0,8 m and their diameter are about similar and range 0,80–0,90 m (Table 1). According to the archaeologist of the excavation, Mrs St. Markoulaki, the progressive decrease observed in the diameter of the drums indicates that they belong to the same column. After the excavation, the drums were transferred for safekeeping and exposure to the garden of the Archaeological Museum of Chania and were placed in a random place on the ground (Fig. 1).

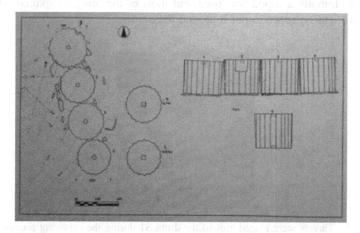

Fig. 4. The five drums were found in second use, in a salvage excavation in the centre of Chania, in Crete, Greece. (Drawing: Ephorate of Antiquities of Chania, Hellenic Ministry of Culture and Sports.)

Nowadays the exhibition of the New Archaeological Museum of Chania is being prepared and the archaeologists are planning to expose the archaic column in reconstruction. The items are very heavy so it is not easy to move them and try to find their original position. The research digitally via the five 3D models of the drums and as a result the digital reconstruction of the column would be valuable and helpful (Parthenios 2017).

The modelling of the drums was done using a photogrammetric Structure from Motion Techniques. Several factors influenced the quality of the models, such as position of the items, accessibility, lighting conditions on all surfaces, size and weight.

For the creation of digital models, the five drums were photographed with two different cameras, Nikon Coolpix P530 and Canon EOS 700D (Fig. 6), following the CIPA 3 × 3 guidelines and rules (Waldhäusl et al. 1994).

Table 1. The dimensions of the five stripped archaic drums

Drum	Dimensions		Stripes
	Diameter	Height	
S1	77 cm	77 cm	17
S2	78 cm	76 cm	20
S3	78 cm	78 cm	20
S4	77,5 cm	78 cm	20
S5	88 cm	75,5 cm	20

It was important to arrange the cameras settings with proper focal length and shutter speed. Some targets were placed around the items to arrange the dimensions (Fig. 5). Cloudy hours were selected, during the process of image recording to avoid shadows on the items. Initially a tripod was used, but most of the images captured without it because there was no space to put it around the drums.

Fig. 5. Targets were placed around the drum S1 during the capturing process.

After photo shooting the appropriate images were selected and the process of the three-dimensional models' production began. The photos, in portrait and in landscape mode, were imported into the Agisoft Photoscan software and the modelling process began (Figs. 7a, b).

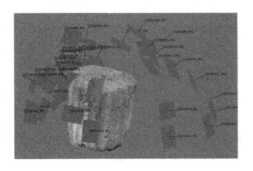

Fig. 6. The images of the drum S1 and the digital camera used for the 3D modeling.

(a)   (b)

Fig. 7. The point cloud and the tiled model of the S1 drum produced in Agisoft Photoscan software.

The sets of images (Fig. 6) used were different for each drum and the process needed a long time, depending on the number of photos used in each model. A number of 3D models were created for each drum (Fig. 8). After carefully comparing them we selected the most appropriate for the next phase of the digital reconstruction of the column.

The cameras used for the modelling of the drums in Agisoft Photoscan, in total were: 42 for drum S1, 28 for S2, 38 for S3, 33 for S4 and 37 for S5.

S1	S2	S3	S4	S5
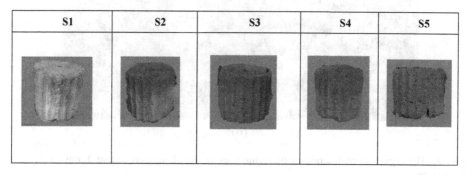				

Fig. 8. The accurate 3D models of the five drums of the archaic column.

Afterwards used Autodesk 3ds Max in order to reconstruct the column. The five obj files, derived from Agisoft Photoscan, were imported into 3ds Max (Figs. 9a, b). Each drum was manually scaled and aligned. The composition was made manually and the position of each drum was chosen according to the perimeter dimensions at the upper and lower sides. First, S5 was placed, followed by S3, S4, S2 and S1 (Figs. 9a, b). Also, the direction and width of the ribs helped. Deterioration of the surfaces did not help because there was a loss of material mainly due to mechanical erosion.

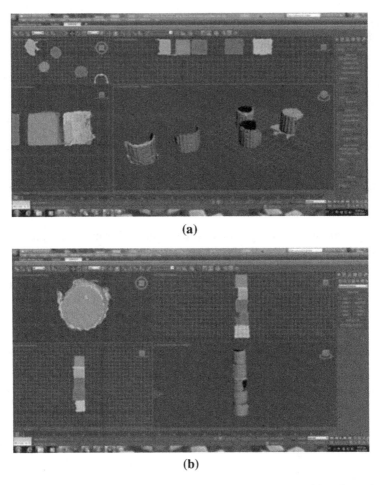

(a)

(b)

Fig. 9. The digital reconstruction of the column using the 3D models of five drums in the 3ds Max software.

Moreover a second experimental path was chosen in addition to the first one in order to test the digital reconstruction of the column. We used HTC Vive along with Tilt Brush software. We took advantage of the software's ability to import 3D geometry (the five .obj files of the drums) and then manipulate each object separately in an immersive 3D environment. We were able to easily grab, move, rotate and finally place each drum on top of the other using the two HTC Vive Controllers (Fig. 10). This offered us the ability to quickly test possible placements for each drum. Furthermore, we were able to sketch in 3D on top of the virtual drums, notating and high lighting points for further research.

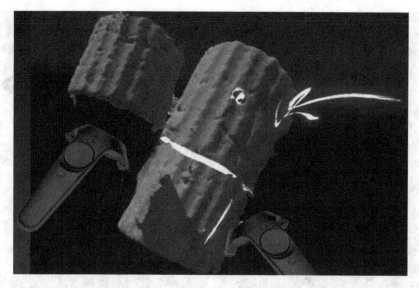

Fig. 10. Using Tilt Brush and HTC Vive for the digital reconstruction of the archaic column from five drums.

7 A Hypothetical Reconstruction of an Archaic Temple

The five sandstone drums were found in 1997, in an excavation by the Greek Ministry of Culture. They were found on the ground, in second use and according to the archaeologist of the excavation, the progressive decrease observed in the diameter of the drums indicates that they belong to the same column. So far there are no more information about the real type, size or location of the ancient construction in which the column belonged. But the type, the form and the dating (c.a. 560 B.C.) of the drums indicate that the column could be part of an ancient temple of Doric order form.

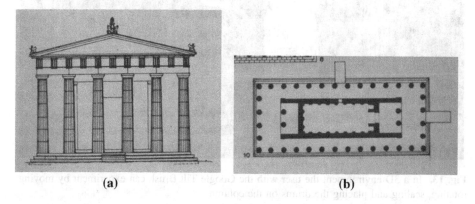

(a)   **(b)**

Fig. 11. The drawings of the temple of Athena Alea at Tegea, Peloponnese. The façade and the plan, of the temple used for the creation of the virtual reconstruction. « Griechisches Bauwesen in der Antike » , Wolfang Müller- Wiener, Fig. 82, p.153 (Müller-Wiener 1988).

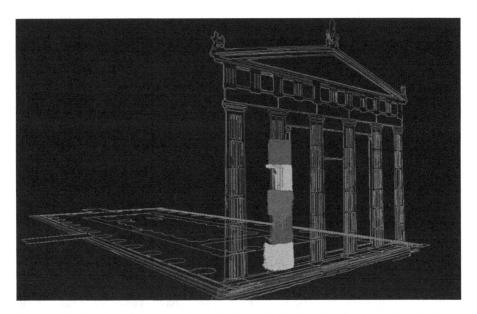

Fig. 12. The façade and the plan view drawings of the Temple of Athena Alea in Tegea, Peloponnese, (traced using Power Trace in Corel Draw) serve as a 3D canvas on which the five drums (modeled in Agisoft Photoscan) are being reconstructed into a column.

Fig. 13. In a 3D environment the user with the Google Tilt Brush can experiment by moving, rotating, scaling and placing the drums on the column.

Using the drawings of a typical Doric Temple (Figs. 11a, b), as documented by Wolfang Müller- Wiener, on « Griechisches Bauwesen in der Antike », Fig. 82, p.153, we experimented with creating a 3D canvas inside the immersive environment on which the 3D reconstruction of the column can take place in real scale. The black and white façade and plan view drawings of the Temple of Athena Alea in Tegea, Peloponnese, created by Wolfang Müller- Wiener, in « Griechisches Bauwesen in der Antike » where traced using Power Trace in Corel Draw (Fig. 12), in order to be converted from scanned images to vectors. The vectors were exported to .dwg files, then imported to 3ds Max and saved as .obj files, which ultimately were imported in Google Tilt Brush in order to serve as the 3D canvas where the user (Fig. 13) can experiment by moving, rotating, scaling and placing the drums on the column.

8   Conclusions

Experimenting with recently commercially available immersive reality applications has the potential of revealing valuable new tools in the field of virtual reconstruction in cultural heritage. Our research demonstrates how simple to use tools can be combined in order to create a virtual reconstruction of an archaic column from its five spare drums, inside a hypothetical 3D environment where the user can interact with the artifacts in real scale and in real time. Future research should focus on improving scaling accuracy of the imported .obj 3D models in the Google Tilt Brush environment.

References

Bruno, F., Bruno, St., De Sensi, G., Luchi, M.-L., Mancuso, St., Muzzupappa, M.: From, 3D reconstruction to virtual reality: a complete methodology for digital archaeological exhibition Digital Heritage, Progress in Cultural Heritage: Documentation, Preservation and Protection. In: 6th International Conference, EuroMed 2016, Nicosia, Cyprus, October 31–November 5, Proceedings, Part I (2016)

Cosmas, J., et al.: 3D MURALE: a multimedia system for archaeology. In: Proceedings of the Conference on Virtual Reality, Archaeology and Cultural Heritage, pp. 297–306 (2001)

De Reu, J., De Smedt, P., Herremans, D., Van Meirvenne, M., Laloo, P., De Clercq, W.: On introducing an image-based 3D reconstruction method in archaeological excavation practice. J. Archaeol. Sci. **41**, 251–262 (2014)

El-Hakim, S.F., Beraldin, J-A.: Detailed 3D reconstruction of monument using multiple techniques. In: ISPRS-CIPA Workshop, Corfu, Greece, 1–2 September 2002

Jones, M.W.: Doric measure and architectural design 2: a Modular reading of the Classical Temple. Am. J. Archaeol. **105**(4), 675–713 (2001)

Kersten, T.P., et al.: The Selimiye Mosque of Edirne, Turkey – an immersive and interactive virtual reality experience using HTC VIVE. In: The International Archives of the Photogrammetry, Remote Sensing and Spatial Information Sciences, Volume XLII-5/W12017 Geomatics & Restoration – Conservation of Cultural Heritage in the Digital Era, 22–24 May, Florence, Italy (2017)

Lippolis, E., Livadiotti, M., Rocco, G.: Architettura greca, Storia e monumenti della polis dale origini al V secolo (2007)

Papaioannou, G., Karabassi, E.-A., Theocharis, Th.: Virtual archaeologist: Assembling the past (2001)

Parthenios, P., Peteinarelis, A., Lousa, S., Efraimidou, N.: Three modes of a monument's 3D Virtual Reconstruction. The case of Giali Tzamisi in Chania, Crete, Digital Heritage International Congress, Granada, Spain, pp. 75–78 (2015). (Conference: 2015 Digital Heritage). https://doi.org/10.1109/DigitalHeritage.2015.7413838

Parthenios, P., Androulaki, Th., Gereoudaki, E., Vidalis, G., Combining structure from motion techniques with low cost equipment for a complete 3D reconstruction of a 13th century church. In: Proceedings 8th International Congress on Archaeology, Computer Graphics, Cultural Heritage and Innovation, 5–7 September, Campus de Vera, Universitat Politecnica de Valencia, Valencia, Spain (2016)

Parthenios, P., Androulaki, Th.: Integrating structure from motion photogrammetry with virtual reality tools as a novel technique for digitally reconstructing an archaic column, CHNT 22, Vienna (2017)

Pujol L.: Archaeology, museum and virtual reality Digit HVM. Revista Digital d' Humanitats (2004). ISSN: 1575–2275 No. 6

Rocco, G.: Guida alla lettura degli ordini architettonici antichi, I. Il dorico, Liguori, Napoli (1994)

Roussou, M.: Immersive Interactive Virtual Reality in the Museum Foundation of the Hellenic World, Athens, Greece (2001)

Roussou, M.: Virtual Heritage: from the Research lab to the Broad Public, Foundation of the Hellenic World, Athens, Greece (2014)

Huang, Q.-X., Flory, S., Gelfand, N., Hofer, M., Pottmann, H.: Reassembling fractured objects by geometric matching. ACM Trans. Graph. (Proc. Siggraph) 25(3), 569–578 (2006)

Scopigno, R., et al.: 3D models for cultural heritage: beyond plain visualization. In: ISTI-CNR, pp. 48–55 (2011)

Müller-Wiener, W.: Griechisches Bauwesen in der Antike, Broschiert (1988)

Waldhäusl, P., Ogleby C.L., Lerma J.L., Georgopoulos A., 3 × 3 rules for simple photogrammetric documentation of architecture (1994)

http://cipa.icomos.org/wp-content/uploads/2017/02/CIPA_3x3_rules_20131018.pdf. Accessed 29 Oct 2018

Zheng, S.Y., Huang, R.Y., Li, J., Wang, Z.: Reassembling 3D thin fragments of unknown geometry in cultural heritage. In: ISPRS Annals of the Photogrammetry, Remote Sensing and Spatial Information Sciences, Volume II-5, 2014, ISPRS Technical Commission V Symposium, 23–25 June, Riva del Garda, Italy (2014)

Reconstruction and Visualization of Cultural Heritage Artwork Objects

Anastasia Moutafidou[1], Ioannis Fudos[2(✉)], George Adamopoulos[1], Anastasios Drosou[1], and Dimitrios Tzovaras[1]

[1] Information Technologies Institute, Center for Research and Technology, Thessaloniki, Greece
[2] Department of Computer Science and Engineering, University of Ioannina, Ioannina, Greece
fudos@cs.uoi.gr

Abstract. Cultural heritage artwork objects usually consist of multiple surfaces with details that become more apparent over time. The most common deformations concern the composition of materials, the use of objects. Reconstruction techniques are used for building 3D models of existing objects from sensor data such as laser scanner and photogrammetry data. Similarly, we can use additional types of sensor data for reconstructing (i) the micro-structure of the object (dents, bumps, cracks) or (ii) the material layers that lie underneath the external surface.

We report on the development of methods for digitally reconstructing and visualizing cultural heritage objects including their material consistency and their micro-structure.

Keywords: Material aging · Visualization tool · 3D reconstruction

1 Introduction

The protection of cultural heritage artifacts is an important but often very tedious and repetitive process. Each object has been crafted using several materials that age differently, a fact that can have a significant impact in its appearance as it ages. Archaeologists and curators bear the brunt of studying the aging process of each material and then apply that knowledge to restore artifacts to a prior condition or to prevent further decay.

The analysis, simulation, emulation and visualization process that is used to model the effect of aging requires highly detailed models of the solid objects and therefore demands a computationally intensive physical modeling simulation.

This work has been partially supported by the European Commission through project Scan4Reco funded by the European Union H2020 Programme under Grant Agreement No. 665091. The opinions expressed in this paper are those of the authors and do not necessarily reflect the views of the European Commission.

© Springer Nature Switzerland AG 2019
A. Moropoulou et al. (Eds.): TMM_CH 2018, CCIS 961, pp. 141–149, 2019.
https://doi.org/10.1007/978-3-030-12957-6_10

3D reconstruction is the process of capturing the boundary surface and the appearance of real objects. This process can be accomplished either by active or passive methods.

Small deformations of the surface structure, and small color variations due to corrosion contribute to a realistic look of objects. Deformations observed in many materials arise due to small-scale interactions among elastic strain, plastic yielding, and material failure. According to Kider [1] the aging process depends on material composition, object usage, weathering conditions and a large number of other physical, biological, and chemical parameters which over long periods of time result in local deformations that are manifested through dents, bumps, cracks and layer peeling. Furthermore, sensor data from ultrasounds and non visible electromagnetic radiation (infrared, ultraviolet and x-rays) are used to reconstruct information regarding multiple material layers.

Subsequently, methods are needed for (i) the analysis of the aging process on sample materials (see Pfaff et al. [2]) the prediction of aging described as a set of surface micro deformations based on the results of the analysis (ii) for the realistic rendering of artwork objects and the visualization of the underlying material layers. Therefore we report on the development of (i) technique for reconstruction of artwork objects with all the macro and micro properties of the surface and the underlying material layers (ii) a powerful visualization tool for browsing a realistic rendered interior and exterior of a reconstructed artwork object. Figure 1 illustrates the reconstruction and visualization pipeline.

2   Related Work

Realistic rendering is one of the most important goals in computer graphics. A key role in realistic rendering is being played by the micro-structure of the surface of the object. This micro-structure is often modeled by local deformations that affect locally the geometry of the boundary surface of the object. Frerichs et al. [3] provide a thorough survey of such methods.

Small deformations such as cracks according to Pfaff et al. [2] are more likely in areas determined by local properties or by physical simulation which result in refining a high-quality triangle mesh. Glondu et al. [4] use a physically-based fracture to create a wide range of crack patterns. Dorsey and Hanrahan [5,6] observed that crack patterns in materials arise due to small-scale interactions between elastic strain, plastic yielding, and material failure.

Lee et al. [7] introduced a method for reproducing visually observable aging deformations by a simplified simulation process. According to Mérillou and Ghazanfarpour [8] determining the effect of an aging period on the visual appearance of an artwork object is a complex process.

El-Gaoudy et al. [9] and Rushmeier [10] noticed that it is of extreme importance the availability of information about the properties of material composition for manufacturing, or restoration.

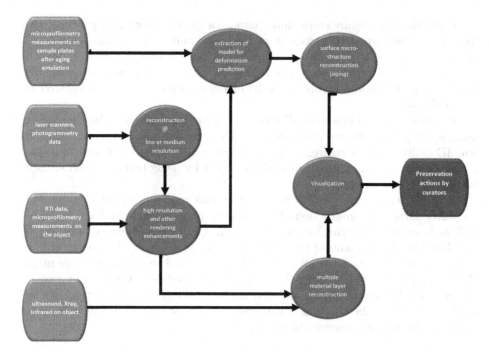

Fig. 1. Reconstruction and visualization pipeline.

Chudnovsky and Preston [11] discuss the background for modeling the kinematics of aging deformations. Various simulation techniques have been used for material aging. For example, Yin et al. [12] studied the surface of wood which is modeled by values assigned to tetrahedral mesh vertices. On the other hand, Paquette et al. [13] present an aging technique that simulates the deformation of an object caused by repetitive impact strikes over long periods of time.

Modern rendering techniques and hardware are capable of capturing aging phenomena efficiently (see e.g. Kider [1]).

Gomes et al. [14] propose a multi step pipeline for the 3D reconstruction technique of cultural heritage artwork objects. This pipeline is based on high res photos and laser scanner data for mesh generation, partial view registration, mesh parametrization and diffuse texture generation. Our pipeline goes beyond this pipeline by incorporating, very high resolution geometry based on micro-profilometry measures, surface micro-structure data (aging), multiple material layers, and detailed material properties using RTI data.

3   3D Reconstruction Techniques

3D reconstruction mainly focuses on reproducing an object by capturing surface geometry information and color. Digital reconstruction of cultural heritage objects is a challenging task since it requires capturing details at the level of micrometers for the external surface.

While there are many steps that compose a 3D reconstruction pipeline, the most important are (i) the acquire of raw color and (x, y, z) information, (ii) the alignment of 3D data into a common reference frame in a process known as registration, (iii) the mesh integration stage where data from all acquired 3D views are combined and finally (iv) a 3D model with its textures is generated as the final result.

3D reconstruction produces 3D models that are used for rendering the object using appropriate software and hardware. These models are composed using 2D and 3D multi-modal measures, which are then analyzed and combined to derive a high fidelity model of the surface of the object by using texturing and surface reconstruction techniques.

Multiple sensors are used to provide the data required for each model. For the outer surface reconstruction of the models, photogrammetry techniques, as well as laser scanning can be used to obtain a low or medium resolution triangulated mesh representing that surface. Micro-profilometry is used to obtain higher resolution data for a surface which can be used to extract texture maps that can reproduce this high resolution information efficiently using the rendering pipeline. Finally texture maps that contain information about the color of an area as well as the specular, roughness and albedo values of a material, are generated by RTI measurements (see Fig. 2).

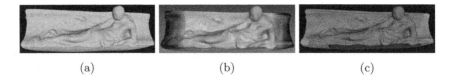

(a)   (b)   (c)

Fig. 2. Multistep high definition 3D reconstruction technique.

4 Reconstructing the Surface Micro-structure and the Underlying Layers

In this section we outline methods for deriving a model for simulating aging based on micro-profilometry measurements taken on material sample plates (Costabel et al. [15]) during an emulated aging process (Moutafidou et al. [16]). We mainly focus on local deformations due to corrosion/erosion and finally cracks by modeling the behavior of displacements locally and observe the results by realistic rendering.

The first type of deformations that we would like to predict are dents or bumps. They commonly occur in areas around a specific point of the surface. According to micro-profilometry measurements on simple metal plates, we have observed that they mainly occur where large deviations from the mean value (distance from the metal plate plane) are obtained and then model their occurrence on real world artwork meshes that we have derives by the methods of Sect. 3.

The second type of deformation that we study is a crack. It's very common for a crack to start occurring where extreme deviations from the mean value are obtained. Since micro-profilometry data from sample plates follow a normal distribution, we can model the frequency of crack occurrence based on the standard deviation of the fitted distribution on sample plates which basically means that cracks are more likely to occur in areas where extreme bumps or dents are present.

For every new aging step the following repeatable process is carried out. Each artificial aging process is being studied by (i) statistically analyzing the micro-profilometry measurements from each sample plate according to maximum likelihood estimation (MLE) which is based on the likelihood function, (ii) finding the parameter values that maximize the likelihood of making the observations given a specific family of probability distribution functions, (iii) determine the set of values for the model parameters that maximize the likelihood function and finally (iv) determine the parameters of Gaussian pdf that best fits each set of micro-profilometry measurements.

The results of the above process is utilized to predict aging on reconstructed artwork objects made by the same material. More specifically Moutafidou et al. [16] focus on the detection of such patterns within sets of data drawn from micro-profilometry measures taken on material samples such as silver, bronze or egg tempera plates during the process of artificial aging.

Finally, we derive not only the surface of the objects but also volumetric information of the interior of the solid object. Multiple sensors are used to provide the data required for each model. Ultrasound, infrared and x-ray measurements provide information for reconstructing details of the inner layers, including geometry and thickness.

5   A Visualization Tool for Artwork Objects

Most archaeologists work is based on the observation of existing cultural heritage artifacts in their current state, combined with the knowledge gathered from experience on how different materials age. The main goal of our work is to adopt a state of art approach from artificial material aging, and rendering so that people involved in curatorial work and in preservation tasks will be able to understand exactly the nature of aging and act accordingly. In this way, it is necessary to have tools that produce and visualize digital representations and models of visual surface appearance and material properties, to help the scientist understand how they evolve over time and under specific environmental conditions.

Material models mainly consist of surface geometry (including normals), material color, reflectance distribution and other light related material properties. Micro-surface modeling on the other hand, is based on a hybrid method that combines material aging models, multiple material layers and physical models. With all this available data for the model there is a need for an appropriate viewer that can visualize all the details. The 3D Viewer needs to be able to render the details provided by all types of sensors while being fast. Figure 3 illustrates the improvement of the rendering result if you added a normal map texture. Figure 4 illustrates the improvement of the rendering result if you added a normal map texture, and other texture maps to improve quality. For that reason, we have developed a rendering tool which is capable of visualizing efficiently all the surface details and the underlying layers.

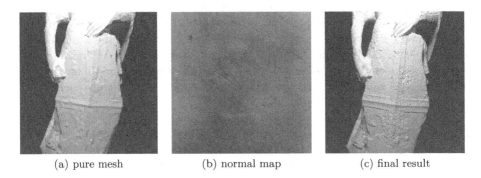

(a) pure mesh   (b) normal map   (c) final result

Fig. 3. Example of adding a normal map texture in a low poly mesh to enhance the final result.

Our visualization tool is built on top of a multi-fragment renderer using vertex and fragment shaders. The outcome is a fast visualization tool that given the scanned geometry and material model is able (i) to portray the appearance of the artifact in the future when specific effects are applied and (ii) to visualize and analyze the underlying material layers.

We have conducted artificial aging experiments with several types of sample plates (bronze, silver, egg-tempera). Furthermore, we have analyzed the micro-profilometry data and derived the distribution function for each pair of (sample plate, time instance). For several artwork objects we have applied an algorithm to create an aging effect such as dents and bumps as well as cracks based on the distribution function. The first algorithm refers to bumps. In this case we conducted bumps and dents in the original object. More specifically based on the original object (Fig. 5a) we derive a confidence interval to compute the number of the deformation. The result is 154 bumps and 99 dents (Fig. 5b). On the other hand we have studied cracks. Based on the same path we conduced cracks based on the previous model (Fig. 5b). In this case we are going to create a new aging step with bumps, dents and cracks. The result is 231 new bumps, 139 new dents and 4 cracks (Fig. 5c).

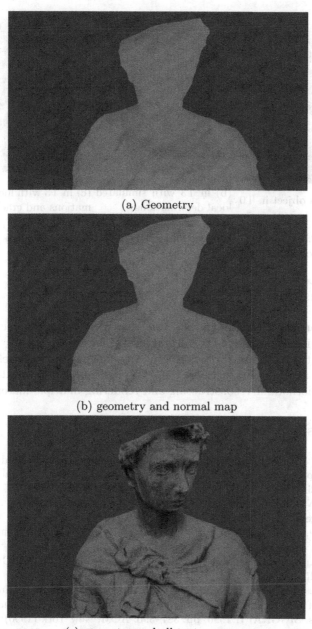

(a) Geometry

(b) geometry and normal map

(c) geometry and all texture maps

Fig. 4. Rendering original geometry, then enhancing using texture maps.

(a) original object in T0   (b) in T3 with simulated local deformations   (c) in T3 with local deformations and cracks

Fig. 5. Simulating local deformations and cracks using data from micro-profilometry measurements on sample plates.

6 Conclusions

We have reported on the design and development of a tool for reconstruction and visualization appropriate for cultural heritage artifacts.

We present a method for predicting and enhancing the aging effect in cultural heritage artifacts. Furthermore, based on multi-modal sensor data we have developed a method for reconstructing and rendering multiple material layers.

References

1. Kider Jr., J.T.: Simulation of three-dimensional model, shape, and appearance aging by physical, chemical, biological, environmental, and weathering effects. Ph.D. thesis, Philadelphia, PA, USA, AAI3542821 (2012)
2. Pfaff, T., Narain, R., de Joya, J.M., O'Brien, J.F.: Adaptive tearing and cracking of thin sheets. ACM Trans. Graph. **33**, 110:1–110:9 (2014)
3. Frerichs, D., Vidler, A., Gatzidis, C.: A survey on object deformation and decomposition in computer graphics. Comput. Graph. **52**, 18–32 (2015)
4. Glondu, L., et al.: Example-based fractured appearance. Comput. Graph. Forum **31**, 1547–1556 (2012)
5. Dorsey, J., Hanrahan, P.: Modeling and rendering of metallic patinas. In: Proceedings of the 23rd Annual Conference on Computer Graphics and Interactive Techniques, SIGGRAPH 1996, pp. 387–396. ACM, New York (1996)
6. Dorsey, J., Edelman, A., Jensen, H.W., Legakis, J., Pedersen, H.K.: Modeling and rendering of weathered stone. In: Proceedings of the 26th Annual Conference on Computer Graphics and Interactive Techniques, SIGGRAPH 1999, pp. 225–234. ACM Press/Addison-Wesley Publishing Co., New York (1999)
7. Lee, S., Kim, J.-W., Ahn, E.: A visual simulation method for weathering progress of stone artifacts. Multimed. Tools Appl. **75**, 15247–15259 (2016)
8. Mérillou, S., Ghazanfarpour, D.: Technical section: a survey of aging and weathering phenomena in computer graphics. Comput. Graph. **32**, 159–174 (2008)

9. El-Gaoudy, H., Kourkoumelis, N., Varella, E., Kovala-Demertzi, D.: The effect of thermal aging and color pigments on the Egyptian linen properties evaluated by physicochemical methods. Appl. Phys. A **105**, 497–507 (2011)

10. Rushmeier, H.: Computer graphics techniques for capturing and rendering the appearance of aging materials. In: Martin, J.W., Ryntz, R.A., Chin, J., Dickie, R.A. (eds.) Service Life Prediction of Polymeric Materials, pp. 283–292. Springer, Boston (2009). https://doi.org/10.1007/978-0-387-84876-1_19

11. Chudnovsky, A., Preston, S.: Geometrical modeling of material aging. Extr. Math. **11**(1), 22–36 (1996)

12. Yin, X., Fujimoto, T., Chiba, N.: CG representation of wood aging with distortion, cracking and erosion. J. Soc. Art Sci. **3**, 216–223 (2004)

13. Paquette, E., Poulin, P., Drettakis, G.: Surface aging by impacts, In: Proceedings of Graphics Interface 2001, GI 2001, pp. 175–182. Canadian Information Processing Society, Toronto (2001)

14. Gomes, L., Bellon, O.R.P., Silva, L.: 3D reconstruction methods for digital preservation of cultural heritage. Pattern Recogn. Lett. **50**, 3–14 (2014)

15. Costabel, M., Dauge, M., Nazarov, S.A., Sokolowski, J.: Analysis of crack singularities in an aging elastic material. ESAIM: Math. Model. Numer. Anal. **40**(3), 553–595 (2006). Version 10.03.2005

16. Moutafidou, A., Adamopoulos, G., Drosou, A., Tzovaras, D., Fudos, I.: Realistic rendering of material aging for artwork objects. In: IEEE International Conference on Image Processing (ICIP 2018), no. 5 (2018)

Vandalized Frescoes' Virtual Retouching

Melina Aikaterini Vlachou ⓘ, Dimitrios Makris⁽✉⁾ ⓘ,
and Leonidas Karampinis ⓘ

University of West Attica,
Campus I, Ag. Spyridonos Street, Egaleo, 122 43 Athens, Greece
melvlachou@hotmail.com, {demak,leokar}@uniwa.gr

Abstract. In this study, we present the results of the application of virtual color restoration on the vandalized frescoes of the Church of Saint John (located in the courtyard of the University of West Attica Campus II). In order to retrieve the monument's interior and exterior, three-dimensional scanning technology and photogrammetry were implemented in two distinct periods. In 2010, scanning through terrestrial laser scanner and capturing the interior were held. documentation via photogrammetry took place in 2016. Due to the extensive vandalism that suffered its interior, the current documentation was limited only to the monument's exterior. For the application of digital color restoration, firstly we decided to individually apply the three retouching techniques in the virtual environment: chromatic abstraction, neutral color and mimetic color restoration technique. As the optical result of the virtual interventions was not aesthetically correct, we decided to combine contemporary aesthetic approaches, as well as basic retouching techniques and apply them in an original way: chromatic abstraction, neutral color and mimetic color restoration technique, which lead to a new mixed method. According to the Venice Charter article 9 retouching aims to restore the aesthetic value of the monument with respect to its historic value and must stop where conjecture begins. Furthermore, the retouching techniques should be visibly distinct by the authentic painting layer and any intervention made should be based on archaeological and historical knowledge. Therefore, the workflow which was decided to follow, complied with the Code of Ethics of Conservation of Cultural Heritage.

Keywords: Virtual retouching · Digital color restoration ·
Three-dimensional scanning technology · Close-range photogrammetry ·
Frescoes · Cultural heritage

1 Introduction

The contribution of wall-paintings in cultural heritage is undeniable, as through them people were able to externalize and transmit their emotions, ethics and traditions to the next generations. The different techniques and materials used indicate their significance. Therefore, the need for documentation, conservation and restoration is mandatory, as their possible damage or loss would affect the international cultural heritage [1].

© Springer Nature Switzerland AG 2019
A. Moropoulou et al. (Eds.): TMM_CH 2018, CCIS 961, pp. 150–170, 2019.
https://doi.org/10.1007/978-3-030-12957-6_11

With the application of three-dimensional (3D) scanning technology in the scientific field of conservation, it is possible to document the minimum detail with the utmost precision, in order to create condition reports for study and research. Moreover, it is possible to approach various aspects of the artifacts, which may no longer be accessible. The current three-dimensional digital models (comparatively more detailed than the ones in the past) have a greater volume of information, making it possible to acquire and analyze through new processes the three-dimensional nature of art, opening up new possibilities for use, [2]. At the same time, interventions can be digitally applied to accept or reject the proposed conservation methods in the real environment, and it is no longer particularly difficult to create purely virtual environments for educational, research, exhibition and even entertainment purposes [3].

The exported data can be collected to create records that include morphological, iconographic, aesthetic characteristics and dimensions of the artifacts, to compare them with other projects of the same or other artists, as well as monitor the wear rate. Besides planning an effective conservation program and studying objects that require special treatment, the three-dimensional scanning can be used to make replicas for display and sale, for safe transport and storage of the original artifacts when necessary. In cases of destruction due to human or natural factor, the exported scan files may allow even the complete morphological restoration of the monument, as we already see in the case of Palmyra.

In this paper the digital three-dimensional recording and documentation of the frescoes of a vandalized monument (temple of St. John of the 14th century located in the olive grove of University of West Attica, Campus II) was attempted with the use of range – based and image – based technologies, (Fig. 1). The recovery of the interior and exterior of the temple was based on terrestrial laser scanning and close-range photogrammetry. Scanning through a terrestrial laser scanner and the corresponding digital photographs of the interior and exterior of the temple were held in 2010, while the documentation via photogrammetry of the exterior was carried out in June 2016. Between these different time periods the monument's interior suffered extensive vandalism with a great part of its temple and frescoes been destroyed at least by half by human factors, (Fig. 2). This is where the purpose and the value of the present work is focused, because it consists the only three-dimensional laser scanning file in which we have the complete recording and documentation of the vandalized temple.

The processing of the resulting digital files was carried out to implement the virtual color restoration on selected three-dimensional parts of the interior of the church. Firstly, we applied the known retouching techniques in the digital environment, so as to accept or reject them. As a result, in order to achieve the perfect digital retouching of the frescoes, approaches of modern aesthetics and basic techniques of color restoration were applied in an original combination, without this implying that basic principles of conservation, such as Articles 3 and 9 of the Venice Charter [4], were not taken into consideration. Finally, it is possible to restore the monument's morphology in the future.

Fig. 1. St. John's church exterior views

Fig. 2. St. John's church interior before and after the unfortunate vandalism

1.1 Research Aims and Objectives

Through this case study, we want to present the positive results of merging the theoretical and applied knowledge of conservation with the use of digital technology for the virtual application of color restoration. Our aim was therefore the virtual retouching of the frescoes of the St. John's church, due to the vandalism that has undergone its interior based only on earlier digital recording. The interdisciplinary nature of the study implies that digital color restoration has been approached according to conservation principles and theory, as it is applied in the context of conservation of cultural heritage [5]. With the help of 3D scanning technologies, we can acquire the virtual representation of artworks, observe and record their state of preservation, and even restore them to their original form. As in our case, on the one hand the entrance to the temple was not allowed after its vandalism, and on the other hand a part of the iconostasis was completely destroyed, we applied color rendering in the digital model of the church, receiving relevant information only from the photographs taken in 2010. The way we worked was divided into two phases: digital processing of the data which derived from the scanning (2010) and digital color restoration directly onto the 3D surface of the digital models.

2   Related Research

Documentation supported on reality-based 3D models is a widespread and effective approach in diverse fields like archaeology, architecture, paintings, sculpture [2, 6]. The application of 3D scanning and close-range photogrammetry enables the growth of study, restoration and dissemination procedures serving cultural heritage researchers and professionals [7]. Examples include, but are not limited to, architectural and building morphology and behavior examination [8, 9], fragmented parts' reassembly and reconstruction [7, 10, 11], quality assessment [12].

Color restoration of wall – paintings and frescoes is a difficult task due to the diverge norms such as rules and directions from international organizations, factors such as visual perception and restorers' objective approaches. Earlier applications contained two-dimensional digital color restoration of cultural heritage artworks. The widespread of 3D digital techniques offered great opportunities for the development of virtual color restoration much closer to the real conditions of conservators theoretical and applied work. Digital color restoration is based on color calibration and color accuracy for recording and fidelity application, especially on cases where the artwork is highly damaged and/or totally destroyed. Most research on digital color restoration focuses on acquiring and retrieving color information. A research team presented a framework of RGB images mapping on detail 3D digital model for both documentation and restoration support that enable restorers to work direly on the colored virtual model, [13]. In [14] is presented a complete workflow of morphological reconstruction and color restoration of the hypothetical appearance of terracotta statues. Digital restoration is applied directly on high-resolution 3D geometry based on MeshLab Paint interface [15]. The author in [16] focus on polychromy studies in order to reanimate classical statuary through the digitization of a Roman portrait for the documentation, study and visualization of the traces of pigments it brought. The efforts' completeness were supported by direct virtual painting over the 3D digital model. A research team in [17] describes the multi-stage physically reassembly and reconstruction of a damaged statue based on digitally – aided restoration techniques and in particular virtual restoration of painted decorations. The team provided an analytical digital model with surface decorations' virtual retouching and restoration applied directly over the digital 3D model. However due to limited knowledge color retouching over 3D model was implemented for didactical purposes. The case study of [18] is a Roman marble sarcophagus's polychromy virtual reconstruction. In particular, the project investigates in which way the digital documentation could be implemented for a realistic compensation of ancient polychromy.

All aforementioned approaches focus on specific digitization frameworks which could contribute to understand and solve particular restoration problems about color and morphology issues. Additionally we should underline the fact that they reside on open source software and applications. A general representative scenario includes the digital restoration based on 2D color intervention that is applied to relevant atlas textures of the 3D model, and followed by the re-implementation on the 3D mesh model.

The alternative path that is followed in the abovementioned studies concerns the use of per-vertex color information mapping. Such scenario tackles many of the limitations of complex dense geometric 3D scanning models with topological problems. In this way the color retouching is applied directly over the surfaces of the 3D model.

The reason why we chose to apply the retouching techniques directly onto the surface of the 3D model was to observe and evaluate the results of restoration on real-time. In this way we were able to reject any unsuccessful interventions made. On the contrary, if we had decided to restore the image and then apply it as a UV – atlas map texture to the 3D model, we would not be able to determine whether or not our attempt was successful. Restorers can receive an important preview of restoration hypothesis based on the particular advantages of 3D representations to provide color, brightness, glossiness, texture, tactile roughness, and material makeup real-time information [19].

3   Historical Research

Initially, historical research was carried out to find information about the architecture of the church and the possible changes that it might have experienced throughout the years. The detailed documentation, as recorded by the historian Orlandos [20], describes in detail the architectural style of the temple and its structural elements. In contrast, no information was found about the color and conservation status of the wall paintings, due to the coating layer covering the interior of the church during that period. The church of St. John is located in the area of Egaleo near the 3rd Cemetery, within the olive grove surrounding the University of West Attica Campus II. The original owner it is thought to be Alexander Benizelos and later it became the property of the Merkati family, descendants of the former Athenian family of the first. Because of the vandalism it suffered in the past years, the entrance to the temple is not allowed, so it is not possible to analyze its current conservation status. Regarding the architectural style of the church, the triconch type with a semicircular apse is encountered during the Byzantine and post-Byzantine periods. Due to the fact that the stonework is irregular without decoration, etc., its construction dates back to 1300 AD, and before the fall of Constantinople (late Byzantine period).

4   Laser Scanner Data Processing

As mentioned above, the prohibition of entering the temple due to the vandalism it suffered, did not allow the recording of its state of preservation and photographing its interior. Therefore, the only way to obtain the color information of the frescoes was to register the photographs taken in 2010, directly to the geometry of the three-dimensional models. In order to achieve the desired result, the laser scanner data was processed, followed by the implementation of the digital intervention.

4.1   Acquiring the 3D Model

In order to acquire the interior and exterior of the temple, range – based and image – based technology were used. The terrestrial laser scanning of the interior of the temple and the according photographs were held in 2010, while the documentation of its exterior using close-range photogrammetry took place in June 2016. The unfortunate vandalism that occurred during these years resulted in the loss of a part of its iconostasis (Fig. 2). Upon completion of the data processing, a detailed three-dimensional model was obtained (26.5M vertices/53.6M faces) [21].

Processing the 3D Model. MeshLab (open source software) was used to create the 3D surface model [15]. The point cloud was imported into the software (Fig. 3) and with the application of the screened Poisson surface reconstruction algorithm [22] it was transformed into a three-dimensional triangular grid (Fig. 4). After cleansing and healing its surface, (Fig. 5), the color from the point cloud was transferred to the vertices of the digital 3D model (Fig. 6), which was then divided into smaller sections [21].

Fig. 3. Two views from the point cloud after its import in MeshLab.

Fig. 4. View of the 3D model after the application of the screened Poisson surface reconstruction algorithm in MeshLab.

Fig. 5. View of the 3D model after the cleaning and healing of its surface in MeshLab.

Fig. 6. View of the 3D model with per vertex color information.

This step was necessary because the large number of triangles, of which the model consisted, made it difficult to handle it. In order to create the textured atlases and simplify the geometry of the final 3D without losing the original analysis quality, Blender was chosen, [23]. After simplifying the geometry of each sub-section and creating the corresponding texture maps, the 3D models were combined to create the texture atlas of the final 3D model [21].

5   Color Restoration – Digital Intervention

As far as frescoes are concerned, the application of color restoration aims at improving the readability of the form and the content of the work, eliminating the color gap, while preserving its historical value and its original condition. It aims at highlighting the authentic artistic, historical and aesthetic elements of the work of art, ensuring its continuity for making it available to the public. The result of the application of the method is considered to be successful when the intervention can be visually separated in close proximity from the original painting layer, while being reversible.

The prevailing theories and techniques converge in the fact that the color restoration should take place depending on the project, its materials, its dimensions, its value (artistic, historical and economic) and the competent authority. The most accepted and often applicable techniques as described by Ornella Casazza [24] are: (a) chromatic selection when vertical dense lines of consecutive colors not only can cover the color gap (with the same final color resulting from the blending of the individual layers), but also can create the lost form; the basic prerequisite is the precise knowledge of the lost form, (b) chromatic abstraction when cross-colored lines of successive colors, create a color grid that varies according to the authentic coloring of the area; forms and shapes are not re-created because we do not have precise knowledge of the lost shapes and forms; (c) neutral color when due to the large size of the unknown loss, a "neutral" color is chosen to emphasize the authentic colors; (d) mimetic color for very specific parts usually motifs or very small losses, by applying a color of lower intensity than the original and (e) mixed techniques, when more than one techniques are applied to a work of art. All of these techniques comply with the basic rules of the Code of Ethics for Conservation, as referred in Articles 3 and 9 of the Charter of Venice. At the ICOMOS conference, which took place in Sri Lanka in 1993 [25], common principles were expressed in which the conservation and restoration interventions should treat the work of art with respect to its historical and aesthetic value.

According to Brandi [26], two are the basic principles governing the science of conservation: reversibility and legibility. Since a digital environment ensures the former, the difficulty lies in ensuring the second principle. This was a major part of this study. Consequently, in order to choose the way in which we would apply the color restoration within the three-dimensional digital environment, we had to study the conservation principles. Our concerns focused mainly on the compliance with the rules of conservation of cultural heritage in the digital environment and the contemporary visual aesthetic effect that the filter of the software would offer, after the application of the digital intervention.

The frescoes which were to be restored were chosen according to their thematic content, their place of significance within the temple, and mainly due to the fact that some of them were not preserved. These were: a part of the dome and the Pantocrator depicted in its center, as well as the frescoes decorating the iconostasis. Color restoration methods were selected based on the type, location, and extent of the losses on the painting layer.

The effectiveness of restoration techniques is based on both critical thinking and knowledge of conservators, as well as on their special skills. For this reason, we were skeptical about the aesthetic result that the MeshLab painting tool [27] would provide. In order to achieve the desired results, we experimented with the hue of the applied colors. We thus came to the conclusion that the hue of the applied colors should be between 40% and 80% compared to the original one and that the colors should be applied in levels so as to blend harmoniously. With the method we created, the colors could be separated from the original, while offering a satisfying and harmonious visual result. In this way we were able to implement the digital intervention in accordance with the Principles and the Code of Ethics of Conservation of Cultural Heritage [1].

6   Digital Retouching

After the processing of the acquired digital files, the application of the virtual color restoration of the selected parts of the interior of the church followed. These parts were selected for their place of significance in the temple, their thematic content and the losses they suffered during its vandalism.

The two maim approaches for color editing – intervention directly on the 3D model geometry (color per vertex mapping, and mapping textured image on the surface geometry) provide many advantages as well as limitations [7]. In general, a feasible parametrisation is not always achievable, mainly due to fact that 3D scanning models are geometrically complex and topologically inconsistent [7, 16]. Therefore we follow an accepted solution of mapping the color information per-vertex [13–18]. Moreover, we rejected the possibility of intervening on the image of the model's texture and re-projecting it onto the mesh, as the result of the restoration would not be directly visible and we could not compare the former and the later state of the frescoes during the digital intervention. The adopted manipulation was consider by the team's restorers efficient and adequate and very close to the real world manipulations.

As described in [21] the resulted color information from the range-based recording was not detailed enough for restorers for mural paintings digital studying and color restoration. Therefore, we decided to registered over the range – based model an available set of images acquired during the same time.

Firstly, the three-dimensional digital models of the parts of interest and their corresponding images were imported in MeshLab. Secondly, the images were registered to the meshes' geometry by means of software's algorithm and visual optimization method [27]. Image projection was required to acquire the chromatic information as the laser used in 2010 did not have the required analysis and we were not permitted to re-enter the interior of the monument to retrieve more data. The complexity of the models' geometry combined with the above mentioned, caused difficulty in the alignment of the photos; the algorithm calculated to a certain extent the focal length and distortion of the camera, requiring manual adjustment of the images, making the process time consuming. In this way, digital color restoration was applied directly onto the three-dimensional surface of the models, with the help of a similar filter provided by MeshLab [15]. At this point, it is should be emphasized that due to the prohibition of entering the temple, the quantitative recording of the frescoes' colors was not possible. For this reason, our knowledge of color information was limited to those provided by the interior photographs and the color restoration was applied according to them. The color restoration of the parts of interest was accomplished using the MeshLab painting tool thanks to the high-resolution colored 3D digital models. The color was applied to the vertices of the meshes, strictly within the limits of the losses, to avoid the over-painting of the original painting layer.

6.1 Application of Digital Color Restoration

Digital color restoration was implemented according to the principles of conservation of cultural heritage, [26, 28]. Retouching aims at the unity of the work of art and must comply with the rules of conservation regardless of whether or not it is applied to a virtual digital environment. Each applied intervention aims at highlighting the historical and artistic values of the work of art, ensuring its unity. The final result not only should be separated from the original painting layer at a close distance and be as neutral as possible, given that there is in fact no neutral color, but should also be reversible. Techniques are selected according to the location, extent and type of damage. In losses where forms have been lost, it is proposed to apply chromatic abstraction in conjunction with neutral color. In small losses, where we do not risk entering a false form or pattern, we can choose the method of mimetic color restoration technique, provided that the applied colors are toned down compared to the original ones. The extensive deterioration of the frescoes' painting layers (discoloration, losses on the painting layer and on the substrate, traces of soot, etc.) determined, as in any case, the choice of the applied color restoration techniques: chromatic abstraction, neutral color and mimetic color restoration technique [24, 26, 28].

Firstly, we decided to apply chromatic selection technique as we would in real conditions. Although this methodology would in fact be ideal, and is applied in real-world restorations (when the forms of missing parts are known). However, such technique is not recommended for greater losses in real-world color interventions. As a case study we decided to apply the chromatic selection with parallel lines in the bigger losses where the color was lost in the Pantocrator. By its implementation in the digital environment we faced an important limitation, (Fig. 7). The manipulation based on chromatic selection technique failed because the aesthetic result could not be acceptable following the fundamental norms of color restoration [1, 4]. Similarly, we applied the chromatic selection technique with parallel lines to the smaller losses of the frescoes whose thematic content was missing, however the resulted aesthetic impression was not satisfied.

Fig. 7. Screenshot of Pantocrator after the application of chromatic abstraction.

Thus we proceed towards the next phase of our approach. Given the fact that we were working in a digital environment, we decided to experiment and combine the above-mentioned techniques to create an original mixed technique for our purposes.

Fig. 8. Screenshot of Madonna and Child before (left) and after (right) the application of digital color restoration.

Fig. 9. Screenshot of the frescoes of the iconostasis before (left) and after (right) the application of digital color restoration.

These techniques, although based on the basic principles and ethics of color restoration, have been adapted to the needs that have arisen in the digital environment. They were thus applied in combination, by way of derogation from what would be acceptable in a real environment, to achieve the desired visual effect. As a result,

Fig. 10. Detail of the fresco depicting Saint John the Baptist before (left) and after (right) the application of the new mixed technique (chromatic abstraction – neutral color).

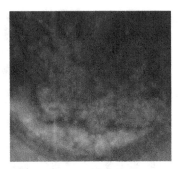
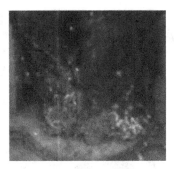

Fig. 11. Detail of the fresco depicting Jesus Christ before (left) and after (right) the application of the new mixed technique (chromatic abstraction – neutral color).

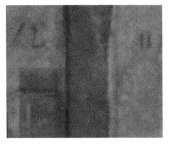

Fig. 12. Detail of the background of the iconostasis' frescoes before (left) and after (right) the application of mimetic color restoration.

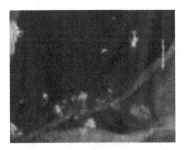
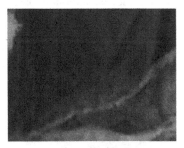

Fig. 13. Detail of Madonna's garment before (left) and after (right) the application of mimetic color restoration.

the technique of chromatic abstraction was applied in combination with the technique of neutral color, creating a new technique for the requirements of the digital environment. In other words, the colors were applied uniformly and not by the use of crossed lines, by simultaneously increasing and decreasing their intensity according to their proximity or not with the original color. In this way the applied colors were visibly separated from the original ones without being projected forward, overshadowing them.

Fig. 14. Screenshot of the fresco depicting Pantocrator before and after (2nd – 60% and 3rd – 80%, row) the application of the new proposed technique.

In areas where there was a loss of mortar and painting layer, an off-white – gray tint was applied to reduce their brightness (Figs. 8 and 9).

According to the above mentioned, the techniques of chromatic abstraction and neutral color were combined and applied on most of the frescoes and in particular where there were losses in the thematic content (Figs. 10 and 11).

In the forms that adorned the iconostasis, especially at the folds of the garments and at the backgrounds, the colors were chosen according to the original, i.e. mimetic color restoration technique, but were applied to a lower hue, in order to separate the new intervention from the original painting layer (Figs. 12 and 13).

In the form of Pantocrator (in the face and the garment) and in the larger part of the surrounding area, the following retouching techniques were chosen to be applied:

(a) the variation of the chromatic abstraction-neutral color retouching technique for the greatest losses, whose boundaries are not of the same color (Figs. 14, 15 and 16);

(b) the variation of neutral-mimetic color restoration technique was applied to the small losses of the dome's decorative motif, (Fig. 14). The applied colors' hue is between 40% and 80% compared to the original one.

Fig. 15. Detail of Pantocrator's face before (left) and after (right) the application of the variation of neutral-mimetic color restoration technique.

In the part of the painting surrounding the form of Pantocrator, the mimetic color restoration technique was applied (Fig. 17). This technique is generally not recommended because of the possibility of incorrect and arbitrary data input, however, in this case it was considered ideal because the pattern of the decoration does not pose a risk of arbitrariness and simultaneously enhances the legibility of the work of art, especially when it is applied to a digital virtual environment.

The restoration of the rest of the dome was approached in a similar way (Fig. 18).

In selected areas of the dome where there were decorative elements and motifs, especially in the background where there were widespread small losses, we chose to apply the technique of mimetic color restoration to facilitate the legibility of the painting layer and its depicted figures (Fig. 19).

Fig. 16. Detail of Pantocrator's garment before (left) and after (right) the application of the variation of neutral-mimetic color restoration technique.

Fig. 17. Detail of the decorative motif surrounding Pantocrator before (left) and after (right) the application of mimetic color restoration technique.

Fig. 18. Screenshot of a part of the dome before (above) and after (below) the application of digital color restoration.

Fig. 19. Detail of a part of the dome before (left) and after (right) the application of mimetic color restoration.

7   Results and Discussion

While at the core of the concept of digital restoration may lay the technical tools [29], the presented approach provides a widening of such concept. In the current study the digital tools in accordance with restorers influence enable the emergence a new color restoration scheme. Although the digital color restoration was applied to selected parts of the temple, the final three-dimensional models are in accordance with the principles of color restoration and modern visual perception. The intervention applied to the particular sections of the temple demonstrates the valuable contribution of scanning technologies to conservation science. We have demonstrated that the implementation of retouching techniques is feasible in the digital environment, but it is possible too being consistent with the principles and ethics of conservation of cultural heritage. All exported 3D digital models have come from precise metrics and can be used in conjunction with the photographic material as a complete record of the previous state of preservation of the monument and as a comparison of the damage it has suffered. It should be emphasized that due to the inability of entering the temple, it was not possible to identify the pigments on the painting layer of the frescoes; something that may happen in the future, provided that the remaining part of the iconostasis will exist when the permission of entering will be granted. However, thanks to the high-resolution photos taken in 2010 which were projected onto the 3D geometry of the digital models, the colors were displayed successfully. It is worth mentioning that the limited implementation of the digital intervention was due to the time-consuming process that was followed; perhaps due to the inability of the algorithm to calculate the angle of the taken photographs. This fact could be eliminated if we could re-enter the interior and take new photographs. Additionally, using the MeshLab software and the z-painting tool that offers, we were able to approach the original restoration conditions, by applying the digital intervention according to the techniques of color restoration and the principles and ethics of conservation. Consequently, we managed to intervene in fully detailed models (300K to 500K vertices) creating the photo-realistic depiction of the frescoes.

By using 3D technology, conservators can apply a variety of restoration techniques to artifacts and monuments that e.g. are not accessible, their bad state of preservation does not allow them to be conserved, they have been vandalized to such an extent that parts of them have been completely lost and destroyed, etc. In our case study, we chose the church of Saint John for two main reasons. Firstly, the entrance to its interior was forbidden and no attempts have been made by the competent authorities for its restoration and preservation. Secondly, the monument has been vandalized throughout the years. Consequently, this church was the most suitable exemplar for our project as our purpose was to demonstrate how 3D visualization can contribute to the field of conservation of Cultural Heritage. We strongly believe that not only can digital color restoration be applied to monuments which have suffered from a variety of catastrophic events, but also in cases where the conservator wants to apply a variety of different color retouching techniques in a digital environment and choose between them the most preferable one, because in a digital environment he/she does not risk of making alterations which will affect the actual artifact.

8 Conclusions

The aim of our study was to present the potential of current technologies in the field of conservation of cultural heritage. The transmission of our tangible and intangible cultural heritage should be a global direction of the next generations, in order to ensure its rescue. Through our multidisciplinary approach and in accordance with the Principles for The Preservation and Conservation-Restoration of Wall Paintings [1], we implemented the digital retouching using the information we collected from the scanning, the photographic material and the existing historical references, thus creating precise copies of the monument's wall paintings. Finally, we improve the iconostasis's readability by restoring the integrity of the church's interior to its condition before the unfortunate vandalism. In order to achieve the desired result, the digital intervention should be implemented in accordance with the rules and the ethics of Conservation and Restoration [26, 28].

Conservators and restorers could implement diverse hypothetical interpretative digital color reconstructions towards virtually experiment and studying artworks. Restorers can have – obtain an important preview of the restoration hypothesis based on the particular advantages of 3D representations (color, brightness, glossiness, texture, tactile roughness, and materials).

The competent multidisciplinary team should have knowledge, critical thinking and specific skills that will determine the success of the intervention. We are sure that, following the above, color restoration can be successfully applied in a digital environment. We chose to apply a combination of the retouching techniques of mimetic restoration, chromatic abstraction and neutral color [26, 28], adapting them to the new requirements of the digital environment. In this way, we created a new original mixed technique that allowed us to experiment without violating the basic principles and theories of conservation (Articles 3 and 9 of the Charter of Venice), [4]. At the end of the digital color restoration, we were able to distinguish the interventions due to the reduced intensity of the applied colors, while at the same time the optical effect was harmonious after the legibility of the works had been restored. We therefore hope that the provided three-dimensional digital model will allow the general public to study and appreciate this monument of Cultural Heritage, but above all that it will be feasible to apply the specific or equivalent methodology and technique to similar Monuments of our cultural heritage; monuments that for different reasons cannot be entirely presented (due to the fact that part of the frescoes or other portable and non-members have been removed for reasons of safety, protection, preservation, etc.) and which could be presented in their restored form, only in a digital environment.

References

1. ICOMOS: Principles for the preservation and conservation-restoration of wall paintings. In: Proceedings of the ICOMOS 14th General Assembly, Zimbabwe, Victoria Falls (2003). http://www.icomos.org/charters/wallpaintings_e.pdf. Accessed 10 July 2018
2. Georgopoulos, A., Stathopoulou, E.K.: Data acquisition for 3D geometric recording: state of the art and recent innovations. In: Vincent, M.L., López-Menchero Bendicho, V.M., Ioannides, M., Levy, T.E. (eds.) Heritage and Archaeology in the Digital Age, pp. 1–26. Springer, Cham (2017). https://doi.org/10.1007/978-3-319-65370-9_1
3. Alliez, P., Bergerot, L., Bernard, J.-F., Boust, C., Bruseker, G., et al.: Digital 3D objects in art and humanities: challenges of creation, interoperability and preservation. White Paper: A result of the PARTHENOS Workshop held in Bordeaux at Maison des Sciences de l'Homme d'Aquitaine and at Archeovision Lab. (France), 30th November–2nd December 2016. PARTHENOS. Digital 3D Objects in Art and Humanities: challenges of creation, interoperability and preservation, November 2016, Bordeaux, France (2017)
4. ICOMOS: The Venice charter for the conservation and restoration of monuments and sites, Venice (1964). http://www.icomos.org/charters/venice_e.pdf. Accessed 10 July 2018
5. Brajer, I.: To retouch or not to retouch? Reflections on the aesthetic completion of wall paintings, CeROArt (2015). http://ceroart.revues.org/4619. Accessed 12 July 2018
6. Scopigno, R., et al.: 3D models for cultural heritage: beyond plain visualization. IEEE Comput. 44(7), 48–55 (2011)
7. Callieri, M., Dellepiane, M., Cignoni, P., Scopigno, R.: Processing sampled 3D data: reconstruction and visualization technologies. In: Stanco, F., Battiato, S., Gallo, G. (eds.) Digital Imaging for Cultural Heritage Preservation: Analysis, Restoration, Reconstruction of Ancient Artworks, pp. 103–132. Taylor and Francis, London (2011)
8. Georgopoulos, A., et al.: Merging Geometric Documentation With Materials Characterization and Analysis of the History of the Holy Aedicule in the Church of the Holy Sepulchre in Jerusalem. The International Archives of the Photogrammetry, Remote Sensing and Spatial Information Sciences, vol. XLII-5/W1, pp. 487–494 (2017)
9. Hatzopoulos, J.N., Stefanakis, D., Georgopoulos, A., Tapinaki, S., Volonakis, P., Liritzis, I.: Use of various surveying technologies to 3D digital mapping and modelling of cultural heritage structures for maintenance and restoration purposes: the Tholos in Delphi, Greece. Mediterr. Archaeol. Archaeometry 17(3), 311–336 (2017)
10. Arabadjis, D., et al.: A novel information system for the automatic reconstruction of highly fragmented objects with application to the reassembly of prehistoric wall paintings and vessels. In: Koui, M., Zezza, F., Kouis, D. (eds.) MONUBASIN 2017, pp. 571–578. Springer, Cham (2018). https://doi.org/10.1007/978-3-319-78093-1_62
11. Tsiafaki, D., Koutsoudis, A., Arnaoutoglou, F., Michailidou, N.: Virtual reassembly and completion of a fragmentary drinking vessel. Virtual Archaeol. Rev. 7(15), 67–76 (2016)

12. Menna, F., Nocerino, E., Remondino, F., Dellepiane, M., Callieri, M., Scopigno, R.: 3D digitization of an heritage masterpiece – a critical analysis on quality assessment. XXIII ISPRS Congress, vol. XLI-B5, pp. 675–683 (2016)
13. Dellepiane, M., Callieri, M., Ponchio, F., Scopigno, R.: Mapping highly detailed color information on extremely dense 3D models: the case of David's restoration. In: Eurographics (Cultural Heritage), pp. 49–56 (2007)
14. Dellepiane, M., et al: Multiple uses of 3D scanning for the valorization of an artistic site: the case of Luni terracottas. In: Eurographics Italian Charter Conference, Salerno, pp. 7–14 (2008)
15. Ranzuglia, G., Callieri, M., Dellepiane, M., Cignoni, P., Scopigno, R.: MeshLab as a complete tool for the integration of photos and color with high resolution 3D geometry data. In: CAA 2012 Conference Proceedings, pp. 406–416 (2013)
16. Graham, C.A.: 3D digitization in an applied context: polychromy research. In: Østergaard, J. S. (ed.) Tracking Colour. The Polychromy of Greek and Roman sculpture in the Ny Carlsberg Glyptotek. Preliminary Report 4, Copenhagen, pp. 64–87 (2012)
17. Arbace, L., et al.: Innovative uses of 3D digital technologies to assist the restoration of a fragmented terracotta statue. J. Cult. Heritage 14(4), 332–345 (2013). https://doi.org/10.1016/j.culher.2012.06.008
18. Siotto, E., Callieri, M., Dellepiane, M., Scopigno, R.: Ancient polychromy: study and virtual reconstruction using open source tools. J. Comput. Cult. Heritage 8(3), 1–20 (2015). https://doi.org/10.1145/2739049
19. Apollonioa, F.I., Gaiania, M., Baldissini, S.: Color definition of open-air architectural heritage and archaeology artworks with the aim of conservation. Digit. Appl. Archaeol. Cult. Heritage 7, 10–31 (2017)
20. Orlandos, A.K.: Dyo anekdotoi naoi tōn Athēnaiōn Mpenizelōn. In: EBBS, pp. 323–328 (1931). (in Greek)
21. Makris, D., Vlachou, M.A., Karampinis, L.: Digital color restoration of vandalized monument's Frescoes. In: Koui, M., Zezza, F., Kouis, D. (eds.) MONUBASIN 2017, pp. 561–569. Springer, Cham (2018). https://doi.org/10.1007/978-3-319-78093-1_61
22. Kazhdan, M., Hoppe, H.: Screened Poisson surface reconstruction. ACM Trans. Graph. 32(3), 1–13 (2013)
23. Blender Foundation. http://www.blender.org/. Accessed 08 July 2018
24. Casazza, O.: Il restauro pittorico: Nell unità di metodologia. Nardini, Firenze (1981)
25. ICOMOS: 10th General Assembly, Colombo Proceedings, National Committee, ICOMOS 1993. Conference Volume. Sri Lanka ICOMOS National Committee, Colombo (1996)
26. Brandi, C., Basile, G.: Theory of Restoration. Nardini, Firenze (2005)
27. Dellepiane, M., Callieri, M.: Visualization of colour information on highly detailed 3D models, ERCIM News 67, Online Edition (2006). https://doi.org/10.1145/2487228.2487237, https://ercim-news.ercim.eu/en67/raampd-and-technology-transfer/vizualization-of-colour-information-on-highly-detailed-3d-models. Accessed 14 July 2018
28. Stoner, J.H., Rushfield, R.: Conservation of Easel Paintings. Routledge, London (2012)
29. Maino, G., Monti, M.: Color management and virtual restoration of artworks. In: Celebi, M., Lecca, M., Smolka, B. (eds.) Color Image and Video Enhancement, pp. 183–231. Springer, Cham (2015). https://doi.org/10.1007/978-3-319-09363-5_7

Idea: Ancient Greek Science and Technology

Panagiotis Ioannidis[1], Angeliki Malakasioti[2]([⊠]) [iD],
and Maria Mavrokostidou[3] [iD]

[1] TETRAGON S.A., 2B, Karaiskaki str., 54641 Thessaloniki, Greece
giotis@tetragon.gr
[2] Athens, Greece
malakasioti@uth.gr
[3] Thessaloniki, Greece
m.mavrokostidou@gmail.com

Abstract. The exhibition IDEA - Ancient Greek Science and Technology presents, in a composite way, the advancement of Greek thought and innovation which created a series of scientific fields, parallel to the discovery of a multitude of technical and technological achievements. The Exhibition highlights all those elements that raised the Ancient Greek world at a wondrous recognizable level, defining the western and consequently the modern world.

The exhibition objectives included: presenting the connection between science and technology in Ancient Greece, associating Greek knowledge and innovation and their contribution to the advancement of new theoretical fields, showcasing the important fields of science, arts and technological achievements in the Greek world, as well as their association with later developments of knowledge till the present, all constituting reminders of the unparalleled contribution of those achievements to the foundation of the western civilization.

The exhibition was inaugurated on September 11, 2016 at the NOESIS Technology Museum. Since March 2018 it has been installed in Hellenic Cosmos Cultural Centre. The modern way of presentation and the use of digital media challenges visitors to engage with the world of ancient Greek science and technology in a familiar and entertaining way. Its 19 thematics include rich visual material, digital and physical models of technological findings, interactive and audiovisual media, and 3d video projections, achieving a multidimensional reconstruction of the technology of the ancient world.

Keywords: Ancient Greece · Science · Technology · Museum ·
Culture · Traveling exhibition · Design · Architecture

1 Introduction

The 'IDEA: Ancient Greek Science and Technology' exhibition consists of the presentation, in one cohesive cultural event, of the evolution of Greek thought that created a series of scientific, technical and cultural fields. These fields, along with their technological achievements, which brought the ancient Greek world in the forefront of scientific interest, laid the foundations of the project's original idea.

© Springer Nature Switzerland AG 2019
A. Moropoulou et al. (Eds.): TMM_CH 2018, CCIS 961, pp. 171–183, 2019.
https://doi.org/10.1007/978-3-030-12957-6_12

The exhibition was initially proposed by Tetragon, a company specialized in exhibition architecture, museology-museography and museum design, to NOESIS – Science Centre and Technology Museum of Thessaloniki in the summer of 2013. The aim of this traveling exhibition was to become a cultural and educational event, directed to the public inside and mainly outside of Greece, showcasing and promoting technological achievements of the ancient Greek world through the prism of philosophy and corresponding fields of early science, which influenced the shaping of the western and worldwide civilization. A complete preliminary design dossier was created that autumn, including the concept, design, structure as well as the exhibition budget.

During next year, the proposal received the approval of financial support for its implementation from Stavros Niarchos Cultural Foundation and the journey of its realization begun – a quite demanding project as its design process had to manage a series of challenges in multiple levels.

1.1   Challenges and Design Issues

At the beginning, a vast field of knowledge had to be summed up in particular and critical conclusions concerning the scientific and philosophical fields, which influenced and fertilized the technological innovations and achievements, giving solutions to everyday problems and human needs since antiquity.

One more fact is that each chosen exhibition section that deals with a distinct area of a subject, constitutes a vast field of temporal and technological evolution, with influences of various functions and adjacent areas of activity, which made the choice of its content particularly challenging. There were fields like Naval and Warfare technology that have overlapping areas and subjects, or others like Architecture and Building, with distinctive contemporary scientific distinction, which therefore suggested a separate presentation.

Another challenge was the required simplification of terms and concepts that risks the distortion of meaning in an effort to incorporate different levels of knowledge, the ultimate goal of being understandable and appealing to younger, international or culturally differentiated audiences.

Responding to a contemporary interpretation of the past, the exhibition's way of presentation had to be formed accordingly, with a thoughtful use of text and image, new digital and interactive media, edutaining exhibition practices, capable of attracting various audiences.

The exhibition design and scenario had to be understandable from the beginning, precise and guiding, as well as offering options for flexible navigation, so that it does not exhaust the perceptual capacity or available time of its visitors.

At the same time, the exhibition, in the extent that its aim is to travel and be hosted in various venues, should have adjustable structure and components as an important parameter in its design, either by increasing or reducing its content, without losing its entity and purpose.

Finally, the necessary validity, useful to anyone that is curious or scientifically interested in additional information and references, was also critical. For this reason, previous exhibitions about Ancient Greek Technology, the permanent one of NOESIS under the care of Prof. Claire Palyvou, as well as those by EMAET (Greek Association of Research on Ancient Greek Technology) [1] made with the contribution and highly important work of its president Prof. Tassios [2], were a valuable background for this exhibition and its contents.

Moreover, contributions from other scientists and researchers of the corresponding scientific fields were requested, and their work and research has formed a point of reference in each exhibition section. All this multitude of research, references, views, scientific and factual criteria, theoretical or even aesthetic differences of the research-ers' approaches, had to be confined to a single narrative - in other words, to the formation of an 'Idea' in reference to the title – with interesting points and conclusions, but also with the necessary cohesion. Thus, the concept of idea and its potential narrative representations enforced all applied methodological design tools.

2   Museological and Museographic Approach

2.1   From the Idea to Creating Space

One of the early questions posed about the exhibition content, but also on a general level, was the relation and influence of science on technology. And this does not relate to the contemporary correlation of any kind of technological innovation and application that has its scientific principle and vision as a given. But to a period where philo-sophical quests start to shape the beginnings of sciences in various fields, and technical solutions often constitute the result of empirical developments and applications.

At the same time, in this period, already since the 6th century B.C., the quests for principles and causes in a series of natural phenomena take a distance from religious beliefs and metaphysical interpretations. Philosophers will indulge in mathematics and physics, while a frame is shaped in the Alexandrian era with personalities that are simultaneously working on many fields, theoretical as well as applied. Archimedes, Ctesibious, Hero and others, expand their activities in technological solutions and applications, while their astronomical observations and theories, along with highly developed techniques, will have as their apex the wondrous Antikythera Mechanism.

The exhibition IDEA: Ancient Greek Science and Technology seeks to display this osmosis and parallelism through time. In this direction, Greek thought, philosophy and science, with their prominent figures and their work as milestones in a timeline, have the form of a lit axis in the exhibition. The visitor walks around it and reads it chronologically, from the beginning of the one side until the end of the other. At the center, two annexes form a small square with two sides in the shape of a Π, where Mathematics and Geometry develop on the one side, while Physics and Biology develop on the other. Theoretical fields, which Greeks advanced and developed on many levels, set the foundations for the evolution in a series of sciences [3].

There is an introductory area at the start of the exhibition space, along with the interactive timeline. This places a series of achievements and technological events in time, defining the range of the periods and their correlations, which the exhibition presents further.

Its main volume is developed on both sides of the central axis in nineteen exhibition sections. It mainly concerns the technological achievements and scientific fields that are realized or grouped as subjects. Sections have different sizes depending on their content and importance, while they unfold in space in positions with correlations and combinations between them regarding their subject, while in some cases, according to the chronological relation. For example, Hydraulics and Ceramics that are placed in the beginning developed as human activities before Theatre or Automata that appear at the end of the technological evolution.

In any case, the exhibition does not suggest a single reading. Visitors can move freely through space, spending time according to their interests, since each exhibition section forms an independent and complete entity.

2.2   Content

The sections bear on their surfaces an aesthetically imposing mythological element and the general introduction that defines the context of each thematic. The multiplicity of gods, divinities and heroes, and their varied expressions and notions, as well as every kind of their representations in art, offer a straightforward introductory gesture in every section of the exhibition.

Each section then presents a text that sums up the historical evolution and the importance it acquired in the course of time. The rest of it is articulated with a selection of the most important technological achievements, with images and concise text and captions. Where applicable, the content includes references to the influence these achievements had on the formation of western and global civilization until today.

In some sections like Hydraulics, Mechanics, Building, Metallurgy, Metrics etc., the previous research and work of EMAET, Prof. T. Tassios and other researchers and contributors of previous exhibitions, symposia and presentations about Ancient Greek Technology, became the thinktank for content choices and documentation. The research work and suggestions by professors and scholars were valuable in triggering the development of the exhibition's narrative[1]. In the rest of them, as well as in parts of the content of all sections, an extended bibliographic research of articles and conference presentations took place to provide more complete documentation and selection of the necessary references.

Afterwards, the team of museologists, archaeologists and architects in Tetragon finally shaped the content of the sections, its extent and the unification of its style so as to acquire a unique cohesive structure.

---

[1] Researchers like Prof. John H. Seiradakis in Astronomy, Prof. Georgios Karadedos in Theatre, Prof. Dimitrios Kalligeropoulos in Automata, Prof. Stefanos Geroulanos in Medicine, ass. Prof Evridiki Kefalidou in Ceramics, Themis Veleni PhD researcher in Music, Prof. Manolis Korres in Architecture and Building, Prof. Evangelos Livieratos in Geography-Chartography, Prof. Vassilis Lamprinoudakis on the Asclepeion of Epidaurus, Kostas Damianidis PhD researcher for Naval Technology and others.

3   Design and Presentation Techniques of the Exhibition

The exhibition, as aforementioned, consists of the main axis of Greek thought and the exhibition sections arranged on either side. Moreover, two more exhibition elements are an introductory interactive timeline and an area for projecting a 15-minute audio-visual production. Part of the axis is formed by touchscreens with educational inter-active games. The Axis of Thought constitutes a point of reference in space running across the exhibition, offering an illuminated dynamic pulse. This luminous appearance of the central element symbolizes the importance of Greek thought and its influence on the evolution of arts, sciences and techniques.

Exhibition sections have, apart from the surfaces information, rich visual material and moving images (3D), dramatizing of ancient personas, holograms, sounds, illus-trations and physical models of technological findings that achieve a multidimensional reconstruction of the ancient world. Some of the screens are interactive, giving the audience the ability to choose between the subjects they include.

All sections have a main object on display, in the form of a physical or digital model, whereas in some cases there is a secondary one. Regarding the form and style of the model's construction, which in most cases are full sized or scaled, the idea was to mainly show the structure and operation of the prototypes rather than try to depict their realistic form. Corresponding to that concept is also their representation on screen - minimal forms with basic information and explanations.

It is important to note that the volume of information that a traveling exhibition contains in text or other material, but also in exhibits, especially when they are not original or of a significant artistic value, should not exceed certain perceptual or temporal limits. When an exhibition deals with multiple and extended scientific fields, it is important to focus on showcasing the most representative elements, their corre-lation and the concise but valid text apposition that accompanies them, so that its essence, and the long-term timeframe of its evolution, will emerge.

Each one of the sections of the IDEA: Ancient Greek Science and Technology exhibition is a vast field, capable of consisting a cohesive frame of documentation and, if accompanied by various exhibits and elements, an autonomous large exhibition by itself. In this case, we aim to focus and showcase the technical achievements of each field, so that the references in the general frame, historical or scientific, should appear through this particular perspective. The main pedagogical role of the exhibition leads to a more concise and comprehensible presentation of the contents.

Even though the exhibition initially appears as of archaeological interest, the aim, the content and identity are destined to be hosted not only in archaeological museums, but in other cultural infrastructures as well. This was also a significant parameter that had to be considered regarding the style and handling of the content or the exhibit design.

One other important element of the exhibition is the 15-minute-long audiovisual production that is shown in a designated space. The scenario is based on the dialogue of two protagonists, a modern young university student and an old man from antiquity. The dialogue takes place while the student falls in a dreamy state. The questions and answers of the hypothetical dialogue cover summarily almost the complete extent of the exhibition thematics. The transcendent nature of the conversation, with a humorous and educational touch, is transferred in contemporary time and space during the introduction and epilogue of the film, while a course on the Antikythera Mechanism is taking place in an amphitheater.

The needs of the traveling exhibition, the necessary adaptability and reinstallation it demands, led to the design of stands using a system of disassembling components. This system is made of supports and frames of aluminium elements, which are filled with back lit printed fabric surfaces, offering a light feel as well as a single, functional whole to the viewer, with a low reuse cost.

The visual identity of the exhibition has as its basic element the choice of the Greek word IDEA, common also in most western languages, with the same meaning, which immediately expresses the beginning of thought, the invention and creation of every achievement. The logo, with its four letters in uneven distances, expresses the evolutionary course of each creation. Circular graphic elements framing events and faces, along with lines linking them, mark the chronological developments, the flow and connections that achievements mentioned in the exhibition sections have between them. Uneven and crooked tracks and paths, correlations and self-contained elements, form the graphics of surfaces, as it happened in the long period the exhibition narrates.

4   Exhibition Sections

The conceptual background of the central axis, as well as of the individual exhibition sections, outline their character and particularities in content and design level.

4.1   The Thought Axis

The development of Ancient Greek technology was initiated by the appearance of the first Greek tribes and continued until the classical era. From the 6th century B.C., an influence and fertilization of technology via the newly born Greek science is observed. The theoretical research contributes so that a series of technical achievements matures, that will apex in the Hellenistic era.

Researching the Greek world, the productive role of large libraries, like the one in Alexandria, and the development of many scientific fields that will in their turn profit from technology, the influence of all these superb achievements will be found since then in various phases of our civilization to this day [4, 5].

4.2   The Thematic Sections

Mathematics/Geometry - Mathematical Knowledge in Ancient Greece. The turning point in relation to previous cultures in the study and development of Mathematics, takes place in the 5th century B.C. in ancient Greece. Then inductive reasoning and productive method are introduced with rigorous proofs of mathematical theorems [6].

With their gradual disconnection from philosophy, after Plato and Aristotle, Mathematics is now a separate science. During the 3rd and 2nd century B.C., in the Hellenistic period, it reaches a golden age with the brilliant works of Euclid, Archimedes and Apollonius.

Its exploration and use in other fields, such as Physics, Astronomy and Geography, will set the overall context of the interaction of sciences. The influence of Greek mathematics will be catalytic at various times in the centuries to come, even to this day.

Physics/Biology - From Observing to Interpreting the World. The ancient Greeks' contribution to Physics was definitive and prompted the discipline's transition to a new era, even though a long tradition already existed in predating civilizations. The difference in the approach of the Greek philosophers-scientists lies in the fact that, rather than simply observing natural phenomena, they tried to interpret and explain them, linking them to logical causes. They challenged traditional and religious beliefs attached to the origins of the cosmos and of natural phenomena by distinguishing between the natural and the supernatural; and claimed that any phenomenon has a natural cause as well as a natural explanation.

Through the implementation of rational critique, arbitrary and unsubstantiated interpretations were substituted by the first natural laws, resulting in a conflict of ideas and theories between philosophers. Thus, the foundations of the science of Physics were laid. This also paved the road for the formation of the first scientifically based worldviews, since Greek philosophers were the first to answer the question of the origin of the cosmos using natural laws presented in a structured form and with logical continuity.

Ceramics – Life and Pottery. Ancient Greek myths, scenes from everyday life and other representations come alive on the surface of vessels, offering invaluable information about the life, customs and traditions of ancient Greeks [7].

The durability of pottery makes these vessels our strongest material link to the ancient world. Since the Archaic period, ceramic vessels, with their characteristic colors and decorations, constitute a symbol of Greek cultural identity worldwide.

The expansion of commerce and the quest for raw materials was one of the causes that gave rise to the Greeks' colonial activity during the Archaic and Classical periods. Commercial amphorae, undecorated and resistant to breakage to ensure the safe transportation of products by sea, are a common finding in many underwater archaeological sites on the Mediterranean and the Black Sea.

Metallurgy – Mining and Metallurgic Technologies in Ancient Greece. The first metals discovered and used by humans were gold and copper. These were used for the first time in Asia, in Persia, Mesopotamia and Asia Minor, where metallurgy first evolved. On the Greek peninsula, the age of metal commenced at the start of the 3rd millennium B.C. Subsequently, there was rapid technological and cultural development.

Noteworthy gold deposits were found in the region of Northern Greece, while silver, copper, lead and iron were mined in mainland Greece and on the islands. It is certain that Greeks also imported their metals from distant lands. During the Classical period, mining production reached its peak, with the Lavrion mines being the most prominent [8].

Music – Ancient Instruments and Evolution. Music in Greece has a long history, dating back to prehistoric times. Depictions of musical performance appear in the Cycladic civilization, as well as in the Minoan and Mycenaean eras.

In Greek Antiquity, Music included Poetry, Melos (Melody) and Dance, which was mainly cultivated as part of the ancient theatre. At the time, learning music was a major part of a youth's overall education.

The greatest evolution of music in antiquity occurs in the Classical era. New instruments were invented, existing ones were perfected, while important mathematical relationships that govern music [9] were discovered.

Naval Technology - Ancient Naval Routes. Navigation, initially within the Aegean Archipelago and later in the Mediterranean, commenced before the 3rd millennium B.C. Its main drives were maritime commerce, colonization, exploration, and communication with distant lands and people.

The Aegean civilization in the 2nd millennium B.C., as noted on the island of Thera (Santorini) and in Minoan Crete, was based on seamanship and on ships that enabled commerce within a wider area.

During the Mycenaean era, the Homeric poems, the Iliad and the Odyssey, mention the naval expedition of the Achaeans to Troy and Odysseus' ten-year wanderings in the sea.

From the start of the first millennium B.C., there is a noted evolution in naval technology and in ancient Greeks' knowledge of navigation. Commerce from city-states and from the colonies [10] that span every coast and niche of the Mediterranean and the Black Sea created the wealth and prosperity of the Hellenic world, while the facilitated communication by sea helped spread knowledge and ideas and further developed Greek civilization.

Theatre – The Birth of Ancient Greek Drama and Theatrical Mechanisms.
Theatre is, in its entirety, a creation of the ancient Greeks; according to Aristophanes, it had infused education in the Athenian democracy of the 5th century B.C. Its origins can be traced back to the chants and circular dances that were rituals in the popular worship of Dionysus. Dramatic performances evolved step by step from religious dances into a form of public art which, in democratic societies, has been watched by ever broader audiences.

This led to the construction of the great stone amphitheaters, solving not just issues of functionality, but technical problems as well, such as auditory and visual comfort. The theatrical atmosphere was created through the use of scenery, costumes, masks, and the use of mechanical equipment [11] for special stage effects.

Hydraulics - Knowledge of Hydraulic Technology. Water use in ancient Greece was related to the satisfaction of everyday needs and to large-scale works. Water networks for domestic use, sewage systems, baths and other sanitary installations are evidence of highly evolved hydraulic technology [12].

At a larger scale, technical works were constructed for water supply of cities, irrigation of fields, drainage of lakes and river-diversion [13].

Astronomy - Observing the Heavens. Already in early antiquity, the Greeks –much like other ancient peoples- turned their eyes to the heavens and engaged in observation and study. Their primary driving force was the need to regulate the calendar and to determine the agricultural year.

Observations by the ancient Greeks in the 6th and 5th centuries BC were made using simple instruments, but they were arduous and methodical. Their observations quickly led to numerous theories and discoveries on the fundamental astronomical problem: the relationships of the basic movements of celestial bodies.

Astronomy received a significant boost in the 4th century BC, when simple observation was replaced by the mathematical and evidentiary methods. This development established Astronomy as a mathematical science.

In the following century, two impressive theories were proposed: Aristarchus's heliocentric theory, which posits the Sun in the centre of the universe, and Apollonius's theory of epicycles and eccentric circles.

The Antikythera Mechanism [14, 15] is one of the rarest and most complex mechanisms that survived from Ancient Greece.

Architecture - The Creation of the City. The works of ancient Greek architecture are the culmination of a long-lasting building and cultural activity that preceded them at the same place and gave birth to one of the greatest revolutions in the history of art.

Greek cities with their democracy acquire a new structure and organization for the needs of their citizens and their administration [16]. The temples are the most important buildings, while the Market, usually surrounded by galleries, is the centre of political and economic life. Around it monuments and public buildings develop, such as theatre-bouleuterion, gymnasium, stadium, aqueducts and baths.

Building – Technical Innovations. The multitude of buildings in the Greek world and their great artistic value would only be possible through a lasting technical development and the introduction of technological innovations in all phases of the building work [17].

The invention of the pulley, the polyspasta, the winches, and the invention of a number of lifting machines such as the monocular and the two-wheeled hoist, helped to transport very large weight stones.

Combined with the creation of intelligent mooring systems, lubricating sliders, wheeled vehicles and advanced tools, they have contributed to the construction of large edifices [18] such as the Parthenon, which today seem to be impossible for that time.

Geography - The Foundations of Geographic Science. The measurement of the Earth and its representation, which concerned many people, acquires its scientific foundation in ancient Greece.

The engagement of Greeks with Geography and Cartography appears at a time of frequent trips, migrations, exploratory excursions and seafaring [19]. Geographic knowledge of the era is enriched from this intense mobility, resulting in the design of more accurate maps that make distant destinations feasible.

Metrics - The First Measurements. The first units of measurement were based on the human body, whole or its parts. The development of trade and contact with other peoples created the need for more stable units of measurement.

The development of metrics gained a special importance in ancient Greece, as the concept of measurement was a basic prerequisite for the development of science and technology. Various measurement methods and tools were used to solve problems of geometry, geography, astronomy, building, engineering, and practical problems of everyday life.

War Technology – Ancient Greek Weaponry. The ingeniousness and the weaponry of the ancient Greeks played a key role on the battlefield. Several stories revolve around ploys such as the mythical Trojan horse, weapons such as the sarissa, the spear of the Macedonian phalanx, and the influential strategic decisions that defined the outcome of crucial battles and determined the course of Greek history [20].

Engineering - Pioneering Machines. The evolution of mechanics and engineering in ancient Greece found a variety of applications in everyday life. Ancient Greeks invented pioneering machines using processed materials, such as copper, tin and iron, and they also invented tools such as the lathe for metal objects.

Telecommunications – Means of Communication in Ancient Greece. Delivering messages in antiquity was important and necessary, since messages typically had military or commercial significance. The original method of transmitting messages was based on messengers, who had to cover large distances in the shortest possible time. Other basic methods to transmit messages with speed and reliability were visual and auditory signals, smoke and fire signals, and encryption systems.

Medicine – Medical Care in Ancient Greece. Medical knowledge has been developed in Mesopotamia and Egypt before the Greeks, but was dominated by magic and superstition. Early medical care in the Minoan and Mycenaean eras was similar. In the 6th century B.C. some city-states, such as Kroton, were famous for their doctors, while during the 5th century B.C. medical schools flourished on the island of Kos (the birthplace of Hippocrates) and in Knidos; these defined the end of a long healing tradition based on religion and dubious practices, and they inaugurated a new era of rational and scientific medicine. All this knowledge that is contained in the Hippocratic Corpus and its apex during the Alexandrian era and up to Galen laid the foundations of Western Medicine and decisively influenced it.

The central point of interest in this section is the rendering of the facilities of Asklepeion of Epidaurus [21], the most important Asclepius of antiquity, in a model scale, so that the visitor can instantly understand the relationship between space, buildings and health services.

Painting/Sculpture - The Arts of Ancient Greek Civilization. In ancient Greece, the arts of Sculpture and Painting produced remarkable works both at the level of technical skill and in artistic value. The canon, the perfection in composition and the technical conquests have functioned through the ages as an archetype, defining not only the method of production but also perceptions of artistic creation to this day.

Elements of ancient Greek art, like dynamic motion and vividness, also appear in Hellenistic art and set the foundations for the creation of Roman art. Reaching its peak during the Classical period, from miniature works to large-scale projects, ancient Greek art dominated and consolidated the concept of representation and ideal imitation of nature in western civilization.

Athletics - The Games in Antiquity. The ancient Greeks devoted as much care to their physical exercise as they did to the cultivation of their intellect. Their systematic occupation with exercise gradually led to the development of specific sports and the organization of Games, such as the Pan-Athenian, Isthmia and Pythia events. The organization of Panhellenic games played a significant role in shaping a common cultural and religious identity among the inhabitants of different city-states.

The most important event was, of course, the Olympic Games, which contributed to the establishment of new institutions and ideals [22, 23]. Their importance is underlined by the Olympic Truce, the suspension of all wars and attacks for the entire duration of the Games.

Automata – The Idea of Automation in the Ancient Greek World. There are three quantum leaps in the history of technology: the construction of tools, the invention of machines that make use of some form of external energy, and the invention of machines that are moved by some form of internal energy – the automata.

During the Hellenistic period, the automata were the innovative technology of the era. Important personalities in the field of automata were the engineers Ctesibius of Alexandria, Philo of Byzantium and Hero of Alexandria [24]. Important written works on Mechanics and Automata are the Pneumatics by Philo of Byzantium, containing fragments of his great opus Mechaniki Syntaxis (Compendium of Mechanics), as well as the works Automata and Pneumatics by Hero of Alexandria (Table 1).

Table 1. Presentation of models per thematic section.

Sections	Models
1. Mathematics - Geometry	Short 3d projection with Archimedean solids
	Pythagorean Theorem
2. Physics – Biology	Short 3d video with the classifications of animals, plants and minerals
3. Ceramics	Techniques for firing and decorating pottery
	Model of commercial ship - pointed amphorae
4. Metallurgy	Lavrion ore mines
5. Music	Hydraulis
6. Naval Technology	Trireme and ram
7. Theatre	Ancient Theatre of Dion – mechanical equipment
8. Hydraulics	Eupalinian Aqueduct
9. Astronomy	Antikythera Mechanism
10. Architecture - Building	Parthenon
	Lifting machine
11. Geography	Hero's Diopter
12. Metrics	Hero's odometer

(continued)

Table 1. (*continued*)

Sections	Models
13. War Technology	Polybolos catapult
14. Engineering	Archimedes screw
15. Telecommunications	Hydraulic Telegraph
16. Medicine	Asclepeion of Epidaurus
17. Painting – Sculpture	Statue of Ephebe of Marathon
	Pantograph for marble sculptures
18. Athletics	Hysplex
19. Automata	Moving automatic theatre

5 Epilogue

The exhibition is placed in a wider framework of interdisciplinarity in a way that is comprehensible and educational for the public. In the context of promoting the Greek civilization, its ideas and achievements, it is necessary to enrich the exclusively archaeological exhibitions and their original exhibits, with new expressions and modern methods of displaying it.

The management of cultural heritage, its peculiarities and its remarkable characteristics can now be explored through education and entertainment (edutainment). At the same time, digital imaging technologies, the creation of 3D representations, as well as sophisticated techniques such as 3D printing, and a range of high-end technologies create nowadays expanding possibilities in displaying aspects of culture, in a more appealing way for younger or broader audiences.

These approaches can be inscribed in an overall contemporary context of cultural meaning making, embracing the potential of broadening accessibility, raising of interest and participation in an intercultural environment.

References

1. Greek Association of Research on Ancient Greek Technology-Thessaloniki Science Center and Technology Museum, Ancients Greek Technology, Thessaloniki (1997)
2. Tassios, Th.: The peak of Hellenistic technology. In: The Antikythera Shipwreck: The Ship, the Treasures, the Mechanism, pp. 17–25. National Archaeological Museum, Athens (2012)
3. Gregory, A.: Eureka: The Birth of Science. Trans. Mavroudis E., University Studio Press (2007)
4. Lloyd, G.E.R.: Early Greek Science: Thales to Aristotle. Trans. Karletsa, P., Crete University Press, Heraklion (2008)
5. Lloyd, G.E.R.: Greek Science after Aristotle. Trans. Karletsa, P., Crete University Press, Heraklion (2008)
6. Tsimpourakis, D.: Mathematical Measurements in Ancient Greece. Aiolos Publications, Athens (2002)
7. Scheibler, I.: Greek Pottery. Production, Trade and Use of the Ancient Greek Vases. Trans. Manakidou E., Kardamitsa Publications, Athens (2010)

8. Tsaimou, P.: Metals in Ancient Times-The Ancient Mining and Metallurgic Technology, 1st edn. Simeon, Athens (1997)
9. Kaimaikis, P.: Philosophy and Music. The Music in Relation to Pythagoreans, Plato, Aristotle and Plotinus. Metaichmio Publications, Athens (2004)
10. Tzalas, H.: The ships of Greeks: from the rafts and dugouts of prehistoric times. In: Vigopoulou, I. (ed.) The Journey: From the Ancient to the Modern Times, pp. 37–46. National Hellenic Research Foundation, Athens (2003)
11. Karadedos, G.: Hellenistic theatre of Dion. In: Veleni, A.P. (ed.) Ancient Theatres of Macedonia, pp. 73–88. Diazoma, National Theatre of Northern Greece, Athens (2012)
12. Palyvou, K.: Sewage and Sanitary installations in the Aegean during the 2nd Millenium BC. In: Ancient Greek Technology, Proceedings of the 1st National Conference on Ancient Technology, pp. 381–389. Cultural and Technological Foundation of ETVA, Thessaloniki (1997)
13. Tassios, Th.: Selected topics of water technology in Ancient Greece. In: Boudouris, K. (ed.) Proceedings of Conference on Hydrotechnologies in Ancient Greece, Thessaloniki, pp. 11–13 (2013)
14. Moussas, X., et al.: The Antikythera mechanism, the oldest known astronomical computer. Rom. Astron. J. **18**(Supplement), 281–282 (2008)
15. Seiradakis, J.H.: The Antikythera mechanism: from the bottom of the sea to the scrutiny of modern technology. In: Proceedings of the meeting 'From Antikythera to the Square Kilometre Array: Lessons from the Ancients' (Antikythera & SKA), Kerastari, Greece, 12–15 June 2012. http://pos.sissa.it/cgi-bin/reader/conf.cgi?confid=170. id.7 (2012)
16. Foka, I., Valavanis, P.: Discover Ancient Greece – Architecture and Urban Planning. Kedros Publishers, Athens (1992)
17. Korres, M.: The construction of ancient columns, Doctorate thesis, National Technical University of Athens, Department of Architecture, Athens (1992). http://thesis.ekt.gr/thesisBookReader/id/8838#page/1/mode/2up. Accessed 18 Feb 2014
18. Korres, M.: From Pentelicon to the Parthenon: The Ancient Quarries and the Story of a Half-Worked Column Capital of the First Marble Parthenon, 1st edn. Melissa Publishing House, Athens (2000)
19. Livieratos, E.: A tour into Cartography and Maps: 25 Centuries from Iones to Ptolemy and Rigas. National Centre for Maps and Cartographic Heritage, Thessaloniki (1998)
20. Rorres, C., Harris, H.G.: Formidable war machine: construction and operation of archimedes' iron hand. In: Proceedings of the Symposium on Extraordinary Machines and Structures in Antiquity, Olympia, Greece (2001)
21. Lamprinoudakis, V., Mastrantonis, T.: The Asclepius of the Epidaurus: The Greatest Sanatorium of the Ancient World. https://bit.ly/2SMWtW5. Accessed 20 Oct 2015
22. Valavanis, P.: Games and Sanctuaries in Ancient Greece. Kapon Editions, Athens (2004)
23. Odysseus Culture Homepage. http://odysseus.culture.gr/h/4/gh41.jsp?obj_id=10841. Accessed 2 Oct 2015
24. Kalligeropoulos, D., Vasileiadou, S.: History of Technology and Automata. Sychroni Ekdotiki, Athens (2005)

Cross-Sector Collaboration for Organizational Transformation: The Case of the National Library of Greece Transition Programme to the Stavros Niarchos Foundation Cultural Center (2015–2018)

Julia Elmaloglou, Georgia Angelaki, and Stephania Xydia[✉]

National Library of Greece, Athens, Greece
jelmaloglou@gmail.com, georgiaangelaki@gmail.com,
stephania.xydia@gmail.com

Abstract. 2018 marked the year of the relocation of the National Library of Greece (NLG) from a centrally located 19th century neoclassical building, where it had been housed since 1903, to its new seaside premises at the Stavros Niarchos Foundation Cultural Center (SNFCC). After being in a state of neglect for almost 20 years, the NLG had to undergo a major transformation before it could move and be able to function in the new building; the collections had to be prepared for safe relocation, new services had to be designed, the public image of the Library needed to be revisited and staff skills had to be developed. The transition of the NLG to the new building was made possible thanks to a designated grant of €5 million from the Stavros Niarchos Foundation (SNF). The Transition Programme was carried out between 2015 and 2018 and was divided in 5 actions: 1. Collections Transition and Development, 2. Digital Services Development, 3. Public Library Department Design, 4. Audience Development and 5. Staff Training. The current paper gives an overview of this large scale, cross-sectoral Public-Private-People Partnership, focusing on the multidisciplinary projects that were designed and implemented, key challenges and successes, innovations and open-ended issues.

Keywords: Transition management · Library innovation · Relocation · Digital transformation · Audience development · Public-Private-People partnerships · Organizational change · Service design

J. Elmaloglou—Head Project Manager/Collections Transition Manager, National Library of Greece Transition Programme 2015–2018. G. Angelaki—Digital Services Manager, National Library of Greece Transition Programme 2015–2018. S. Xydia—Audience Development Manager, National Library of Greece Transition Programme 2015–2018.

A. Moropoulou et al. (Eds.): TMM_CH 2018, CCIS 961, pp. 184–202, 2019.
https://doi.org/10.1007/978-3-030-12957-6_13

1 Introduction

1.1 The Stavros Niarchos Foundation Cultural Center (SNFCC)

2018 marked the year of the historic relocation of the National Library of Greece (NLG) to its new premises at the Stavros Niarchos Foundation Cultural Center (SNFCC) [1, 2], the new architectural landmark of Athens. The SNFCC houses the National Library of Greece, the Greek National Opera (GNO) and a 210 acre park. The construction of the SNFCC was funded exclusively by the Stavros Niarchos Foundation (SNF) at a cost of ca. €620 million. After its completion in 2017, the ownership and management were fully transferred to the Greek State. SNFCC S.A. is the company responsible for facility management and park management, with its board of directors referring to the Ministry of Finance. Each of the two institutions (NLG, GNO) is managed independently, referring to the Ministry of Education and the Ministry of Culture respectively, and represented in the SNFCC S.A. board of directors.

Since the first months of operation, the SNFCC has become an attraction for millions of visitors and has contributed to repositioning Athens as a top city-break destination. The SNFCC endows the National Library with 22,000 m^2 of additional space and 8 new reading rooms, serving the needs of a constantly evolving library, with respect to its collections and services, aiming to contribute to its further emergence as an institutional pillar for scientific research and knowledge diffusion into society.

1.2 The National Library of Greece (NLG)

The NLG was founded in 1832; it is the primary custodian of the nation's intellectual print and increasingly electronic output. It hosts the biggest print collection in the country, primarily due to the legal deposit mandate, with holdings including more than one million items such as books, journals, manuscripts, maps, newspapers, music scores and various artworks. The Manuscripts collection consists of ca. 5,400 bound codices, the oldest of which dates back to the 9th century AD, being one of the largest collection of Greek manuscripts in the world. The NLG also houses 170 archives, the most prominent of which being the 1821 Greek Revolution Fighters Archive. The Rare Printed Books collection comprises ca. 17,500 invaluable volumes; the oldest book of the Library, the first printed book written in the Greek Language and published by a Greek, dates back to 1476 [3].

The main Library departments are: ISBN/ISSN/ISMN [4], Reading Rooms, Cataloguing Department, Manuscripts and Treasures, Journals and Newspapers, Conservation, Digitization, Secretariat (human resources & protocol), Accounting and IT.

Partly because of the transition, a new law was voted [5] in 2017, expanding the mandate of the Library, while enabling to synchronize with advances in digital technology and supporting the enhanced capacities emerging from the new building. Besides its long established services, the law assigned to the NLG the responsibility of aggregating all born digital and digitized publications via digital legal deposit and Web archiving. Under this new law, the NLG became the coordinator of the national network of libraries (which to date includes 220 municipal and public libraries all over the country), and the national coordinator regarding electronic access to items created or published in Greece for people with disabilities. Additionally, the new law established a Public Library department, something rather uncommon, since most national libraries are first and foremost research libraries.

1.3   A Pressing Need for Change

In order to evaluate the scale of organizational change that the relocation of the NLG involved, it is important to take into consideration the starting point and the conditions that previously prevailed within and around the organization. The period before the Transition Programme was a period of neglect for the NLG, with the Library being underfunded, understaffed and cut off from the national and international environment.

In 2014, the NLG was based in three different buildings that had not been maintained for years, providing insufficient space for hosting its collections. In 2014, the operating budget was €865,654[1], among the lower of European National Libraries, prohibiting any investment in collections or service development. For almost 15 years, no cleaning or conservation was made to the collections, and no development occurred apart from the continuance of the legal deposit. Additionally, there was no acquisition of international bibliography and no access to electronic resources, apart from the Hellenic Academic Libraries Link Consortium resources [6]. IT infrastructure was outdated and there was no department focusing on communications or programming, thus keeping the institution in an introvert modus operandi and only able to serve a limited number of researchers. There was only 100 staff, 48 of which permanent employees with 17 of them close to retirement, and 52 teachers seconded on an annual basis, providing limited opportunities of investment in training and skills development. The organizational structure was outdated and insufficient to cover the operational needs. Furthermore, the NLG was left with only acting directors for the period 2005–2014, at a time in which the Greek State entered into negotiations with the SNF for the construction of the SNFCC. As a result, the NLG had limited capacity to contribute with inputs for the development of strategic frameworks or concrete specifications for the new building.

2   The Transition Programme Overview

2.1   The Transition Programme Vision

In order to enable the NLG to meet the challenge of a multilayered operation in the new building, meet the increased expectations of its users and fulfill the requirements of the new law, a major organizational transformation was required. This was enabled through the design and implementation of a Transition Programme, funded exclusively by the SNF with a designated grant of €5 million.

The vision of the Transition Programme was, ultimately, to bring the Library to the 21st century, redefining its institutional role in relation to its collections, services and

---

[1] This amount excludes salary costs, since all NLG staff is directly reimbursed by the Ministry of Education.

audiences. While the actual relocation of the collections would be covered by the State Budget, the NLG Transition Programme intended to operationally modernize the institution. This involved reorganizing and developing collections, upgrading existing and designing new services in the context of a digital transformation, training staff in order to respond to the ever-increasing operational needs and expanding its outreach, reconnecting the NLG with different user groups.

2.2   Programme Structure

The NLG was the sole beneficiary of the SNF grant, although funds were not directly diverted to the institution due to existing legal constraints and limited managerial capacity. Consequently, the contractual and financial management was handled by two intermediary not-for-profit organizations, which would ensure that the funding provided by the SNF would cover the breadth of works required, based on the needs of the NLG.

The SNF grant was managed by Future Library from the Programme commencement until July 2016, when management was transferred to the Friends and Patrons of the SNFCC (F&P SNFCC), up until the completion of the Programme in March 2018. Implementation was supervised by a Management Board comprising, in the first phase, Dr. Filippos Tsimpoglou (Director General, NLG), Georgios Bokos (Professor Emeritus, Ionian University), Rob Davies (Independent Library Consultant) and Dimitris Protopsaltou (Future Library Director). The latter two were replaced in the second phase by Christos Skourlas (Professor, University of West Attica).

Following an open selection process, the authors of this paper were appointed as Action Managers, in order to implement the Transition Programme and to supervise the timely completion of the projects within the given budget. This Transition Task Force was in charge of strategic planning, process design, coordination of the development of technical/functional specifications and procurement procedures, management and monitoring of projects and coordination of all pertinent working teams consisting of NLG employees and external contractors.

Over a period of less than three years, the Transition Task Force coordinated the implementation of 50 projects for the NLG, working with no less than 150 Library staff and 167 contractors including universities, medium-sized businesses and large multinational companies, international and domestic publishers, ICT companies, creative agencies and individual experts. In total, 400 professionals were employed and worked in parallel with preparations for the historic relocation. The mixed model of collaboration between the SNF, the NLG and the two intermediaries (Future Library & F&P SNFCC) allowed for the necessary flexibility, with regards to project management, while ensuring transparency and accountability for all stakeholders involved.

2.3   Management Structure

To effectively bridge internal needs with external know-how within a coherent strategic framework, the Programme implementation employed a cross-sector collaboration structure, with internal decision-making processes designed from scratch. The Transition Task Force worked in tandem with NLG staff: a 3-membered Transition Team formed by the NLG Director General assumed a broader coordination role, handling the necessary communications with the SNF, the SNFCC S.A. and the NLG Supervisory Council. Its responsibilities included organizing meetings for monitoring project progress, and handling all team and staff assignments by the Director General. For each of the 5 actions, one NLG Liaison was appointed by the Director, working closely with the respective Action Manager, overseeing the implementation of all corresponding projects. A weekly Coordination Meeting with the Steering Committee was the central point of reference for monitoring the overall progress of the Programme.

Each project was overseen by a Project Team comprising NLG staff who worked directly with the respective Action Manager and the selected contractor's team. Due to the interdisciplinary nature of the projects, the NLG Project Teams comprised members of different Departments within the institution, fostering cross-departmental collaboration and generating input from diverse areas of expertise and practical experience. The Project Teams were responsible for conducting initial requirements analysis, drafting technical specifications and evaluation criteria for tenders, evaluating technical offers, proposing contractor selection, participating in regular meetings and providing data to the selected contractor, evaluating and approving deliverables. The following Table 1 depicts the structure of the Transition Programme.

During the years 2015–2018, over 30 Project Teams were formed within the NLG in order to oversee the 50 projects of the Transition Programme. This structure introduced the Library to an experimental, project-based organizational model which broke the silos between previously disconnected departments and enabled the NLG staff to perceive workflows and user journeys across services and collections, both online and offline. This organizational model ensured the completion of multiple projects in parallel timelines, achieving coordination at an unprecedented pace, scale and breadth.

Table 1. Transition programme structure

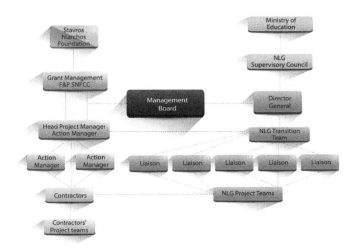

3 The Transition Programme Implementation

The Transition Programme implementation was organized in 5 separate actions:

1. Collections Transition and Development
2. Digital Services Development
3. Public Library Department Design
4. Audience Development
5. Staff Training.

3.1 Collections Transition and Development

Action 1 focused on preparing and re-organizing all collections for their transfer to the new premises ensuring safe relocation and minimal detriment, as well as permanent and efficient control and future inventorisation of all material. Three categories of projects were implemented, encompassing preparation, processing and development projects.

Preparation projects included the Rare Printed Books collection expansion, cleaning and conservation of the collections, as well as documentation and classification of archives. A specialized team went through the entire collection, book by book, following predefined scientific protocols, to identify and select hidden treasures that had remained neglected and part of the general collection. More than 12,000 volumes of valuable incunabula and rare printed books were selected and relocated to the new vaults at the SNFCC. Considering that the NLG collections had not been cleaned for almost 15 years, one entire year was dedicated to specialized cleaning using non-chemical materials. An experienced team of conservators followed a precise and systematic process for the surface cleaning of the collections. This process involved the use of specialised vacuum cleaners with mounted HEPA filters (Museum Vac), microfibre cloths and soft brushes. The stacks were also cleaned with no use of solvents or water [7]. The cleaning process ran through approximately 16 km of shelves, providing care to ca. 800,000 items. As part of the preparation projects, the institutional (in-house) archive of the NLG was classified according to the General International Standard Archival Description, ISAD(G) [8], comprising 40,000 documents and covering 143 years of operations[2], unveiling precious material about the institution's history. Finally, an extensive study was completed to determine specifications for the relocation process. This required granular mapping and measurement of the collections, detailed relocation planning according to the category and natural size of all material, shelf coding according to the building management system of the SNFCC in all relevant spaces, storage options outline, rare and fragile material handling outline, identification of multiple start-to-end transfer points through all stack levels at both buildings, ensuring accuracy and consistency of the transfer.

Processing projects implemented as part of Action 1 of the Transition Programme ensured that each item entering the new building had a unique ID, acquired through a 7-step specifically designed procedure. This comprised preventive conservation treatment for sensitive items prior to placing RFID tags, using specialized protective

---

[2] The project covered the period 1832–1975, since after that year the archive is considered active.

materials such as Melinex™ strips, cotton tapes and autoclaved bags[3]. This step was followed by placing RFID[4] tags on every item, choosing among three different positions; upper, lower or middle part of the inner book backboard, in order to facilitate simultaneous scanning for multiple items in the future. Subsequently, items were matched with their respective description in the integrated library system (ILS) using predefined searching criteria, and these descriptions were corrected when necessary, according to international library standards so as to improve the online catalogue access points and thus the end user's experience. Further steps included creating new bibliographic records for previously acquired material, in accordance with the Anglo-American Cataloguing Rules (AACR2), thus optimizing electronic access to the collection; classification followed the Dewey Decimal Classification (DDC) [9] and the Library of Congress Classification (LCC) [10] systems. Book covers or title pages for all material published prior to the year 2000 were digitized and the thumbnails were subsequently integrated in the online public access catalogue. The final step involved reclassification of material in the book stacks when necessary, so that the entire collection became ready for transfer. Within one year, preventive conservation was applied to 80,000 sensitive items, 720,000 RFID tags were placed, 528,000 items were identified in the ILS, 372,500 bibliographic records were corrected, 75,000 new bibliographic records were created, and 400,000 cover or title pages were digitised.

Furthermore, the Library's foreign-language collections were expanded for the first time after 13 years. 2,000 books and 500 e-books were acquired from the international market, focusing on the humanities and Hellenic studies. Moreover, a substantial investment on perpetual access in 11 electronic databases was completed, with an emphasis on Ancient Greece, the New Testament and Byzantine Studies.

A designated project was designed based on the findings of the Rare Printed Books collection expansion, aiming to select items of great historic and/or aesthetic value, which will constitute a special collection titled "The Treasures of the National Library of Greece"[5]. Each of the 252 selected documents was accompanied by a concise text, which provided brief information as to where, when and by whom it was created, and where, how and when it reached the NLG. This collection[6] will be used for multiple purposes (communication applications, permanent exhibition and training programmes) in order to highlight the value of the NLG material, to attract special audiences and to meet specialized educational needs.

---

[3] The decision to apply these materials was based on similar practices in other libraries and was made under the basic principle that all interventions should remain reversible and in particular that the RFID tag should be secure but in no direct contact with the book.

[4] Radio Frequency IDentification (RFID) is the latest technology used in libraries; unlike Electro-Mechanical (EM) and Radio Frequency (RF) systems, which have been used in libraries for decades, RFID-based systems move beyond security and theft detection systems to become and also provide efficient material tracking throughout the library, including easier and faster charge and discharge, inventorying, general handling and user statistics.

[5] A similar project was completed at the NLG in 1999 and resulted in an exhibition and a relevant publication; however since important material was unveiled it was considered appropriate for the concept of the treasures to be revisited.

[6] This selection can by no means be viewed as definitive since the search and recording of items will continue and the NLG collections will be constantly enriched.

It was the first time in the Library's history that all these projects were completed with each material item on-hand, posing several challenges and a high level of complexity. On a daily basis, more than 5,000 items were moving within two buildings, thus requiring continuous workflows and appropriately experienced work-teams, ensuring minimal invasion and damage of the collections. The often inappropriate weather conditions along with the restrictions of the existing buildings posed substantial risks and demanded methodical and standardized processes with the ability to immediately respond to changing schedules. Time was limited thus requiring very close production monitoring and constant troubleshooting on the spot. Action 1 comprised the most labour-intensive projects involving more than 130 external professionals and 30 Library staff from 11 different disciplines and areas of expertise, thus requiring flexible production chains and efficient teams integration, in order to minimize risks for the collections and optimize the preparation for the historic relocation of the NLG collections.

3.2   Digital Services Development

The vision for Action 2 was to design and develop the first version of an integrated electronic infrastructure of national importance that could deliver the new services mandated by the law and support the enhanced Library operations at the new building.

At the beginning of the Programme, a few systems were in place: an external company had developed and had been maintaining the ISBN/ISSN/ISMN service, the Koha Integrated Library System was used for the cataloguing of print material and VuFind was used for searching the Library collections. The old website was not supported for a long period of time. The IT department comprised three staff members, none of which had software development skills. Users' handling was done offline on printed cards and overall administration relied on heavy paperwork. The Library's relocation to the new building demanded an upgrade of the existing digital services, the development of new services for end-users from scratch, the modernisation and expansion of digital storage and networks, as well as staff training, in order to be able to support the enhanced functions at the new building.

All developed services needed to be integrated into a state-of-the-art, modular, scalable and secure electronic infrastructure. In this context, ensuring sustainability beyond the life of the individual projects was a major issue which was addressed early in the Programme, in the following ways:

- Open source solutions were adopted wherever possible, as well as open data and open content policies, in order to lower costs, to avoid vendor lock-in and to allow for maximum re-usability of data and content. Wherever possible, ownership of the developed software and data was ensured by the Library through contractual arrangements. Strategic partnerships were forged with individual partners and contractors were selected on the basis of their unique expertise and capacity to provide cost-efficient and customised support in the years to come.
- International standards and best practices in the library and information science domains were investigated extensively and adopted in order to provide future-proof solutions. A data-centric approach was embedded in the design of the systems

architecture, taking into account modern developments like OCRing, text and data mining, semantic web and artificial intelligence. As the Library moves into web harvesting, digitization of legacy collections and digital legal deposit, data will become increasingly important. Therefore, the architecture catered for the data being independently stored, standardised, interlinked, processable, reusable and openly available through multiple channels to the extent possible.

- Detailed technical and functional specifications were developed for each service following a multidisciplinary approach which involved analysis and integration of the technological, business, bibliographic, communication, UX and legal aspects of each service. This allowed for close monitoring of project implementation by cross-departmental teams which were gradually formed in the course of the project. Consequently, this resulted into enhanced ownership of projects' results by the Library staff and important transfer of knowledge.

In less than 2.5 years, in the framework of Action 2, 16 projects were designed and implemented, involving 50 contractors, 110 external collaborators and 30 Library staff. The average development time frame for each digital services project was 6 months and the average maintenance and support contract was pre-paid by the grant for 2 years. This meant that numerous projects run in parallel.

Action 2 Digital Services projects are divided in 4 main groups:

Main Library Operations. These projects included the upgrading of existing software and services that underpin the core library operations, namely the handling of the print and digital collections. Services which underpin the management of the lifecycle of print collections, such as the Koha ILS, the ISBN/ISSN/ISMN and Vufind, were upgraded to include enhanced functionality such as material circulation and lending functions. The new Digital Collections Platform was designed and developed to handle the whole digital content lifecycle from digitization to online delivery, using the IIIF image delivery standard. Ebsco's Electronic Discovery Service (EDS) was acquired to provide indexing and an integrated point of access to the entire Library's distributed catalogues and electronic resources and finally, the BiblioConserv CDS software was acquired and extensively customized, in order to meet the specific needs of the Conservation department.

Research and Development (R&D). These projects aimed at piloting innovative services related to the Library's core mission. The first version of the National Web Archive Infrastructure was developed using the widely adopted open source components such as Heritrix for harvesting, Solr for indexing and the Wayback Machine for searching and for webpage reconstruction and display. The initial crawls produced already the biggest available corpus in Greek, which exists in a machine readable form to date and there is already a lot of interest to develop linguistic tools to be further used in machine learning like stemmers, stop words, etc. [11]. The second R&D project regarded the development of an e-Lending service for trade e-books for the first time in the country, in order to promote e-reading as this market is still in its infancy in Greece. Fourteen publishers provided 2,899 electronic copies allowing users to borrow a book in the same way one can borrow a print book by downloading it on their phone or tablet, for a limited period of time [12]. Finally, the Greek Libraries Union Catalogue was developed, which uses participatory cataloguing practices and advanced algorithmic techniques to produce "master" records which can then be used for copy-cataloguing, thus, significantly lowering costs of cataloguing and creating economies of scale among participating libraries.

Enterprise Operations. These projects aimed at modernizing and automating the Library's administrative operations which lie with the Secretariat and the Protocol Departments. A commercial HRMS SaaS system was acquired and customised to correspond to the needs of the Library, the employees' archive and part of the Protocol archives were digitised and the data were inserted in the System. The use of the e-Protocol system was generalised to serve as the main Document Management System which helped significantly improve automation and control of the enterprise operations. One important aspect of the HRMS is that it acts as the authentication source for single-sign-on of the library staff and automatically assigns roles for them to the different services. Additionally, the new Website Platform was developed to provide a one-stop-shop for the user regarding the Library's collections, events and services. Furthermore, a commercial SaaS payment gateway solution was acquired and adapted to interoperate with all digital services that involve online financial transactions such as the ISBN service, Koha, etc. Last but not least, a horizontal User Identity Management Infrastructure (UIM) was developed allowing users to be registered once and to have access to all the digital services via a single sign on service. The infrastructure implements an innovative authentication process as it automatically verifies the user data with two important public registries- the tax registry for physical persons and social security registry and it has been presented as a Greek public sector digital transformation best practice case in e-government fora [13]. All the aforementioned digital services interoperate and depend on the UIM for the authentication of their users.

Digital Storage, Networks and Equipment. These projects aimed, on one hand, at remediating the lack of appropriate datacenter at the Library and, on the other, at complementing the state-of-the-art digitisation infrastructure at SNFCC with the acquisition of 3 high-end archival and end-user digitisation units.

Additionally, specialized legal services and IT expertise was provided by individual experts while several projects in Action 4 (Audience Development) framed the development of the digital services and in particular the design of the new visual identity and copyrighting. The more or less consistent visual design, the integrated workflows and the single sign-on service provide a unified navigation experience from the user's point of view, through the different digital services.

Finally, the specifications for the Library's Integrated Collections Management System were designed following a thorough analysis and documentation exercise that involved all Library departments with the aim to systematize, integrate and align all workflows related to the complete lifecycle of print, electronic and digital items in a single integrated infrastructure that would also include digital legal deposit and OAIS-compliant digital preservation requirements, providing the basis for the development of the Library's infrastructure v2.0. However, with an average timeframe of development of 6 months for the Transition projects, the project was finally procured via the State Budget [14].

In terms of development, most services had been completed in a beta form by the end of December 2017, however, there were several pending issues including stress and security testing, interoperability testing with other services, finalisation of communication texts and translations while at the same time, the Library's actual move was starting in January 2018. The services were going to gradually launch in the course of 2018 as the Library was preparing to open also its physical spaces to the public.

We can conclude that thanks to the Transition Programme, the Library established the first instance of an integrated, modular and secure electronic infrastructure that can serve diverse needs and can easily scale in the future to incorporate new developments and needs. Digital services and infrastructures can only continue to develop and expand as the actual use will come in place, digitisation will be systematised and new user needs will rise. Following the operationalisation of the services, version 2.0 of the infrastructure should cater for the development of value-added services for targeted communities such as the research community, the visually impaired, etc.

3.3   Public Library Department Design

Action 3 focused on the development of the Public Library Department. The operation of a Public Library had been embedded in the architectural approach of the SNFCC from the very beginning; however, its organic integration as a new department according to the upgraded institutional role of the NLG was far from secured. In the context of Action 3, service design aimed at converting the ground floor into the main reception area and access point for all NLG services, and defining a development strategy that would gradually transform the 5,000 m^2 open space into a living hub of exploration, experimentation, learning and creativity. The respective study was based on an analysis of international best practices, intending to develop a quality lending collection, design modern services for diverse user profiles, use and adapt existing constructional and technological infrastructure to further facilitate the user journey and provide innovative programming options. The study maintained the separation of the ground floor in three main areas serving children, teenagers and adults respectively, while defining four concentric circles of activity around the iconic Book Castle, encompassing relevant services: a research zone (book stacks), a quiet zone (reading rooms), a discovery zone (open activity spaces) and a creation zone (closed studios/workshops spaces/makerspace). In addition, the study provided an Operations Plan for the new department, suggesting specifications for appropriate staffing, guidelines for the management of collections and programmes, optimal internal workflows and an Internal Policies & Regulations draft, to be finalized by the NLG following its pilot operation at the new premises.

The Public Library Department required a new, diverse and attractive collection. To this end, a participatory approach was adopted, appointing 21 experts from diverse fields to select the best books according to a common set of criteria and to submit comprehensive lists that were subsequently used to procure 15,000 books from the domestic and international book market. Another 25,000 items were hand-picked by this interdisciplinary team of experts from the NLG collection of second copies at the Votanikos building which remained unused. The lending collection thus comprises quality and awarded publications in the following categories: children literature, adult literature, philosophy and psychology, religion, social sciences, natural sciences, technology, arts, general reference books, geography, history and local history; 30% of the collection is in non-Greek languages (English, French, German) aiming to address the international visitors of the SNFCC.

To date, more than 30,000 items have been bibliographically processed by an external team of 17 experienced librarians and placed on the shelves following the Dewey Decimal Classification System (DDC), while the collection has been enriched with educational materials, board games and video games, subscriptions of Greek and foreign language magazines. In the framework of Action 3, four educational programs for children and teenagers were designed in order for the NLG to receive regular school visits, and a comprehensive annual programming strategy was developed in order to provide unconventional series of workshops, exhibitions, lectures etc. that promote literacy, life-long learning and cultural creation.

Furthermore, a series of communication applications were designed to differentiate the Public Library Department as a sub-brand among the traditional umbrella of NLG services, suggesting playful aesthetics to its signage system, digital presence and print material. Despite the lack of sufficient and experienced staff for the operation of the Public Library Department, the external know-how provided during the Transition Programme enabled a small team within the NLG to become familiarized with its needs and coordinate its pilot operation since July 2018.

Taking into account the importance of the Public Library space as the first touchpoint of all NLG users and especially new audiences at the SNFCC, appropriate staffing, capitalization of established partnerships and long-term funding of programming activities remain a prerequisite in order to build upon the studies and projects completed during the Transition Programme, ensuring its gradual development into a vibrant public space and community hub providing equal, free, open access to knowledge and entertainment to all.

3.4 Audience Development

Action 4 aimed at broadening the outreach of the NLG at local, national and international level, overcoming the introvert and exclusive character of its services in past decades, and embracing the global challenge of libraries to reinvent their role in society. The new facilities offer the possibility to serve a larger number of citizens and to provide services that reach new user groups (children, teenagers, families, tourists, minority groups), thus enhancing its institutional mission and social impact. Considering the lack of experienced staff or relevant departments related to Communications and Engagement within the NLG, it became evident that the institution was in need of the development of a comprehensive Communication Strategy from scratch, which would align and promote its upgraded services, while radically redefining its relationship with the Greek society and the international research community. Design thinking methodologies [15] and participatory leadership approaches were adopted throughout the implementation of Action 4, taking into account the growing importance of libraries as catalysts of social innovation and enablers of community action [16]. The five stages of Design Thinking were used throughout the implementation of Action 4, enabling multiple stakeholders to empathise, define, ideate, prototype and test new ideas and practices in an open, experimental framework. This strategic approach to NLG Communications surfaced key challenges and solutions regarding the identity and mission of the NLG, which informed ongoing projects across all 5 Actions and enhanced user-centered design throughout the Transition Programme implementation.

The Communication Strategy was developed over two years, ensuring that the NLG would acquire the necessary tools, know-how and partnerships for the establishment of a Communications Department and the development of a consistent outreach strategy once operating at the SNFCC. The process was initiated with an extensive Audience Research, which engaged over 1500 citizens in defining needs, aspirations and ideas for the future of the National Library of Greece through quantitative and qualitative research. Design thinking labs with key stakeholders, questionnaires with NLG staff, in-depth interviews with leading thinkers from the library sector, surveys and focus groups with diverse user groups were some of the tools used to harness human potential and distill a Strategic Brief for the NLG. This document served as the basis of an open, nation-wide design competition engaging Greece's creative community to reconsider the visual identity of a leading national cultural institution. This open ideation process, the first of its kind by a Greek Public Sector organization, was completed in three phases; 40 design studios submitted portfolios in the first phase, and 5 creative teams were selected in the second phase, presenting publicly 10 proposals for the NLG brand at the final stage. The winning team questioned the need for a static logo and instead developed a comprehensive visual language, which focuses on typography as imagery valorising the Greek and Latin alphabet, tackles content as form and defines the Library as a "intraliminal" space; a place where the past meets the future, the analogue meets the digital and the national meets the global. After a 12-month experimentation period with the NLG content, producing pilot print and digital communication applications, the design team delivered an open Brand Manual which serves as the basis for the NLG Communication Strategy. According to it, future designers of the NLG become authors of a visual, perpetual story written in a recognisable, unconventional visual language, and NLG communications focus on the institution's gradual transformation from a collection of books to a community of creators, researchers and explorers. The new visual identity of the NLG won the Golden Prize at the Greek Design Awards, creating a precedent for a radical redefinition of the image and outreach strategy of public institutions.

As part of Action 4, the Transition Programme enabled the establishment of a Press Office and the training of an internal Communications Team, the setting up of multiple communication channels and user databases, as well as the development of creative content (copyrighting, translations, photography, videos), all of which are essential for the upcoming establishment of a Communications Department and the launch of the initial communication campaigns.

Considering the scale of the ongoing transformation within the organisation, the development of the NLG Communication Strategy inevitably led to strategic questions on service design, requiring redefinition of the 5Ws of Communication (who – what – where - when - why) for every single service, with an additional emphasis on the "how". A series of service design workshops and simulation exercises was thus organised, engaging NLG staff to map user journeys, identify touchpoints in the physical and digital realm, align processes across different departments and provide relevant content for diverse communication channels. This process led to the development of an MVP study for the pilot operation of the NLG at the SNFCC, which linked outputs across all actions of the Transition Programme and prioritised implementation-ready services.

Beyond the development of the Communication Strategy, a key success for the Audience Development of the NLG was the establishment of the Greek Libraries Network (GNL) [17]. Building on previous work developed by Future Library, the NLG took over the coordination of the network, providing to its members efficient networking and collaboration tools, regular training and nation-wide programmes for children and teenagers. To date, the NLG brings together 220 public and municipal libraries, in a joint effort to upgrade the services offered to local communities and to create a nation-wide programming framework for the promotion of reading, creativity and civic engagement. Since 2015, the NLG coordinates the iconic Summer Reading Campaign for children aged 4–14, designing a series of creative workshops which attract over 60.000 children across the country on a yearly basis. Centralised design and content development, continuous online support of librarians and integrated communication campaigns enable a multiplier effect that converts the NLG premises at SNFCC as a physical hub of a vibrant community of librarians, educators, volunteers and young readers.

This community was developed further through the pilot events' programming that was produced during the Transition Programme. By establishing strategic alliances, the NLG undertook the organization of pilot events in the new premises in anticipation of the official inauguration, including scientific conferences, guided tours, original educational activities and literary events which left its own unique cultural mark within the wider SNFCC complex. The public's response to the above pilot initiatives exceeded every expectation, making it probably one of the first libraries to be filled with people before it was filled with books.

3.5 Staff Training

Action 5 focused on preparing Library employees in order to efficiently respond to the increased and multilayered operational requirements at the new premises. As previously mentioned, there was only 100 staff when the Transition Programme commenced; 26 new employees were hired in spring 2016, all of them educated to postgraduate level, which substantially increased capacity.

"Any management's ability to achieve maximum benefits from change depends in part on how effectively they create and maintain a climate that minimizes resistant behavior and encourages acceptance and support" [18]. In order to facilitate organizational change, transition workshops were organized on a monthly basis, giving the opportunity to expose problems, address interpersonal issues, discuss ongoing projects, propose solutions and improve alignment around goals. Furthermore, weekly mutual learning sessions were organized featuring presentations by different contractors and partners in order to keep all staff sufficiently informed about projects' implementation, outcomes and overall progress, to familiarise them with the new services under development and provide a platform where they could express questions and interest in joining specific Project Teams.

The Transition Programme implementation itself turned out to serve as a "learning-by-doing" tool, an invaluable training process for the NLG staff and for all external professionals involved since it offered a unique opportunity of exposure to interdisciplinary collaboration and co-creation processes. Taking into account that the programme required an organizational change which diverged from the pre-existing institutional status quo, particular focus was given in training and educating change agents within the NLG that could "distance themselves from their existing institutions and persuade other organization members to adopt practices that not only were new, but also broke with the norms of their institutional environment" [19]. Throughout programme implementation, early adopters with the adequate hard and soft skills were identified within the organisation, in order to diffuse innovations [20] in terms of process and service design across the NLG system.

This approach strengthened cross-sectional communication and increased direct involvement of all NLG employees participating in the individual project groups, resulting in them gradually gaining increased ownership of the Transition Programme. In an effort to respond to the ever-increasing future operational requirements, a general mapping of immediate and future training needs of the staff was completed, in order for the Library to identify relevant national and international conferences worth attending, to schedule field trips related to the new services and launch targeted training programs.

Upon completion of the Transition Programme, it becomes evident that continuous training, empowerment of change agents, capacity building across departments and team building are of the utmost importance in order for the NLG to capitalize on the programme's outcomes and successfully integrate future additional human resources.

4   Process Design for Organisational Transformation

The NLG Transition Programme implementation proved to be a highly demanding experiment in transition management using private funding to transform a public organization. With regards to the process design the Transition Task Force adopted a human-centred approach combining organisational change principles and a strong community engagement component, underpinned by the overall digital transformation of the organisation. User needs were put at the core of the service design process and participatory leadership was tested in practice, in an attempt to update the definition of "public services" within the NLG. Pulling together a constellation of partners and experts from multiple disciplines mobilised by a noble cause and a shared purpose, the Transition Programme enabled participating actors to embrace complexity, unleash creativity and co-create new possibilities. In this framework, the adoption of a lean management methodology was essential in order to successfully tackle unpredictable parameters arising throughout implementation. A visionary leadership on behalf of the NLG and a supportive funding mechanism on behalf of the SNF significantly enhanced the potential for systemic transformation.

As there was no exact precedent in the collaboration between a public organisation, a private donor and a non-profit as the intermediary for a grant of this size, the management process was designed in a way that would ensure transparency, accountability and the least bureaucracy possible, taking into consideration an extremely tight timeline. Process design was based on international best practices in the library domain, using Greek public tendering guidelines that ensure the protection of public interest, while at the same time exploiting the flexibility provided by private procedures.

Financial and project monitoring involved the following steps for most of the projects: The Transition Task Force performed extensive needs analysis within the NLG, delivered detailed technical and functional specifications, followed by thorough market research performed for each tender, in order to attract the most suitable possible candidates. Tender specifications and technical offers were formally evaluated by the NLG Project teams; the selected contractors were approved by the Management Board and subsequently endorsed by the donor (SNF).

All project progress was monitored closely, to ensure smooth collaboration between the Library and the contractors and transfer of knowledge between external experts and Library staff. Deliverables were submitted by the Action Managers to the relevant Project Team through the NLG Protocol, with the latter reviewing and providing formal approval for the grant manager (F&P SNFCC) to proceed with invoicing and payment. Final control of all monthly-issued invoices was undertaken by the donor (SNF). The aforementioned safeguards were essential to securing quality assurance throughout Programme implementation.

Despite the initial resistance to change, the Transition Task Force succeeded in creating a productive framework of cross-sector collaboration, testing in practice participatory leadership methods that fostered relationships of trust and establishing the use of co-working tools throughout Programme implementation. Besides daily presence and regular meetings with staff across NLG buildings, online project management and communication tools (basecamp, google docs and skype) were used to enable remote and collective work between the contractors and the library staff and to speed up decision-making processes. Taking into consideration that such tools had never been used in the NLG at organisational level, the contribution of the Transition Programme in upgrading internal processes and building project management capacity is of paramount importance.

5   Conclusions and Future Challenges

The Transition Programme brought the National Library of Greece at the threshold of the SNFCC and concluded at the time when the actual relocation of the collections and the staff started in spring 2018. The immediate successes of the Programme are obvious: Collections are "in order", technical infrastructures are expanded, staff skills are enhanced, a new bold image for the Library has been carved and a new set of services and programs for the public have been developed. The financial resources provided by the SNF empowered the NLG with the necessary human resources, advanced tools and expert know-how, to make the leap from past to future, from analogue to digital, from national to global, paving the way for re-establishing relevance, increasing reach and securing resilience.

The Transition Programme, undoubtedly, catalyzed culture change and opened up an unprecedented amount of possibilities for the Library, triggering a substantial increase in the state budget allocated to the NLG, and attracting numerous collaboration offers from major stakeholders. These include opportunities to participate in national and European projects, organize scientific workshops, host presentations and art exhibitions, receive donations of private archives and collections, train interns and co-organise first class international events such as the UNESCO Book Capital festivities and the 2019 World Library and Information Congress of IFLA.

Upon Programme completion, the Transition Task Force provided the NLG leadership with a strategic framework for further fundraising and organisational development, focusing on three main pillars:

1. The development of the NLG as a global Center of Excellence for Hellenic Studies
2. The establishment of the NLG as the leading library of the country and a Competence Center for librarianship and information science
3. The cultivation of an open community for life-long learning and cultural creation.

Nevertheless, completion of the Transition Programme funding raises several open-ended issues, regarding operationalization of new services developed and sustainability of results. The NLG is currently faced with the immediate challenges of pilot operation in the new building: user testing, beta operation and further adaptation of workflows are required before the services can be launched in the new premises and online. The development of a comprehensive Strategic and Business Plan for the NLG constitutes a prerequisite in order to seize the momentum and build on established strategic alliances.

The Transition Programme raised the bar, requiring increased human and financial resources for the Library operation. A significant raise in staff numbers and expertise across multiple disciplines, and the establishment of a new organogram encompassing new departments will be needed for the Library to go into full operation[7]. Increased administrative capacity within the NLG will be required to manage diverse revenue streams via participation in EU and national programmes, event sponsorships, private donations and commercial activities, making the most of its collections and built infrastructures, e.g. ensuring the restoration and strategic exploitation of the Vallianeio and the Votanikos buildings to overcome SNFCC limitations.

In conclusion, the extent to which the potential generated through the Transition Programme 2015–2018 will be unleashed for the benefit of the Greek society and the international community remains to be seen. Yet this case study has undoubtedly contributed to the field of cultural management, providing new insights into how to harness human potential from public, private and civil society actors to drive growth. At a time of inefficient top-down reform efforts within the Greek State and increased polarization between the public vs. private narrative, the NLG Transition Programme serves as a unique case study for a cross-sector collaboration based on a Public-Private-People partnership model. Although this model still requires long-term testing, further research, analysis and impact assessment, the outcomes of the NLG Transition Programme demonstrated a window of opportunity worth exploring further.

---

[7] The agreement of the SNF with the Greek State (Law 3785/2009) set the minimum number of employees at the beginning of operations at 286; however staff in 2018 is only 140, whereas existing constraints of the public service hiring procedures do not guarantee timely and sufficient manning of all established and foreseen departments.

Acknowledgements. First and foremost, we would like to thank Dr. Filippos Tsimpoglou, NLG Director General, for his trust, constant support and excellent collaboration throughout the Transition Programme implementation. We would also like to thank Stavros Zoumboulakis, President of the NLG Supervisory Council, for his valuable advice and the members of the Transition Programme Management Board, particularly Professors George Bokos and Christos Skourlas for their valuable insights and continuous help.

Special thanks to the NLG Transition Team (Evi Stefani, Chrysanthi Vasiliadou, Varvara Moula) and the NLG Liaisons (Vasiliki Tsigouni, Panagiotis Paloukos, Eirini Pavlakou, Grigoris Chrysostomidis, Georgios Parlavantzas) for their important contribution to programme implementation. Additionally, we thank all NLG Project Teams and all NLG staff for embracing the vision for a new National Library.

We would also like to thank Eleni Liveriadou (Chief Operations Officer, F&P SNFCC) for her excellent collaboration and substantial contribution to the grant management.

Last but not least we would like to thank the Stavros Niarchos Foundation for making the Transition Programme possible and especially Eva Polyzogopoulou (Assistant Director of Programs & Operations, SNF) for her valuable contribution to overall implementation.

References

1. The Economist. https://www.economist.com/prospero/2018/05/14/libraries-that-speak-loudly?fsrc=scn/fb/te/bl/ed/librariesthatspeakloudlyshelfawareness. Accessed 18 Sept 2018
2. The Guardian. https://www.theguardian.com/travel/2017/jul/05/athens-cultural-centre-stavros-niarchos-renzo-piano-design. Accessed 18 Sept 2018
3. Laskaris, K.: Epitomi ton okto tou logou meron kai allon tinon anagkaion. Damilas and Paravicino, Milan (1476)
4. The International ISBN Agency Homepage. https://www.isbn-international.org/. Accessed 17 Sept 2018
5. Hellenic Parliament, Law 4452 (2017). https://www.e-nomothesia.gr/kat-ekpaideuse/nomos-4452-2017-phek-17a15-2-2017.html. Accessed 17 Sept 2018
6. Hellenic Academic Libraries Link Homepage. https://www.heal-link.gr/. Accessed 17 Sept 2018
7. Gkinni, Z., Tsaroucha, C., Sarris, N.: The National Library of Greece: Moving into a New Era, IFLA WLIC 2017 Libraries. Solidarity. Society. http://library.ifla.org/1654/1/170-gkinni-en.pdf. Accessed 17 Sept 2018
8. The International Council on Archives, the International Standard Archival Description. https://www.ica.org/en/isadg-general-international-standard-archival-description-second-edition. Accessed 17 Sept 2018
9. The OCLC Dewey Homepage. https://www.oclc.org/en/dewey.html. Accessed 19 Sept 2018
10. The Library of Congress Classification. https://www.loc.gov/catdir/cpso/lcc.html. Accessed 19 Sept 2018
11. Presentation of the Web Archive Project by the Athens University of Economic and Business at Naftemporiki Newspaper. https://www.naftemporiki.gr/story/1347787/protopora-prospatheia-glossologikon-poron-gia-tin-elliniki. Accessed 17 Sept 2018
12. Kathimerini Newspaper. http://www.kathimerini.gr/962504/article/epikairothta/ellada/hlektroniko-anagnwsthrio-apo-thn-e8nikh-vivlio8hkh-2500-titloi-sto-diadiktyo. Accessed 17 Sept 2018

13. Project Presentation by Dr. Eftihia Vraimaki, 9th OTS e-government Forum, 17 November 2017. https://www.youtube.com/watch?v=Faa0e4-rGt4. Accessed 17 Sept 2018

14. Public Tender on the Integrated Collections Management System. https://diavgeia.gov.gr/decision/view/%CE%A17%CE%9F%CE%9646%CE%A8%CE%962%CE%9C-%CE%9F%CE%A41. Accessed 18 Sept 2018

15. Design Thinking for Libraries. http://designthinkingforlibraries.com. Accessed 18 Sept 2018

16. De Moor, A., Van Den Assen, R.: Public libraries as social innovation catalysts. In: CONFERENCE 2013, Prato Community Informatics Research Network (CIRN) C, Monash Centre, Prato, Italy. https://communitysense.files.wordpress.com/2013/10/cirn2013_de-moor-van-den-assem_publ.pdf. Accessed 18 Sept 2018

17. Greek Libraries Network Homepage. https://network.nlg.gr/. Accessed 17 Sept 2018

18. Judson, A.: Changing behavior in Organizations: Minimizing Resistance to Change. Blackwell Publishings, Cambridge (1991)

19. Battilana, J., Casciaro, T.: Change agents, networks, and institutions: contingency theory of organizational achange. Acad. Manag. J. **55**(2), 381–398 (2012). https://doi.org/10.5465/amj.2009.0891

20. Hoffman, V., Christinck, A., Probst, K.: Farmers and researchers: how can collaborative advantages be created in participatory research and technology development? Agric. Hum. Values **24**(3), 355–368 (2007). https://doi.org/10.1007/s10460-007-9072-2

Volos in the Middle Ages: A Proposal for the Rescue of a Cultural Heritage

Konstantia Triantafyllopoulou[✉] [iD]

University of Thessaly, Volos, Greece
nadtriadafil@yahoo.gr

Abstract. In this paper the history of the 'unknown' Medieval Volos will be studied and an attempt will be made to promote the cultural heritage of the Byzantine era. The effort will be completed with the proposal to create a thematic museum in the area of the Castle of Palea and a cultural route through the points of interest we have identified. The aim of the survey is to create a tourist attraction and thus to improve cultural tourism. Furthermore, knowing the history of our region means to know the evolution followed by the society. This is important because many people ignore the history of this period. For the region of Volos we have a lot of historical evidence for earlier historical periods, such as Neolithic and Classical. However, it remains unknown what happened in the region during the medieval period from the 4th to the 15th century AD.

Keywords: Middle Ages · Castle of Palea · Cultural tourism · Thematic museum · Volos

1 Introduction

Cultural heritage is important in understanding the past and preserving the identity in the future. Archaeological sites and monuments provide information on the cultures, customs and beliefs of the past. They help us understand who we are and where we can go. If these sites are destroyed, contextual data is lost [1]. Tourism and Culture are interrelated phenomena and the desire of tourists to get acquainted with the specific features of the place they visit is intertwined with their will to travel for touristic purposes. The particular cultural heritage of each region is the basis for further development and is an element of tourist attraction. A prerequisite for the economic development of the touristic area is the proper management to obtain value as a place to visit [2]. It is important for national identities and cultural differences to attract the attention of other cultures by creating investment, developing the economy and enhancing the identity and image of a country [3].

Through this work, an attempt is made to collect historical data concerning Volos and the wider region in the obscure medieval period (4th–15th century AD). Although research and excavations have taken place, the history of the place of this era is not known to the general public. The purpose of this study is to highlight the medieval history of the region and to use it for the development of Cultural Tourism which will contribute to the economic development of the area in modern ways. At the same time, the inhabitants of the area will become aware of their own history.

© Springer Nature Switzerland AG 2019
A. Moropoulou et al. (Eds.): TMM_CH 2018, CCIS 961, pp. 203–221, 2019.
https://doi.org/10.1007/978-3-030-12957-6_14

The chosen methodology was the following: the first part of the research is a review of the international literature and commentary on secondary data such as maps and designs. The second part demonstrates the proposal of creating a thematic museum which is accompanied by a marketing plan where the strategy for its implementation is presented.

2 Historical Information

2.1 Early Period - Paleochristian Demetriada

Due to its favorable geographic location, Magnesia has created important cities and harbors during the Early Christian and Byzantine period. The ancient settlements that are preserved in the same place in the Early Byzantine period are Demetriada, Iolkos and Fthiotides Thives [4, 5].

In the location of ancient Demetriada, in Pefkakia, the city continues its existence, and becomes an Early Christian city where a number of basilica churches was built [6–8]. The existence of two ports used to serve the transit trade within the Thessalian area as well as with ports far from Greece [9]. The city was divided into two sections from which its center was in the northern port. There, and to the northeast of the Ancient Theater in the excavations of 1912, by Arvanitopoulos and in 1969-1972 by Milojcic a 4th-5th century Early Christian church was discovered under the name Basilica A' of Damocratia and it is the most important monument of the early Byzantine era in Thessaly [4, 9–12]. A secular building with an arch and mosaic floor that is supposed to be the house of Damocratia was revealed in the West side of the basilica A' [5].

The second part of the city was the southern harbor at the foot of Prophet Elias, in the area of today's New Pagasais [11, 12]. There, the excavations of Theocharis and Milojcic from 1961 to 1972 brought to light the Basilica B' Cemetery church, a 4th century three-aisled with narthexes and frescoes, which is the oldest Christian temple in Thessaly [4, 6, 9, 12].

In 520 AD Demetriada moved to the hill of the Palea area some 2 km away and was fortified in the attempt of the Byzantine Emperor Justinian to protect cities who were strategically located from the catastrophic invasions of the barbarians Goths and Huns [13, 14].

In the pre-existing position of Demetriada, the Velegizite Slavs lived in the 7th century AD, having appeared in the 6th century AD in central and southern Balkan Greece. The lack of border guarding both onshore and sea by the Byzantine army made it easy to settle in the area [9, 14, 15]. According to St. Demetrius' chronicle, the Slavs arrived from the land and from the sea by single-handed ships carved on tree trunks [16].[1] The preservation of Demetriada at the same location was a precondition for safety

---

[1] The city, as described by Kordatos [17], after the destruction of the barbarian genders, did not remind the once-glorious Demetriada. However, as Avramea points out, the city of Demetriada, unlike Phthiotides Thives, was neither occupied nor besieged by the Slavs, who seemed to live there peacefully and were engaged with agriculture [9].

from both the land and the sea; on the contrary, the hill of Palea was a safer area [13, 18]. In addition, excavations did not reveal any new building activity in ancient Demetriada beyond the 6th century AD, nor the repair of its walls [13].

The great technical work of the aqueduct was constructed in the 4th century AD and was used throughout the medieval life of the city [11, 19, 20].

Early Christian Fthiotides Thives (Nea Anchialos)

To the south coast of Pagasitikos Gulf, in Homeric Pyrasos, which was renamed Thives, today's Nea Anchialos is one of the largest centers of the Early Byzantine era, the magnificent harbor town of Magnesia, Fthiotides Thives [7]. The excavations that took place in Fthiotides Thives in 1923 by G. Sotiriou, continued in 1959 by Lazaridis and revealed the flourishing of the region during the Roman and Early Christian years. The findings of the excavations give us a complete picture of the Early Christian settlement since the excavations brought to light seven basilica churches of Early Christianity [5]. Part of the architectural decoration was brought directly from Constantinople [21]. The earthquakes of the 6th century destructed the thriving city of Fthiotides Thives with the elaborate basilicas. The region, after its extinction in the 6th-7th century AD, never developed again, nor was referred to any source [5, 15].

2.2   Middle Period – the Castle of Volos

The Castle of Volos is known as the Castle of Palea. The hill of Palea has not stopped been inhabited since 3000 BC [14, 22]. As mentioned, during the 6th century AD the hill was fortified by Justinian and turned into a fortress. The inhabitants of Demetriada moved to the hill of Palea. The settlement inside the powerful Castle, although no longer big in size, was powerful, and preserved its existence throughout the Byzantine period until the Ottoman period [7, 9, 23–25].

Early Christian Walls

The Castle had a quadrilateral shape and dimensions 190 m in the south and 320 m in the west. The surface of the Castle covered sixty acres, consisted of four sides and the perimeter of its walls reached one kilometer. Northeast of the wall there was the citadel that separated the wall from the city, while there was an opening gate for communication. The walls at the northern points were accompanied by quadrilateral towers, and in the southern points where better protection was required, the towers were circular in shape. In the fortress there was a trench of 15 m wide. The width of the walls with additions over the centuries reached about three meters [7, 14]. Stone monuments and statues of ancient Demetriada of the Hellenistic period were used as materials for the fortification of the Castle of Palea [26].

The history of Demetriada in the castle of Palea over the centuries is depicted on its walls with repairs and renovations by the owners and conquerors. During the founding of Agioi Theodoroi church in 1889, an early Christian basilica church showed up. The excavations that took place in 1973-1974, brought to light the mosaic floor and architectural members of the same church appeared on the south side of the church of Agioi Theodoroi [9, 14, 25, 27].

Two baths from the Early Christian era came to light from the excavations. The first one is located on Souliou Street and dates back to the 4th century and the second to the 6th century a part of which is visible from the actual Louli's complex [14].

At the same time that the Castle was fortified, in the 6th century there were also created signage stations - towers with permanent guard for the transmission of bright passwords for the timely information of the inhabitants about any dangers, relayed signals were transmitted at the next stations, their remains are now found on Pelion shores (Pyrophoric vigles -fryktories) [9].

After the earthquakes of the 6th century and the invasions of the Slavs who hit the area, the Saracen pirates appeared.[2] The year 896 AD seems to be the most likely year in which Demetriada falls into the hands of the Saracens after a lands siege raid [25, 28, 29]. It is speculated that for the conquest of the fortified city, it became a wall battle [4, 7].[3] The consequences of the Saracen conquest were devastating to Demetriada, led the city to complete decay and decadence [9, 13, 28]. In the 11th century, the pirates reappeared in Demetriada [14]. In 1040 AD, Demetriada becomes occupied by the Bulgarians, who repaired the walls of the Castle when they themselves caused the major damage from the use of siege machines. However, their stay was short [4, 9].

Demetriada remains an important port until the 12th century. In 1082 AD Alexios Komnenos granted maritime privileges to the Venetians for free trade and renewal of the years 1126 AD, 1148 AD, 1187 AD and 1198 AD, the occupation of the port passes on to them, serving their own interests [4, 7, 14].

2.3   Late Middle Ages: Franks and Catalans

During the late middle ages, a period of Frankish domination, the available information is mainly concerned to Pelion and Magnesia [13]. The most famous feudal family who ruled the region were Melissenos (1207–1320 AD).[4]

In 1205 AD Frankish rule began in Greece with Constantinos Melissenos [13]. Melissinos's main concern was to protect its territory from pirates raids and built small fortress stations (pyrgovigles-view towers), ruins rescued on rugged shores of Mount Pelion and mountain ridges.[5] He himself lived in the area of Palaiokastro of Lechonia (Methoni) and was titled Lord of Demetriada and Lechonia [30, 31]. Palaiokastro is within walking distance of Ano Lechonia on a hill 250 m high [5, 32]. At the same

---

[2] The Saracens were Arab Muslim pirates who, starting raids from northern Africa and Aqaba, arrived in the Mediterranean. The raids of the Saracens in Greece began before the 9th century AD, lasting until the 10th century AD, causing the terror and desolation of the Aegean islands. In 824 AD, they conquered Crete, reaching Thessaloniki, the second capital of the Byzantine Empire, dominating it in 904 AD [28].

[3] For the conquest of a city were used wooden constructions with floors of a height corresponding to the walls known as woodcutters. Within it were the soldiers who were about to attack. The movement was done with oxen, horses and pedestrians on wheels. Series of archers protected those who were moving it, throwing their arrows to the wall [28].

[4] Members of the Melissenos family were distinguished throughout the Byzantine territory and emerged as senior entitles of Byzantium, such as generals, princes, ecclesiastical lords, toparchs [30].

[5] An enemy ship warning system was the flashing of tower tops using intermittent smoke [31].

time, the Benedictine monks of the city of Amalfi, built the monastery of St. Andrew, in the place where the monastery of St. Lavrentius is today.

The Melissian feudals of Demetriada from 1207 AD to 1423 AD founded many monasteries in Pelion and the monastic life flourished [5, 33]. Konstantinos Melissenos founded the Monastery of Oxeias Episkepseos in Makrinitsa in 1214 AD, in the name of the Virgin Mary Yperagnou Theomitiros, 70–80 m east of today's metropolis of Makrinitsa [30, 31].

In 1271–1272 AD, Nicholas Melissenos and his wife, princess Anna Melissenos founded the monastery of Prophet John the Baptist of Nea Petra in Portaria (Dryani-nouna) [25, 30, 34]. In 1275 AD, under the command of toparch Ioannis Melissenos, a naval battle took place in the Pagasitikos Gulf, near the old Demetriada between Byzantines who won the battle, and Franks [13, 31].

In 1310, Demetriada was ruled by the Catalans [14]. The Greek-Catalan family Melissenon - Novelles (1320–1423) was created after a marriage relationship with Catalan Othon De Novelles and Anna Melissenis.[6] From 1320 until the first conquest by the Turks in 1396, the Hellenic-Catalonian family governed the place, and the area of Demetriada and Lechonia flourished [30, 31].

Ottoman Occupation

The final conquest of Magnesia by the Ottoman Turks took place in 1423 AD by Turahan-Beni and the Castle turned into an Ottoman fortress [30]. The Turks chose the Byzantine Castle as a military center by expelling its Christian inhabitants to Pelion and the Castle was inhabited exclusively by them [5]. Until 1840 the Castle was the only settlement in the area except from the new district of merchants, which was established in 1841 opposite the Castle [7].

In 1889 the castle walls were demolished by the Greek authorities. However, part of its walls, as mentioned above, is maintained on the west side, the NE corner and sections in the north, east and west [7, 13, 14].

3   A New Museum

According to ICOM, the *Museum is defined as a non-profit, permanent institution in the service of society and its development, open to the public which acquires, conserves, researches, communicates and exhibits the tangible and intangible heritage of humanity and its environment for the purposes of education, study and enjoyment* [35]. Museums are a place of preservation of cultural heritage, they constitute recreational activity in people's free time and at the same time a lever of sustainable development of a region [36]. Hein [37], defines the museum as a multi-role communication system. The more effectively it communicates with the general public, the more accurately it fulfills its mission [38].

Although museum has been known to society as an encyclopedic institution since the 19th century, has begun to change form and becoming more anthropocentric only in recent

---

[6] Daughter of Dimitriadas toparch Gabriel Melissenos and sister of the 5th in the order of succession, Stefanos Melissenos Gavriilopoulos [31].

years. The visitor is no longer a passive receiver, but participates in it communicatively and ideologically in order to make the museum a competitive choice in the use of leisure time [39, 40].

Despite museum still has the priority to protect the exhibits, it changes the way they are presented, because it focuses on visitors. Today's museum mission is to become accessible to all, to enlarge its audience and to self-maintain its existence [41]. The booming of museums began in the 1970's, with the Pompidou Center in Paris as the most exemplary museum that combines education, entertainment and library with the cultural center [42, 43]. Also, the Guggenheim Museum in Bilbao managed to change the industrial image of the region as a significant attraction for thousands of visitors and boost its economic growth [42, 44]. While the Louvre Museum in Abu Dhabi, which cost 1 billion pounds, was recently inaugurated to transform the city's image [45]. The most successful museums offer a variety of experiences and attract different audiences reflecting the diverse needs of each visitor [46].

The challenge for the new museums is to give priority to designing an integrated strategy that is identical to its vision as determined by its management [47]. The museum, that communicates both with public and modern society, always inspires us [48, 49].

Technology has greatly facilitated the shift of the museum to live and attractive presentations to the public in an effective way. Texts, icons, videos, sound, 3D graphics, virtual reality they all complement the exhibits in many museums. The virtual and 3D exhibits often have the same objectives as the real ones [50]. In the last decades, virtual reality technologies have been used in the field of cultural heritage. Protecting and restoring past remnants has become a powerful tool thanks to the capabilities of three-dimensional reconstructions of archaeological sites and finds, virtual and augmented reality systems, creating realistic, high-performance three-dimensional models. The museums are interested in presenting their collections in an attractive way [51, 52].

There are many types of museums, characterized by the objects they collect and exhibit vary in size, pores, collections, architecture [53]. Museums can be divided into three categories: historical, technological and thematic. Nevertheless, museum has become so diverse and its categorization changes as it changes itself [54].

The Thematic Museum

The term "thematic" constitutes of a product with special supply and demand characteristics that attract specific tourists, being an alternative form of cultural entertainment. This category includes the "thematic" museum, the thematic park and the mega events [55]. The "thematic" museum is of special interest and category, presents a special collection, which the visitor admires and is taught by e.x. Museum of musical instruments, Museum of Arts, Nomismatic Museum etc. The exhibitions of the thematic museums are often inspired by a specific person, group or natural artifacts which have an important role in realizing why something is valuable and the reason of its maintenance for the next generations [56].

Cultural Tourism

According to the World Tourism Organization, cultural tourism involves the movement of people with cultural motivations, including educational, cultural tours, festivals, visits to historical monuments, popular culture and pilgrimage [57]. Cultural tourism is also recognized as a major cultural activity that evolves as a form of tourism of special interest that participants have different travel motives from others [58, 59]. The increase of income and free time has been the basis for tourism development over the last 50 years. From 25 million in 1950, international tourism has surpassed 1bn in 2014. Europe is the primary destination of half of international arrivals with 608 million in 2015 [60]. In 2016, international arrivals reached 1.2 billion with a forecast of 1.8 by 2023 [61]. In the US, cultural tourism reaches 81% [62]. Tourism is a very important for the global economy and, despite the occasional crises in recent decades, it is resilient with continued expansion and diversification. However, the distinction between random cultural tourists and those who have as their primary objective is difficult [63].

Cultural tourism, according to the World Tourism Organization, is defined as "trips aimed primarily at visiting sites and events that their cultural and historical value has transformed as part of the cultural heritage of a community" [64]. The World Tourism Organization predicts that by the year 2020 cultural tourism will have a leading role in the world. It is a qualitative alternative to the problems posed by mass tourism as it prioritizes the protection of the natural environment and the cultural heritage to be maintained for future generations [48, 65]. Cultural tourism increases competitiveness, creates job opportunities, restricts rural immigration, generates income for investment, promotes sense of pride and self-esteem in host societies [66].

3.1   European Examples of Medieval Thematic Museums

We mention indicatively examples of museums with medieval and Byzantine collections, each one illustrating a different scale. The Grand Palace Mosaic Museum, in Constantinople, Turkey presents part of the preserved mosaic floor of the Byzantine Palace [67]. The Mosaic Museum, Tamo, in Ravenna, Italy is also a big, innovative, exhibitional center of multiple use, since the innovative presentation of the exhibits consists of interactive applications, multimedia and new technologies [68]. On the other hand, the Archiepiscopical Museum, in Ravenna, Italy is distinguished through its exceptional artifacts that belong to the old Cathedral of the city [69]. A distinguished category are the Medieval torture Museums such as in San Gimignano, Italy, and the Rothenburg ob der Tauber, Germany, where mechanisms of torture are exhibited in order to inform the public about the civil rights [70, 71]. The Musêe de Cluny in Paris, France, is dedicated in medieval art and symbolizes the gothic Past of the country [72]. Royal Armoires, in Leeds, England is the biggest and best designed museum of arms and armoires in the world. The exhibits are interpreted in a way to bridge and define the military Past [73]. In addition, the Medieval Mile museum in Ireland, which was established in 2017 represents a huge treasure of work and life in Ireland [74]. The Medieval Cyprus Museum (Castle of Limassol) demonstrates exhibits from all medieval period, the Byzantine and Christian Museum, in Athens, Greece has more than 25000 artifacts and exhibits and the Museum of Byzantine civilization in Thessaloniki, which is the most important in Greece, hosts periodical exhibitions and offers educational programs [38, 75, 76]. Exploiting Medieval Heritage –Churches is another example of the evolution of museology and technology introducing new methods of attraction. Applied cases such as Santa Maria Maggiore di Siponto, Apulia, Italy, where the old basilica church was recreated in a permanent installation made of wire (mesh technique) [77]. The Cluny Abbey in Burgundy, France, where the roman-gothic church was digitally recreated using augmented reality [78]. Also, the Arian in Piazza exhibition that took place in 2013 in Ravenna, Italy demonstrated how technology could be applied in order to maintain the inheritance [79].

4 Case Study: Municipality of Volos

Volos is located in eastern central Greece in Thessaly and is the capital of the Magnesia regional unit. The united municipality occupies an area of 387,14 sq. Km. The permanent population is 144,449 inhabitants according to the 2011 ELSTAT census [80, 81]. The city of Volos at the edge of the Pagasitic Gulf is one of the oldest inhabited areas of Greece. From prehistoric times until today it has not stopped being inhabited and in 21th century has become one of the largest industrial centers in Greece. Today, the de-industrialized city shapes its modern character following the trend of the era [7, 82, 83].

As far as tourism is concerned, the total arrivals in Greece in 2013 with 16mn arrivals kept constantly rising to 25mn arrivals in 2017. In Volos, the annual arrivals are more than 200,000 with most of them being made by nationals, as opposed to the whole country [84].

There are many 3-star hotels in the city of Volos, including two 4-star hotels (Park, Volos Palace), one 5-star hotel (Xenia) and 3 accommodation with rooms to let. There are more than 10 museums in the city of Volos. The Athanasakio Archaeological Museum of Volos is the most important museum of the city, founded in 1909 and for many years was the only organized museum in Thessaly [85]. The annual visits reach 30,000 persons. The most visited months are April, March and May. The fact that maximum visits don't take place during the summer months means that they were carried out by locals, or even by schools, and not by foreign tourists [86].

5 Proposal: Creation of a Thematic Museum

According to the cultural heritage as mentioned above, we propose to create a Thematic Medieval Museum in the area of the Castle of Palea, in the core of the old Byzantine settlement. More specifically, Giannitson Street with Lachana was chosen to host the new museum. There is open space suitable for the development of such idea. According to the Municipal Property of Volos, it is owned by the Municipality of Volos and is called 'Youth Square'.

The proposed location is adjacent to the archaeological site of the Castle where a large part of the wall is visible. The choice of this location also has the following advantages: it is functional, it has open space, connects to the Castle, the walls are visible, has the potential to merge with the Castle, there are no houses nearby, parking spaces can be created, easy accessibility and just across the street is the Tsalapatas multi-site with all the infrastructure needed. The proximity of the archaeological site of the Castle with the Tsalapatas, and the other two City and Railway museums can create a cluster of museums (Fig. 1).

Fig. 1. The area chosen for the new museum in Giannitson – Lahana street (Source: own processing)

We propose the creation of a building of 300 m^2. There and mainly through the use of new image and sound technologies, the history will be narrated and the medieval heritage of the region will be presented. The Museum will have an amphitheater for the presentation of 3D projections, documentary films which will demonstrate the history of Demetriada and its change from a Hellenistic city to an early Christian one. In addition all the historical events that followed the relocation to the hill of Palea until the Ottoman conquest will be vividly presented. A virtual reality movie experience will be offered including important historical events like battles, wall battles, naval battlements, etc. which took place in the area of Pagasitikos and near the Castle.

Indicatively, I would like to mention: A proposal for the representation of a naval battle by the Byzantine fleet of Emperor Michael H. Palaiologos with the warships [87]. Moreover, historical events that took place initially in Demetriada and then at the Castle, such as the settlement of the Slavs in Demetriada with monoliths, the construction of Castle walls, the transfer of settlements, the attacks of Saracens pirates with battlefields, Bulgarians, Turks and Battleships.

The life in the city of Fthiotides Thives and the historical events will be presented through a 3D animation movie. Presentation of the history of other neighboring settlements, their evolution and the development of Pelion in the late Middle Ages will be also offered.

Moreover a recreation of the damaged monuments, as they were in their original form, will be available to visitors through 3D digital models, such as the Castle walls, the Early Christian basilica church of Damocratia, the Cemetery basilica, the Byzantine baths and the great technical work of the Aqueduct that worked throughout the Byzantine era etc. The visitor will be able to get useful information about each monument and explore the texture, the interior and exterior decoration, the evolution over the years, the materials of the construction etc.

In the following section we present some architectural recreaction of the *Basilica A' church of Damocratia* made by the architect Ebrahim Abnar (Figs. 2, 3 and 4).

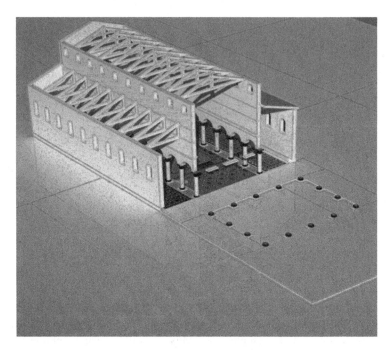

Fig. 2. Architectural recreation of *Basilica A' church of Damocratia* (Design: Ebrahim Abnar)

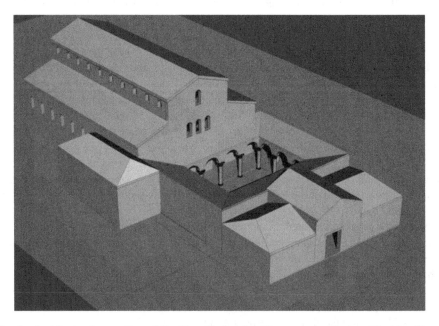

Fig. 3. Architectural recreation of *Basilica A' church of Damocratia* (Design: Ebrahim Abnar)

Fig. 4. Interior architectural recreation of *Basilica A' church of Damocratia* (Design: Ebrahim Abnar)

Another optional offer for the visitor will be the access to information about the history, architecture, art (mosaics) trends and styles that prevailed at any time in the region but also in the Byzantine Empire in general by an application. Information about the Bishops of Demetriada, the actions and the work of each Emperor and the changes that the Empire suffered will be also available. In addition, the thematic Museum of Palea will also intergrate a Virtual and Augmented reality room where the visitor will be able to tour virtually in the monument, city, etc.), an Installation theme, which will allow the visitor to 'feel' the history in an artistic way and to participate actively is also proposed, Video Projections on the adjacent to the museum Castle walls will also take the Museum experience out of the museum and bring to life Medieval times.

In the museum, we suggest real Byzantine objects, such as portions of preserved mosaics, coins and other artefacts, should be transferred which are scattered in various collections.

Holograms (3D digital clones of objects) will complete the displays.

Educational programs and activities will be available to schools, tourists and families with special packages of activities with interactive, educational applications and programs which will offer a creative experience with medieval art (eg, draw, paint, built the Castle, the church, the mosaic, etc.). A café area, a showroom and an event venue will be available at the museum. Information brochures as well as a book with representations of the original and current form of the monuments will be designed.

A Medieval Cultural Route is proposed to create a medieval cultural path commensurate with the course and evolution of society as it has been formed over the centuries and the conquerors. The starting point is the early Christian Demetriada with the first stop at the Basilica of the Democracy, then the Aqueduct and the Cemetery Basilica. There is a visit to Fthiotides Thives and a tour of the preserved basilicas as their flourish is defined at the same time. Then return to the city of Volos and visit the archaeological site of the Castle. From there we go to the area of Agios Konstantinos to see the surviving basilicas. Then we reach Lechonia and climb up to Palaiokastro and Agios Lavrentios. On the return we go to Portaria and Makrinitsa to see the positions of the Melissenian monasteries. In return we visit the hill of the bishopric (Fig. 5).

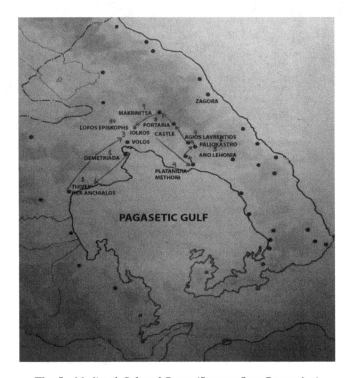

Fig. 5. *Medieval Cultural Route* (Source: Own Processing)

6 Marketing Plan

In order to ensure the viability of the museum, it is necessary to create a marketing plan. This is defined as follows:

Analysis of the current situation: There are 30 Byzantine museums in Greece [88].

Competitive environment: The medieval and Byzantine museums exist in Greece in Athens (324 km) and Thessaloniki (217 km) and Veria (200 km) as well as in the Mediterranean countries. Particularly:

1. Byzantine and Christian Museum of Athens, 2. Byzantine Museum o Thessaloniki, 3. The Medieval Museum of Rhodes, 4. Byzantine Museum of Veria.

6.1 SWOT Analysis of the New Museum

With the implementation of SWOT analysis, we aimed at identifying opportunities, exploiting the advantages and addressing the drawbacks of creating the new museum.

STRONG POINTS	WEAK POINTS
• Archaeological site of Castle	• Small number of tourists
• Easy accessibility	• Cannot be a large-scale Museum
• Within walking distance: Tsalapatta Brickworks Museum, City Museum, Volos Railway Museum	
• Touristic area	
• Restaurants	
• University	
• Widening tourist season	
• Economic development	
• Development of cultural tourism	
• New job positions	
• Unobstructed visual position of the Castle	
CHANCES	THREATS
• Promotion of Cultural Tourism	• Lack of funds
• Increase tourism	• Economic crisis
• Network of Museums (Berlin Museum).	• Indifference
• Innovative	• Incorrect management
• New job positions	• Incomplete advertising
• Tourist attraction	• Other Medieval museums (Thessaloniki, Mystras, Rhodes, Malta).
• Economic development	• Competitive Cities

6.2 Marketing Objectives

Estimated Benefits

The success of the museum, its recognition and its acceptance by the local community are the direct goals of our proposal. The creation of the museum will act as a level of economic and touristic development of the region. The attraction of cultural tourists from all Greece and mainly from abroad is an important factor and goal for the success of the museum. The promotion of the Thematic Medieval museum will be associated with the region's Brand name through the development of cultural tourism. In the context of promoting cultural tourism, the local economy will flourish, by the creation of new job's opportunities and by the improval of the quality of life. Historic preservation creates benefits for the local economy, strengthening links between the community and citizens, creating a sense of pride. Also it encourages private investment and sustainable business activities.

Product

As a product we define the Medieval Museum, the logo and the Brand name that accompany it as follows: The decorative calligraphy letters of the medieval alphabet we chose form the word mean and refers us to the English word medieval meaning Middle Ages and the medievolos to the Medieval Volos. The medieval mean and the word Volos in calibri font symbolizes the coexistence of the old with the new. The aim is that the visitor, through the slogan, realizes the 'theme' of the museum (Fig. 6).

Fig. 6. Brand name – proposed Logo (Source: Own processing)

6.3 Target Market/Group

The target market is cultural tourists. This category of tourists explores the cultural heritage of the place by visiting museums, archaeological sites, monuments, towns, historical or natural sites, cathedrals, castles, churches, historic houses with in-depth presentations and interpretations by specialized tour guides etc [55]. The cultural tourist is the "oldest" type of "modern" types of tourism [58]. People have been traveling for cultural reasons since ancient times, and nowadays are called alternative tourists. Cultural tourists usually have more money to spend and stay in the area more [41, 89]. They also have a high level of education and high professional positions. They want to enrich their lives with new experiences and combine vacations with cultural enjoyment and are younger than they once were [90].

Positioning

The ultimate product is a Thematic Medieval Museum. Its competitive advantage is its differentiation due to its unique medieval character and innovation due to the use of new technologies, thus differentiating it by giving it the attributes of becoming a brand name.

Promotion

The promotion of the museum will be achieved through public relations, tour operators, travel agencies where it will be advertised and will be part of holiday packages. Social media, direct mail, local television, radio, Sunday newspapers, Art Newspaper, magazines, advertising at airports, hotels, brochures, road transport, internet, banners and posters. A museum website will be created where its programs will be demonstrated.

Place

The distribution of the museum product is through distribution channels, either directly from the museum itself and addressed to the public or intermediary intermediaries and indirectly made with tour operators through holiday packages.

7 Conclusions

The aim of this study was to explore the history of the city of Volos and the wider region in order to suggest new visiting opportunities. Cultural Tourism (which is an alternative form of tourism chosen by those who want to discover culture of a country) was chosen to attract quality tourism by increasing its rates and attracting tourists of different nationalities. At the same time, the inhabitants of the area will become acquainted with their history. As a result of the innovative investment, the city will improve its image and reputation. The success of the museum through its recognition and emergence as a Brand name will help to improve the quality of life of residents and create new businesses in the region. The use of new technologies has been deemed necessary both for tourist purposes and for the understanding and interpretation of ruins and finds, to protect their degradation and to be able to examine in depth elements such as the meaning and purpose of their creation, their original meaning and history, allowing them to be restored. The representation of the past and the monuments harmoniously combines the old with the new giving to the museum a comparative advantage creating a competitive destination equal to other medieval well-known museums attracting a large number of visitors. The way we handle, promote and harness our cultural heritage attributes elements of our education but above all to our quality [48]. Culture is important for tourism and for the attractiveness and competitiveness of destinations. An essential prerequisite for a successful destination is the positive synergy between tourism and culture and, in order to be considered successful, cooperation between the public and private sectors and local communities is needed such as the Municipality of Volos, the Hellenic Ministry of Culture or any other official institution. One should look forward to implement this proposal.

Aknowledgments. This proposal was the subject of the MSc thesis with the title 'Medieval Volos, a proposal for the creation of a thematic museum', carried out under the supervision of Professor Michel Zouboulakis (Department of Economics, University of Thessaly) in the master program 'Tourism and Cultural Planning and Development' of University of Thessaly.

References

1. Renfrew, C., Bahn, P.: Archaeology: Theories, Methods, and Practice. Thames and Hudson, New York (2011)
2. Cocossis, C., Tsartas, P.: Sustainable Tourism Development and Environment (Βιώσιμη Τουριστική Ανάπτυξη και Περιβάλλον). Kritiki, Athens (2001)
3. Virginija, J.: Interaction between cultural/creative tourism and tourism/cultural heritage industries. In: Tourism-from Empirical Research Towards Practical Application. InTech (2016). https://doi.org/10.5772/62661

4. Avramea, A.: H Byzantini Thessalia mechri tou 1204: Symboli eis tin historiki geografia (Η βυζαντινή Θεσσαλία μέχρι του 1204: συμβολή εις την ιστορικήν γεωγραφίαν) (Doctoral dissertation). Ethnikon kai Kapodistriakon Panepistimion Athenon (1974)

5. Asimakopoulou Atzaka, P.: Palaioxristianiki kai byzantini Magnisia. In: Hourmouziadis, G., Asimakopoulou Atzaka, P., Makris, K.A. (eds.) Magnisia. To Hroniko enos Politismou (Μαγνησία. Το χρονικό ενός πολιτισμού) Ancient Magnesia. Kapon, Athens (1982)

6. Ephorate of Antiquities of Magnesia. Early Christian Monuments of Demetriada (2016). http://efamagvolos.culture.gr/Palaioxristianika_Demetriadas.html. Accessed 6 Sep 2017

7. Chastaoglou, V.: To portraito ths polhs apo ton 19° eos shmera (Το πορτραίτο της πόλης πό τον 19ο αιώνα έως σήμερα). Di.Ki, Volos (2007)

8. Konidaris, G.: Mitropolis Demetriados (Μητρόπολις Δημητριάδος). TEE, 5 (1964)

9. Papathanasiou, A.: H Byzantini Demetriada 431-1204 (Η Βυζαντινή Δημητριάδα 431-1204). OMIROS, Volos (1995)

10. Arvanitopoulos, A.S.: Anaskafai kai Erevnai en Thessalia kai Makedonia kata to etos 1912 (Ανασκαφαί και έρευναι εν Θεσσαλία και Μακεδονία κατά το έτος 1912). PAE (1913)

11. Makri-Skoteinioti, T., Mpatziou-Efstathiou, A.: Protasi Anadeiksis tou arxaiologikou horou Dimitriados. In: Mnhmeia ths Magnhsias (Μνημεία της Μαγνησίας). Conference proceedings Anadeiksi tou diahronikou mnhmiakou ploutou ths magnhsias kai ths evriteris periohis, Volos, pp. 11–13 (2002)

12. Ntina, A.: Palaiohristianika Mnhmeia Dhmhtriados En Volo (Εν Βόλω), vol. 12, pp. 12–15. Dimos Volou (2004)

13. Liapis, K.: To Kastro tou Volou mesa stous aiones (Το Κάστρο του Βόλου μέσα στους Αιώνες). Ores, Volos (1991)

14. Anastasiadou, A., Ntina, A.: The Castle of Palea (Το Κάστρο στα Παλαιά του Βόλου) Hellenic Ministry of Culture and Sports 7th Ephorate of Byzantine Antiquities. Larissa, Volos (2014)

15. Mango, C.A.: Byzantium: The Empire of New Rome. Weidenfeld & Nicolson, London (1980)

16. Lemerle, P.: The oldest collections of the miracles of St. Demetrius and the penetration of the Slavs in the Balkans. I. The text (Les plus anciens recueils des miracles de saint Démétrius et la pénétration des Slaves dans les Balkans. I. Le texte) (1979)

17. Kordatos, G.: Istoria tis eparchias Volou kai Ayias (Ιστορία της Επαρχίας Βόλου και Άγιας), Athens (1960)

18. Papachatzis, N.: I periohi tou Volou apo apopsi istoriki kai arhaiologiki (Η περιοχή του Βόλου από άποψη ιστορική και αρχαιολογική), Volos (1967)

19. Intzesiloglou, H.: To archaio theatro Demetriados. In: Mnhmeia ths Magnhsias (Μνημεία της Μαγνησίας). Conference proccedings Anadeiksi tou diahronikou mnhmiakou ploutou ths magnhsias kai ths evriteris periohis, Volos, pp. 11–13 (2002)

20. To archaio theatro Demetriados (2012). http://www.theatreofdemetrias.gr/theatre/index.php?option=com_content&view=article&id=19&Itemid=137&lang=el. Accessed 5 Sep 2017

21. Krautheimer, R.: Early Christian and Byzantine Architecture. Penguin books Ltd, Harmondworth (1979)

22. Giannopoulos, N.I.: Palaiochristianiki epigraphi kai palaiochristianiko nekrotafeio Volou (Παλαιοχριστιανική επιγραφή και παλαιοχριστιανικό νεκροταφείο Βόλου). EEVS 12 (1936)

23. Marzolff, P.: Zur Stadtbaugeschichte von Demetrias. In: Proceedings of an International Conference on Ancient Thessaly in memory of Dr R. Theochari, Athens (1992)

24. Giannopoulos, N.I.: To frourio tou Volou (Το φρούριο του Βόλου). EEVS 8 (1931)

25. Tsopotos, D.: Istoria touVolou (Ιστορία του Βόλου). KODV, Volos (1991)

26. Kravaritou, S.: Isiac Cults, Civic Priesthood and Social Elite in Hellenistic Demetrias (Thessaly), Notes on RICIS 112/0703 and beyond. Tekmeria 12 (2014). https://doi.org/10.12681/tekmeria.305

27. Ntina, A.: To kastro tou Volou: Istoria -Archaiologia- Anadeiksi. In: Mnhmeia ths Magnhsias (Μνημεία της Μαγνησίας). Conference proccedings Anadeiksi tou diahronikou mnhmiakou ploutou ths magnhsias kai ths evriteris periohis, Volos, pp. 11–13 (2002)

28. Papathanassiou, A.: Thessaliko hmerologio (Θεσσαλικό ημερολόγιο), tomos 19, trito afieroma. Larissa (1991)

29. Continuatus, T.: Ioannes Cameniata. Symeon Magister, Georgius Monachus, Bonn ed (1838)

30. Papathanassiou, A.: Melissinoi tis Demetriadas, ktitores I. Monon (Οι Μελλησινοί της Δημητριάδας κτήτορες I.Μονών), Papazisis (1989)

31. Papathanasiou, A.: H Magnesia kai to Pilio ston istero mesaiona (1204–1423) (Η Μαγνησία και το Πήλιο στον ύστερο Μεσαίωνα), Volos (1998)

32. Liapis, K.: Ta Lechonia kai to Paliokastro tous (Τα Λεχώνια και το Παλαιόκαστρο τους). Thessaliko hmerologio, tomos 39, Larissa (2001)

33. Dimou, D.: Vizitsa Piliou. To hthes kai to simera (Βυζίτσα Πηλίου. Το χθές και το σήμερα) (2014)

34. Mamaloukos, S.: Oi Byzantinoi naoi ths Magnhsias En Volo (Εν Βόλω), Dimos Volou, vol. 12, pp. 32–39 (2004)

35. ICOM. International Counsil of Museums (2007). http://icom.museum/the-vision/museum-definition/. Accessed 3 Dec 2017

36. Mortaki, S.: Cultural tourism is a local business: the example of the maritime museum of Litochoro in Greece. Am. Int. J. Contemp. Res. (2013)

37. Hein, G.E.: The role of museums in society: education and social action. Curator **48**(4), 357 (2005). https://doi.org/10.1111/j.2151-6952.2005.tb00180.x

38. Bounia, A., Gazi, A.: Ethnika mouseia sth notia Evropi (Εθνικά μουσεία στη νότια Ευρώπη). Kaleidoscopio, Athens (2012)

39. Stamatelou, A.: International Museum Day 1992–2005: New Trends, New Challenges (Διεθνής Ημέρα των Μουσείων 1992–2005: Νέες Τάσεις, Νέες προκλήσεις). Museum, Revised edition of the museum research center of the University of Athens, Issue 5, pp. 18–22 (2007)

40. Scott, C.: Branding: positioning museums in the 21 st century. Int. J. Arts Manage. 35–39 (2000)

41. Black, G.: The Engaging Museum: Developing Museums for Visitor Involvement. Psychology Press, Abingdon (2005)

42. Van Aalst, I., Boogaarts, I.: From museum to mass entertainment: the evolution of the role of museums in cities. Eur. Urban Reg. Stud. **9**(3), 195–209 (2002). https://doi.org/10.1177/096977640200900301

43. Deffner, A., Metaxas, Th.: Tourism development, industrial heritage and special museums: The case of the Kavala tobacco museum (Τουριστική ανάπτυξη, βιομηχανική κληρονομιά και ειδικά μουσεία: Η περίπτωση του μουσείου καπνού Καβάλας). Pan-Hellenic and International Geographical Conferences, Collection of Practices, pp. 581–589 (2010)

44. Metaxas, T., Petrakos, G.: Regional competitiveness and cities competition. In: Proceedings of Greek Department of European Regional Science Association with subject 'Regional Development in Greece: Trends and Perspectives', (University of Thessaly). University Thessaly Press, pp. 207–230 (2004)

45. Trend, N.: The Telegraph, Louvre Abu Dhabi: first look inside the £1 billion art museum in the desert (2017). http://www.telegraph.co.uk/travel/destinations/middle-east/united-arab-emirates/abu-dhabi/articles/louvre-abu-dhabi-first-look-review/. Accessed 5 Dec 2017

46. Kolter, N., Kotler, P.: Museum Strategy and Marketing. Designing Missions: Building Audiences (1998)
47. American Association of Museums. Commission on Museums for a New Century. Museums for a new century: A report of the commission on museums for a new century. Amer Assn of Museums (1984)
48. Bounia, A.: Cultural Policy in Greece, the Case of the National Museums (1990–2010): an Overview. Museum Policies in Europe 1990–2010: Negotiating Professional and Political Utopia, 127 (2012)
49. Pournou, A., Jones, A.M., Moss, S.T.: Biodeterioration dynamics of marine wreck-sites determine the need for their in situ protection. Int. J. Nautical Archaeol. **30**(2), 299–305 (2001). https://doi.org/10.1111/j.1095-9270.2001.tb01377.x
50. Lepouras, G., Katifori, A., Vassilakis, C., Charitos, D.: Real exhibitions in a virtual museum. Virtual Reality **7**(2), 120–128 (2004)
51. Bruno, F., Bruno, S., De Sensi, G., Luchi, M.L., Mancuso, S., Muzzupappa, M.: From 3D reconstruction to virtual reality: a complete methodology for digital archaeological exhibition. J. Cult. Herit. **11**(1), 42–49 (2010). https://doi.org/10.1016/j.culher.2009.02.006
52. Wojciechowski, R., Walczak, K., White, M., Cellary, W.: Building virtual and augmented reality museum exhibitions. In: Proceedings of the Ninth International Conference on 3D Web Technology, pp. 135–144. ACM, April 2004. https://doi.org/10.1145/985040.985060
53. Noesis. Science Center and Technology Museum (2017). http://www.noesis.edu.gr/%CF%84%CE%B9-%CE%B5%CE%AF%CE%BD%CE%B1%CE%B9-%CE%AD%CE%BD%CE%B1-%CE%BC%CE%BF%CF%85%CF%83%CE%B5%CE%AF%CE%BF/. Accessed 25 Oct 2017
54. Categorization of museums. Historical, technological, thematic (2017). http://poieinkaiprattein.org/culture/museums/new-use-of-media-by-museums-study-for-volos/categorization-of-museums-historical-technological-thematic/. Accessed 20 Nov 2017
55. Cocossis, C., Tsartas, P., Gribba, E.: Special and alternative forms of tourism: Demand and supply of new tourism products (Ειδικές και εναλλακτικές μορφές τουρισμού: Ζήτηση και προσφορά νέων προϊόντων τουρισμού). Kritiki, Athens (2011)
56. Galllagher, B.: The Thematic Museum, Interpretation & Archive (2012). http://barcelona.b-guided.com/en/noticias/b-ing/the-thematic-museum-interpretation-archive-35.html. Accessed 6 Dec 2017
57. OECD: The Impact of Culture on Tourism. OECD, Paris (2009)
58. McKercher, B., Du Cros, H.: Cultural Tourism: The Partnership Between Tourism and Cultural Heritage Management. Routledge (2002)
59. Rizzo, I., Mignosa, A. (eds.): Handbook on the Economics of Cultural Heritage. Edward Elgar Publishing, UK (2013)
60. Cocossis, H.: Cultural Heritage and Sustainable Tourism: The challenges. InHerit, Athens (2017)
61. World tourism organization. Annual report (2016). http://cf.cdn.unwto.org/sites/all/files/pdf/annual_report_2016_web_0.pdf. Accessed 3 Jan 2018
62. Selier, N.: Cultural Tourism Leads the Growth of Travel Industry (2017). http://www.solimarinternational.com/resources-page/blog/item/228-cultural-tourism-leads-the-growth-of-travel-industry. Accessed 12 Dec 2017
63. Noonan, D.S., Rizzo, I.: Economics of Cultural Tourism: Issues and Perspectives (2017)
64. World tourism organization. Annual report (2012). http://cf.cdn.unwto.org/sites/all/files/pdf/annual_report_2012.pdf. Accessed 3 Jan 2018
65. Sdrali, D., Chazapi, K.: Cultural Tourism in a Greek Insular Community: The Residents Perspective (2007)

66. UNWTO/UNESCO. World Conference on Tourism and Culture gathers Ministers of Tourism and Culture for the first time (2015). http://media.unwto.org/press-release/2015-02-06/unwtounesco-world-conference-tourism-and-culture-gathers-ministers-tourism. Accessed 3 Jan 2018

67. İstanbul - The Great Palace Mosaic Museum (2017). http://www.kultur.gov.tr/EN-113952/istanbul—the-great-palace-mosaic-museum.html. Accessed 19 Dec 2017

68. Museo Tamo (2017). http://www.tamoravenna.it/. Accessed 20 Nov 2017

69. Opera di Religione della Diocesi di Ravenna. Archiepiscopical museum, Ravenna mosaic (2017). http://www.ravennamosaici.it/musei/museo-arcivescovile/?lang=en. Accessed 16 Nov 2017

70. Museo della Tortura e della penna di morte (2017). http://www.torturemuseum.it/musei-permanenti/san-gimignano/. Accessed 16 Dec 2017

71. Medieval crime museum (2017). https://www.kriminalmuseum.eu/about-us/history-of-the-museum/?lang=en. Accessed 19 Dec 2017

72. Museê de Cluny- Cluny Museum (2017). http://www.musee-moyenage.fr/. Accessed 29 Oct 2017

73. Sturtevant, P.: The public mediavalist. Medieval Tourism Bucket list: The Royal Armouries Museum (2014). https://www.publicmedievalist.com/royal-armouries/. Accessed 20 Nov 2017

74. Medieval Mile Museum (2017). https://www.medievalmilemuseum.ie/. Accessed 18 Nov 2017

75. Mesaioniko Mouseio Kyprou (Kastro Lemesou). http://www.mcw.gov.cy/mcw/DA/DA.nsf/0/61E651EA3D6F3976C2257199001EFB8F?OpenDocument. Accessed 30 Sep 2017

76. Mouseio Byzantinou politismou Thessalonikis. https://www.mbp.gr/. Accessed 28 Dec 2017

77. Artist Edoardo Tresoldi creates a phantom basilica in Italy's Puglia (2016). https://thespaces.com/artist-edoardo-tresoldi-creates-a-phantom-basilica-in-italys-puglia/. Accessed 8 Nov 2017

78. Landrieu, J., Père, C., Rollier-Hanselmann, J., Castandet, S., Schotté, G.: Digital rebirth of the greatest church of Cluny Maior Ecclesia: From optronic surveys to real time use of the digital Model (2011). https://doi.org/10.5194/isprsarchives-xxxviii-5-w16-31-2011

79. Zaccarini, M., Iannucci, A., Orlandi, M., Vandini, M., Zambruno, S.: A multi-disciplinary approach to the preservation of cultural heritage: a case study on the Piazzetta degli Ariani, Ravenna. In: Digital Heritage International Congress (DigitalHeritage) 2013, vol. 2, pp. 337–340. IEEE, October 2013. https://doi.org/10.1109/digitalheritage.2013.6744775

80. Hellenic Statistical Authority (2017). http://www.statistics.gr/el/statistics/-/publication/SAM03/2011

81. Municipality of Volos (2013). http://dimosvolos.gr/?p=880#more-880

82. Chastaoglou, V.: Volos, h dhmiourgia ths neas polhs. E Istorika, 66 (2001)

83. Panagopoulos, B.: O Volos ths viomihanias ton grammaton kai ton agon, E Istorika, 66 (2001)

84. Hellenic Statistical Authority (2017). http://www.statistics.gr/el/statistics/-/publication/STO12/

85. Municipality of Volos (2013). http://dimosvolos.gr/?page_id=249

86. Hellenic Statistical Authority (2017). https://www.statistics.gr/statistics/-/publication/SCI21/

87. Papathanassiou, A.: Arheio thessalikon meleton, tomos 8, Volos (1988)

88. Ministry of Culture and Sports. Byzantine and Post-Byzantine museums and collections (2012). http://odysseus.culture.gr/h/1/gh110.jsp?theme_id=22

89. Sharpley, R., Telfer, D.J. (eds.): Tourism and Development: Concepts and Issues, vol. 63. Channel View Publications, Bristol (2014)

90. Richards, G.: Cultural Tourism: Global and Local Perspectives. Psychology Press, London (2007)

Information Technology, Smart Devices and Augmented Reality Applications for Cultural Heritage Enhancement: The Kalamata 1821 Project

Vayia V. Panagiotidis[✉], George Malaperdas, Eleni Palamara,
Vasiliki Valantou, and Nikolaos Zacharias

Laboratory of Archaeometry, University of the Peloponnese,
Old Camp, 24133 Kalamata, Greece
vayiap@gmail.com

Abstract. The purpose of this paper is to present the development of a modern web multimedia application named "Kalamata Action Map". The application provides a map based environment for users to exploit the Messenian landscape of the early 19th century thus enhancing our understanding on temporal and spatial interactions of that Era and towards the present. Designed for educational and touristic purposes, using historic material as a starting point "Kalamata Action Map" takes advantage of the intriguing urban surroundings, avoiding the limitations imposed by a closed museum environment. The application will be developed for use over the internet with open access. Data and images of landscape and architectural features, artifacts, everyday life etc. associated with historical information in reference to the War for Greek Independence and the establishment of the Modern Greek State will be integrated in a friendly and enticing way offering a unique user experience.

The application provides interactive maps of the city of Kalamata, serving as an exciting and modern depiction of the historic events that led to the beginning of the War for Greek Independence in 1821. Moreover, and additionally to the various historic, folklore and architectural elements, the application will also include digital representations of artworks and everyday objects and suggested routes for nearby destinations in the Prefecture of Messenia, therefore creating a digital network of interconnecting locations of historic interest, improving their commercial and touristic connections. The use of the application has the potential to serve as a significant reference point both for residents and visitors.

Keywords: GIS · Kalamata · Cultural heritage management ·
Digital applications

1 Introduction

Geographic Information Systems (GIS) in collaboration with location based application software provide a friendly, digital platform for the management and enhancement of archaeological and cultural heritage environments. The work introduces a topobased application using GIS named "Kalamata Action Map" that provides a map based environment for users to view the Messenian landscape of the early 19th century, portrayed by Finlay [1] and Christou [2].

© Springer Nature Switzerland AG 2019
A. Moropoulou et al. (Eds.): TMM_CH 2018, CCIS 961, pp. 222–231, 2019.
https://doi.org/10.1007/978-3-030-12957-6_15

The study area is located in the south west prefecture of Greece mainly Messenia with some references to neighboring areas of Laconia (West Mani) and Arcadia (SW). The application is developed to complement the user experience of visitors to the "Kalamata 1821: Roads of Freedom" website. "Kalamata 1821: Roads of Freedom" is a co-financed by Greece and the European Union project under the program "Research, Creation and Innovation" ESPA 2014–2020. The "Kalamata 1821: Roads of Freedom" project aims to study and exhibit to the wider public, important local parameters of the 1821 era and their interconnections with the present, through roads of history, culture and trade. Emphasis will be given to specific aspects of the 1821 Revolution, which began in Kalamata, highlighting the historic significance and the strong interrelation with the wider area (e.g. preparation for the uprising, economic life of the region, beginning of the revolution, Navarino naval battle), in order to describe the 200 years of history and the historic correlations. By using new vision and multimedia technologies, the project will attempt to interactively depict the historic moments to the citizens/visitors of Kalamata, on a permanent basis.

The project website contains plethora of information in reference to the city of Kalamata, its history, culture, legends, and economy and how they all evolved to the present. The "Kalamata Action Map" application compliments the website offering a spatial and temporal view of this information with images, links, videos etc. Via the "Kalamata Action Map" application users can participate in trails, encouraging users to discover the city though a network of historic buildings and significant landmarks, which will come to life via intriguing storytelling, pictures and Augmented Reality [3]. These insights, historic and cultural, will offer an alternate aspect of the city providing tourists information and motivation to visit buildings, landmarks and sites of historic significance, enhancing cultural heritage and sustainability and making "the Kalamata Action Map" a tool for protecting the routes and scenes where the events of the Greek Revolution took place while promoting Cultural Heritage of the 19th century.

2 The Historical Base

The Peloponnese is one the most restless regions of the subdued Greek areas. Following the failed Orlov revolt in the Peloponnese and later in Crete that began in February 1770 the region remained in an electrified state. In the following years, significant efforts will lead to the liberation of Kalamata, the first liberated city in the occupied by the Ottomans Greece.

In early January 1821 Theodoros Kolokotronis pursued by the Ottomans, arrives in Kardamili to find shelter in the tower of the Mourtzinos (Troupakis) family. In Early February Grigorios Dikaios (aka Papaflessas) arrives in Mani after the Vostitsa meeting with Paleon Patron Germanos seeking allies. On the Ottoman side, Hoursit Pasha the Moras Valesi is transferred with his troops from Tripolitsa to Epirus to assist in the battle against Ali Pasha, the local ottoman ruler who was fighting to gain his independence from the authority of the High Porte, leaving the Peloponnese without sufficient military forces [4].

The Ottoman authorities in Kalamata suspecting efforts of revolution from the Greeks send their families away to find safe haven in near castles. Locally, the "Voevode" of Kalamata, official ottoman ruler, Souleiman Agas Arnaoutoglou calls for conference with the Greek authorities, ("Proestoi"), to discuss his concerns. They reassure him for their loyalty and having spread rumors about outlaws wandering in the region, they propose that he reinforces his guard of only 150 Ottoman soldiers with capable Greek men from Mani.

In the middle of March 1821, a large supply of ammunition arrives to the small harbor of Almiros from Smyrna, one of the largest commercial ports in the Ottoman Empire, sent by the "Philiki Eteria" (Society of Friends), a secret organization faithful to the revolution. Papaflessas informs his allies and Nikitas Stamatelopoulos, aka Nikitaras, and Christos Papageorgiou (aka Anagnostaras), are summoned to the harbor to move the shipment safely. With the assistance of villagers and mules the ammunition was transported from the harbor to the Monasteri of Mardaki outside the Taygetos Mountain village of Megoloanastasova (modern Nedousa). ("Dromos Baroutiou"-Road of Gunpowder) In order to transport the shipment Papaflessas received customs documents from Petrobei Mavromichalis, Greek ruler of Mani who was occupied with his family for many years in commercial transactions, involving him in the scheme.

The fact that the shipment during the transportation was escorted by armed men, drew attention of Ottoman authorities. The "Proestoi" were summoned for interrogation and explained that it is a simple shipment of olive oil but due to the constant attacks from outlaws, armed forces are required to escort all goods transports. They also convinced Arnaoutoglou during the audience that he must take action to protect them and the Messenian region which is under his authority.

The events escalate quickly after that audience; Arnaoutoglou summons Petrobei Mavromichalis to send reinforcements and the latter jumps to the call sending his son Elias and 150 trained in war Maniates to protect the castle. Elias Mavromichalis convinces Arnaoutoglou that the incoming threat of Kleftes is too strong to be fought off by 150 Ottoman soldiers and 150 Maniates and that he must call for more troops from his father. Arnaoutoglou takes this advice seriously and immediately calls for more reinforcements from Mani. That was the signal for the Greek troops gathered in the small harbor of Kitries to take over the city.

Already on March 17th 1821 the Maniates of Tsimova (Modern Areopoli) had risen the revolutionary flag. Approximately 2000 Maniates of "Western Sparta" headed by Petrobei Mavromichali, the Mavromichalis clan, the Mourtzinous and Christaioi families and Theodoros Kolokotronis approach the city from the South East taking over the hills surrounding the castle.

At the same time from the west and the north, Greek forces from the neighboring villages of Messenia approached the city under the lead of Papaflessas, Nikitaras and Anagnostaras. From the Taygetos villages of Sabazika (modern Akovos) and Megoloanastasova the forces moved in to Mardaki and from there west to Velanidia.

From Sitsova (modern Alagonia), Tsernitsiova (modern Artemisia) Mikroanastasova (modern Piges) with the others from Megaloanastasova and Sabazika together marched south to Dipotama and Tourles, hills around the castle of Kalamata. At the same time men from Lada and Karveli moved to Ai Lia and met the other forces from Taygetos at Dipotama on their way to Kalamata. All routes and paths connecting Kalamata with its rural region were taken over by Greek revolutionary forces leaving no way out.

Elias Mavromichalis announces to Arnaoutoglou on March 23rd 1821 that the castle of Kalamata is surrounded by Greek forces and the wisest thing for him to do is to resign. Arnaoutoglou realizing the situation surrendered and the city of Kalamata was liberate on the same day.

3 The Application

Since the first major connection of mapping to a specific event, that took place in 1854 when Dr. John Snow connected the cholera spread in London by mapping the way the epidemic moved spatially, visualizing data spatially has developed into its own scientific field. In the 1960s mapping was digitized, coordinates were stored on mainframe computers, and map graphics were the output via line printers. By the 1980s GIS had evolved into a multileveled tool available to a vast number of scientific fields. By analyzing data through space and time, either it be population fluctuation, property allocation or crisis prediction, GIS offered a unique spatial visualization. Archaeologists acknowledged quite early the role of GIS in data collection and management and in the 1990s GIS turned into a significant tool for archaeologists providing spatial and temporal analysis of the archaeological information and later on combined with remote sensing and modeling technologies. GIS in collaboration with location based application software provides a friendly, digital platform in a technologically innovative manner for the management and enhancement of archaeological and cultural heritage environments thus enhancing our understanding on temporal and spatial interactions [5, 6]. Through mixing digital technologies, web applications and GIS with the field of Cultural Heritage Management, the upcoming field of virtual – digital cultural heritage emerges. The use of technological means to present, sustain and conserve cultural heritage, whether it be visualization strategies, teaching tools or marketing features, is achieved through the integration of modern technology with existing tools [7].

In the case of the Kalamata 1821: Roads of Freedom project, "Kalamata Action Map" focuses on the city of Kalamata and its surrounding villages in a maximum radius of 60 km. This wide area includes the western Mani villages, east of Kalamata, as well as the villages west of Kalamata, such as Messini, Eva, as well as the Taygetos villages to the north. The maps for the application where created using GIS software,

ArcGIS. In order to correlate between the 19th century landscape and the present geography two historic maps of the area were used. These maps provided a base for villages and city names as well as road networks of the early 19th century, and the means to create trails, transportation routes and to determine the "hidden" routes used before the revolution. Many such pre-revolutionary roads can be recognized even today in the area at the foot of Taygetos mountain overlooking the Kalamata shoreline, the Messenian Gulf (Fig. 1).

Fig. 1. Southern Greece, with the adjacent Islands, A. Arrowsmith (1828) [II] [8]

The routes, trails and points that were imported to ArcGIS create an image of communication, commercial networks and transportation routes. By placing these historic points on a modern map of the area an alternate aspect of the world is laid open to the user. Routes following historic roads and trails for commerce or even for the march to the liberation of Kalamata can be relived and visited. These routes offer users the opportunity to physically follow the footsteps of the past incorporating visual aids such as pictures, paintings, videos etc.

From the development of the maps used, based on the bibliography, the Greek forces taking action in the March 23rd events, moved through coarse terrain in the area surrounding Kalamata. The largest distance was covered by the Mani clans 57.8 km via Kardamyli to Kalamata. The liberating forces surrounding the Castle and Center of Kalamata holding defense lines at Tourles Hill 560 m NE, Koumari Hill 710 m N, Nedontas Bridge 350 m W and Fragolimna Hill 245 m SE (Fig. 2).

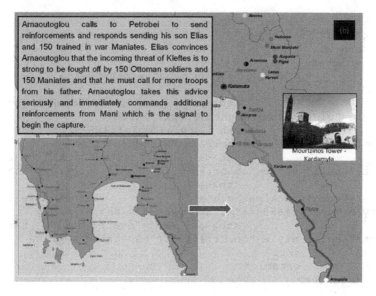

Fig. 2. Screen portraying the route the Maniates followed to Kalamata.

"Kalamata Action Map" presents points of interest situated on a basemap providing the platform for additional information regarding the spatial data to be presented. The integration of the GIS spatial analysis to the website was realized using HTML, JavaScript and CSS (Cascading Style Sheets) for web applications. Specifically, open source libraries were incorporated into the website's HTML code for the app page. CSS provides the necessary stylization for the pages (colors, background, alignment etc.) and Leaflet's open source JavaScript library provides the necessary tools for the map implementation. Leaflet supports most mobile and desktop platforms, the libraries used where adapted to the parameters needed for the "Kalamata Action Map".

```
<html>
<head>
  <title> Καλαμάτα 1821 - Δρόμοι Ελευθερίας </title>
  <metahttp-equiv="Content-Type"  content="Text/html;
charset=ISO-8859-7" />
  <metaname="keywords"  content="Kalamata1821,  Kala-
mata, history" />
  <meta name="description" content="..." />
  <meta name="robots" content="follow, index" />
  <meta  name="copyright"  content="Laboratory of Ar-
chaeometry    -    University    of    the    Peloponnese
(http://ham.uop.gr/en/research/labs/archaeometry)" />
  <meta    name="author"    content="vayiap@gmail.com
(Vayia Panagiotidis)" />
  <link rel="stylesheet" href="leaflet.css" />
  <script src="leaflet.js"></script>
```

```
    </head>
    <body>
     <center>
             <table id="logo_menu" width="90%">
             <tr>
                     <td><img src="images/logoK21.jpg"
style="width:89px;height:78.5px;"></td>
                     <th     style="color:#6699ff;    font-
family:Arial; font-size:22px;">Ιστορία</th>
                     <th><a   href="culture.html"   tar-
get="_self" style="color:#6699ff">Πολιτισμός</a></th>
                     <th><a
href="traditionalpaths.html"           target="_self"
style="color:#6699ff">Διαδρομές</a></th>
                     <th><a   href="events.html"   tar-
get="_self" style="color:#6699ff">Δράσεις</a></th>
             </tr>
             </table>
     </center>

     <center>
             <div id="mapid"></div>
     </center>
     <br>
     <center>
             <div                      id="espa"><img
src="images/logoESPA.jpg" height="10%"></a></div>
     </center>

     <script>
             var          mymap              =
L.map('mapid').setView([37.040, 22.117], 14.5);

     L.tileLayer('https://server.arcgisonline.com/ArcGI
S/rest/services/World_Imagery/MapServer/tile/{z}/{y}/{
x}', {maxZoom: 18}).addTo(mymap);

             var  marker_1  =  L.marker([37.04458266,
22.11402595]).addTo(mymap);
             mark-
er_1.bindPopup("<b>ΙερόςΝαόςΑγίουΙωάννη</b><br><a
href='images/thumps/             ai_giannis.html'
target='_blank'><img       src=images/ai_giannis.jpg
height=20%></a>").openPopup();

             var  marker_2  =  L.marker([37.04429066,
22.1162169]).addTo(mymap);
             mark-
er_2.bindPopup("<b>ΙερόςΝαόςΥπαπαντήςτουΣωτήρος</b><br
><a   href='images/thumps/  naiskos_ypapantis.html'
target='_blank'><img  src=images/naiskos_ypapantis.jpg
height=20%></a>").openPopup();
```

```
        var    marker_3    =    L.marker([37.0435481,
22.11302076]).addTo(mymap);
        mark-
er_3.bindPopup("<b>ΙερόςΝαόςΑγίωνΑποστόλων</b><br><a
href='images/thumps/          agioi_apostoloi.html'
target='_blank'><img    src=images/agioi_apostoloi.jpg
height=20%></a>").openPopup();

        var    marker_4    =    L.marker([37.046354,
22.116767]).addTo(mymap);

    marker_4.bindPopup("<b>ΚάστροτηςΚαλαμάτας</b><br><
a           href='images/thumps/kalamata_castle.html'
target='_blank'><img    src=images/kalamata_castle.jpg
height=20%></a>").openPopup();
    </script>
    </body>
</html>
```

[Excerpt from code written in History.html file]

Our approach to presenting the liberation of Kalamata through the Action map comprises of a base map centered on the castle of Kalamata and its surroundings were the city center was, before, during the ottoman occupation and the years after the liberation. As the city was fortified most of the houses, of the small rural type one next to another following the ground shape, some of them still inhabited today, were enclosed within the fortification. Economic improvement from the 18th century established a Greek "elite" that built great luxury houses outside the walls, such as the tower-house of Panagiotis Benakis, a rich and powerful man who had a significant role in the Orlov revolt. After the failure of the Orlovs, his tower-house was demolished by the Ottomans, but not his family chapel "Taxiarchaki" still existing in the Mavro-michali Square. Another important building of that period, located south of Ypapanti church near the castle, shows a typical rural architecture were small blocks of stones are bind with any kind of material in an asymmetrical way. The house belonged to the Korfiotaki family and later became the residency of the local Ottoman ruler Souleiman Arnaoutoglou. After the liberation, the square of Ayioi Apostoloi chapel became the city center were today all celebrations for the liberation take place and were a lot of neoclassical buildings of that era still remain [9].

The Map is organized into four sections or themes, History, Culture, Routes and Activities (Fig. 3). Each section includes its own map with selected relevant points and their connection. Points are initially collected using GPS and include the locations in the area that have been selected to present the story of Kalamata from the four different aspects. The points are clickable; on first click a balloon window appears containing a thumbnail and short description of the site. Additionally, the points work as hyperlinks to other pages where relevant information regarding the location and visual aids are presented [10, 11].

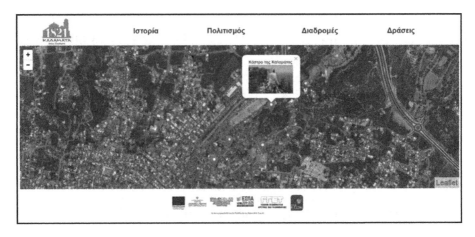

Fig. 3. Screenshot "History" tab "Kalamata Action Map"

The "Culture" section provides a compact review of the areas that play a part in molding Messenian culture from the 19th century to the present. A visitor can learn about ceremonies, traditions, trade and financial practices etc. Similarly, the "History" tab provides historic information of the period (Fig. 4). "Routes" and "Activities" are innovative features, creating in the first case actual networks connecting points of interest depending on the visitor's interests. For example, if someone wanted to hike the route the Mani clans followed from Kardamyli to Kalamata they would check into this page and click Kardamyli which in the Routes page is linked to the trail from Kardamyli to Kalamata as travelled in March 1821. "Activities" will provide insight over the area's map on events, concerts, reenactments etc. all connected to the Kalamata 1821: Roads of Freedom project.

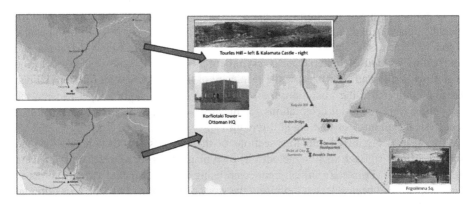

Fig. 4. Screen presenting the routes from Taygetos, West Messenia and East (Mani) and forces that surrounded the city.

4 Future Aims

Presented above is the evolution of the topobased application as initially given in short at the 42nd International Symposium in Archaeometry [12]. The implementation focused on only the GIS maps and the data they produced such as distances covered by the Greek revolutionaries. The goal is to create an overall application that will contain the data previously mentioned but will also incorporate further visual and simulative digital tools. Images will be enhanced with 360° image viewer, augmented reality for specific uses for example people praying in the Ayioi Apostoloi chapel or a view of the castle gate. Documentation will be enriched from the results of the historic research of the Academic team of the Kalamata 1821 project.

The potential seems endless when using digital technologies in cultural heritage. The Kalamata 1821: Roads of Freedom project aspires to become a paradigm for relevant projects not only in projecting historic information but by making that information come alive and make the reader, visitor, and student part of the experience.

References

1. Finlay, G.: History of the Greek Revolution, vol. 1 (2017). (Original work published 1861)
2. Christou, A: Πολιτικές και Κοινωνικές Όψεις της Επανάστασης του 1821 [The Political and Social Aspects of the Revolution of 1821]. Papazisis, Athens (2013). (in Greek)
3. Papagiannakis, G., Magnenat-Thalmann, N.: Virtual worlds and augmented reality in cultural heritage applications. In: Baltsavias, M., et al. (eds.) Recording, Modeling and Visualization of Cultural Heritage, pp. 419–430. Taylor & Francis Group (2006)
4. Brewer, A.: The Greek War of Independence, The Struggle for Freedom and from Ottoman Depression. The Overlook Press, New York (2011)
5. Ioannidis, M., Thalmann, N.M., Papagiannakis, G.: Mixed Reality and Gamification for Cultural Heritage (2017). https://doi.org/10.1007/978-3-319-49607-8
6. Malaperdas, G., Zacharias, N.: A geospatial analysis of mycenaean habitation sites using a geocumulative versus habitation approach. J. Geosci. Environ. Prot. **6**, 111–131 (2018). https://doi.org/10.4236/gep.2018.61008
7. White, M., Petridis, P., Liarokapis, F., Plecinckx, D.: Multimodal mixed reality interfaces for visualizing digital heritage. Int. J. Archit. Comput. **5**(2), 322–337 (2007). http://www.multi-science.co.uk/ijac.htm
8. Arrowsmith, A.: Orbis Terrarum Veteribus Noti Descriptio. A Comparative Atlas of Ancient And Modern Geography, From original Authorities, and upon a New Plan For The Use Of Eton School (1828). London, Published by the Author, 10, Soho Square
9. Σπηλιοπούλου, Ι.: Η πολεοδομική και αρχιτεκτονική εξέλιξη της Καλαμάτας από τα προεπαναστατικά χρόνια έως την περίοδο του μεσοπολέμου. Η τύχη της πόλης μετά τους σεισμούς. Στο: Γ. Ξανθάκη-Καραμάνου (επιστημονική εποπτεία), Α. Δουλαβέρας – Ι. Σπηλιοπούλου (επιστημονική επιμέλεια) Μεσσηνία: Συμβολές στην ιστορία και στον πολιτισμό της. Εκδόσεις Παπαζήση, Αθήνα (2012). (in Greek)
10. Foni, A.E., Papagiannakis, G., Magnenat-Thalmann, N.: A taxonomy of visualization strategies for cultural heritage applications. ACM J. Comput. Cult. Herit. **3**(1), Article 1 (2010). https://doi.org/10.1145/1805961.1805962
11. Droj, G.: Cultural Heritage Conservation by GIS, Társadalom – térinformatika – kataszter * GISopen konferencia (2010)
12. Panagiotidis, V.V., Malaperdas, G., Zacharias, N.: Digital enhancment of cultural heritage landscapes: the Kalamata 1821 project. In: 42nd ISA Book of Abstracts, p. 368 (2018)

DiscoVRCoolTour: Discovering, Capturing and Experiencing Cultural Heritage and Events Using Innovative 3D Digitisation Technologies and Affordable Consumer Electronics

Constantin Makropoulos[1], Dimitra Pappa[1(✉)], René Hellmuth[2],
Alexander Karapidis[3], Stephan Wilhelm[3], Vassilis Pitsilis[1],
and Florian Wehner[4]

[1] National Centre for Scientific Research (NCSR) "Demokritos",
Agia Paraskevi, Greece
dimitra@dat.demokritos.gr
[2] Institute for Human Factors and Technology Management,
University of Stuttgart, Stuttgart, Germany
[3] Fraunhofer Institute for Industrial Engineering IAO, Stuttgart, Germany
[4] University of Applied Sciences of Schmalkalden, Schmalkalden, Germany

Abstract. Recent years have seen the growing digitisation of cultural heritage, leveraged by innovative information technologies (imaging technologies, multimedia, virtual reality etc.). Advanced digitisation technologies have been instrumental in transforming conservation and scientific research methods regarding cultural heritage, as well as people's experience of cultural heritage relics, monuments and events, thus paving the way for novel consumer services.

The present paper revolves around the use of advanced 2D/3D digital scanning of large scale objects and surroundings and the valorisation of the digital spatial models produced, in order to advance preservation efforts, to enhance scientific research work and to create unique, immersive cultural experiences, using affordable consumer electronics. With regards to the latter, the proposed DiscoVRCoolTour prototype specifically targets the production, marketing and consumption of cultural tourism. Digitisation technologies are already in use in the context of cultural tourism (e.g. in museums and monuments). However, limited research and solutions can be found with respect to the interaction between cultural heritage, scan/photo and immersive technologies, potential customers' and visitors' experiences in the cultural tourism locations, events and attractions. Physical as well as virtual customer services based on digitisation technologies for cultural tourism attractions, locations and entire destinations are still not exploited properly.

Overall, a manifold of applications and services can be generated from the adoption and adaptation of relevant 2D/3D digital scanning technologies already applied in other sectors (e.g. construction industry). In this context, the paper first presents relevant digital technologies for digital data acquisition of large scale objects and surroundings and discusses critical aspects of the proposed solution, namely with regards to digital imaging, scan/photographing methods, virtual

© Springer Nature Switzerland AG 2019
A. Moropoulou et al. (Eds.): TMM_CH 2018, CCIS 961, pp. 232–249, 2019.
https://doi.org/10.1007/978-3-030-12957-6_16

reality experience, secure metadata storage, etc. Subsequently, the applications and expected benefits of the DiscoVRCoolTour prototype for cultural heritage conservation and valorisation are discussed, including new emerging forms of cooperation and novel "technology-induced" business models.

Keywords: Cultural heritage · 3D digitisation · Laser scanning

1   Introduction

Advanced digitisation technologies have been instrumental in transforming conservation and scientific research methods in cultural heritage, as well as people's experience of cultural heritage relics, monuments and events ([1–4]), thus paving the way for new scientific epistemologies and novel consumer services. Cultural heritage artefacts that traditionally were presented in two-dimensional form, as drawings and/or in digital two-dimensional (2D) form (CAD artefacts), are increasingly captured, modelled and visualised in three dimensions and/or in 3D virtual environments [5]. New technologies and techniques are emerging (e.g. photogrammetry, laser scanning etc.), allowing for more accurate digital capture of 3-dimensional objects and surfaces. Early efforts included modelling and rendering of artefacts and architecture from photographs (e.g. [6, 7]). Current applications employ advanced non-contact close or long range scanning, modelling, analysis and computer-based visualisation tools to produce:

- three-dimensional (3D) recordings of archaeological sites and buildings (e.g. [8–10]) and of small objects (e.g. [11]);
- three-dimensional visualisations of cultural heritage sites, using airborne scanning and imaging ([12]) or from geospatial information ([13]).
- four- and five- dimensional models of the evolution of 3D spatial reconstructions over time and from different perspectives ([7, 14]).

Digital technologies hold the promise of enhancing the experiential and interpretive dynamics of the cultural heritage representations, in a way that can transform a person's understanding of a cultural artefact. Digitisation technologies are already in use in the field of heritage (e.g. in museums or monuments). Limited research and solutions can be found in the interaction between cultural heritage, scan/photo and immersive technologies, potential customers and visitors' experiences in the cultural tourism locations, events and attractions. The use of advanced 2D/3D digital scanning of small and large-scale objects and surroundings and the valorisation of the digital spatial models produced has the potential to advance preservation efforts, enhance scientific research work and create unique, immersive cultural experiences, using affordable consumer electronics.

Acknowledging the potential of 3D digitisation technologies, the present paper presents the design of an innovative system for discovering, capturing and experiencing cultural heritage. The DiscoVRCoolTour system thus promotes the creation of new value chains for tourism. To produce the system design, the analysis follows a problem-centred approach that builds on the principles of the Design Science Research Methodology (DSRM) for Information Systems development ([15, 16]). In the design-science paradigm, knowledge and understanding of a problem domain and its solution are achieved in the building and application of the designed artefact. The present paper focuses on the early stages of the DSRM methodology, starting from the identification of opportunities for innovation to the design of the DiscoVRCoolTour system. In this process, three distinct pillars were investigated: (a) the requirements of the application area (cultural tourism); (b) the current state-of-the-art and emerging directions in digital data acquisition of large buildings, environments and objects; and (c) 3D digitisation applications in cultural heritage.

The remainder of this paper is structured as follows: Sect. 2 discusses the new opportunities created by advanced digitisation technologies in the field of cultural heritage, including an introduction to the specific needs and requirements of the targeted application domain of the DiscoVRCoolTour system: cultural tourism, and briefly summarises related work. Section 3 summarises the findings of our survey of relevant digital technologies for digital data acquisition of large scale objects and surroundings. Section 4 presents the design of the DiscoVRCoolTour system, outlining the core elements of the architecture. Section 5 concludes the paper.

2   3D Digitisation Applications in Cultural Heritage

Advanced technologies (2D/3D digital acquisition, three-dimensional modelling, as well as new data processing and management technologies etc. can bring significant benefits to cultural heritage communication, preservation, and scientific research. In recent years, numerous initiatives have been launched, involving the modelling and rendering of digital cultural heritage in 3D for research and preservation and/or communication purposes.

2.1   Preservation & Scientific Research in Cultural Heritage

Digitisation technology has significantly contributed to the advancement of **preservation & scientific research.** New archaeological epistemologies have emerged from the marriage of archaeology, computer science, and engineering novel applications for 3D scanning, photogrammetry, and computer modelling, coupled with global positioning systems (GPS) and geographic information systems (GIS), and virtual collaborative environments, stand at the core of modern **cyber-archaeology** and **cyber-archaeometry** ([17–20]).

In this context, digital artefact modelling is gradually becoming an indispensable instrument of archaeology [21]. The three-dimensional recording and archiving of archaeological sites is imperative for the **documentation** of cultural heritage for future generations [8]. Modern technologies allow for the best documentation and facilitate novel digital survey methodologies for high-quality research (e.g. metric surveys of historic buildings, sites and landscapes). The three-dimensional (3D) way of recording allows for more realistic and less biased capturing, compared to two-dimensional archives. The 3D models produced can be used for sharing accurate information about a specific cultural heritage artefact for future study and revision. Garstki [22] notes that a 3D model may capture the visual appearance of the original artefact and can be manipulated in three dimensions or investigated from numerous perspectives. Advanced digitisation technologies can provide the means for archaeological practitioners to conduct rapid and cost-effective **surveys** [23]. Virtual 3D artefact models can significantly advance scientists understanding of the past [4], for example, in order to provide new insight into the construction history of the Tomb Chamber of the Holy Aedicule of the Holy Sepulchre in Jerusalem [24], or in order to facilitate the investigation of the geometric decorative apparatuses of Alhambra [25]. A combination of laser scanning and photogrammetric approaches have been used to generate 3D models and virtually reconstruct parts of Stonehenge, allowing researchers to discover new evidence [13].

Digitisation technologies emerge as key enablers of **preservation** and for future **in situ monitoring** of the archaeological remains, providing detailed and contextualised knowledge of the position, size, shape and identity of the components of a historic site. This is particularly important in excavations when no in situ preservation for structures can follow, or in the case of open-air heritage, which typically remains exposed to weathering by the elements of nature (such as weather conditions, earthquakes etc.), and human destruction (e.g. in the case of underwater cultural heritage [26], rock art [27] etc.). Doulamis et al. [14] advocate the need for 5D modeling for the comprehensive reservation and assessment of outdoor large scale cultural sites. This involves the enhancement of 3D geometry models through the addition of the "time" and "level of detail" dimensions, to assess the spatial and temporal diversity of cultural heritage objects, and accommodate the needs of the different actors involved, respectively.

2.2 Communication of Digital Heritage and Cultural Tourism

The later has revolutionised the concept of "museum", leading to the emergence of novel museum services powered by the **3D digitisation of museum artefacts**. In recent decades many GLAM organisations (i.e. Galleries, Libraries, Archives, and Museums) have launched digitisation projects to improve representation, engage visitors with content in new, innovative ways and support cultural revitalisation. Advances in digitisation technologies provide the means for new forms of engagement with museum-held heritage via 3D multimedia-rich museum websites, online "walk-through" museum explorations [28], virtual museum exhibitions ([29, 30]), virtual environment system installed within a real museum [31], etc. This has also led to novel context-aware representations of cultural heritage, produced from merging 3D models of artefacts, like in the case of the MUVI - Virtual Museum of Daily Life [32].

It has also helped enhance the cultural experience onsite, giving birth to **digital applications for cultural tourism.** Tourism touches upon people's connections with other peoples, places and the past. One of the defining principles of tourism practice lies in the fact that people physically travel to spaces outside of their daily life and experience other or different environments [33]. Through particular types of setting, these spaces allow us to live out various types of touristic realities [34]. In this context, the encounter with, and participation in, what are usually defined as "different cultures", are important elements in most forms of tourism "Modern consumers want context related, authentic experience concepts and seek a balance between control by the experience stager and self-determined activity with its spontaneity, freedom and self-expression" [35].

Cultural locations and spaces can be enriched by scanning and overlaying virtual annotations on top of these places. Devices can be directed at the point of interest, and 2D/3D, e.g. texts, sounds, icons, videos, and digitised 3D artefacts, are added to the users' view. Moreover, these kinds of applications provide cultural tourists with fast knowledge acquisition of immediate cultural location-based information of specific points of interest [36], allow them to explore personal cultural locations and points of interest [37] and give them opportunities to discover new or unknown knowledge ([7, 38]). From the provider view, new business models and opportunities can be identified and initiated by new digitisation technologies used [39] and the competitiveness for a cultural location or site will increase [40]. Every location, monument and tour has a specific "wow-factor" that has to be captured and transported via a digital technology and a digital communication channel. This represents the principal objective of the DiscoVRCoolTour system: Its aim is to explore innovative digitisation technologies to create an environment for consumers to discover, capture and experience cultural heritage and events in an innovative way, using innovative 3D digitisation technologies and affordable consumer electronic equipment.

3   Technologies for Digital Data Acquisition of Large Buildings, Environments and Objects

This section first presents the results of a technology analysis. The technology analysis includes various technologies to create digital twins. It deals with the categories laser scan, photogrammetry, BIM model, Software that creates BIM models from a point cloud, Smartphone Apps and Panorama Photos which are created by trolleys, drones or manual modelling. The table in Fig. 1 is separated in the parts software and hardware. In addition, technologies are presented, mainly from start-ups that form hybrid forms of the above-mentioned technologies or that combine different technologies. The technology analysis is based on an international search across different industries, with a focus on the construction industry. Many solutions can be assigned to the division of

	Software							
Laserscan	Bentley Pointools	Faro (Scene, WebShare)	Leica	NavVis	PointCab	RealWorks (T	Scalypso	Z+F
Photogrammetry	AgiSoft (Photoscan)	ContextCapture	Photomodeler X	Pix4D	Reality Capture			
BIM-Model	Atudodesk Revit	Vectorworks	ArchiCAD	AutoCAD	Allplan	MicroStation	Dynamo	Rhinoceros
Laserscan to BIM-Model	Snapkin	viatechnik	Faro VirtuSurv	EdgeWise	IMAGIniT			
Smartphone App	Canvas	Magic Plan	Cam to Plan					
Panorama Photo	PanormaStudio	easypano	AutoStitch	Hugin	Panorama Tools	NaVVis	LMStitch	Panorama Factory
	Hardware							
Laserscan	Faro Focus 350	Faro Freestyle	NavVis Trolley M6	Leica ScanStation P50	Leica BLK 360	Trimble TX5	Surphaser 105HSX	Topcon GLS 2000M
Photogrammetry	Scarabot X8 (drone)	Intel Falcon 8+ (drone)	DJI Matrice 200 (drone)	RKM 8X (drone)	(reflex cameras)			
BIM-Model								
Laserscan to BIM-Model								
Smartphone App								
Panorama Photo	Giraffe360	Google Gigapixel	(reflex cameras)					

Fig. 1. Overview of current software and hardware technologies

plan and build [41]. The goal is to evaluate existing technologies from different industries and to assess the usability and experience, through digital models, for cultural heritage.

It should be mentioned at this point that the hardware and software solutions listed above are not exhaustive. Rather, the listing is seen as the current state of current solutions that offer interesting approaches and services.

3.1 Laser Scan

Laser scanning can be subdivided into different application areas. Terrestrial laser scanning (TLS) means the use of equipment from a fixed standpoint (tripod). TLS is increasingly being supplemented or even replaced by mobile laser scanning (MLS). In the latter case, the object detection is carried out by moving platforms, such as cars, trolleys, trains or boats, but also by drones. The trend is increasingly in the direction of multi-parameter detection by multisensory systems, as geometric, spectral and/or thermal information. These technologies can be used in a wide variety of areas, from surveying in plant construction, architectural surveying, archaeological documentation to visualisation and reconstruction [42]. These technologies can also be used to record cultural heritage and landmarks. After a rapid development of the scanners, a consistent expansion of the software offers is now on the agenda. Based on various applications in the field of architectural surveying and surveying of historical structures, the state of the art will be presented and illustrated and explained by means of practical examples.

3D Laser Scan technology is a fast method to obtain three-dimensional information of buildings, outdoor facilities and other monumental objects [43]. The laser scanning can be divided into five phases [44]. These phases are shown in Fig. 2.

Fig. 2. Phases of laser scanning

For the exemplary scans of this paper, the BLK360 from Leica was used. The choice fell on the BLK360 because of its easy handling and good transportability. The scanner weighs less than 1 kg and is 10 cm × 16 cm in size. One measurement is accurate to 5 mm (0.25"). The range is 60 cm–60 m. A high-resolution scan can easily capture between 50 and 80 million points in 3 min and be over 1 GB in size.

Figure 3 shows image renderings of the meshed 3D objects of the laser scan in a test room. In the area of the laser scanning that is marked in red, it is noticeable that the laser images the surface of the masonry in a constant, consistent quality over the entire area. However, the laser captures the smooth masonry surface much rougher than it really is.

Figure 4 shows a photograph of the masonry, illustrating the actual surface structure of the material. This is actually smoother than the scan result captured by the laser.

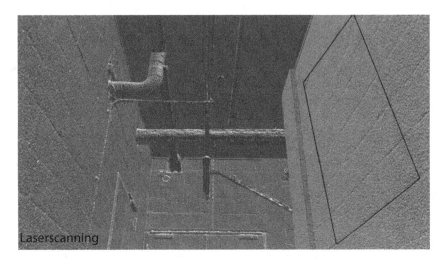

Fig. 3. Properties of the laser scan phases of laser scanning (Color figure online)

Fig. 4. Photograph of the actual masonry surface

Except for the indicated inaccuracies, the laser scan is very qualified to capture buildings or other objects digitally and to get a good virtual impression.

3.2 Photogrammetry

Photogrammetry is about making measurements from photographs, especially for recovering the exact positions of surface points. It is a measurement technique that is used to extract the geometry, displacement, and deformation of a structure using photographs or digital images. The concept of photogrammetry is related to perspective concept and goes back to Leonardo da Vinci. You can divide photogrammetry into two types of subjects: Interior captures and exterior [45]. For the recordings the Nikon D810 with the lens Sigma type 24 mm was used in this experiment. In addition, the software RealityCapture is used. For photogrammetry, the same phases apply as for laser scanning (see Fig. 2). In the planning phase, a photography scheme is developed in photogrammetry to cover the test room (see Fig. 5) with photos wherever possible. The special nature of the test room must be taken into account.

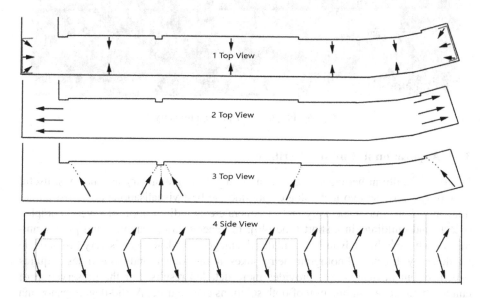

Fig. 5. Photography scheme

When looking at the image rendering of the photogrammetry from Fig. 6, it is noticeable that the surface structure of the scan at the bottom provides the more realistic result, but deteriorates qualitatively towards the top. The qualitative deterioration upwards is due to the photography scheme. While the lower part of the marked area was photographed in parallel from the opposite wall, only the camera angle was adjusted upwards for the upper part of the area. A better result for the upper part of the marked area would have resulted in an increase of the camera tripod.

Depending on the field of application, the paper helps to decide which method is more suitable for different cultural sights. The Leica BLK360 laser scanner used is more suitable for everyday use than photogrammetry, as without experience, in a shorter time in everyday situations, sufficiently good results can be achieved. Despite the higher quality potential for everyday life, the photogrammetric method is too costly in many aspects and requires skilled personnel.

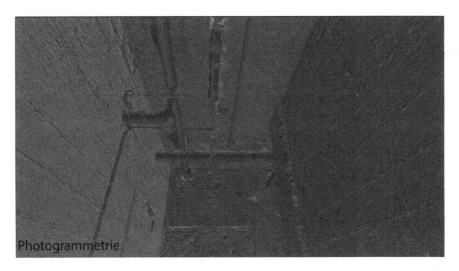

Fig. 6. Properties of photogrammetry

3.3 Application in Cultural Heritage

In terms of cultural heritage, laser scanning and photogrammetry are the most useful technologies. These two technologies provide the best visualisations of tourist attractions in a short time. Especially laser scanners create a digitisation of several complex objects and buildings in a short time [46]. In general, laser scanning and photogrammetry each have their advantages and disadvantages. The laser scan is very fast and can be done by laymen. Photogrammetry takes a bit longer and sometimes requires expertise, but in many cases provides more detailed results. For the preservation of cultural heritage, a combination of both solutions can be used. A 360-degree panorama picture with a point cloud deposited. This makes it possible to take measurements in the panoramic image and store information at certain, freely selectable points. The unique selling point is that the points of interest are visible from all perspectives, but only need to be created once [47]. The points of interest are stored in the point cloud. These are thus visible from every view in the panorama picture. The system of NavVis [48] is shown in Fig. 7. Another not to be underestimated application concerns the restoration of missing building parts. By transferring the scan data into vectorised geometry data (Scan to BIM), parts can be reproduced very efficiently and restored on demand. If the original state is to be retained for special reasons, the missing part can be made visible with augmented reality. For reproduction, more and more 3D printing processes are used.

The top two images in the collage are the panoramic images. The two pictures in the lower area are the representations of the point cloud. The blue point (point of interest) is viewed from two different perspectives. It should be clear that the point of interest is fixed to the point cloud, not to the panoramic image. For virtual tours of cultural sites, this technology is very useful to provide information about e.g. a monument to the "virtual tourist". The method is referred by us as the star method.

Fig. 7. Panorama Picture and point cloud with point of interest

Information is fixed and deposited at the individual points (stars) of the point cloud. From each viewing angle, the information is thus apparent. Another perspective and point cloud view is shown in Fig. 8.

The point cloud is switched to transparent in Fig. 8. The point of interest is also visible in this perspective. For the "virtual tourist", the three-dimensional models can be provided in various applications. An up-to-the-minute and easy way for the user are browser viewers specialised for digital models. Explorations can be made via smartphone, tablet or PC. As mentioned above, three-dimensional models are obtained using

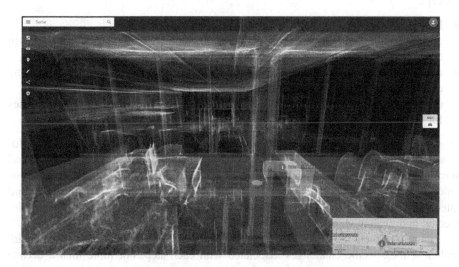

Fig. 8. Point cloud with point of interest

laser scanning and photogrammetry. The photogrammetry is more accurate, but this is negligible for a visual impression without the aim of a survey. The use of a point of interest on the other hand is very helpful for the use case cultural heritage. Especially because of the integration of additional information, the application provides in context with people's experience of cultural heritage relics, monuments and events a new way of sightseeing and visitation.

4   The Proposed Solution: DiscoVRCoolTouR

The DiscoVRCoolTouR system specifically targets cultural heritage digitisation to promote the creation of new value chains for tourism. Emphasis is placed on the creation of authentic and rich user experiences. To this end, DiscoVRCoolTouR experiments with innovative digitisation technologies to enable a manifold of novel cultural heritage services and applications. The objective is to develop new technology-based solutions for virtual immersive experience of cultural events, arts and heritage. For this purpose, innovative technologies will be adapted, combined and deployed, in order to support the digitisation and 3D representation of cultural heritage. The digital spatial models produced will build a digital vault, in the form of an output format-independent database. The ubiquitous availability of output devices of HMDs, hand-helds, tablets and projections, as well as the broadband availability of digital content, will enable future commercial as well as non-commercial applications of the stored data. Content editing tools and processes will be made available to facilitate the creation of new applications on new devices, for different audiences and for different purposes.

The design of the DiscoVRCoolTouR system is based on well-defined objectives, identified and elaborated following the DSRM approach. The following specific objectives have been identified during the development of the DiscoVRCoolTouR concept:

- Discover new ways for users to experience cultural heritage;
- develop/adapt and deploy state-of-the-art digitisation technologies for application in the area of cultural heritage and cultural events;
- describe reference workflows for new cultural heritage services, with specific qualification demands and use of technology and tools;
- evaluate the usefulness and efficiency of scanning and digital imaging technologies for cultural heritage valorisation.

Overall, DiscoVRCoolTouR will study, adapt, implement and evaluate technologies, tools and methods for the 2D/3D digital scanning of large scale objects and surroundings and the valorisation of the digital spatial models for the creation of unique, immersive cultural experiences. New "technology induced" business and interaction models, both commercial and noncommercial, will be developed, implemented and evaluated.

Figure 9 depicts the DiscoVRCoolTouR cultural heritage digitisation value chain, including the specific challenges associated with each step. For more than a decade there have been extensive measures for the digital processing and visualisation of collections. While the digital processing of cultural objects in 2D has long been one of the core tasks of libraries, archives and museums, 3D digitisation of three-dimensional artefacts is still a challenge. In order to meet the growing demand for optimised scanning in 3D for large rooms and environments, DiscoVRCoolTouR intends to close this gap both for the acquisition of raw data and the use and utilisation in new co-operation forms and business models.

Fig. 9. DiscoVRCoolTouR cultural heritage digitisation value chain and challenges

The architecture of DiscoVRCoolTouR (Fig. 10) will consist of several subsystems that are integrated together to provide the user with a comprehensive, motivating experience. Critical aspects of the proposed solution include digital imaging, scan/photographing methods, virtual reality experience, secure metadata storage, etc.

Fig. 10. DiscoVRCoolTouR architecture

Content Acquisition and Digitisation by Industrial Image Capture

Industrial image processing is widely developed in the field of non-destructive technologies for Quality Assurance. The demands on process and product quality have made the precision of the optical test methods more and more extensive. The field of metrology covers a wide range of applications. It is possible to accurately measure shape and position tolerances, angular misalignments, bores and wall thicknesses.

Industrial image data acquisition and processing is extremely precise but also correspondingly expensive and can only be used to a limited extent economically for the mass market or applications with less stringent requirements.

New Integrated Technology Carriers (Laserscanners, High-Resolution Cameras)

The digitisation of rooms and environments is simplified by new integrated technology carriers. Mobile (i.e. mobile/flyable) platforms integrate a large number of optical sensors in addition to HD cameras as well as laser scanners and additional measuring transducers to simultaneously measure field strength measurements of earth's magnetic field and WLAN access points provide a multimodal dataset as a digital spatial model and build a starting point for a wide range of utilisation concepts for new business models. These mobile technology carriers (e.g. NavVis) are to be used within the framework of the project.

Visualisation by Immersive VR Technology

Walk-through digital spatial models on a 1:1 scale are now used especially for architecture and building visualisations. The viewer enters a multi-sided projection system in which he can experience photorealistic computer models mostly on the basis of CAD/BIM (Building Information Models) Data. The tracking and the corresponding perspective data projection allow a spatial quality of experience, which is only surpassed by reality. For the development of the final experience of the user, a system named CAVE will be used as a scalable multi layered projection room for immersive experiences ([49, 50]), making use of the models produced with 3D scanning technologies and other existing content that will be demonstrated in the virtual space, such as text, videos, audio or other 3D models that will be provided. Through the DICE platform [51] the user will be able to navigate in the 3D space, interact with the additional content and information included in it and be part of a holistic and engaging VR experience. Being in the centre of the event, the user will have access to sights, sounds, feelings and emotions that offer an advanced level of understanding of the demonstrated experience.

Head Mounted Displays (HMD's) as a Cost-Effective Mass Medium for Interactive, Tracked 3D Play Platforms

Since about 4 years visual output devices conquer the mass market with Oculus, HTC, Microsoft and Sony, more and more players, also in connection with game consoles penetrate the end user and wait for appropriate content. In many cases it can be seen that the available solution is still looking for an adequate problem. Nevertheless, a serious bottleneck is being lifted by falling prices and the widespread penetration of these devices.

360 Degree Camera-Technology

In photography, an omnidirectional camera (from "omni", meaning all) is a camera with a 360-degree field of view. 360-degree cameras are important in areas where large visual field coverage is needed, such as in panoramic photography and in combination with VR-HMD´s to create cutting-edge, dynamic environments. For the average consumer, a 360-degree camera offers an unbelievably effective method of capturing moments and memories in a more detailed and vivid sense than any photograph ever could!

Innovation-Driven Consumer Market
With declining hardware prices and extremely high-resolution cameras as well as cloud-based processing algorithms, straightforward and robust applications are created that are especially suited for the imaging of large-scale objects with characteristic surfaces (materiality and patina). The first applications with panoramic photography on the web were already shown in the early 1990s as web applications (Quicktime VR). The acquisition, processing and representation of the digital 2D image data were reserved for experts and software developers. Latest developments in photogrammetry i.e. by realities.io show significant potential for further applied research. The most mesmerising places around the world can be picked as a spot on the globe and virtually be travelled by the push of a button.

5 Benefits, Results and Conclusions

The DiscoVRCoolTouR approach shows that digitisation also impacts the tourism sector enabling different new business opportunities for tourism-oriented cultural locations and monuments. On the technological level, the DiscoVRCoolTouR approach will result in significant enhancement of knowledge and experience in the areas of content generation, business model development and system integration, particularly on 3D display technology and virtual/augmented reality that could be applied in future customer projects in the tourism sector. Using these technologies will enable tourism destinations to differentiate tourism products significantly.

In detail, the adapted technology implementation in the area of cultural heritage will facilitate the usage of the mentioned technology/methodology of the DiscoVRCoolTouR solution in large scale cases enabling the preservation of cultural heritage and providing comprehensive and motivating experience in an innovative way by combining 3D digitisation technologies and affordable consumer electronic equipment.

Development of socio-cultural exchange platforms: Platforms enhancing the school class dialog through students communicating the culture of their country in live sessions and explaining individual content with visual examples based on visual models located on internet portals (e.g. SteamVR) are meant to convey and strengthen the dialogue between nations and the mutual appreciation of cultural life and origin. Approaches such as the project "Multaka – Meetingpoint Museum" in Berlin may serve as a possible model for implementation.

Tour planning and detailed highlight presentation: Provision of a planning basis for special interest groups i.e. disabled people, elderly people, students, scientists. The VR digital Tour can provide a very realistic view and representation on the tour logistics and object access. So, the complete set of digital representation for a tour can answer questions that are previously unknown. Based on this scenario, new business opportunities for tour providers (value-added information), tourism associations (marketing) or travel agencies (information source for customer consulting) can be developed. Furthermore, today's job profiles of employees working in these areas will be reshaped according to AR/VR enriched technologies provided in their companies. New competences could be derived from the new work tasks to enable employees to cope with the needs of their digitally enriched work environments.

Marketing of cultural events, historical objects and gastronomic/culinary facilities: In VR applications not only an optical space effect arises since the ambience and sense perception in combination with the MP3 surround sound format offer a previously unavailable world of experience for potential customers. The opportunity of data enrichment e.g. the user case-oriented combination with information and third-party media content can create new forms of interaction between customers and suppliers, to purchase tickets directly, etc. The decentralized marketing of cultural tourism facilities for potential target groups will be massively facilitated by the digitisation of entire locations.

New technology solutions: The digital object models of the event locations provide the organizers with high-quality data bases for planning. For example, the planning of logistic processes can be simulated as well as rescue and emergency concepts, and scenarios can be planned more clearly on the basis of the digital models. On the other hand, guests and customers can take their seats in almost live images and assess the quality of the position in the room.

Use of the scanned location/event for the travel industry: Travel guide, travel agent for "cultural tours", tourism organisation and marketing of the destination, an amphitheatre in original size with an almost original soundtrack can make the experience quality that a "hot spot" has to offer. An avatar or a real guide can use the dialog to provide details about the local attractions which provide an intensive journey preparation and anticipation of the experience in reality.

In summation, DiscoVRCoolTouR is not merely a technological approach. It is the alignment of new digital technologies combined with concrete application fields where business models with emerging new/re-shaped competence-based job profiles to cope with the work tasks can be derived from. So, let the future begin …

References

1. Bianchini, C., Ippolito, A., Bartolomei, C.: The surveying and representation process applied to architecture: non-contact methods for the documentation of cultural heritage. In: Handbook of Research on Emerging Digital Tools for Architectural Surveying, Modeling, and Representation, pp. 44–93. IGI Global (2015)
2. Brusaporci, S. (ed.): Digital Innovations in Architectural Heritage Conservation: Emerging Research and Opportunities: Emerging Research and Opportunities. IGI Global (2017)
3. Neuhofer, B., Buhalis, D., Ladkin, A.: A typology of technology-enhanced tourism experiences. Int. J. Tour. Res. **16**(4), 340–350 (2014)
4. Liritzis, I., Al-Otaibi, F.M., Volonakis, P., Drivaliari, A.: Digital technologies and trends in cultural heritage. Mediterr. Archaeol. Archaeometry **15**(3), 313–332 (2015)
5. Georgopoulos, A., Kontogianni, G., Koutsaftis, C., Skamantzari, M.: Serious games at the service of cultural heritage and tourism. In: Katsoni, V., Upadhya, A., Stratigea, A. (eds.) Tourism, Culture and Heritage in a Smart Economy. SPBE, pp. 3–17. Springer, Cham (2017). https://doi.org/10.1007/978-3-319-47732-9_1

6. Debevec, P.E., Taylor, C.J., Malik, J.: Modeling and rendering architecture from photographs: a hybrid geometry-and image-based approach. In: Proceedings of the 23rd Annual Conference on Computer Graphics and Interactive Techniques, pp. 11–20. ACM (1996)
7. Ioannides, M., et al.: Online 4D reconstruction using multi-images available under open access. In: ISPRS Annals of the Photogrammetry, Remote Sensing and Saptial Information Sciences, vol. II-5 W, 1, pp. 169–174 (2013)
8. De Reu, J., et al.: Towards a three-dimensional cost-effective registration of the archaeological heritage. J. Archaeol. Sci. **40**(2), 1108–1121 (2013)
9. González-Aguilera, D., Muñoz-Nieto, A., Gómez-Lahoz, J., Herrero-Pascual, J., Gutierrez-Alonso, G.: 3D digital surveying and modelling of cave geometry: Application to paleolithic rock art. Sensors **9**(2), 1108–1127 (2009)
10. Remondino, F., El-Hakim, S., Girardi, S., Rizzi, A., Benedetti, S., Gonzo, L.: 3D virtual reconstruction and visualization of complex architectures-The 3D-ARCH project. In: International Archives of the Photogrammetry, Remote Sensing and Spatial Information Sciences, vol. 38(5/W10) (2009)
11. Boehler, W., Heinz, G., Marbs, A.: The potential of non-contact close range laser scanners for cultural heritage recording. In: International Archives of Photogrammetry Remote Sensing and Spatial Information Sciences, vol. 34(5/C7), pp. 430–436 (2002)
12. Lasaponara, R., Masini, N.: Full-waveform Airborne Laser Scanning for the detection of medieval archaeological microtopographic relief. J. Cult. Herit. **10**, e78–e82 (2009)
13. Bryan, P., Dodson, A., Abbott, M.: Using geospatial imaging techniques to reveal and share the secrets of Stonehenge. Int. J. Herit. Digit. Era **3**(1), 69–81 (2014)
14. Doulamis, A., et al.: 5D modelling: an efficient approach for creating spatiotemporal predictive 3D maps of large-scale cultural resources. In: ISPRS Annals of Photogrammetry, Remote Sensing & Spatial Information Sciences (2015)
15. Peffers, K., Tuunanen, T., Rothenberger, M.A., Chatterjee, S.: A design science research methodology for information systems research. J. Manage. Inf. Syst. **24**(3), 45–78 (2007)
16. Hevner, A.R., March, S.T., Park, J., Ram, S.: Design science in information systems research. MIS Q. **28**(1), 75–105 (2004)
17. Forte, M.: Cyber archaeology: 3D sensing and digital embodiment. In: Forte, M., Campana, S. (eds.) Digital Methods and Remote Sensing in Archaeology. QMHSS, pp. 271–289. Springer, Cham (2016). https://doi.org/10.1007/978-3-319-40658-9_12
18. Levy, T.E., Smith, N.G., Najjar, M., DeFanti, T.A., Kuester, F., Lin, A.Y.M.: Cyber-Archaeology in the Holy Land. California Institute for Telecommunications and Information Technology (Calit2), UC San Diego (2012)
19. Liritzis, I., Volonakis, P., Vosinakis, S., Pavlidis, G.: Cyber-archaeometry from cyber-archaeology: new dynamic trends in archaeometric training and research. In: Virtual Archaeology (Methods and benefits). Proceedings of the Second International Conference held at the State Hermitage Museum, pp. 38–40. The State Hermitage Publishers, Saint Petersburg (2015b)
20. Liritzis, I., et al.: Delphi4Delphi: first results of the digital archaeology initiative for ancient Delphi, Greece. Antiquity **90**(354) (2016)
21. Historic England. 3D Laser Scanning for Heritage: Advice and Guidance on the Use of Laser Scanning in Archaeology and Architecture. Historic England, Swindon (2018)
22. Garstki, K.: Virtual representation: the production of 3D digital artifacts. J. Archaeol. Method Theor. **24**(3), 726–750 (2017)
23. McCarthy, J.: Multi-image photogrammetry as a practical tool for cultural heritage survey and community engagement. J. Archaeol. Sci. **43**, 175–185 (2014)

24. Moropoulou, A., et al.: Five-dimensional (5D) modelling of the Holy Aedicule of the church of the Holy Sepulchre through an innovative and interdisciplinary approach. Mixed Reality and Gamification for Cultural Heritage, pp. 247–270. Springer, Cham (2017). https://doi.org/10.1007/978-3-319-49607-8_9

25. Marco, R.: Unfolding geometry from unity: digital survey and 3D modeling of islamic decorative apparatus in Generalife Palace, Alhambra. In: Cocchiarella, L. (ed.) ICGG 2018. AISC, vol. 809, pp. 664–676. Springer, Cham (2019). https://doi.org/10.1007/978-3-319-95588-9_55

26. Kan, H., Katagiri, C., Nakanishi, Y., Yoshizaki, S., Nagao, M., Ono, R.: Assessment and significance of a world war II battle site: recording the USS Emmons using a high-resolution DEM combining multibeam bathymetry and SfM photogrammetry. Int. J. Nautical Archaeol. (2018)

27. Horn, C., Ling, J., Bertilsson, U., Potter, R.: By all means necessary–2.5D and 3D recording of surfaces in the study of Southern Scandinavian rock art. Open Archaeol. **4**(1), 81–96 (2018)

28. Proctor, N.: The Google art project: a new generation of museums on the web? Curator Mus. J. **54**(2), 215–221 (2011)

29. Patel, M., White, M., Walczak, K., Sayd, P.: Digitisation to presentation: building virtual museum exhibitions. In: Vision, Video and Graphics (2003)

30. Lepouras, G., Katifori, A., Vassilakis, C., Charitos, D.: Real exhibitions in a virtual museum. Virtual Reality **7**(2), 120–128 (2004)

31. Charitos, D., Lepouras, G., Vassilakis, C., Katifori, V., Charissi, A., Halatsi, L.: Designing a virtual museum within a museum. In: Virtual Reality, Archeology, and Cultural Heritage: Proceedings of the 2001 Conference on Virtual reality, Archeology, and Cultural Heritage, vol. 28, no. 30, p. 284 (2001)

32. Chiavarini, B., Liguori, M.C., Guidazzoli, A., Verri, L., Imboden, S., De Luca, D.: On-line interactive virtual environments in Blend4web. The integration of pre-existing 3d models in the MUVI-Virtual museum of daily life project. In: Proceedings of Electronic Imaging and the Visual Arts-EVA, pp. 117–124 (2017)

33. Selänniemi, T.: On holiday in the liminoid playground: play, time, and self in tourism. In: Bauer, T.G., McKercher, B. (eds.) Sex and Tourism: Journeys of Romance, Love, and Lust, pp. 19–34. Haworth, New York (2003)

34. Shields, R.: Places on the Margin – Alternative Geographies of Modernity. Routledge, London (1991)

35. Binkhorst, E., Den Dekker, T.: Agenda for co-creation tourism experience research. J. Hosp. Mark. Manage. **18**(2/3), 311–327 (2009)

36. Yovcheva, Z., Buhalis, D., Gatzidis, C.: Empirical evaluation of smartphone augmented reality browsers in an urban tourism destination context. Int. J. Mob. Hum. Comput. Interact. **6**(2), 10–31 (2014)

37. Leue, M.C., Jung, T., tom Dieck, D.: Google glass augmented reality: generic learning outcomes for art galleries. In: Tussyadiah, I., Inversini, A. (eds.) Information and Communication Technologies in Tourism 2015, pp. 463–476. Springer, Cham (2015). https://doi.org/10.1007/978-3-319-14343-9_34

38. Charitonos, K., Blake, C., Scanlon, E., Jones, A.: Museum learning via social and mobile technologies: (How) can online interactions enhance the visitor experience? Br. J. Educ. Technol. **43**(5), 802–819 (2012)

39. Nägele, R.: Verfahren zur technisch-induzierten Gestaltung von Geschäftsmodellen, Dissertation Uni Stuttgart 2017 (2017)

40. Neuhofer, B., Buhalis, D., Ladkin, A.: Smart technologies for personalized experiences: a case study in the hospitality domain. Electron. Markets **25**(3), 243–254 (2015)

41. Hellmuth, R.: Research of the Potentials of a BIM model for building technology. Master thesis. University of Stuttgart, Stuttgart. IGE (2017)
42. Mettenleiter, M., Härtl, F., Kresser, S., Fröhlich, C.: Laserscanning–Phasenbasierte Lasermesstechnik für die hochpräzise und schnelle dreidimensionale Umgebungserfassung, München: Verlag Moderne Industrie (Die Bibliothek der Technik, Band 371) (2015)
43. Choi, S.P., Shin, M.S., Yang, I.T., Acharya, T.D.: Application of data mining techniques for the development of 3D laser scan data management program. Int. J. Appl. Eng. Res. **12**(14), 4658–4662 (2017)
44. Wehner, Fl.: Comparative investigation of laser scanning and photogrammetry for interior reconstruction. Bachelorthesis. University of applied sciences Schmalkalden, Schmalkalden. Informatik (2018)
45. Baqersad, J., Poozesh, P., Niezrecki, C., Avitabile, P.: Photogrammetry and optical methods in structural dynamics – a review. Mech. Syst. Sign. Process. **86**, 17–34 (2017). https://doi.org/10.1016/j.ymssp.2016.02.011
46. Cooper, J.P., Wetherelt, A., Zazzaro, C., Eyre, M.: From Boatyard to museum: 3D laser scanning and digital modelling of the Qatar Museums watercraft collection, Doha, Qatar. Int. J. Nautical Archaeol. (2018)
47. Wilhelm, S.: Visions and developments for buildings in the digital age. Corp. Real Estate J. **2017**(7), 51–62 (2017)
48. Huitl, R., Schroth, G., Hilsenbeck, S., Schweiger, F., Steinbach, E.: TUMindoor: an extensive image and point cloud dataset for visual indoor localization and mapping. In: 19th IEEE International Conference on Image Processing (ICIP) 2012, pp. 1773–1776. IEEE (2012)
49. Ihrén, J., Frisch, K.J.: The fully immersive cave. In: Bullinger, H.-J., Riedel, O. (eds.) 3rd International Immersive Projection Technology Worskhop, 10–11 May 1999, Center of the Fraunhofer Society Stuttgart IZS (1999)
50. Seiler, U.T., Koch, V., von Both, P.: Immersive virtual simulation of spaces. In: Proceedings of the 33rd eCAADe Conference, Vienna University of Technology, Vienna, Austria, 16–18 September 2015, vol. 1, pp. 77–88 (2015)
51. Thomopoulos, S.C.A., et al.: DICE: digital immersive cultural environment. In: Ioannides, M., et al. (eds.) EuroMed 2016. LNCS, vol. 10058, pp. 758–777. Springer, Cham (2016). https://doi.org/10.1007/978-3-319-48496-9_61

The Digitization of the Tangible Cultural Heritage and the Related Policy Framework

Konstantina Siountri[1,2,3(✉)], Evangelia Vagena[3],
Dimitrios D. Vergados[3], Joseph Stefanou[3,4],
and Christos-Nikolaos Anagnostopoulos[2]

[1] Hellenic Ministry of Culture, Athens, Greece
[2] Cultural Technology and Communication Department,
University of the Aegean, Mytilene, Greece
{ksiountri, canag}@aegean.gr
[3] Department of Informatics, University of Piraeus, Piraeus, Greece
{ksiountri, vergados}@unipi.gr,
evangelia.vagena@gmail.com
[4] School of Architecture,
National Technological University of Athens, Athens, Greece
joseph@central.ntua.gr

Abstract. Nowadays, the digital revolution has changed the conventional way of acquiring images and reproduction of the existing or the imaginary world, leading to new forms and dimensions of the "reality". The perception of the user is expanded through 360° technology, Augmented Reality (AR), Mixed Reality (MR), Virtual Reality (VR) platforms that originally sprang through the gaming industry and have influenced, among other things, the field of culture, both in terms of its production and its management and enhancement. Our research work aims to present the issues of the related policy framework that arise from the use of the new digital technologies which in our days drastically change the way of management of the Greek tangible cultural heritage. The paper will concentrate at the interaction between archaeological law, copyright law and the use of digital media of creation, promotion and diffusion of tangible cultural heritage.

Keywords: Virtual environments · Digital culture ·
Cultural heritage management · Legal framework · Copyright ·
Personal data · Privacy

1 Introduction

Cultural heritage, either tangible[1] or intangible[2], is unique and valuable and it is our society's obligation to preserve, protect and pass it on to future generations. Furthermore, cultural heritage is a driving force for the creative sectors and an important

---

[1] For example buildings, monuments, artefacts, clothing, artwork, books, machines, historic towns, archaeological sites, https://europa.eu/cultural-heritage/about.

[2] For example practices, representations, expressions, knowledge, oral traditions, performing arts, social practices and traditional craftsmanship https://europa.eu/cultural-heritage/about.

© Springer Nature Switzerland AG 2019
A. Moropoulou et al. (Eds.): TMM_CH 2018, CCIS 961, pp. 250–260, 2019.
https://doi.org/10.1007/978-3-030-12957-6_17

resource for economic growth, employment[3], social cohesion and elimination of social differences, religious and political contradictions.

The diffusion of cultural heritage, undergoing considerable changes as a result of increased digital technologies, is important in order to raise public awareness and involvement. For this reason, the Hellenic legislation refers to the promotion of cultural goods as a state's responsibility through the spread of knowledge and through the operation of cultural institutions, museums, etc. (Law 3028/02, Article 39 and Article 44).

Nowadays, the use of new technologies contributes to the recording, analysis, planning and monitoring of cultural projects through appropriate interactive and multimedia tools. As the physical presence of the audience is not required e.g. Virtual Museums, more and more people are getting global free access to cultural goods. The education of young people is also important, since digital applications have a great impact in youth.

Virtual environments deliver through a three-dimensional graphics a representation of a real-life or an "artificial" world, where the user can interact with the appropriate interface equipment. The experience of interacting with a virtual environment is characterized by the illusion of participation in a digital place with full or partial immersion, where often the reality is synthesized or replaced by the fantastic.

Virtual reality is based on polygonal models, which are created (a) with software, (b) through three-dimensional scanners, (c) by photogrammetry or (d) with combinations of these three methods. The results of the production or processing of new digital illustration can be presented through technologies such as 360°, augmented reality (AR), mixed reality (MR) and virtual reality (VR).

It is obvious that all of the above that originally emerged through gaming have affected, among others, the area of culture, both in its production as well as its dissemination to the general public. Especially regarding the promotion and commercialization of cultural heritage, digital technologies change drastically the traditional ways of management.

This study aims to present the applicable Greek institutional framework and the legal limitations imposed on digital cultural projects, with emphasis on tangible cultural heritage, based on issues related to the accuracy of the scientific content, the obligations of the involved stakeholders and all copyright issues.

2   Cultural Heritage and Legal Framework

In order to present the Greek institutional framework related to Cultural Heritage, the use of digital media and the obligations of involved parties, for the purposes of this study, we need to define the concept of Contemporary Culture.

---

[3] Cultural and creative industries (CCIs) are estimated to be responsible for over 3% of the EU's gross domestic product and job (https://ec.europa.eu/culture/policy/cultural-creative-industries_en).

Cultural Heritage[4] is the legacy of physical artifacts and intangible attributes of a group or society that are inherited from the past generations, maintained in the present while granted in the future to benefit future generations.

Contemporary Culture[5] is the combination of physical, intellectual, technical achievements and performance, which is the result of creative forces and human capabilities as expressed in modern times.

The above definitions are necessary because we strongly believe that any digital display of the past is a contemporary project that reflects the artistic contribution and the technological abilities of its creator. And therefore, the legislation considered will be a combination of legal provisions (e.g. Law. 2121/1993, as applied on "Copyright, related rights and other cultural issues") in conjunction with Law 3028/2002, as applied, "For the Protection of Antiquities and Cultural Heritage in general".

3 The Archaeological Law 3028/2002

In Greece, the protection of cultural heritage is under the auspices of the State[6] as applied under law 3028/2002 "For the protection of Antiquities and Cultural Heritage in general". The Authorization of the use of monument images on the Internet and in digital applications, whether commercial or artistic, educational or scientific reasons, is given after examination of the issue and approval of the Ministry of Culture as defined in Article 46 of that law. More specifically, Article 46 paragraph 4 and 5 provides the following:

"4. For the production, reproduction and dissemination to the public, for direct or indirect economic or commercial purpose, casts, copies (see footnote 6) or monuments illustration, belonging to the State or property situated in archaeological sites and historical places or individually or cell found in museums or collections of the State, in any manner and medium, including electronic and digital, the Internet (Internet), of telecommunication or other connection networks and creating databases with images of above, other entities or persons other than the State of TAPA and the Greek Culture Organization SA require prior authorization. Permission is granted for a charge, in favor of TAPA natural or legal persons, with the Minister of Culture, which is defined, and the duration of the license, the conditions under which and the fee payable.

5. Production, reproduction and use of these products for other purposes, such as artistic, educational or scientific use is allowed, against payment of a fee in favor of TAPA from which it can be exempted by the Minister of Culture."

Based on Law 3028/2002 and the relevant Joint Ministerial Decision (JMD) YPPOT/ 126463/28.12.2011 (- GG 3046/B/30.12.2011) the fees to the Greek State are provided in relation to the following:

---

[4] From Wikipedia, the free encyclopedia (https://el.wikipedia.org/ recovering 26.6.2018).

[5] Govaris, Ch. (1999), Introduction to Intercultural Education. Athens, Publications Path.

[6] The reference to "copies", under our interpretation of the law, covers also the technique of 3D printing.

- photo shoot
- filming
- use as illustration in printed and electronic publications (e-book)
- use for internet illustration
- use in audiovisual or electronic guides
- use for the production of audiovisual works
- holding of events

The Directorate of the National Monuments Record (DEAM), which functions under the Directorate General of Antiquities and Heritage Ministry of Culture, is competent for the application of the relevant provisions, regarding both licensing and control of the above content.

More specifically, the DEAM, among many other duties, is responsible[7] for the *"definition of specifications, terms and conditions and making available through the Internet and other information and communication technologies derived from the above digital content, including data and metadata that accompany it"*. And the *"authorization of use by third parties and similar monuments illustration of audiovisual material, the use of Information and Communication Technologies to the internet and digital applications other than those authorized by the EFA and the Special Regional Services"*.

The term *"images"* in the above Joint Ministerial Decision (JMD) Chapter 2, Art. 4, Par. 1 (Government Gazette (GG) 3046/B/30.12.2011) is defined as the *"displays of monuments belonging to the State or property situated in archaeological sites and historical places or individually, or moveable monuments found in museums or collections of the State"*.

Nevertheless, in Chapter 2, Art. 5, Par. 2 of the above Joint Ministerial Decision (JMD), the term "photographs" is also inserted and it is linked directly to the amount of fees (depending on their number), creating a confusion whether the displays generated by design software or three-dimensional scanning (laser scanning) are finally subject to fees. Also, the reference to "online" does not specify the number of sites, making the web to be treated as a single entity and it is not clear whether the fee should be paid in respect of each site separately or whether once paid it covers any form of online display by the user.

Furthermore, according to Ch. 2, Art. 6 of the above Joint Ministerial Decision (JMD) on charges for production of audiovisual works, the question arises whether the charge per second of display may be applied to VR tours digital applications. The virtual tour is an audiovisual work and it is likely to be "implemented" in areas being subject to the provisions of 3028/2002. But the "digital time" of touring is selected by each user separately. Moreover, it is not clear whether the display relates only to monuments, as they exist today or it also concerns iconic reproductions and restorations produced by software.

Finally, according to Ch. 2, Art. 7 of the above Joint Ministerial Decision (JMD) on fees using illustration or audiovisual works in digital electronic applications guides that are distributed via mobile telephone or other wired or wireless electronic networks, a fee is provided amounting to 20% of the retail selling price of each digital application. In the

---

[7] P.D.4 "Ministry of Culture and Sport Agency," Official Gazette 7/A/22.01.2018.

current digital landscape and the widespread use of personal electronic devices (e.g. Smart phones) the application of this provision tends to be very difficult, if not possible at all.

Cultural displays involve often content illustrated on social media (like facebook, instagram etc), on digital platforms such as the platform Open Heritage by Google Arts & Culture or platform 3dwarehouse[8] and on crowd sourcing popular platforms like Wikimedia commons[9], YouTube etc in which users from all over the world can add content.

In these cases, there may be copyright issues since the terms of use of these sites include clauses under which users who add - upload content (images, audiovisual works) give both to other users of the media as well as to the companies who own these platforms the permission to use the material. In some cases, the users who upload the content may not have even the right to provide such a license and finally the content uploaded may be illegally reproduced and made available to the public but without the users realizing it. At the same time the control of the work is lost because of the license provided and it may be difficult to enforce the law. It is also noteworthy that the scope of application of those provisions covers also the activity of Cultural Services Department of the Ministry (e.g. Antiquities, Museums etc.), which maintain pages on social networks and often post pictures of monuments, parts of their collections or actions e.g. excavations and findings.

We should also remember that any license obtained under L. 3028/2002 does not in principle entitle the licensee to any further transfer to a third party nor to authorize a third party to use the material obtained. The only case, under which one may use contents protected by copyright without asking for permission from the original beneficiary of protection, would be in case of application of one of the limitations provided by the Greek Copyright Law and more specifically in Articles 18-28 C of law 2121/1993, as applied today.

4   Copyright

Since the digital cultural projects can combine heritage and contemporary creation, copyright law issues emerge.

The main copyright law governing the law of copyright is Law 2121/1993[10] on "Copyright, related rights and other cultural issues" (Government Gazette A/25/1993),

---

[8] https://artsandculture.google.com/project/cyark, https://3dwarehouse.sketchup.com (Recovery: 06/25/2018).

[9] https://commons.wikimedia.org/ (Recovery: 06/25/2018).

[10] Under Article 2 of N.2121/1993 a protected work is "any original intellectual creation speech, art or science, expressed in any form, notably written or oral texts, musical compositions, with or without text, plays, with music or without, choreographies and pantomimes, audiovisual works, works of fine art, including drawings, paintings and sculptures, engravings and lithographs, works of architecture, photographs, works applied arts, the illustrator in the maps, three-dimensional works relative to geography, topography, architecture or science". To protected works also belong to translations, the adaptations, arrangements other works or expressions conversions of folklore and works collections or expressions collections of folklore or simple facts and data, such as encyclopedias and anthologies, as the selection or arrangement of their content is original. The protection of the works of this paragraph is without prejudice to the rights to pre-existing works, which were used as object conversions or collections.

as amended and in force, is the basic copyright law in Greece as combined with the recent law 4481/2017 (Government Gazette A/100/07.20.2017) for the collective management of copyright and related rights, granting of multi-territory licenses for online use of musical works and other Ministry of Culture and Sport competence.

Law 4481 provides for the establishment and the "Committee for the Notification of Copyright and Related Rights Infringement on the Internet (EDPPI)".

Copyright is a right of "double nature" as it includes both property as well as moral rights. As provided in ar. I of law 2121/93 *"Authors have, with the creation of the work, the right of copyright in that work, which includes, as exclusive and absolute rights, the right to exploit the work (economic right) and the right to protect their personal connection with the work (moral right)."*

The economic right allows the author to benefit financially from the exploitation of his work. It confers upon the authors notably the right to authorize or prohibit inter alia the fixation and direct or indirect, temporary or permanent reproduction of their works by any means and in any form, in whole or in part, the arrangement, adaptation of other alteration of their works, concerning the original or copies of their works, the distribution to the public in any form by sale or by other means. If someone does any of the above without the permission of the author, he infringes copyright economic rights of the latter unless one of the restrictions in Articles 18-28C of Law 2121/1993 may be applied.

The moral rights confer upon the author notably the following rights:

(a) to decide on the time, place and manner in which the work shall be made accessible to the public (publication)
(b) to demand that his status as the author of the work be acknowledged and, in particular, to the extent that it is possible, that his name be indicated on the copies of his work and noted whenever his work is used publicly, or, on the contrary, if he so wishes, that his work be presented anonymously or under a pseudonym
(c) to prohibit any distortion, mutilation or other modification of his work and any offence to the author due to the circumstances of the presentation of the work in public
(d) to have access to his work, even when the economic right in the work or the physical embodiment of the work belongs to another person; in those latter cases, the access is effected with minimum possible nuisance to the right holder
(e) in the case of a literary or scientific work, to rescind a contract transferring the economic right or an exploitation contract or license of which his work is the object, subject to payment of material damages to the other contracting party, for the pecuniary loss he has sustained, when the author considers such action to be necessary for the protection of his personality because of changes in his beliefs or in the circumstances.

It should also be noted that the moral rights are independent from the economic rights and remain with the author even after the transfer of the economic rights.

According to article 28 of Law 2121/1193 museums which own the physical carriers into which works of fine art have been incorporated are entitled, without the consent of the author and without payment, to exhibit those works to the public on the museum premises, or during exhibitions organized in museums. Also, the presentation of a fine art work to the public, and its reproduction in catalogues to the extent necessary to promote its sale, is permissible, without the consent of the author and without payment.

The Law sets time limits to the duration of copyright protection, so that is becomes a public good for the benefit of the society. In particular article 29 of Law 2121/1993 provides that Copyright shall last for the whole of the author's life and for seventy (70) years after his death, calculated from 1st January of the year after the author's death. After the expiry of the period of copyright protection, the State, represented by the Minister of Culture, may exercise the rights relating to the acknowledgment of the author's paternity and the rights relating to the protection of the integrity of the work deriving from the moral rights.

According to Law 2121/1993, Article 2, Par. 5, the following are not protected by copyright:

- Expressions of folklore, news and simple events, mathematic formulas, procedures and methods.
- Projects created by civil servants in the context of the official duties of the employee (the property right is transferred automatically to the State, unless otherwise agreed).

On this basis copyright arising from the photographing or filming of protected works, such as monuments, are first recognized for the originator - creator while the rights of the content depicted belong to the Ministry of Culture (which is mandatory, and according to par. 3 of the Joint Ministerial Decision (JMD) 2199/12.09.2005 each photo should be stated that: "the Copyright of the depicted ancient objects belongs to the Greek Ministry of Culture (Law 3028/2002)".

It should also be reminded that according to article 26 of Law 2121/1993, the Greek law allows only the occasional reproduction and dissemination by the mass media works that are legally in a public place. Therefore, any other use case and reproduction photographic or electronic forms should have the necessary licenses to ensure both the archaeological and the law of copyright in relation to immovable monuments.

There is also the alternative of the contractual schemes such as Creative Commons, aim to change the way right holders exercise their rights without modifying copyright's exclusive character which can be used. Creative commons is a nonprofit corporation dedicated to making it easier for people to share and build upon the work of others, consistent with the rules of copyright. The organization provides free licenses and other legal tools to mark creative work with the freedom the creator wants it to carry, so others can share, remix, use commercially, or any combination thereof. Creative commons movement defines the spectrum of possibilities between full copyright and the public domain since it gives author the possibility to opt from all rights reserved to no rights reserved. It is also possible to keep your copyright while allowing certain uses of your work—a "some rights reserved" copyright. The six main licenses offered when you choose to publish your work with a Creative Commons license all include the basic element of attribution by which you let others copy, distribute, display, and perform your copyrighted work - and derivative works based upon it - but only if they give credit the way you request. Creative Commons also has two public domain tools: CC0 and Public Domain Certification. CC0 enables authors and copyright holders to dedicate their works to the public domain. Public Domain Certification facilitates the discovery of works already in the public domain. A person using CC0 (called the "affirmer" in the legal code) waives all of his or her copyright and neighboring and related rights in a work, to the fullest extent permitted by law. If the waiver isn't effective for any reason, then CC0 acts as a license from the affirmer granting the public an unconditional, irrevocable, non exclusive, royalty free license to use the work for any purpose.

5   Examples of Problematic Implementation of Legal Provisions

In Ancient Corinth, during the last two years, two digitization projects were produced for training purposes, one of a graphic restoration and one of a documentation of the existing condition of the site, which despite their high quality or their global impact, they did not apply for the necessary license given by the competent Services of the Hellenic Ministry of Culture. Both of the above productions are on the Internet.

More specifically, we are referring to the 3D animation video of a Graphic restoration of Ancient Corinth in the 2nd century AD (YouTube, duration 6:18 min[11]), production of George Terzis archaeologists, Danila Loginov, Andrey Zarov, and Vyacheslav Derbenev group «History in 3D Creative Team». In March 2017, the American School of Classical Studies at Athens (ASCSA) presented it on its website[12] with positive comments on the accuracy and the amount of the given information.

Also on April 18, 2018 on the occasion of World Heritage Day the Google Arts & Culture[13] announced its partnership with the nonprofit CyArk for a new digital project that offers users "a chance to experience some of the greatest wonders of the world in 3D", e.g. Ancient Corinth[14], The Kingdom of Thailand Ayutthaya, Palace Al Alzem in Syria, Berlin's Brandenburg Gate and the ancient Maya metropolis of Chichén Itzá in Mexico. Despite the reference on the site of CyArK[15] to the Hellenic Ministry of Culture and ASCSA as partners, the Ministry of Culture remains firmly opposed to Google's practices.

In the web there are many digital applications that offer under payment or free guided tours to a various archaeological sites, through virtual and augmented reality. They are also enriched with information, historical pictures, videos, stories etc. The majority of them have not applied for the necessary permission and their historical accuracy is under question.

In Athens for example we can distinguish the cases of Athens Time Walk (Company Diadrasis Ladas I & Co GP) and the Walk the Wall Athens (Dipylon Non Profit Organisation). These applications are available through Google Play platform.

---

[11] https://www.youtube.com/watch?v=dEHPfMIyLfc.

[12] http://www.ascsa.edu.gr/index.php/news/newsDetails/3d-animation-brings-new-life-to-roman-era-corinth.

[13] https://news.gtp.gr/2018/04/18/ancient-corinth-goes-3d-google-cyark-project/.

[14] https://artsandculture.google.com/exhibit/1QLCbLZyVC8kKg.

[15] https://artsandculture.google.com/exhibit/1QLCbLZyVC8kKg.

Athens Time-Walk[16] won the Award for Best Application for Culture in the media contest Heritage in Motion Awards 2017[17], the joint initiative of the European Museum Academy and Europa Nostra and is authorized by Europeana. The Walk the Wall Athens was awarded the silver award in this year's Mobile Excellence Awards in the Tourism/Culture category[18]

Finally, the case of video games is worth mentioning, especially the case of the adventure game Assassin's Creed Odyssey produced by Ubisoft company, which recently announced the release on October 5[19] of a version that takes place in ancient Athens 5 c. B.C. The appeal of the game to millions of players around the world and the enthusiastic reviews for the quality of graphics and the historic information has already raised the debate about the boundaries between imagination and reality and whether it involves issues that concern the implementation of Law 3028/2002.

6 European Union Initiatives

Greece is a state member of European Union and therefore we should also take into deep consideration of the initiative of EU for the diffusion of the cultural "stock" of Europe through the digitization, primarily for conservation purposes as well as the disposal of digital content[20].

In this context it is worth mentioning the main provisions of the European Commission Recommendation of 27 October 2011 on the digitization and online accessibility of cultural material and digital preservation. The Recommendation highlights that the Digital Agenda for Europe seeks to optimize the benefits of information technologies for economic growth, job creation and quality of life of European citizens, as part of Europe 2020. The digitization and preservation of Europe's cultural memory which includes print (books, journals, newspapers), photographs, museum objects, archival documents, sound and audiovisual material, monuments and archaeological sites ("cultural material"), it is one of the main areas of intervention of the Digital Agenda.

The Recommendation makes particular reference to Europeana, "*The European digital library, archive and museum*" which started to operate on November 20, 2008. According to the text of the Recommendation further development of Europeana platform will depend largely on how the Member States and their cultural institutions upload content and in this respect should be encouraged measures to achieve this result. To achieve that aim, among other things, it is recommended that Member States inter alia in relation to intellectual property issues improve conditions for digitization and online accessibility of the material copyright including: (a) rapid and correct transposition and implementation of the provisions Directive on orphan works, following its

---

[16] https://www.heritageinmotion.eu/himentry/slug-e725d48429303680a4948577fc397609.

[17] https://heritageinmotion.eu/pressrelease/winners-heritage-in-motion-2017.

[18] https://dipylon.org/en/2018/09/20/an-award-for-walk-the-wall-athens/.

[19] http://www.greece-is.com/assassins-creed-odyssey-stuns-incredible-recreation-ancient-athens/.

[20] http://s3platform.jrc.ec.europa.eu/digitisation-of-cultural-heritage.

adoption, in consultation with interested parties for the purposes of adoption in order to facilitate a rapid implementation; a close monitoring of the implementation of the Directive after its adoption; (b) creating the conditions of the legal framework for strengthening licensing mechanisms identified and agreed by the stakeholders to digitize large-scale and cross-border access to works out of commerce; and (c) to contribute to the promotion and availability of databases with info on the rights status, connected at European level, such as ARROW.

Taking into consideration the above, the project "National Archive of Monuments" of the Hellenic Ministry of Culture is necessary to be completed. This archive will provide a unified documentation system for Monuments and Collections Management that can be linked directly to other platforms and include all the checked by the Greek State data and metadata.

7   Conclusions

Therefore, we may conclude that the application of the relevant Greek legal framework for digital imaging and online distribution of protected cultural resources will become increasingly difficult in the future, requiring taking a lot of extra technological measures on the part of the Competent Services, since technology seems to be constantly ahead of the law.

For that reason, the establishment of a special department dedicated to the continuous web and electronic monitoring to avoid trafficking of illegal or not proper content seems to be necessary, since the market does not seem to follow the existing laws of licensing.

Moreover, the contracts within designers, photographers, computer applications producers, etc. as regards to the production, reproduction and further online promotion of their work and the scope of the concession or not of their property rights, should be established.

However, the critical issues relating to the enrichment of the relevant legal framework, the introduction (or the clarification) of regulations for the online crowd sourcing platforms and social media and the copyright protection and management point out that we need a serious national digital cultural strategy.

It is also under consideration that in this new digital era the involved institutional stakeholders would benefit more by the adoption of the Open Data and introduction of a new system of "certification". That means that instead of trying to find out in this global market who is illegal, the Greek State could introduce an Organization for Standardization that would be concerned with "rewarding" and promoting the best practices.

Acknowledgements. The publication of this paper has been partly supported by the University of Piraeus Research Center (UPRC).

References

1. Kallinikou, D.: Copyright and Digital Libraries. Sakkoulas Publications
2. Vagena, E.: Technological Protection and Digital Copyright Management. Nomiki Vivliothiki, Athens (2010)
3. Website: The Hellenic Copyrights Organization. http://www.opi.gr. Accessed 2 Feb 2018

Novel Educational Approach for the Preservation of Monuments

Authentic Learning to Better Prepare for Preservation Work

Pieter de Vries[1(✉)] and Antonia Moropoulou[2]

[1] Delft University of Technology, 2628 BX Delft, Netherlands
pieter.devries@tudelft.nl
[2] National Technical University of Athens, 15780 Athens, Greece

Abstract. Cultural heritage preservation has developed from an insular vocation into a field of innovative scientific methodologies characterized by a holistic approach combining a range of scientific fields. Unfortunately, preservation education has not been able to keep pace with all these developments. In this paper authentic learning is analyzed as a possible educational scenario to help improve preservation education to connect to the state of affairs at the preservation workplace. The purpose is to sketch an educational framework based on knowledge and experiences with authentic learning in other engineering fields as a primer for the design and implementation of 'authentic learning for preservation'. From the analysis it becomes clear that authentic learning can support the selection of valuable unknown experiences and support the design and development of 'authentic learning for preservation' experiments to help closing the gap between preservation education and cultural heritage practice. The authentic learning model as presented here clearly supplies a framework to consider in this endeavor. As such the paper can be helpful in the discussion about the usefulness and feasibility of this approach.

Keywords: Preservation education · Authentic learning ·
Learning technologies

1 Introduction

Cultural heritage preservation has developed from an insular vocation into a field of innovative scientific methodologies characterized by a holistic approach combining a range of scientific fields. Students are prepared to handle several fields, but to cope with the number and heterogeneity of the different sciences involved in the preservation profession, other skills are needed to adequately deal with the daily reality of preservation work. This paper is about the development of an authentic learning approach that supports the use and appreciation of other and new skills in the context of a higher education institution. Good engineering education relies on: motivation, real life learning and the ability to communicate and collaborate, which demands students to be self-developers in a continuous improvement process [1]. An 'authentic setting' provides opportunities to respond in an adequate way on the demand of continuously changing requirements.

© Springer Nature Switzerland AG 2019
A. Moropoulou et al. (Eds.): TMM_CH 2018, CCIS 961, pp. 263–272, 2019.
https://doi.org/10.1007/978-3-030-12957-6_18

Authentic learning is an instructional approach that allows students to explore, discuss, and meaningfully construct concepts and relationships in contexts that involve real-world problems and projects that are relevant to the learner. It is making use of multiple educational and instructional techniques focusing on connecting the curriculum with real-world issues, problems, and applications. The basic idea is that students are more motivated in what they are learning when it is about new concepts and skills, and this exposure to a real-life context prepares them with practical and useful skills, and with topics that are relevant and applicable in their prospective working environment. The emergence of the internet and other new tools for communication, visualization, and simulation, have lowered the threshold for organizing and experiencing a more authentic learning environment. So, authentic learning allows for blending different kinds of learning and to give students opportunities to think and act like professionals but in a learning or training context. In a physical sense it is easy to organize, but in a pedagogical sense it needs a clever design to deal with the tasks and the limitations imposed by the curriculum and institutional constraints [2].

This paper describes the essence of an authentic learning approach in relation to the developments in cultural heritage preservation. The purpose is to sketch an educational framework based on knowledge and experiences with authentic learning in other engineering fields as a primer for the design and implementation of 'authentic learning for preservation'.

2 Research Context

2.1 The Bigger Picture

One of the issues high on the agenda in Europe has been skill shortages in the engineering field. Looking at the progress made it seems an issue that requires permanent attention as it is firmly related to the ongoing changes in society. In general shortage emerge where the skills required are unavailable in the workforce and the opportunities for change insufficient. It might be that people are over - or under skilled, whatever their qualification level, but apparently their skills do not match the job [3]. One of the actions taken to deal with the mismatch is to forge stronger links between institutions and industry with structured partnerships as 'knowledge alliances' to adapt curriculum to demands of society. This notion intertwines with the perception that good engineering education relies on real life learning and the ability to communicate and collaborate, which requires that students are capable of self-directed learning in a continuous improvement process [1]. Operating in an Authentic learning environment can help to enlarge the understanding of innovation and entrepreneurship in real life and can function as a catalyzer for the incorporation of authentic learning in the field of preservation education.

Against the backdrop of the skills agenda, the skills shortage issue is clearly high on the agenda of the cultural heritage sector. The European Year for Cultural Heritage 2018 initiative called 'Skills for Heritage: enhancing education and training for the traditional and new professions' [4]. During this year the European institutions also focus on the 'support of the development of specialized skills and improve knowledge management and knowledge transfer in the cultural heritage sector, considering the implications of the digital shift'. Europe is renowned for its exceptional skills in the field of heritage preservation and conservation, but not well prepared to improve the transmission of heritage knowledge and skills to the younger generations. There is a lack of high-level professionals in "traditional" heritage occupations and therefor it is important to explore possible responses. Not only is the cycle of professional preparation often very long, also the 21st century innovation and digitalization require an appropriated response as to make better use of new technologies for heritage preservation [5].

2.2   The Focus of This Paper

The focus is on exploring authentic learning as an educational format that could lower the threshold between formal preservation education and the workplace. It is believed that improving the relationship will bring real-world problems and project within reach, bringing new concepts and skills in a format that can be dealt with in a school environment. It is an instructional approach that supports the opportunities for students to acquire knowledge, skills and attitudes directly related to a workplace setting. This paper should clarify the authentic learning concept, while using a related case study as an example of how this format can be applied in daily practice within an engineering context and deduce lessons for HEE. In this way the paper should show to be helpful to fuel the discussion about the usefulness and feasibility of this approach.

The next section is about positioning authentic learning in the broader field of learning for innovation. Innovation is very much tied to technological development and as such any change in education should have to be considered with the digital transformation in mind. In Sect. 3 the origins of authentic learning will be described including the basic structure of the way this concept has been applied in the case study. Section 4 is about the experiences with in addition a final section reflecting on the outcomes and results and the potential possibilities and relevance authentic learning might have for preservation education.

3   Authentic Learning

Authentic learning is an umbrella term for pedagogical strategies with the purpose to connect learners to environments where they gain practical knowledge and experience and lifelong learning skills including metacognitive reflection and self-awareness as being essential [6]. Strategies include vocational training, apprenticeships and scientific inquiries [7]. An increasing number of institutions have started to connect to the world outside campus, but the real-world problems and work situations do not yet have the urgency they deserve in relation to the skills issues we face.

Most teaching and learning are taking place in class rooms, laboratories, libraries and the study at home. Learning though becomes increasingly virtual through the use of ICT and such activities as job-shadowing, project-based learning, apprenticeships, Erasmus+ programmes. These are key elements of authentic learning. One such element on the micro-level is 'Learning by doing'. In this paper we focus on this 'practical' approach in relation to preservation education. The connect is to focus on the position of the student and to see what authentic learning could add and requires.

The argument is that students often come to the workplace where their competences may not always match with the demands. Demands change rapidly as a result of innovation, and the demands vary across Europe given the heterogeneity of industry, HEE and development policies. Furthermore, students often are not well-prepared to deal with the demands of innovation in organisations. It is of paramount importance that students are better prepared and have a firm understanding of practice when they leave their schools. An authentic learning environment that mimics real life in the workplace is an effective way to improve the student's capability to cope with learning demands at the workplace. Therefor the connect between the school and the future work environment needs to improve to also stay informed about changes that may take place at a rate and in ways schools are not able to cope with given their current institutional organization.

Exposure to a real-life context with topics that are relevant and applicable in their prospective working environment prepares students not only with practical and useful knowledge and skills, but has a beneficial effect on motivation [5]. Herrington and Oliver [8] state that authentic learning environments enable students to feel involved in a project as part of a larger whole with tasks that could never be carried out individually and with a stimulus for (higher) thinking processes through communication and discussion. The emergence of the internet and other new tools for communication, visualization, and simulation have lowered the threshold to develop and use authentic learning environments. As a matter of fact, ICT-use has grown immensely in the day-today practice of preservation work, but has not yet affected the approaches to teaching and learning in Cultural Heritage Education enough [4, 9].

3.1 The Theory

The origins of authentic learning draw largely on the theoretical constructs of situated learning and cognitive apprenticeships. One of the findings was that meaningful learning will only take place if embedded in the social and physical context within which it will be used [10]. Therefor one of the questions was how can situated learning be operationalized or in other words, what are the critical characteristics of a situated learning environment in higher education. According to Herrington [11] the following framework summarises the characteristics for the design of such a learning environment:

Authentic learning characteristics

1. An authentic context that reflects the way the knowledge will be used in real life
2. Authentic activities
3. Access to expert performances and the modelling of processes
4. Multiple roles and perspectives
5. Collaborative construction of knowledge
6. Reflection
7. Articulation
8. Coaching and scaffolding
9. Authentic assessment

These critical elements have been used to design and evaluate learning environments. In addition, authentic tasks have been developed as an integral component of an authentic learning environment [2]. These tasks were derived from paper reviews and were used to select cases for investigation. In the case descriptions that follow, these tasks have been used for the design of the course and the evaluation.

Authentic learning tasks

1. Authentic tasks have real-world relevance
2. Authentic tasks are ill-defined, requiring students to define the tasks and sub-tasks needed to complete the activity
3. Authentic tasks comprise complex tasks to be investigated by students over a sustained period of time
4. Authentic tasks provide the opportunity for students to examine the task from different perspectives, using a variety of resources
5. Authentic tasks provide the opportunity to collaborate:
6. Authentic tasks provide the opportunity to reflect: Activities
7. Authentic tasks can be integrated and applied across different subject areas and lead beyond domain-specific outcomes
8. Authentic tasks are seamlessly integrated with assessment
9. Authentic tasks create polished products valuable in their own right rather than as preparation for something else
10. Authentic tasks allow competing solutions and diversity of outcome.

Further research explored the conditions and factors that contributed to the successful use of the tasks. The most successful applications were customer-oriented and could be considered using education more as a process than a product. The applications did not necessarily provide real experiences, but provide 'cognitive realism'. Student support was accepted as a need to get accustomed to a totally different way of learning and to convince the students that this approach in the end would be fruitful.

4 Authentic Learning Case Description

For the case description we selected an engineering course from the Faculty of Technology, Policy and Management at the Delft University of Technology. This course 'E-learning in Corporations' was designed using the authentic task framework next to design principles derived from the theories of constructivism and connectivism [12–15]. It was an elective course on the use of e-learning in corporations for bachelor and master students with mainly an ICT background. The course has been operational during three consecutive years and a more detailed description can be found here [2].

The objective of the course was to confront students with the day to-day reality of managing learning demands in a market organization. Students worked in small groups and could opt for the role of educational, technical or organizational e-learning consultant. The three roles were specified in a competence matrix to show the different roles and tasks in the project, so students knew they were developing different knowledge and skills. Hence, sharing knowledge and learning experiences, presenting findings and challenges to fellow students and discussing were crucial in ensuring that everyone benefits from the knowledge gained. ICT tools played a very important role in all interactions that took place and was a strong enabler in the process.

The topic of the course depended on the questions and demands from the company. In one case this was about 'E-learning for On-boarding Programs' and comprised a comparative analysis to create a benchmark for this company 'X', which is an international marketing organization. Company 'X' has been transformed from a decentralized organization into a more centralized organization and decided to implement an On-boarding Program for newly hired employees. The main objective of the student project was to shorten the induction period of new hires and bring employees together to collaboratively share knowledge.

4.1 The Course Cycle

The course cycle gives a concise overview of the course and the learning activities taking place. The course cycle of E-Learning in Corporations consisted of five steps and is graphically represented in Fig. 1. Step 1 and 2 are about the exploration of the subject e-learning. Students work on individual assignments and the outcomes were discussed in the group. Next was the devise of learning goals. Students formulate their individual learning goals in addition to the generic ones. Students work in small groups, have different roles and gain different knowledge and skills. Sharing knowledge and learning experiences, presenting findings and challenges play a very important role.

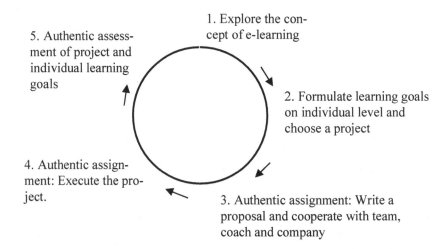

Fig. 1. The course cycle of e-learning in corporations

Step 3 is about writing a proposal. The students develop a proposal on the basis of available information and interviews and make an analysis of the company's learning demand. Issues that play a role in this process are: asking the right questions, interpretation of the data, understanding and meeting customer expectations, effective information exchange, and so on. The final outcome of this phase is a proposal containing the business problem, desired results, related activities, the planning of these activities, the division of tasks and responsibilities and a financial overview. This offer needs to be approved by the company. The core of working from an authentic learning perspective is the endorsement of learning in a natural way where motivation and initiative matter [2]. The students use a variety of different tools during the planning and executing of the project.

The final step in this cycle is Authentic assessment. Continuous review of what is going on is an important element of the authentic learning environment. Such a review focuses on how the group operates, but also on the perception students have of their own performance. Reflection on cooperation within the group and with the client in the different stages of the project is crucial. Therefore, it is important for the students to reflect on progress during the project. The summative evaluation is a combination of a process and a product assessment and takes into consideration the students' individual performance and their contributions to the project. The students personal learning goals and their personal perception of all the activities in the course are discussed at the end of the course using an 'after action review'. This 'after action review' is also used to assess the group performance: students need to answer questions like:

- What did we want to achieve?
- What happened?
- Why did it happen?
- How would we do it next time?

On the basis of this all, students receive an individual grade.

4.2 Results and Experiences

A new course with a non-traditional format certainly has its specific challenges. The course design and the achievements are discussed below using the format of the Course Cycle in relation to the characteristics and tasks of the authentic learning environment.

Organization

The design and preparation of the course logically took more time than usual. In addition to the pedagogical format it is the interaction with the business organization that accumulates time. Not all companies are interested in this kind of projects because of the expected time investment and there is no guarantee that there will be a useful final product. Also, not all assignments fit in the scope of the course. Most assignments were found through the use of the teacher's personal network. This requires targeted actions and a continuing effort. This also means that during the project the teacher needs to monitor the students closely, to help them identify and overcome the problems that have to do with a lack of experience. The assessment phase of the project requires a lot of attention, because many sources of information serve as input for the final assessment which is time consuming and there are always issues of validity.

Process

Most students became knowledgeable enough about e-learning to handle their position as junior consultant, but with little in-depth knowledge about the matter. This often led to a misjudgement of the real business situation, which surfaced mostly during the careful analysis of the interviews (2×30 min). Particularly when it comes to substantive advice, e.g. the selection of tools for improving internal communications, the students were confronted with a complex thinking process they are barely trained for. Because the students themselves co-determine the process. They are a key factor in this student-centred approach, but miss the experience to fully understand the consequences for the project. It is rather difficult for the students to monitor their learning objectives and act accordingly. Most miss the overview, due to the complexity of the business case. The dependence on third parties is sometimes difficult to digest, such as the repeated postponement of an appointment by a manager and thus increasing the time spent on less important issues. On the other hand, most students have experience with different types of e-learning, blended learning and problem-based education. This knowledge and experience were used to speed up the knowledge transfer of the students. A lack of knowledge was mostly quickly solved using their ICT skills to find answers and solutions, which also functioned as a degree of self-correcting power.

Content: Technology and Business Model

In general the students had good knowledge of ICT-related issues and little trouble with the e-learning technology itself. However, e-learning technology usually plays a minor role in the IT policy of a company and suggestions that transcend the existing framework were not well received. E-learning technology has its own limitations and therefore the opportunity to really make a difference is not likely. On the other hand, the boom in new technologies which are not necessarily familiar to all students makes it difficult to choose an appropriate solution. A good business model would definitely help in selecting tools, but the dominant business model of most training departments relies fully on formal training courses. This made it more difficult for students to come up with real e-learning solutions.

Just an example of the students' feedback taken from the experiences at Company X on E-learning for Onboarding:

- The assignment was very informative and instructive as well, because it taught me some aspects that have never come up doing other, non-real cases. Having several interviews and interactions with companies, structuring and preparing these, preparing an outline for research, experiencing collaboration issues with companies and within my group, the role of e-Learning within companies, and the e-learning literature were all very valuable for me."
- "Overall I really learned a lot from this course. It gave me a good impression what it is like to be a Consultant, even though we only scratched the surface of this profession. Especially on communicating aspects, verbally as well as written, I see where I can improve my skills."
- "I also got to see the importance of informal learning. I am a student who used to work alone a lot. I now see the link between the speed of the study process and collaborative ways of working. The pace of organization processes can also be enhanced by collaborative practices."

5 Conclusions and Discussion

The focus in this paper was on exploring authentic learning as an educational format that could lower the threshold between formal preservation education and the workplace. Therefor the authentic learning model has been analyzed from a 'learning by doing' perspective to see if the concept would supply the theory and ingredients for the application in preservation education. So far, no concrete examples from that field have been available, so another case study was used from the engineering field to reflect on the experiences and the question of usefulness. The general answer can be that authentic learning does supply a format that could support preservation education to become more up to date. As in all other fields, ICT is a significant enabler and if properly used can contribute to innovate, channel and improve educational interventions. The most interesting element is that in combination with new educational scenarios ICT offers a range of possibilities to also better connect to the use of ICT as applied in the professional field like museums and others.

From experiences in the educational field it is also clear that ICT does not make a difference if the educational scenario is not accurately designed and planned by people who are willing and able to make effective educational use of technological tools [16]. This calls for further research in the field, in order to find interesting experiences and to design and develop pilot projects aiming at closing the gap between preservation education and cultural heritage practice. The authentic learning model as presented here clearly supplies a framework to consider in this endeavour.

References

1. Moropoulou, A., De Vries, P.: On the problems, challenges and prospects for the european higher engineering education arising from the global economic crisis. In: Come, F. (ed.) SEFI@40 Driving Engineering Education to Meet Future Challenges, Brussels, pp. 81–85 (2013)
2. De Vries, P., Van den Bogaard, M.: A contemporary educational model for lifelong learning practices. In: Lappalainen, P., Oy, L. (eds.) European Continuing Engineering Education, Conceptualizing the Lessons Learned, Finland, SEFI and TKK Dipoli, pp. 209–218 (2009)
3. Cedefop: Skills, Qualifications and Jobs in the EU: the making of the perfect match. Luxembourg: Publications of the European Union (2015)
4. Homepage 'The European Year of Cultural Heritage 2018' Europa EU. https://europa.eu/cultural-heritage/about. Accessed 15 Aug 2018
5. European Initiative no 8: Skills for Heritage: enhancing education and training for the traditional and new professions INNOVATION PILLAR 2018, 10 July 2018. http://www.ecco-eu.org/fileadmin/user_upload/8_Skills_for_heritage_for_NC.pdf
6. Lombardi, M.: Authentic learning for the 21st century: an overview. In: Oblinger, D.G. (ed.) Educause Learning Initiative (2007)
7. Becker Adams, S., et al.: Horizon Report: 2018 Higher Education Edition, Reading (2018). ISBN 978-0-9906415-8-2
8. Herrington, J.: Authentic e-learning in higher education: design principles for authentic learning environments and tasks. In: Reeves, T.C., Yamashita, S. (eds.) Proceedings of E-Learn 2006, pp. 3164–3173. AACE, Chesapeake (2006a)

9. Ott, M., Pozzi, F.: Towards a new era for cultural heritage education: discussing the role of ICT. Comput. Hum. Behav. **27**(4), 1365–1371 (2011). https://doi.org/10.1016/j.chb.2010.07.031
10. Brown, J.S., Collins, A., Duguid, P.: Situated cognition and the culture of learning. Educ. Res. **18**(1), 32–42 (1989)
11. Herrington, J.: Design principles for authentic learning environments and tasks. In: World Conference on E-Learning in Corporate, Government, Healthcare, and Higher Education (2006b)
12. Jonassen, D., Mayes, T., McAleese, R.: A manifesto for a constructivist approach to uses of technology in higher education. In: Duffy, T.M., Lowyck, J., Jonassen, D.H., Welsh, T.M. (eds.) Designing Environments for Constructive Learning. NATO ASI Series, vol. 105, pp. 231–247. Springer, Heidelberg (1993). https://doi.org/10.1007/978-3-642-78069-1_12
13. Wertsch, J.V.: Vygotsky and the Formation of the Mind. Cambridge (1997)
14. Siemens, G.: Connectivism: a learning theory for the digital age. Int. J. Instr. Technol. Distance Learn. **2**(1), 3–10 (2005)
15. Siemens, G.: Knowing Knowledge. Lulu Publishers (2006). www.lulu.com
16. Lytras, M.: Teaching in the knowledge society: an art of passion. Int. J. Teach. Case Stud. **1**(1/2) (2007). https://doi.org/10.1504/ijtcs.2007.014205

20 Years of the N.T.U.A. Interdisciplinary Post Graduate Programme "Protection of Monuments"

Irene Efesiou[1], Eleni Maistrou[1], Antonia Moropoulou[2],
Maria Balodimou[1], and Antonia Lampropoulou[2(✉)]

[1] School of Architectural Engineering, National Technical University of Athens,
Athens, Greece
[2] School of Chemical Engineering, National Technical University of Athens,
Athens, Greece
labrop@mail.ntua.gr

Abstract. The National Technical University of Athens Postgraduate Master Program "Protection of Monuments", organized by the School of Architectural Engineering with co-organization by the School of Chemical Engineering and collaboration with the Schools of Civil Engineering and Rural-Surveying Engineering has offered to the Greek society and to the field of preservation of cultural heritage, 20 years of creative interdisciplinary education of engineers, archaeologists, conservators, natural scientists and other scientists of related disciplines. It provides advanced education and specialization, develops competencies and skills to solve complex problems, in order to educate researchers and professionals in conservation and restoration of architectural heritage and historic materials. The interdisciplinary approach is divided in two Directions of Studies: Direction A "Conservation and restoration of historic buildings"; Direction B "Materials and conservation interventions". The curriculum of multidisciplinary courses is constantly updated and renewed according to scientific-technological advancements, innovative holistic digital methodologies and international standards. It includes core subjects aiming to scientific specialization and optional subjects aiming to extend and deepen knowledge. During the last 20 years of continuous operation, it has created new successful trends in the development of an interdisciplinary educational system and the infusion of innovation in the process of creative and sustainable preservation of Cultural Heritage. It has also helped to create a new generation of engineers and other experts technologically and scientifically skilled, open minded and collaborative with a creative and visionary approach to the Protection of Monuments. In the near future the program will also be offered to international students.

Keywords: Protection of Monuments · Preservation of cultural heritage · Interdisciplinarity · Engineering education

© Springer Nature Switzerland AG 2019
A. Moropoulou et al. (Eds.): TMM_CH 2018, CCIS 961, pp. 273–284, 2019.
https://doi.org/10.1007/978-3-030-12957-6_19

1   Introduction

The Year 2018 is celebrated by the European Commission as the "European Year of Cultural Heritage", with the main slogan "Our Heritage: Where The Past Meets The Future". It also coincides with the 20th anniversary of the foundation of the Interdisciplinary Post Graduate Master Program on "Protection of Monuments" by the National Technical University of Athens (NTUA), which was one among the very first Master Programs founded in NTUA (Fig. 1).

Fig. 1. 2018 European Year of Cultural Heritage-20 years of N.T.U.A. Interdisciplinary Post Graduate Program "Protection of Monuments"

This Master Program has offered to the Greek society and to the field of preservation of cultural heritage 20 years of creative interdisciplinary education of architect engineers, chemical engineers, civil engineers, rural & surveying engineers, archaeologists, conservators, natural scientists and other scientists of related disciplines.

Cultural internationalism and transdisciplinary cooperation, as we know them today, are ideas which emerged after World War I but mostly after World War II. It took them many years to flourish, to be practically implemented, to be able to break the nationalistic mentality and the barriers of strict professional disciplines that ruled the procedures for the Protection of Cultural Heritage in every country.

Unfortunately, during the last years, Nationalism is rising in some European countries, endangering - among others - this valuable initiative.

The Athens Charter Conference on the restoration of historic buildings (1931), the Venice Charter (1964), the creation of UNESCO, ICCROM and ICOMOS, the Amsterdam Declaration (1975), the Granada Convention (1985) had a crucial role in this, at the same time with the new fertile European University education of the younger generations regarding the Protection of our shared Cultural Heritage.

1.1   Protection of Architectural Heritage in Greece

Although Greece has a very rich Cultural Heritage, considered as the cradle of European civilization, with monuments, settlements, natural landscapes, intangible heritage dating back from prehistoric times to the 21st century, the education concerning the protection of our built heritage has started quite late.

The massive reconstruction of Athens and other big cities, after World War II with modern multi storey buildings, due to the interior migration in the 50's from the provinces to the capital city, led the traditional settlements to deterioration and abandonment, thus, outshining the protection of Architectural Heritage.

The declaration of 1975 as year of the European Architecture, by the Council of Europe under the general motto "A Future For Our Past", was crucial for Greek Universities. It raised awareness, and sensitivity to the academic world, regarding our built and natural heritage which gradually led the Higher Educational Schools of Architectural Engineering to start offering courses in the field of "Restoration and Conservation". Until then, the education of young engineers was mostly oriented towards contemporary building methods and new materials, modern architecture and urban planning of new neighborhoods.

In the meantime, the protection of vernacular architecture and natural landscapes, issues regarding restoration methods for built heritage, structural strengthening against seismic risks, as well as a viable reuse of traditional houses, villages and industrial heritage, all started to become critical issues. As a result, a necessity for specialized engineers in restoration and conservation emerged.

Passionate young architects and civil engineers, willing to follow this discipline, were obliged to study abroad in other European Post Graduate Programs, in order to return and face the Greek reality.

2   Post Graduate Master Program: "Protection of Monuments"

In the late 90's, NTUA was in the process of introducing new interdisciplinary Post Graduate Master Programs to address the emerging needs of advanced engineering education. In 1998, the Interdisciplinary Post Graduate Master Program "Protection of Monuments" was founded to provide education for engineers in the field of restoration and conservation of monuments and sites and offer a brighter future to the Architectural Heritage of the country, which was endangered by massive reconstruction, natural hazards and devaluation.

This Master Program was based on the principles of the international Charters and Conventions and on the strong will, scientific knowledge and determination, of visionary NTUA Professors as Professors Ch. Bouras, D. Zivas from NTUA School of Architecture and Prof. A. Moropoulou from NTUA School of Chemical Engineering. After the Senates' approval, NTUA organized an innovative and integrated curriculum, adopting an interdisciplinary approach to cover all aspects in the diverse field of Monuments' protection.

Within this framework, the close collaboration of the four NTUA Schools of Architecture, Chemical Engineering, Civil Engineering and Rural & Surveying Engineering was very decisive for the interdisciplinary character of the Master Program.

The holistic approach in the field of protection of monuments, due to the diversity of values, scopes and the required methodology, is divided in two Directions of Studies:

- Direction A: "Conservation and restoration of historic buildings and sites", organized by the NTUA School of Architectural Engineering
- Direction B: "Materials and conservation interventions", organized by the NTUA School of Chemical Engineering

Many Professors contributed significantly to the organization of the course curriculum and to keep it updated throughout the years. Former Directors of the Master Program, whose leadership was instrumental in the successful implementation of the educational curriculum, were Professors Emeritus N. Kalogeras, M. Biris, M. Korres, E. Maistrou, Professor I. Efesiou, from the School of Architecture. The contribution of Prof. A. Moropoulou, who was amongst the founders and Studies Director of Direction B since 1998, from the School of Chemical Engineering, must also be underlined. Apart from the Directors, many acknowledged Professors contributed to the Master Program's successful evolution, such as Professors Emeritus K. Mylonas, F. Goulielmos, P. Touliatos, I. Kizis, from School of Architecture, Th. Skoulikidis and G. Batis from School of Chemical Engineering, Th. Tassios, P. Karydis and K. Syrmakezis from School of Civil Engineering, and D. Balodimos from School of Rural & Surveying Engineering. Today, the Special Interdepartmental Committee that is responsible for the administrative and scientific issues of the program, comprises of Ass. Prof. C. Caradimas (Director), Prof. G. Marinou, Prof. A. Moropoulou, Prof. E. Vintzilaiou and Prof. C. Ioannidis.

The Master Program was initially running on funds from the Ministry of Education and partly from the Greek Ministry of Culture. International experts and Professors were invited to give lectures. It provides advanced education and specialization, develops competencies and skills to solve complex problems, in order to educate researchers and professionals in conservation and restoration of monuments, historic materials and architectural heritage. The post-industrial era of sustainable development, demands highly qualified scientific and technological support for its evolution.

The students attending the Master Program cover a wide range of disciplines: Architect engineers, Chemical engineers, Civil engineers, Rural & Surveying engineers, Art Historians, Archaeologists, Conservators and other scientists with education relevant to the field of restoration and conservation. In order to attend the Master Program candidates must have a university diploma equivalent of 5 years of studies for engineers and 4 years of studies for natural scientists and archaeologists.

Initially the Master Program's duration was 2 semesters. Since the curriculum's major evaluation and reformation in 2003 it became 3–4 semesters until the recent re-establishment in 2018 with a final duration of 4 semesters. It awards a Post Graduate Specialization Diploma, Master of Science in the scientific field of "Protection of Monuments" in the direction of "Conservation and Restoration of Historic Buildings and Sites" or "Materials and Conservation Interventions", according to the direction that the students enrol and attend.

The curriculum of multidisciplinary courses is constantly updated and renewed according to continuous research, theoretical, scientific, technological, socioeconomic advancements, holistic digital methodologies, sustainable approaches, international expertise and standards. It includes core subjects aiming to open-minded scientific specialization based on transdisciplinary collaboration and optional subjects aiming to extend and deepen knowledge and capacities. This successful combination has achieved an effective framework, merging multidisciplinary theoretical and practical education, through laboratory and on-site surveys, experimental courses and educational visits and the elaboration of innovative postgraduate theses (Table 1).

Table 1. Post Graduate Program "Protection of Monuments" courses' curriculum

Mandatory courses for both directions	
1.1. History and theory of restoration	
1.2. Introduction to the pathology & restoration of monuments & building materials	
1.3. Legislation and management of monuments	
Direction A Conservation and restoration of historic buildings and sites	Direction B Materials and conservation interventions
Core courses	Core courses
2.1. Methodology of analysis and documentation	2.1. Science and engineering of building materials and materials of architectural surfaces
2.2. Methods of conservation and restoration	2.2. Science and engineering of materials and conservation-restoration-protection interventions
2.3. Urban conservation and reuse of historical buildings	2.3. Environmental management for the preservation of monuments
Optional courses	Optional courses
3.1. Special issues concerning archaeological research	3.1. Corrosion and conservation of metal objects of art and constructions
3.2 Monuments and museums - museology	3.2. Specific techniques of materials and conservation interventions with emphasis on earthquake protection
3.3. Protection of the industrial heritage – protection and rehabilitation	3.3. Pilot conservation applications in monuments
3.4. Contemporary architectural heritage	3.4. Environmental management planning for the preservation of monuments and complexes
3.5. Geometrical documentation of monuments	3.5. Environmental management for the protection of museum exhibits
3.6. Digital recording methodologies for the documentation and revealing of monuments	3.6. Materials, techniques and technologies for cultural heritage objects' conservation and preservation
3.7. Technical infrastructures and technological applications	3.7. Archaeometry and methods of revealing archaeo-environment
3.8. Specific techniques of materials and conservation interventions with emphasis on earthquake protection (Specialized course for civil engineers)	3.8. Interdisciplinary documentation, diagnosis, revealing and protection of cultural heritage in the direction of sustainable development
Final Dissertation Thesis for both directions	

A special mention must be given to the dissertation thesis of both directions, as they cover a big range of research topics, such as the following:

- Conservation, consolidation and restoration of monuments.
- Restoration and reuse of historical buildings from ancient to traditional and more recent reinforced concrete constructions.
- Conservation and reuse of industrial buildings and complexes.
- Protection and enhancement of archaeological sites, ancient paths, Underwater Cultural Heritage and other Cultural Landscapes.
- Maintenance, consolidation and enhancement of fortifications.
- Protection, rehabilitation and enhancement of historic cities, towns and villages.
- Specific issues on strengthening of historical structures.

These topics demonstrate the interdisciplinary character of the dissertation theses of this Master Program. In many cases, students of different disciplines collaborate on the same project, during the elaboration of their theses, highlighting the value of inter-disciplinarity in hands-on education (Fig. 2).

Fig. 2. Work in situ and in labs, presentations, educational visits of students

The very important role using resources of active Research Programs in Education concerning monuments' protection must be underlined. Incessant research with the collaboration of enthusiastic students is the only way to progress knowledge in the field of Protection of Monuments and form a new generation of experts with hands'-on experience, that will always explore new paths in order to succeed in their difficult task (Fig. 3).

Fig. 3. Characteristic examples of dissertation theses in 2017

The interdisciplinarity of the field of cultural heritage is further emphasized by Direction B in the approach adopted elaborate simultaneously dissertation theses in the same case study (monument or historic building) by many students of different disciplines. This approach is exemplified in the analysis and restoration projects, including the following case studies:

- Archaeological sites: Temple of Hephaestus (Thission, Athens), Temple of Pythian Apollo (Rhodes)
- Byzantine monuments: Hagia Sophia (Istanbul, Turkey), Holy Aedicule of the Holy Sepulchre (Jerusalem), Kaisariani Monastery (Athens), Monasteries in Serbia, Monastery of Varnakova (Nafpaktia)
- Historic buildings and structures: National Archaeological Museum (Athens), Buildings of National Bank of Greece (Athens and Piraeus), Kapodistrian Orphanage (Aegina), Historic buildings of the Municipality of Athens (Villa Klonaridi, Villa Dourouti), Historic Towers in Mani, Plaka bridge (Epirus)
- Historic cities: Medieval City of Rhodes, Fortezza of Rethymnon, Corfu Fortress

Concurrently, a very important research field for Engineers is the case of Transdisciplinary Multispectral Modeling. Multispectral modeling is a methodology that is rapidly becoming a very important factor for engineers in the analysis and restoration procedure.

In many dissertation theses of this Master Program, students apply multispectral modeling methodologies in their projects, according to the special needs of each study.

This methodology is rapidly becoming a very important and almost indispensable factor in the analysis and restoration procedure. All the collaborating NTUA Schools are working hard in order to enrich the modeling methodology with all data required, intending to gain a useful tool easily manageable by all disciplines in the different survey and restoration cases. In this research field, the scientific input of the School of Rural & Surveying engineers is very important, due to the special technological-digital background needed, leading even to innovative interdisciplinary virtual reality projects.

The transdisciplinary approach for the protection of monuments not only encompasses cooperation among scientists of various disciplines, such as engineers and archaeologists, which has always been a catalytic factor for an interdisciplinary approach, but also actively engages common research by Humanities disciplines, such as Theology, Sociology and others, within a digital environment that acts as a fulcrum for a common new and international language that covers all aspects and engages diverse scholars for an holistic monuments' protection.

Among the most recent innovative transdisciplinary NTUA research projects, where the Master Programme "Protection of Monuments" has contributed, are: The restoration of the historic Bridge of Plaka in Epirus, the reuse and conservation of Villa Klonaridi and Villa Dourouti in Athens, the reuse and restoration of the French industrial metallic wharf in Lavrion-Attica.

In addition, during the last years, Direction B has focused in Transdisciplinary Multispectral Modeling research, with innovative approaches in emblematic monuments, such as the Holy Sepulchre's Holy Aedicule in Jerusalem and the Temple of Pythian Apollo in Rhodes.

Direction A also researches in this area, trying to cover the vast range of data regarding the modeling from ancient ruins to historical buildings, complexes and sites.

The creative collaboration and exchange of expertise with other European Postgraduate programs - especially with École de Chaillot - must be noted. Three very successful workshops were organized between the program's Direction A and École de Chaillot in 2009 (Nauplion), 2011 (Kastania, Mani) and 2015 (Vatheia, Mani), the results of which were presented in Paris and in Athens. The concept of the workshop was the collaboration of the students, initially 10 days in Greece working on the survey and proposals for the restoration of selected monuments and a few months later 10 days in Paris, during which the final proposals were concluded and presented to the Greek and French Professors who participated and guided the students through the workshop (Fig. 4).

Direction B has a long standing cooperation with Princeton University (USA) and Boğaziçi University (Turkey) regarding earthquake protection of built cultural heritage with the organization of keynote speeches in NTUA and Princeton University and two workshops in Hagia Sophia in Istanbul (2000 and 2011). University of Ca' Foscari (Venice, Italy) has provided contribution to the program's curriculum through lectures and co-supervision of Master Thesis.

Fig. 4. Collaboration and exchange with École de Chaillot

Both Directions have emphasized the importance of cooperation with social partners and municipalities via the elaboration of Master theses and projects in historic settlements and monuments all around Greece. More specifically, Direction B has successfully performed the elaboration of case study educational visits and postgraduate Master Theses in the Medieval City of Rhodes and historic buildings of the Municipality of Athens, with the support of the Municipalities of Rhodes and Athens, respectively.

3   The Programme's Evolution and Its Future Perspectives

During the last 20 years, the field of Cultural Heritage protection has evolved, in response to new knowledge that has been developed, the easier availability of innovative and state of the art technologies, tools and techniques, and the wider involvement of disciplines other than the traditional ones.

The concept of the built cultural heritage has been subject to constant evolution, particularly in the second half of the 20th century and the beginning of the 21st, extending from individual monuments and archaeological sites to historic urban areas, rural settlements and cultural landscapes.

Nowadays, digital technology has invaded the cultural heritage domain, reorientating completely traditional tools, methods and techniques. The use of digital technology is now indispensable, in order to create simple drawings which can be used by all engineers, as in geometrical surveying, recording, modelling, historic data filing and mapping, on site advanced diagnostics and inspection by non-destructive techniques, robotics, on line monitoring of monuments etc. However, digital technology cannot be a substitute for the experience and expert knowledge needed to analyse, understand, evaluate, assess and to propose the appropriate restoration solution and methodology that should be followed. On the other hand, digital technology has revealed new horizons to transdisciplinary cooperation, by opening the barriers between different scientific fields, creating common languages, tools and methodology, even enabling them to collaborate equally in innovative projects.

In addition, the field of cultural heritage protection has to take into account issues related to compliance with contemporary requirements, guidelines and legislation regarding buildings' resilience to environmental risks (earthquakes, floods, fire), safety (war), energy efficiency, which have emerged as critical factors in a changing environment within climate change, global socio-economic crisis and path towards sustainability.

However, an evolution utilising latest technologies and tools should not be achieved at the expense of the historic fabric, but instead should transdisciplinary engage the issues of compatibility, reversibility, sustainable reuse, active participation and social involvement towards an efficient restoration, protection, rehabilitation and management of built cultural heritage.

The Master Program has been positively evaluated twice, has organized two conferences and published two booklets with students thesis. Since its foundation in 1998, approximately 500 students have graduated and the alumnae usually face no unemployment, even during the current economic crisis. They are engaged in different departments of the Ministry of Culture and in different positions in the public or private sector.

Experience from the last 20 years of the Master Program has emphasized the advantages of close collaboration with the relevant authorities, such as the Ministry of Culture, Municipalities and Cultural Heritage organizations. In fact, as a result of this collaboration, many graduates of this program are employed by these authorities, serving as a "nucleus" of Innovation and Change-of-Mentality within these authorities, introducing new ideas, know-how and expertise. To this end, this Master Program serves more than purely an academic tool and has created Greek Public services with highly equipped engineers and other scientists that can manage and give solutions to big issues concerning the Protection of Cultural Heritage in Greece.

A large number of our graduates participate in conferences and seminars, being always up to date in their field of interest and many continue on a PhD level.

The Master Program is offered in Greek. In the near future the Master Program will be also taught in English and offered to international English speaking students, in order to widen the horizon of the education in the field of "Protection of Monuments" to the World Cultural Heritage.

4   Conclusions

During these 20 years of continuous operation the Post Graduate Master Program on "Protection of Monuments" has created new successful trends in the development of an interdisciplinary educational system and the infusion of innovation in the process of creative and sustainable preservation of Cultural heritage.

It has also helped to create a new generation of engineers and other experts technologically and scientifically skilled, open minded and collaborative with a creative and visionary approach to the Protection of Monuments of our shared European and World wide Heritage.

References

1. ICOMOS: Guidelines for Education and training in the conservation of monuments, ensembles and sites. ICOMOS General Assembly, Colombo (1993)
2. Moropoulou, A.: Innovative education and training for the conservation of cultural heritage. In: PACT, vol. 58, pp. 71–80 (2000). J. European Study Group on Physical, Chemical, Biological and Mathematical Techniques Applied to Archaeology
3. Maistrou E.: Higher education on conservation and restoration in Greece. The case of National Technical University of Athens. In: Strategies for the World's Cultural Heritage - Preservation in a Globalized World: Principles, Practices, Perspectives. ICOMOS General Assembly, Madrid (2002)
4. Moropoulou A.: 5 years experience on European perspectives of the relevant NTUA MSc direction for materials and conservation interventions. In: 7th International Symposium of the Organization of World Heritage Cities. OWHC, Rhodes (2003)
5. Cakmak, A.S., Freely, J., Erdik, M., Moropoulou, A., Labropoulos, K.: Byzantine Istanbul as an open lab for international interdisciplinary courses of Princeton University, Bogazici University and National Technical University - Materials Science and Engineering Section. In: 7th International Symposium of the Organization of World Heritage Cities. OWHC, Rhodes (2003)
6. Moropoulou, A.: Educational and specialization needs of engineers in the protection of cultural heritage. In: Workshop on Educational and specialization needs of engineers in the protection of Cultural Heritage. Creation of new roles and jobs for Engineers. Technical Chamber of Greece, Athens (2006)
7. UNESCO: Vienna Memorandum on "World Heritage and Contemporary Architecture – Managing the Historic Urban Landscape". UNESCO General Assembly, Paris (2005)
8. Moropoulou, A.: From national to European and international research and education programmes. In: 7th European Conference CHRESP on Safeguarded Cultural Heritage: Understanding and Viability for the Enlarged Europe, Prague (2006)
9. Moropoulou, A., Konstanti, A., Kokkinos, Ch.: Education and training in cultural heritage protection: the Greek experience. In: International Conference Heritage Protection: Construction Aspects, Dubrovnik (2006)
10. Kalogeras, N.: Post graduate studies and the practicing the engineering profession: post graduate program "Protection of Monuments". In: Workshop on Postgraduate Studies in Engineering Education: Experiences and Perspectives. Technical Chamber of Greece – NTUA, Athens (2006)
11. Moropoulou, A., Labropoulou, A., Konstanti, A., Kiousi, A.: European Master level education on protection of cultural heritage: national experiences and European perspectives. In: 8th European Conference on Research for Protection, Conservation and Enhancement of Cultural Heritage. CHRESP, Ljubljana (2008)
12. Balodimou, M.: The contribution of the Post Graduate Programme Protection of Monuments, to the protection and projection of the Greek Industrial Heritage. In: TICCIH Bulletin, Period C, Broch. 1, Athens, pp. 75–79 (2010)
13. Maistrou E.: Evaluation of Heritage as Basis for Planning Strategy in the Post Graduate Program of NTUA Protection of Monuments: Chronicity Sensitive Interventions in Historic Environments, vol. 7. Alinea editrice (2011)
14. CIF ICOMOS: Principles for Capacity Building through Education and Training in Safeguarding and Integrated Conservation of Cultural Heritage. Submitted to CIF Members (2013)

15. Chéhrazade, N., Maistrou, E.: Formations et échanges de l'École de Chaillot et de l'École d'Architecture de l'Université Technique Nationale d'Athènes. In: 5th International Meeting on the Mediterranean Architectural Heritage. RIPAM, Marseille (2013)
16. Moropoulou, A., Lampropoulou, A.: Evaluation of the interdisciplinary program of postgraduate studies "Protection of Monuments" – Direction "Materials and conservation Interventions". In: THALES-COMASUCH Conference on Scientific Support to Decision Making for the Compatible and Sustainable Protection of Cultural Heritage. NTUA, Athens (2015)
17. Caradimas, C., Efesiou, E.: University education concerning European cultural/architectural heritage. In: Safeguarding the Values of European Cultural Heritage. ICOMOS Hellenic Under the Auspices of UNESCO Hellenic National Commission, Athens (2018)

Education and Training for the Preservation of Cultural Heritage, as a Strategic Aim of the Department of Architecture, Frederick University Cyprus

Marios Pelekanos[✉] and Byron Ioannou

Department of Architecture, Frederick University,
7, Y. Frederikou Street, 1036 Nicosia, Cyprus
{art.pem, b.ioannou}@frederick.ac.cy

Abstract. The Department of Architecture at Frederick University Cyprus offers a five-year program which leads to the professional degree of an Architect Engineer. The aim is to educate future architects, providing them with knowledge and sensitivity on the built environment of the European and Mediterranean region and to become an outstanding academic center for studies in the wider European context. There are 4 compulsory courses that directly deal with architectural history and cultural heritage and another 13 that put cultural heritage as a base for their learning outcomes. In addition, research and diploma project courses give also the option to students, to deal with the analysis of historical buildings and constructional systems, history of architecture, urban restoration projects, integration of contemporary architecture into a heritage environment and intervention proposals. Many courses bring students in direct contact with selected historical buildings and urban complexes. Two compulsory courses include investigation on site, constructional analysis and intervention proposals on selected monuments. The Department offers a Master's Degree in "Conservation and Restoration of Historical Structures and Monuments". The aim of the program is to instruct students, not only in the methodology concerning the protection and restoration of historical buildings, but also in the practices, which are nowadays internationally applied in restoration projects. The course accepts students with degrees in architecture, civil engineering, archaeology, history etc. Finally, the Department's strategic goal is the dissemination of research on heritage conservation to the local society through public lectures, high school student events, cooperation with local authorities and publications.

Keywords: Education · Training · Architecture · Cultural heritage · Preservation

This paper was based on information that was prepared by Frederick University Cyprus. Part of this information is available through the University's website. Another part derives from the Course Information Packages of the relevant undergraduate and postgraduate courses. Tables were prepared by the authors and photos belong to the authors and their colleagues in the Department of Architecture. Plans and other documentation on the constructional analysis of historical buildings are part of the relevant undergraduate and postgraduate research programs.

© Springer Nature Switzerland AG 2019
A. Moropoulou et al. (Eds.): TMM_CH 2018, CCIS 961, pp. 285–308, 2019.
https://doi.org/10.1007/978-3-030-12957-6_20

1 Introduction

Frederick University is a private university operating in the Republic of Cyprus, a member state of the European Union [1]. Frederick University was established after a decision by the Council of Ministers of the Republic of Cyprus on 12th September 2007. Although the establishment of the University is relatively recent, the organisation has a long history of more than 50 years in higher education. Frederick University operates from two campuses, in Nicosia and Limassol.

The University offers a broad range of academic programs of study in the areas of Science, Engineering, Business, Arts, Architecture, Media, Humanities, Health, and Education. The University has a strong focus on academic research, being one of the leading research organizations in the country.

The mission of Frederick University is the provision of learning opportunities through teaching and research in the areas of science, technology, social sciences, and the arts, as well as a systematic contribution to the wider social context. The main objectives are the promotion of science, knowledge and education through teaching and research aiming at the enhancement of society in general, the dissemination, application and scientific exchange of knowledge, and the provision of high-quality, internationally-recognized undergraduate and postgraduate education.

The Department of Architecture offers a five-year program which leads to the professional degree (Diploma) of an Architect Engineer [2]. The degree is accredited by the Architects' Council of Europe [3]. The aim of the Department is to educate future architects, providing them with knowledge and sensitivity on the built environment of the European and Mediterranean region and to become an outstanding academic center for studies in the wider European context.

The Department is functioning in one of the seven main buildings of the University in Nicosia campus. The building comprises of two main lecture rooms, seven architectural studios, one computer lab, administration and faculty offices for the Department and a special fabrication laboratory for laser-cutting and model making.

2 Undergraduate Program in Architecture

The Undergraduate Program is based on the ECTS credit accumulation mode of study. Students can be awarded the BA in Architecture upon completion of 240 credits and/ or the Professional Degree of Architect Engineer upon completion of 300 ECTS. These credits are allocated to 43 compulsory and 4 elective courses.

Among the compulsory courses, there are 4 that directly deal with architectural history and cultural heritage and another 13 that put cultural heritage as a base for their learning outcomes [4] (Table 1).

In addition, research and diploma project courses give also the option to students, to deal with cultural heritage subjects, such as the analysis of historical buildings and constructional systems, history of architecture, urban restoration projects, integration of contemporary architecture into a heritage environment and intervention proposals. Selected projects were exposed at Heritage European Network event at CAM – Chania, June 2018.

Table 1. Undergraduate courses with content and objectives based on cultural heritage education and training.

Sem.	Course	Content and objectives
1	APX131 Architectural Technology I	The course is an introduction to architectural technology and it attempts an initial general review of the basic principles and knowledge of the laws of nature concerning structures both historical and contemporary. Identification of basic building materials and construction systems, the actions and stresses of the structural components and also the deformation, durability and stiffness of the basic components and types of structures (historical and contemporary)
3	APX221 History of Architecture I	The aim of the course is to analyze the architecture and art of every cultural period, from the renaissance and baroque to the 21st century, and show how within each, culture-art and architecture creates within his own space and time, but also enriches the architectural structure with the pre-existing knowledge of older or parallel cultures
3	APX231 Architectural Technology III	The course deals with the technology of the light structures, i.e. those whose bearing structure is made out of wood or steel, both historical and contemporary. Analysis of behaviour of bearing members under load, the pathology and fire resistance of the wooden structure, the study of joints and construction design guidelines in the past and today
4	APX202 Architectural Design 4	Contemporary architecture design project in a cultural landscape. A recent project for the students was to design a small housing project in a heritage neighborhood of Old Nicosia
4	APX222 History of Architecture II	The aim of the course is to analyze architecture, from the ancient Mesopotamian civilization to the gothic period, and show how within each, technology, culture-art and architecture creates within its own space and time, but also enriches the architectural structure with the pre-existing knowledge of older cultures
4	APX233 Architectural Technology IV	Techniques adopted in traditional structures and historical monuments, compared to the techniques proposed for contemporary structures, usually stress the importance of a systematic analysis of the design guidelines that achieve resistance to dynamic loads, mainly earthquakes. The course deals with the correlation between the laws of Nature and design with emphasis on the durability and stiffness of the structures and the relation between dynamic loads and the behavior of the main types of bearing systems

(continued)

Table 1. (*continued*)

Sem.	Course	Content and objectives
5	APX301 Architectural Design 5	Integrated urban design project at the edges of selected heritage sites (Walled Nicosia, Kaimakli Core, Pallouriotissa Core), in 2014 in cooperation with the Lebanese American University it focused on "Walled Nicosia Catalyst", in 2016 in cooperation with the National Technical University of Athens it focused on "Re-suturing Buffer Zone", a divided and abandoned neighborhood of early 20th century colonial dwellings
5	APX332 Architectural Technology V	Identification and measurement of a selected traditional building leads to its constructural analysis, a basic prerequisite for the understanding of its behaviour in loads and stresses through the ages. The course deals with traditional building systems, their pathology and vulnerability and especially their ability to undertake dynamic loads
6	APX302 Architectural Design 6	Architectural design project integrated in a heritage site (Mitsero Quarries and Industrial Heritage Structures, Aglantzia cemetery, Pera Orinis heritage settlement, Galata heritage settlement)
7	APX431 Architectural Technology VI.	The findings of the constructural analysis of a historical building are used to identify the causes of its deterioration and damages and implement immediate measures for protection. The course deals also with the distinction of the construction phases of a historical structure, the methods and the necessity of monitoring actions and the design guidelines of an appropriate restoration proposal
7–8	APX401-402 Architectural Design 7–8	Integrated studio. Holistic Interventions inside or on the edge of a heritage site (Recent projects were located at Old Nicosia Moat, Chania Walled City, Famagusta Ghost City) in 2014–15 the cooperation with Chania Municipality has led to a bilateral workshop on interventions in heritage sites
8	APX423 Urbanism III	Applied integrated urban projects, involving heritage sites (Moni Power Station Industrial Site, 20th century suburbs around Nicosia medieval moat) 2016 cooperation with the University of Genoa "Moni Power Station Regeneration", 2018 cooperation with the German University of Cairo "NiLe Project"

Many courses demand activities which bring students in direct contact with selected historical buildings and urban complexes. Two compulsory courses (APX332 and APX431), during the 3rd and 4th year of studies, include investigation on site, constructional analysis and intervention proposals on selected monuments.

Within this framework, every year, several historical structures or monuments are selected, in collaboration with the Department of Antiquities - Ministry of Communications and Works, Local Authorities and Local Bishoprics, in order to be measured, documented, investigated and analysed, to record their vulnerability, history of erection and alterations and to point out any necessary immediate rescue measures.

The 3rd year course APX332 deals with the constructional analysis of historical buildings. The content and objectives of the course are to:

1. Identify, by on site measurement, an existing traditional building structural system. Analyse its basic principles of design from the constructional point of view.
2. Review traditional building practices. Record the pathology and vulnerability of historical buildings and their ability to undertake dynamic loads.
3. Relate the various effects on the constructional system through ages with its behaviour during dynamic load.
4. Manage information from multiple sources. Analyse monuments by interpretation of findings. Identify primary and secondary structural members, connections and components. Distinguish additions, repairs and modifications. Analyse the loading bearing system.
5. Identify causes of deterioration and damages. Distinguish construction phases. Develop immediate measures for protection and restoration proposals.

The related 4th year course APX431 deals with the same project, where the findings of the constructional analysis of the historical building are used to identify the causes of its deterioration and damages and implement immediate measures for protection. The course deals also with the distinction of the construction phases of a historical structure, the methods and the necessity of monitoring actions and the design guidelines of an appropriate restoration proposal (Figs. 1, 2, 3, 4, 5 and 6).

Fig. 1. Students measuring the timber roof basilica of Panagia Moutoulla.

Fig. 2. Team coordinators are giving instructions to students while working with them on site.

Fig. 3. Students transferring measurements on their scaled sketches at the church of Panagia Araka.

Fig. 4. Team coordinators supervising the transfer of the manual measurements on digital drawings.

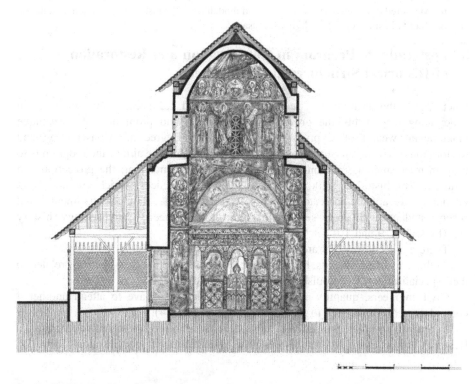

Fig. 5. Scaled plans based on measured drawings, made by hand (Panagia Araka cross- section). Scale of original drawing 1 = 20.

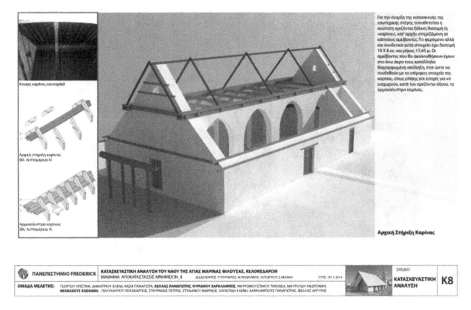

Fig. 6. Measured drawings serve as basis for 3d modelling and constructional system simulation (Timber roofed basilica of Agia Marina Filousas).

3   Postgraduate Program in Conservation and Restoration of Historical Structures and Monuments

Since 2013, the Department, recognizing the need for specializing scientists in undertaking responsibly the preservation, protection and promotion of the unique monumental wealth of Cyprus, offers a Master's Degree in "Conservation and Restoration of Historical Structures and Monuments" [5]. The aim of the program is to fully instruct students, not only in the methodology concerning the protection and restoration of historical buildings, but also in the practices, which are nowadays internationally applied in restoration projects. The course is inter-departmental and accepts students with degrees in architecture, civil engineering, archaeology, history etc. (Figs. 7, 8, 9 and 10).

There are 4 courses that are compulsory to all students (Table 2):

Architects, archaeologists, historians etc. have to attend another 4 courses related to their specialisation (3 compulsory and 1 elective) (Table 3):

Civil engineers, quantity surveyors, topographers etc. have to attend another 4 compulsory courses related to their specialisation (Table 4):

In order to complete the program all students have to prepare, submit and present a Research Project and a Masters Dissertation, related to their specialisation (Table 5):

Table 2. M.Sc. in "Conservation and Restoration of Historical Structures and Monuments" - Common courses.

Sem.	Course	Content and objectives
1	MACOM01 Restoration Theory	History of conservation// Philosophy and Ethics for the protection of monuments and restorations• International Agreements, Contracts and Tenders// Current concepts of recognition, protection, restoration, and use of monuments, Ensembles and Historical Construction.// Problems and Practical implementation of the policy of Protection and Rehabilitation.// Legislation Protection of Cultural Heritage in Cyprus and Europe.// Statutory instruments for recording, protection and Restoration of Monuments and Historical Constructions.// Ethics and Methodology Management of Cultural Heritage.// European Union and International Organizations on the Protection of Cultural Heritage
1	MACOM02 Constructional analysis of historical building systems I	Identification, Documentation and Analysis of Historical Building Construction Systems.// Pathology and Vulnerability of Historical Building Construction Systems (Structures and Sites).// Description and identification of the typically encountered Historical Building Construction Systems (H.B.C.S) in the wider region of the Eastern Mediterranean Basin.// References to traditional building materials and the corresponding structural members.// Local Historical Construction Systems (L.H.C.S.).// The particularities of (H.B.C.S) and (L.H.C.S.) with reference to their typical anti-seismic behavior from prehistory to the 20th century.// Definition of the terms 'Pathology' and 'Vulnerability' of (H.B.C.S) and their correlation with the occurrence of damages. The process of weakening during the structure's life
1	MACOM03 Methods for documentation and pathology	Documentation (geometry, photography, photogrammetry, historical etc.) of the historical structure.// Geometrical documentation methods (methods by hand and by using special devices and their combination).// Documentation and general assessment of pathology.// General assessment of vulnerability.//

(*continued*)

Table 2. (*continued*)

Sem.	Course	Content and objectives
1	MACOM04 Restoration methods and techniques I	Basic principles and scientific ethics of restoration.// Analysis and design methodology of interventions on monuments and traditional structures.// Intervention fundamentals: Compatibility, durability), reversibility.// Framework for the intervention on historic structures so as to conform to the applicable safety standards without harming the integrity, authenticity and originality.// Design proposals for the restoration or repair of monuments and historic structures.// Stone bearing structures during the classical, Byzantine and post-Byzantine years.// Materials properties and proper use (mortar, grout, stone, metal, brick, wood, reinforced concrete)

Table 3. M.Sc. in "Conservation and Restoration of Historical Structures and Monuments" - Courses for specialization in Architecture.

Sem.	Course	Content and objectives
2	MACOM05 Constructional analysis of historical building systems II	Identification, Documentation and Analysis of selected case studies of Historical Building Construction Systems (HBCS).// Pathology and Vulnerability of specific HBCS Systems (stone, brick, wooden and mixed bearing systems).// Description and identification of selected encountered HBCS in the wider region of the Eastern Mediterranean Basin (Cyprus, Greece, Italy etc.).// Thorough study on the basic traditional building materials, their properties and the expected behavior of the corresponding structural member.// Case studies reference of special Local Historical Construction Systems (L.H.C.S.).// Typical examples of HBCS and LHCS with emphasis on their anti-seismic mechanisms and behavior.//

(*continued*)

Table 3. (*continued*)

Sem.	Course	Content and objectives
2	MACOM06 Restoration methods and techniques II	Implementation of basic principles and scientific ethics of restoration.// Analysis and design methodology of interventions on specific case studies.// Intervention fundamentals: Compatibility, durability, reversibility, implemented in actual projects.// Case studies analysis on the intervention on historic structures so as to conform to safety standards without harming the integrity, authenticity and originality.// Design proposals and implementations concerning restoration and/or repair of specific historic structures.// Thorough study and case study analyses on stone bearing structures.// Elaboration on materials properties and proper use (mortar, grout, stone, metal, brick, wood, reinforced concrete)
2	MACOM07 Restoration interventions in building complexes	Identify the history of architectural conservation of historic centers.// Recognize the various charters of conservation and the Cypriot laws on the subject.// Develop the ability to discuss philosophical ideas on conservation of historic centers or towns.// Develop critical thinking.// Construct personal viewpoints and ideology regarding new problems that arise in architectural conservation.// Express a personal ideology and standpoint in regards to architectural conservation.// Show argumentative skills in speaking and writing
2	MACOM08 Management and promotion of archaeological sites (Elective)	Familiarity with the meaning of the principle of organization and management of the archaeological site with the usual needs protection, coverage, collection products excavations, groundwater management and ground water, etc. mounting excavation slopes.// Study of ways to optimize and use of found and excavated antiquities.// Report to the ways of traffic and circulation within the site, methods of information, guidance and information to visitors, the organization of auxiliary spaces etc.
2	MACOM09 Protection - reuse of industrial heritage buildings (Elective)	The introduction to industrial and building shells pre-industrial heritage in Cyprus, Greece and parts of the eastern Mediterranean Basin. Influences from Central and Western Europe.// Rehabilitation, recovery and reuse of industrial facilities eggshell characteristics and historical equipment.// Examples of applications

Table 4. M.Sc. in "Conservation and Restoration of Historical Structures and Monuments" - Courses for specialization in Civil Engineering.

Sem.	Course	Course title and content
2	MACOM10	Masonry engineering and materials technology
2	MACOM11	Static and earthquake analysis of historical building systems
2	MACOM12	Foundations of historical buildings
2	MACOM13	Specialised matters on maintenance and restoration

Table 5. M.Sc. in "Conservation and Restoration of Historical Structures and Monuments" - Research Project and Master Thesis.

Sem.	Course	Content and objectives
3	MACOM14 Research Project	Define the problem under investigation.// Writing of a Task Scheduler and Methodology approach and solving the problem.// Bibliographic research and information.// Approaching the subject of investigation through the rules of science and arts.// Finding the parameters of the problem.// Investigation of the influence of individual parameters on the overall problem and evaluation of the results.// Execution requiring fieldwork.// Execution requiring computational, laboratory and design work.// Preliminary results presentation and corrections.// Completion of Research Project
3	MACOM15 Master Thesis	Definition of the question, the relevant parameters, the brief and the aim of the project.// Development of a Task Scheduler and Methodology approach.// Bibliographic research and information.// Execution of the required fieldwork (on site measurements and survey).// Execution of the required computational, laboratory and/or design work (Proposal).// Preliminary results presentation.// Final project presentation

The M.Sc. in Conservation and Restoration of Historical Structures and Monuments is developed to meet the needs for education in the field of conservation and restoration, by providing the skills needed to key professionals, who are or will be active in the private sector, in governmental institutions and local authorities and as consultant engineers.

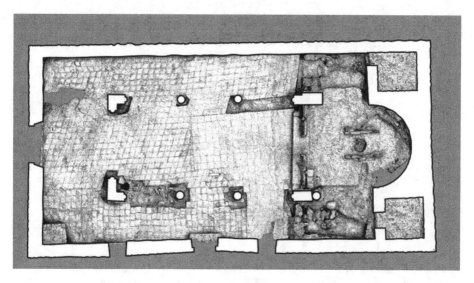

Fig. 7. Photogrammetric plan of the timber roofed basilica of Panagia Iamatiki Arakapa.

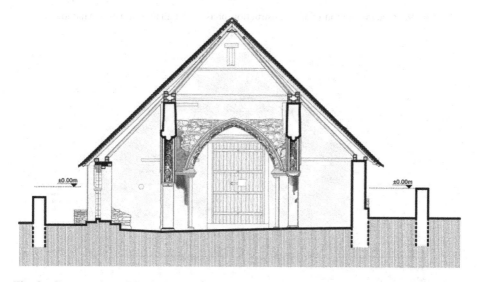

Fig. 8. Cross section of the timber roofed basilica of Panagia Iamatiki Arakapa, drawn by hand.

Β΄ ΕΚΔΟΧΗ:

1η φάση, Β΄ εκδοχή: Βυζαντινή περίοδος (330 - 1191):

Υπάρχει η πιθανότητα τα ίχνη της λιθοδομής που αποκαλύφθηκαν από την μέχρι σήμερα ανασκαφή στην περιοχή του ιερού να ανήκουν στην αρχική ανέγερση του ναού κατά τη βυζαντινή περίοδο (βλ. φωτ. 62, 63 και σχέδιο 17).

2η φάση, Β΄ εκδοχή: περίοδος φραγκοκρατίας (1191-1489μ.Χ.), τρίκλιτος καμαροσκέπαστος ναός:

Δεν μπορεί να αποκλειστεί το ενδεχόμενο και η αρχική φράγκικη εκκλησία να αποτελούσε τρίκλιτο καμαροσκέπαστο ναό. Η συνέχιση των εσκαφικών θα καταδείξει κατά πόσον υπάρχουν ίχνη παλιάς τοιχοποιίας κάτω από το νότιο εξωτερικό τοίχο ή κατά πόσον θα αποκαλυφθούν σπαράγματα τοιχογραφιών στο υπόστρωμα του δαπέδου του νότιου κλίτους. Ήδη στην περιοχή της σήραγγας στο νότιο μέρος του ιερού διαφαίνεται ίχνος λιθοδομής, αλλά δεν είναι σίγουρο η αυτό αποτελεί.

Η πιθανότητα ο αρχικός ναός επί φραγκοκρατίας να ήταν τρίκλιτος και καμαροσκέπαστος ενισχύεται από τις ενδείξεις ύπαρξης γυναικωνίτη, λόγω λίθινης κατασκευής η οποία βρίσκεται στη Ν/Δ εσωτερική γωνία του νότιου τρίτου κλίτους, και η οποία πιθανόν να αποτελούσε τη βάση ξύλινης σκάλας που οδηγούσε προς το ανώγειο μέρος του γυναικωνίτη (βλ. φωτ. 74 και σχέδιο 18). Άλλες ενδείξεις για την ύπαρξη γυναικωνίτη σε ανώγειο χώρο αποτελούν και ξύλινα σπαράγια εντοιχισμένα σε ψηλά επίπεδα στο δυτικό εξωτερικό τοίχο του ναού (βλ. φωτ. 75). Το επίπεδο του δαπέδου του γυναικωνίτη αναγνωστικά θα ήταν πάνω από την κορυφή της τοξοστοιχίας (βλ. φωτ. 14,15) οπόταν αναγνωστικά ο ναός ήταν καμαροσκέπαστος για να υπάρχει ικανοποιητικό ύψος στο ανώγειο. Στο ανατολικό μέρος του δυτικού τόξου στους Γ/αμματόσχημους πεσσούς υπάρχουν ίχνη υποδοχών στερέωσης ξύλινου – πιθανόν καφασωτού – διαφράγματος (βλ. φωτ. 71). Σε αυτό το στάδιο θα πρέπει η δυτική είσοδος του ναού να λειτουργούσε για τις γυναίκες και η νότια για τους άντρες.

Ο ναός αυτή την περίοδο πιθανόν να ήταν καμαροσκέπαστος κρίνοντας επίσης από την καμπυλότητα της τοιχοποιίας και των τοξογράφων στο βόρειο σκέλος της τοξοστοιχίας.

3η φάση Α΄ και Β΄ εκδοχή: κατάρρευση ή κατεδάφιση γύρω στο 150C μέρους του ναού με εξαίρεση το βόρειο και δυτικό σκέλος της τοξοστοιχίας και περιοχή της βόρειας πύλης του ιερού.

Αργότερα, πιθανόν ο φράγκικος ναός να κατέρρευσε, με εξαίρεση το βόρειο και δυτικό σκέλος της τοξοστοιχίας ή να ήταν σε κακή κατάσταση (βλ. σχέδιο 19).

Αναφέρεται ότι ο Αρακάπας βρίσκεται πάνω σε ρήγματα (βλ. χάρτη 2). Στην Κύπρο σημειώθηκαν ιστορικά δύο σεισμοί οι οποίοι επηρέασαν και τη Λεμεσό, ένας μέτριος ισχύος το 1303μ.Χ. και ένας ισχυρός το 1491μ.Χ. Δεν υπάρχουν όμως αναφορές ότι προκλήθηκαν ζημιές στην περιοχή της υπό μελέτη εκκλησίας.

Σχέδιο 17 - 1η φάση, Β΄ εκδοχή: Βυζαντινή περίοδος (330 - 1191)

Σχέδιο 18 - 2η φάση, Β΄ εκδοχή: φραγκοκρατία (1191-1489μ.Χ.), τρίκλιτος καμαροσκέπαστος ναός:

Σχέδιο 19 - 2η φάση Α΄ και Β΄ εκδοχή : 1500μΧ - Κατάρρευση ή κατεδάφιση το ναού - διάφωση βόρειου και δυτικού σκέλους της τοξοστοιχίας και περιοχή βόρειας πύλης ιερού.

Fig. 9. Documentation of the construction phases of Panagia Iamatiki Arakapa.

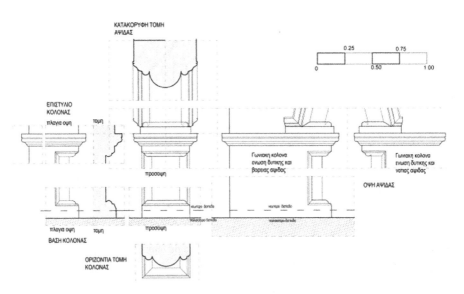

ΚΑΤΑΚΟΡΥΦΗ ΤΟΜΗ ΑΨΙΔΑΣ

ΕΠΙΣΤΥΛΙΟ ΚΟΛΟΝΑΣ
πλάγια όψη   τομή
προσόψη

Γωνιακή κολόνα ένωση δυτικής και βόρειας αψίδας

Γωνιακή κολόνα ένωση δυτικής και νότιας αψίδας

ΟΨΗ ΑΨΙΔΑΣ

πλάγια όψη   τομή   προσόψη
ΒΑΣΗ ΚΟΛΟΝΑΣ

ΟΡΙΖΟΝΤΙΑ ΤΟΜΗ ΚΟΛΟΝΑΣ

Fig. 10. Documentation of construction details (Panagia Iamatiki Arakapa).

4   Research Unit on Historical Constructional Systems

After all these years of intensive work on the subject, based on all the above mentioned research projects, the Department of Architecture has established a Research Unit on Historical Constructional Systems. Director of the Unit is the Head of the Department, Professor Panos Touliatos. The Unit undertakes research projects that cope with special local or global constructional systems, including documentation, constructional analysis and experimental implementation. Research is carried out by the academic staff of the Department, undergraduate and postgraduate students, graduates and scientific associates. Some of the projects that the Unit has undertaken until today include:

1. Research on the phenomenon of ascending moisture at the Katholikon of the Monastery of St. Nicholas at Orounta, including experimental implementation (in collaboration with the Department of Antiquities).
2. Research on the phenomenon of ascending moisture at the Church of St. Charalampos at Deneia (the project is expected to be implemented during 2019, in collaboration with the Department of Antiquities).
3. Research on the phenomenon of ascending moisture at the Timber-Roofed Church of St. Mary at Kourdali, including experimental implementation (in collaboration with the Department of Antiquities).
4. Constructional Analysis of the Dome Roofing of the Katholikon of the Monastery of Docheiarios, Mount Athos, Greece, including proposal for the restoration of the affected roofing (study approved by the authorities, expected to be implemented in the near future) (Figs. 11 and 12).

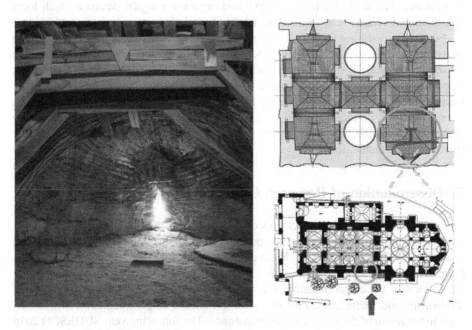

Fig. 11. Constructional Analysis of the Dome Roofing of the Katholikon of the Monastery of Docheiarios, Mount Athos, Greece.

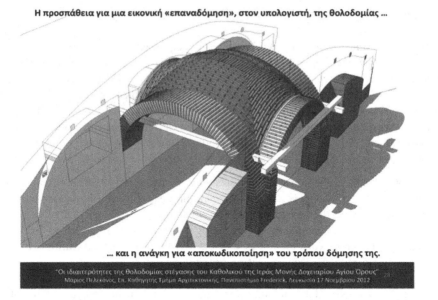

Fig. 12. 3D-Simulation Analysis of the Dome Roofing of the Katholikon of the Monastery of Docheiarios, Mount Athos, Greece.

The Unit recently founded a database of local historical constructional systems, which is based, until today, on the research carried out by the Department during the past decade. The final aim is to establish and organize a digital database of all local constructional systems and to disseminate this knowledge to researchers worldwide.

Furthermore, during the recent years the Unit has established a close collaboration with Politechnico di Torino, Italy. The collaboration includes presentations on the constructional analysis of historical structures by the academic staff of the Department, as part of the relevant Masters course of PoliTo, exchange of students through the Erasmus program and co-supervising of Final Projects carried out by Italian students on Cyprus' cultural heritage. Recently, this collaboration has extended to the Sapienza Universita di Roma, Italy and a third agreement is expected to be signed with Università di Catania, Italy.

5   Dissemination of Research Outcomes

The dissemination of all research outcomes is realized through specific actions, i.e. workshops, public presentations and conferences. Examples of workshops in the past years are the European Design Workshop SUDESCO 2013, which dealt with the Industrial heritage and public space in Limassol [6]. The workshop central theme was the renewal of the costal industrial site of the early 20th century, a unique complex of modernistic and traditional industrial buildings, aiming to the integration of this heritage in the realm of the city's coastal promenade. The following year SUDESCO 2014 dealt with medieval architectural heritage and public space in Agia Napa [7]. Agia

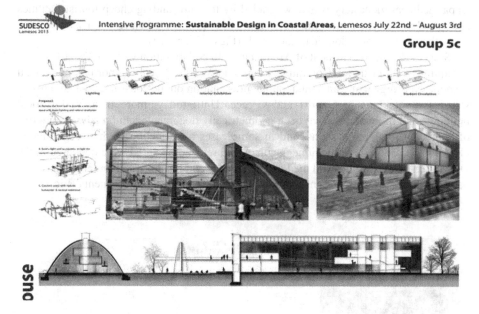

Fig. 13. Sudesco 2013 student projects which dealt with the Industrial heritage and the surrounding public space in Limassol.

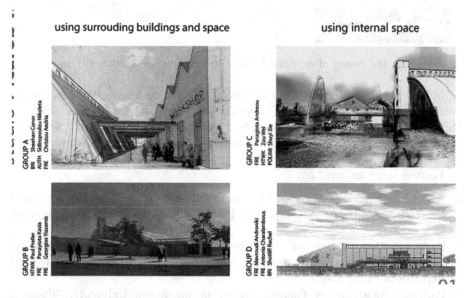

Fig. 14. Sudesco 2013 student proposals based on the renewal of the costal industrial site of Limassol.

Napa medieval monastery was downgraded by the surrounding cheap tourist facilities and visual noise. The aim of the workshop was to propose urban renewal schemes that make the monument a dominant landmark (Figs. 13 and 14).

Since 2012 the Department of Architecture, in collaboration with the Department of Antiquities - Ministry of Communications and Works, Local Authorities and Local Bishoprics, is organising an annual public presentation on the constructional analysis of selected monuments in Cyprus and Greece. Until today, more than 20 historical buildings and monuments have been geometrically documented and analysed in terms of their constructional systems, and in some cases a distinction of their construction phases was proposed. All documents, plans and relevant material were delivered to the Department of Antiquities and the local authorities and bishoprics, in order to serve as a base for the future acts of conservation and restoration of the monuments (Fig. 15).

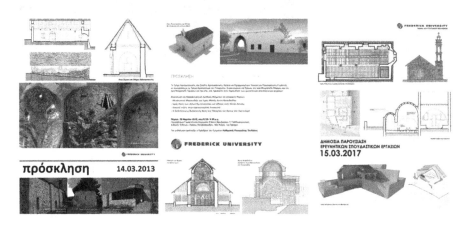

Fig. 15. Posters of the Annual Public Presentations (2013, 2015, 2017).

In 2012, within the framework of the official events of the Cyprus Presidency of the Council of the European Union, Frederick University and the Cyprus Scientific Technical Chamber (ETEK) organized four lectures and two Pancyprian Seminars, which were included in the program "Contribution of Monuments to World Culture and Importance for European Integration".

Lectures were delivered by:

1. Professor Theodosios Tasios, Emeritus Professor of the National Technical University of Athens: "Assessment of Problems in the Structural Restoration of Monuments".
2. Professor Kleri Palyvou of the Department of Architecture at the Aristotle University of Thessaloniki: "From Outdoor Worship to the Temple. Architectural Expressions of the Sanctuary".
3. Professor Anastasia Sali-Papasali, Rector of the Ionian University: "Building Thematic Museums - The Case of Achilles in Corfu".
4. Bishop of Morphou, Mr. Neophytos: "Ecclesiastical Monuments of Cyprus as a Relationship of History and Theology. Monuments or Tombs?".

The First Pancyprian Seminar was organised around a broad thematic, "The Documentation of Monumental Structures". The seminar was attended by a large number of historians, archaeologists, architects, civil engineers and students. In the seminar, a series of topics and practical applications were presented and examined, which are part of the process of analysing monuments and relate to their historical, archaeological, structural and architectural documentation, as a single process, starting with the training of the scientists and students involved and ending with scientific research at each monument separately.

The papers focused mainly on ecclesiastical and other monuments of Cyprus and the wider region during the Byzantine, Frankish, Venetian and Ottoman times. Special reference was made to the importance of constructional analysis as a tool to address the pathology and vulnerability of monumental structures. Speakers were the members of Academic Staff of the University of Frederick Panagiotis Touliatos, Nasso Chrysochou and Marios Pelekanos and the Officers of the Department of Antiquities Eleni Prokopiou and Giorgos Filotheou.

The second Pancyprian Seminar's thematic was the "Structural Restoration of Monuments". As in the first Seminar, the event was attended by a large number of architects and civil engineers, and it was organised under the auspices of the Minister of Education and Culture, George Demosthenous (Figs 16 and 17).

Fig. 16. Photo during the First Pancyprian Seminar on "The Documentation of Monumental Structures".

Fig. 17. Lecture by Professor Theodosios Tasios, on "Assessment of Problems in the Structural Restoration of Monuments".

As mentioned during the seminar, the main objective of Building Restoration of Monuments is, as a first step, to address their existing pathology and to provide them with such strength and security characteristics that they can take on the foreseeable loads and strains without to show failures or collapse. It has therefore been stressed that structural restoration is a prerequisite for the rescue and protection of Monuments, as well as for their re-use and promotion.

Finally, the Department plans and intensively works on the publication, during next year 2019, of a special 400-pages edition, presenting over twenty research projects on constructional analysis of monuments, that were carried out during the past decade by the Research Unit on Historical Constructional Systems of the Department of Architecture.

6   Conclusion

Frederick University is a private university operating in Cyprus, established after a decision by the Council of Ministers of the Republic of Cyprus on 12th September 2007. The organisation has a long history of more than 50 years in higher education and it operates from two campuses, in Nicosia and Limassol (Fig. 18).

The Department of Architecture offers a five-year program which leads to the professional degree (Diploma) of an Architect Engineer. The aim of the Department is to educate future architects, providing them with knowledge and sensitivity on the built environment of the European and Mediterranean region and to become an outstanding

Fig. 18. Frederick University Cyprus, Graduation Ceremony 2018.

academic center for studies in the wider European context. The Department's strategy for promoting Education and training for the preservation of cultural heritage is consisted of:

- High volume of relevant undergraduate courses at the five year Diploma Degree of Architect Engineer,
- The design and run of an interdisciplinary MSc program in Conservation,
- Coordination and participation to international workshops and events related to heritage issues,
- Invited lectureships of esteemed academics and practitioners on conservation issues,
- Publication of books, articles and conference papers on these issues,
- Permanent bilateral cooperation with institutions of excellence in heritage issues from Greece and Italy,
- Dissemination of heritage conservation awareness to the local society through local events, high school student events, publications and cooperation with local authorities along the island and abroad.

The Undergraduate Program is based on the ECTS credit accumulation mode of study. Students can be awarded the BA in Architecture upon completion of 240 credits and/ or the Professional Degree of Architect Engineer upon completion of 300 ECTS. These credits are allocated to 43 compulsory and 4 elective courses. Among the compulsory courses, there are 4 that directly deal with architectural history and cultural heritage and another 13 that put cultural heritage as a base for their learning outcomes.

Many courses demand activities which bring students in direct contact with selected historical buildings and urban complexes. Two compulsory courses (APX332 and APX431), during the 3rd and 4th year of studies, include investigation on site, constructional analysis and intervention proposals on selected monuments (Figs. 19 and 20).

Fig. 19. Educational visit from the Department of Architecture, Frederick University Cyprus in Genova, 2016.

Fig. 20. Project Study by the Department of Architecture, Frederick University Cyprus concerning the re-use of an abandoned industrial heritage site, 2016.

Since 2013, the Department, recognizing the need for specializing scientists in undertaking responsibly the preservation, protection and promotion of the unique monumental wealth of Cyprus, offers a Master's Degree in "Conservation and Restoration of Historical Structures and Monuments". The aim of the program is to fully instruct students, not only in the methodology concerning the protection and restoration of historical buildings, but also in the practices, which are nowadays internationally applied in restoration projects. The course is inter-departmental and accepts students with degrees in architecture, civil engineering, archaeology, history etc.

The dissemination of all research outcomes is realized through specific actions, i.e. workshops, public presentations and conferences. Finally, the Department plans and intensively works on the publication, during next year 2019, of a special 400-pages edition, presenting over twenty student research projects on constructional analysis of monuments, that were carried out during the past decade by the Research Unit of Historical Constructional Systems of the Department of Architecture (Fig. 21).

Fig. 21. Photo of one of the two pavilions of the Research Unit on Historical Constructional Systems of the Department of Architecture, during the "Researcher's Night", 2018.

All the above, clearly state that educating and training young architects on the preservation and restoration of cultural heritage, is an important strategic aim of the Department of Architecture at Frederick University Cyprus. This aim is realized through numerous courses, workshops, seminars, public presentations and publications, which involve practically all stakeholders around the importance of preserving the built remnants of our cultural heritage treasure.

References

1. Frederick University Cyprus. http://www.frederick.ac.cy/
2. http://www.frederick.ac.cy/school-of-engineering-undergraduate-programs/diploma-in-architect-engineer
3. Architects' Council of Europe. https://www.ace-cae.eu/
4. http://www.frederick.ac.cy/diploma-in-architectural-program-structure/diploma-in-architectural-courses
5. http://www.frederick.ac.cy/school-of-engineering-postgraduate-programs/msc-in-conservation-and-restoration-of-historical-structures-and-monuments
6. SUDESCO 2013. http://www.cy-arch.com/sustainable-design-in-coastal-areas/
7. SUDESCO 2014. http://www.cy-arch.com/sustainable-design-in-coastal-areas-sudesco/

Cultural Heritage
and Education/Training/Occupational Activity
of Engineers in Greece

Stamatia Gavela[✉] and Anastasia Sotiropoulou

Department of Civil Engineering Educators, School of Pedagogical
and Technological Education, 14121 Maroussi, Greece
matina@gavela.gr

Abstract. Culture heritage importance has many aspects. From the influence of our sense of identity to the economic aspects as tourism is a leading industry in culture heritage sites. The purpose of this study is to investigate whether engineers perceive the relation of their education, training and experience received or delivered by them, to activities that include the aspect of cultural heritage. The survey was conducted with an online questionnaire that was filled in by 133 engineers. The demographic, academic and occupational characteristics of the respondents were examined for their impact on the responses. Although the role of engineers in the field of cultural heritage is undoubted, the basic answer was that their studies did not include lessons in this field and they did not prepared them to have professional activity related to cultural heritage. The survey could be extended with interviews for further investigation. Such a survey could give directions for the upgrade of education and training of engineers.

Keywords: Cultural heritage · Education · Training

1 Introduction

The current year, 2018, is the European Year of Cultural Heritage [1]. The objective of the European Year of Cultural Heritage is to raise awareness of the social and economic importance of cultural heritage, as cultural heritage has a universal value for us, either being individuals or communities and societies.

Cultural heritage is a major and growing industry of great social significance with a huge economic impact. It consists of many forms. Firstly, it is divided into tangible (e.g. buildings, monuments, clothing, artwork, books, machines, historic towns, archaeological sites) and intangible (e.g. practices, knowledge, skills, oral traditions and expressions, festive events). Cultural heritage also includes landscapes, flora and fauna, as well as resources that were created in digital form or that have been digitalized as a way to preserve them [2]. Except from the intrinsic value of cultural heritage itself, there is also the importance of heritage for the social and economic development. Cultural tourism accounts for a big growing percent of global tourism. The profitability of this type of tourism bases not that much on the admittance fee

© Springer Nature Switzerland AG 2019
A. Moropoulou et al. (Eds.): TMM_CH 2018, CCIS 961, pp. 309–318, 2019.
https://doi.org/10.1007/978-3-030-12957-6_21

charged to gain access to a museum or a site, but mostly on the commercialization of products related to visit, on the economic benefits to the area in which the museum or site is located and on the job creation. "Heritage" implies a sense of continuity and duty.

Conservation and restoration of cultural heritage is a field of interdisciplinary collaboration among professionals with different educational backgrounds and occupational experience. The role of engineers in this field is undoubted. The case of the rehabilitation of the Holy Aedicule of the Holy Sepulchre is an example of a project that succeeded through the interdisciplinary planning and support, using methods and perspectives from different disciplines, namely from the scientific fields of architecture, civil engineering, surveying engineering, chemical engineering, materials science and engineering, information technology, archaeometry and archaeology [3].

Objects and monuments have been created with a wide range of materials and manufacturing techniques, whose study require extensive research in order to preserve them from the speeding degradation that may be caused when they are uncovering from the ground or when they are exposed to pollution, or other factors straining them. The examination and condition assessment of findings and structures and the development of treatment methods for their protection and conservation for future generation bring numerous career paths that lead to specializations [4]. According to Golfomitsou et al. [4] the strong interdisciplinary nature of this subject area makes the formation of a common degree difficult if not impossible. At postgraduate level, students from diverse disciplines acquire a common language and set of skills while at the same time they develop their own professional identity following their individual interests and background. The truth is that graduates must have already acquired the culture and sensitivity to choose a path in cultural heritage field.

The present study investigates whether engineers perceive the relation of their education, training and experience received or delivered by them, to activities that include the aspect of cultural heritage. There are no previous similar studies analyzing the specific topic.

2   Method

A questionnaire-based survey was conducted to acquire data from people in Greece, owning an engineer degree. 133 on line filled questionnaires were collected. The questionnaire content is presented in Table 1. It was created with Google Forms and the participation was anonymous. It was tried to keep short, as the data reliability can be influenced by a long and time consuming questionnaire which make the respondents to lose their concentration or interest before finishing the questionnaire [5].

Due to the restrictions posed by General Data Protection Regulation (GDPR) we are unable to keep record of personal communication data such as emails. A survey based on personalized selection of questionees so to assure identical distribution with the real distribution of the study group, would create legal problems. Consequently a procedure of blind and open questionnaires population procedure was preferred. Nevertheless, the distribution of the finally gathered questionnaires is not notably different from the real one. Lefever et al. [5] pointed out that on line data collection

Table 1. Questionnaire.

Code	Question	Type of answer
Q1	Gender	Female/Male
Q2	Age	5 year intervals (>20)
Q3	Basic studies	Short-answer text
Q4	Institution of basic studies	Short-answer text
Q5	Post graduate studies	Short-answer text
Q6	Institution of post graduate studies	Short-answer text
Q7	PhD	Yes/No
Q8	PhD field	Short-answer text
Q9	Are you an owner of a pedagogical competence certificate?	Yes/No
Q10	From which institution was it issued?	Short-answer text
Q11	Occupational activity	Short-answer text
Q12	Occupational activity location (prefecture)	Short-answer text
Q13	Status of activity	Public sector/private sector/free lancer
Q14	Did your studies (undergraduate and postgraduate) include lessons related to cultural heritage?	Ordinal scale (1 = not at all – 5 = fully)
Q15	Do you consider that your studies made you more sensitive about cultural heritage?	Ordinal scale (1 = not at all – 5 = fully)
Q16	Do you believe that your studies prepared you to have professional activity related to cultural heritage, according to your expertise?	Ordinal scale (1 = not at all – 5 = fully)
Q17	How much do you intend to have a professional activity related to cultural heritage, according to your expertise?	Ordinal scale (1 = not at all – 5 = fully)
Q18	Do you believe that professional activity related to cultural heritage, according to your expertise, can preserve you financially?	Ordinal scale (1 = not at all – 5 = fully)
Q19	Are you interested in training related to cultural heritage subjects, according to your expertise?	Ordinal scale (1 = not at all – 5 = fully)
Q20	If you would like to receive training on cultural heritage, which way would you choose?	Seminar/Postgraduate studies/conferences/Other
Q21	If you are involved in education activities, please indicate in which of the following category(s) you are dealing with	Secondary education/higher education/private vocational training/public vocational training

(*continued*)

Table 1. (*continued*)

Code	Question	Type of answer
Q22	The courses that you are teaching include sections on cultural heritage?	Yes/No
Q23	Which of the following do you use within the courses you are teaching?	Photos/3D illustrations/visits to monuments and buildings of cultural heritage/laboratory tests/Other
Q24	In the context of the courses you are teaching, do you believe that you achieve the interest of students in cultural heritage?	Yes/No

gives participation rates that are low compared to those generally seen in traditional pencil-and-paper surveys, but the respondents are freely choose to participate, which is a positive factor with regard to the quality of the responses. The sample is considered adequate as it corresponds to a group of people freely responding to the survey so any statistically significant result emerging from such a sample is considered to be valid. Furthermore forcing the survey to extent to a wider sample may have resulted to the inclusion of answers provided by people not really concerned on the subject of the survey, taking into account that the questionnaire was available for filling in for a proper period of time.

Since there are no previous similar studies and validated questionnaire on the exact topic, the questionnaire was designed from scratch. The first questions refer to demographic characteristics. Questions 14 to 20 refer to the perception of engineers' education, training and experience received by them, to activities that include the aspect of cultural heritage. Questions 21 to 24 refer to those who are involved to education activities.

Statistically significant correlations among the mean responses of questions 14 to 20 were investigated for subgroups that were formatted according to the answers of the demographic questions. The investigation used the T-student confidence intervals for the mean.

3   Results and Discussion

From the 133 filled in questionnaires obtained, 61% were filled in by male and 39% were filled in by female. The percentages are compatible with the percentages of male and female engineers in Greece. The ages of the respondents was mostly from 26 to 55 years old (Fig. 1). Age and gender distribution of the respondents was proved to be non related thus providing an unbiased sample. That means that both male and female respondents are evenly distributed the age scale. The sample was not biased in terms of age distribution according to gender distribution as these two parameters was observed that was not statistically correlated for the collected sample.

Most of the respondents (38%) were civil engineers. The rest were architecture engineers (8%), mechanical engineers (22%), electrical engineers (6%), electrical and computer engineers (3%), chemical engineers (8%), rural and surveying engineers (4%), electronic engineers (5%), one mining and metallurgical engineer and one production and management engineer. The engineers were both from university level (64%) and technological level (36%). 19% of them (25 out of 133 respondents) are graduates from the School of Pedagogical and Technological Education. 35% of the respondents are owners of PhD degree.

Table 2 presents the distribution of engineers' specialization area in Greece. Data was collected from Technical Chamber of Greece. A prerequisite for being a member of the Technical Chamber of Greece is to be licensed as a qualified engineer and to be a graduate of engineering schools of Greek Universities, or of equivalent schools abroad. At present, not graduates of technological level are able to be members.

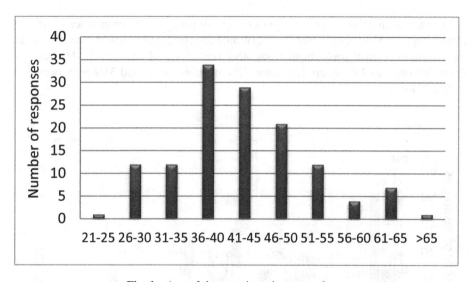

Fig. 1. Age of the questionnaire respondents.

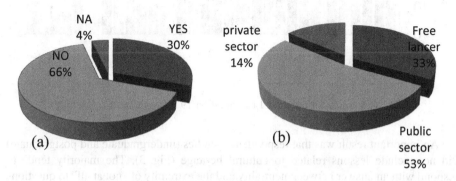

Fig. 2. (a) Answers to question 9 (Are you an owner of a pedagogical competence certificate?) and (b) answers to question 13 (Status of activity)

Table 2. Distribution of engineers' specialization (Data: Technical Chamber of Greece, 2018).

Specialization area	Percentage (%)
Civil Engineers	27.0
Architecture Engineers	17.1
Mechanical Engineers	15.5
Electrical Engineers	16.7
Mechanical and Electrical Engineers	1.1
Rural and Surveying Engineers	5.6
Chemical Engineers	9.1
Mining and Metallurgical Engineers	2.3
Naval and Naval Mechanical Engineers	1.7
Electronic Engineers	3.9

There was an adequate percentage of owners of pedagogical competence certificate (30%) (Fig. 2). Most respondents' occupational activity location was Attica (67%) but there were also small percentages from other prefectures of Greece too. The respondents' activity was 14% in private sector, 53% in public sector and 33% were occupied as free lancers.

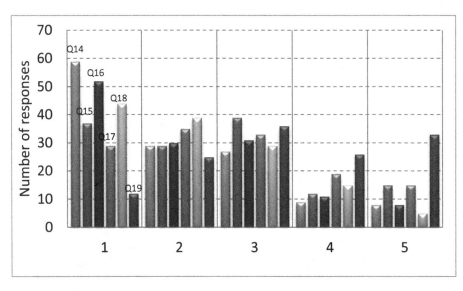

Fig. 3. Responses for question 14 to question 19 (1 = not at all – 5 = fully).

An important result was that respondents' studies (undergraduate and postgraduate) did not include lessons related to cultural heritage (Fig. 3). The majority tends to respond with an answer between neutrality and the extremity of "not at all" to questions Q14 to Q18. They do not consider that their studies made them sensitive about cultural heritage, they do not believe that their studies prepared them to have professional activity related to cultural heritage and they do not believe that professional activity related to cultural heritage can preserve them financially. On the other hand they seem interested in receiving training related to cultural heritage subjects, according to their expertise, mostly (80%) in the forms of seminars and/or conferences (40%) rather than postgraduate studies (19%) or other. In questions 14 to 19 there is a statistically significant linear correlation in any pair of them. That is, anyone who tends to answer one of these questions with a higher score tends the same with the rest of the questions. Therefore those six questions could be also considered in overall as different aspects of the same issue, for example the survey participants' perception of relation with cultural heritage. An engineer could involve to many aspects of cultural heritage field, as building restoration, digitization, materials science etc. So, this perception should be enhanced. Gender and age did not appear to be significantly correlated to the answers of questions 14 to 19. Status of activity (private sector/public sector/free lancer) was not found to be significantly correlated to the answers of questions 14 to 19.

The perception of graduates of School of Pedagogical and Technological Education for the relation of their studies, training and experience to activities that include the aspect of cultural heritage was investigated, since the pedagogical nature of the School could provoke a difference in the answers of the graduates of this School in comparison with others. A statistically significant difference was found between the mean answers from graduates of School of Pedagogical and Technological Education and the mean answers from all the others in question 15. Those graduates consider that their studies made them more sensitive about cultural heritage. Although this difference is not verified when examining the overall answers to questions 14 to 19. Figure 4 shows the mean score (1 = not at all – 5 = fully) and 95% confidence limits for questions 14 to 19 for two subgroups of respondents, according to whether they are graduates of the School of Pedagogical and Technological Education or not. The results are based on 95% confidence intervals estimation assuming a T-student distribution.

Out of 133 respondents, 78 answered that are involved in education activities. From them, 28% are involved in secondary education, 63% in higher education, 18% in public vocational training and 10% answered that are involved in private vocational training (Fig. 5). 23 respondents answered that the courses they are teaching include sections on cultural heritage. The majority of respondents within their courses uses photos (77% of the respondents that answered question 23). Other used in courses is laboratory tests (49% of the respondents that answered the question), 3D illustrations (32% of the respondents that answered the question) and visits to monuments and buildings of cultural heritage (31% of the respondents that answered the question). 56% of those who are involved in education believe that they do not achieve the interest of students in cultural heritage.

A statistically significant correlation (Table 3) in the responses to question 22 ("The courses that you are teaching include sections on cultural heritage?") has been observed according to the gender of the respondents, showing that female respondents tend to answer yes to that question more than male respondents. This could be attributed to either of the following: (a) Women prefer to be involved in lessons having relation to cultural heritage issues, (b) women tend to recognize and identify cultural

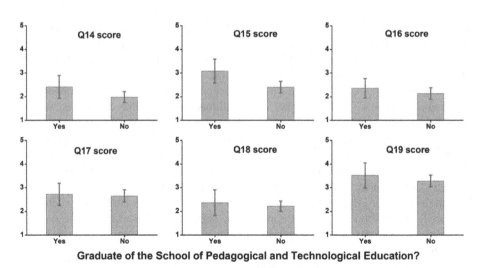

Fig. 4. Comparative results for question 14 to 19 concerning whether the respondent was a graduate of the School of Pedagogical and Technological Education or not (1 = not at all – 5 = fully).

heritage issues in the scope of lessons even when men do not recognize the same in the same lesson, and (c) women tend to, anyway, involve cultural heritage issues when they apply the scope of a lesson they take over.

Table 3. Correlation between responses of question 22 and gender.

Type of answer to question 22	Number of male answers	Number of female answers
Yes	9	14
No	51	18

Developments in computer science enable us to create immersive, interactive and photorealistic experiences that could provide learning experiences that would otherwise be difficult to obtain and could enhance the interest in cultural heritage. Information and Communication Technologies (ICT) have been used mainly in the tourism sector to bring people closer to cultural heritage [6] but they could also be widely used in education. ICT, if it is properly and suitably used, can contribute to innovate and improve educational interventions in the field of cultural heritage [7].

Cultural heritage education could be considered as a pedagogical or teaching process in which people are enabled to learn about cultural heritage. This education should not stop to a level of education but continue throughout the whole education process. The importance of cultural heritage education has already been underlined in 1998 through Recommendation No. R (98)5 of Europe Committee of Ministers to Member States: *"educational activities in the heritage field are an ideal way of giving meaning to the future by providing a better understanding of the past"*.

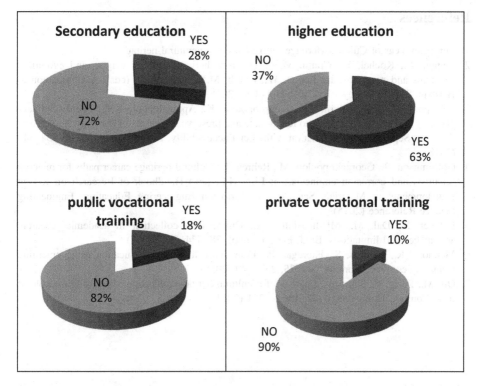

Fig. 5. Responses to question 21 ("If you are involved in education activities, please indicate in which of the following category(s) you are dealing with").

4 Conclusions

Although the role of engineers in the field of cultural heritage is undoubted, the basic answer from the respondents of the questionnaire was that their studies did not include lessons in this field and they did not prepared them to have professional activity related to cultural heritage. They do not consider that their studies made them sensitive about cultural heritage and they do not believe that professional activity related to cultural heritage can preserve them financially.

Respondents that are involved in education, within their courses, use photos, laboratory tests, 3D illustrations and visits to monuments and buildings of cultural heritage 56% of those who are involved in education believe that they do not achieve the interest of students in cultural heritage.

The survey could be extended with interviews for further investigation. Also a systematic validation of the questionnaire could be performed according to adequate techniques. Such a survey could give directions for the upgrade of education and training of engineers.

References

1. European Year of Cultural Heritage. https://europa.eu/cultural-heritage
2. Logan, W., Kochel, U., Craith, M.N.: The new heritage studies: origins and evolution, problems and prospects. In: Logan, W., Craith, M.N., Kochel, U. (eds.) A Companion to Heritage Studies. Wiley Blackwell, Hoboken (2015)
3. Moropoulou, A., Farmakidi, C.P., Lampropoulos, K., Apostolopoulou, M.: Interdisciplinary planning and scientific support to rehabilitate and preserve the values of the Holy Aedicule of the Holy Sepulchre in interrelation with social accessibility. Sociol. Anthropol. **6**, 534–546 (2018)
4. Golfomitsou, S., Georgakopoulou, M., Rehren, T.: Cultural heritage career paths for material scientists and corrosion engineers. In: Lim, H.L. (ed.) Handbook of Research on Recent Developments in Materials Science and Corrosion Engineering Education. Engineering Science Reference (2015)
5. Lefever, S., Dal, M., Matthíasdóttir, A.: Online data collection in academic research: advantages and limitations. Br. J. Edu. Technol. **38**, 574–582 (2007)
6. Mendoza, R., Baldiris, S., Fabregat, R.: Framework to heritage education using emerging technologies. Proc. Comput. Sci. **75**, 239–249 (2015)
7. Ott, M., Pozzi, F.: Towards a new era for cultural heritage education: discussing the role of ICT. Comput. Hum. Behav. **27**, 1365–1371 (2011)

The Historic Centre of Vimercate: Investigation, Education, Community Involvement

Stefano Della Torre[✉] [iD], Rossella Moioli [iD], and Lorenzo Cantini [iD]

Department of Architecture, Built Environment and Construction Engineering,
Politecnico di Milano, Via Ponzio 31, 20133 Milan, Italy
{stefano.dellatorre,rossella.moioli,
lorenzo.cantini}@polimi.it

Abstract. In the last three years multidisciplinary investigations have been carried out on a number of historic buildings in the town of Vimercate, near Milan, whose origins date back to the times of the Romans. The red ribbon that ties together all the analyses, carried out in the frame of university courses, is the continuity of the layered structures and the slow evolution of building techniques.

Beyond the interest of some new achievements in history of architecture, as it is the case for the medieval S. Antonio's church and the baroque Trotti Palace, the main value of these activities has been a transfer of knowledge to trigger a changing attitude in preservation and planning policies.

In the last decades, the local Municipality did not put the preservation and valorization of built heritage as a leverage for development and quality of life. History was felt just as a burden. Therefore, relevant knowledge was just frozen in old sentences, as if research and innovation had no role in the development of the local community.

The organization of activities with schools, associations and people tried to disseminate curiosity and awareness. Some of the activities were also framed in granted projects on preventive conservation. The first tangible outputs have been observed both in the recognition of values in urban regeneration projects, e.g. in the area of the former hospital, and in the principles adopted for the future urban plan.

The implementation of multidisciplinary technologies proved to be effective in feeding communication and developing audience. The learning experience in the realm of real processes proved to be effective in improving students' attitudes.

Keywords: Diagnostic techniques · Heritage education ·
Community involvement

1 Introduction

Vimercate is a small town not far from Milan, which in its urban structure still conserves the footprints of its foundation, dating back to Roman times [1]. The historic centre embraces both outstanding buildings spanning from the Middle Ages to 19th

© Springer Nature Switzerland AG 2019
A. Moropoulou et al. (Eds.): TMM_CH 2018, CCIS 961, pp. 319–328, 2019.
https://doi.org/10.1007/978-3-030-12957-6_22

century, and layered dwelling houses built by techniques slowly developing through the centuries. For these reasons, the town was perfectly suitable in the last three years as a didactic laboratory for architecture students from the Politecnico di Milano. The didactic project, offered to international students, encompasses the geometrical survey of buildings or urban sectors, implementing both traditional direct techniques and instrumental acquisition with digital 2D and 3D modelling, and also GIS for some cases, the recognition of historical layers and building techniques, the technological analysis of buildings and their decays, the evaluations for a sustainable adaptive reuse, according to the planned conservation vision. It is therefore an articulated didactic project, inspired by the debates on education to heritage conservation and restoration [2–9].

The exercise is supported by archaeometric and on the field diagnostic techniques, with the aim of enabling students to get accustomed with these practices, as well as of enhancing the quality of the outputs. Unfortunately, up today it was impossible to implement also laboratory analyses, not for choice, but because of the practical problems, in the given didactic organization, of getting permissions to sample, paying the laboratories and getting results on time. The Authors hope to be able in the future to program also direct sampling and chemical-physical analyses, in order to complete the students' experience and to disambiguate some issues.

The analysed buildings include the churches of S. Antonio [10] and the Holy Virgin of the Rosary, still open to worship, and the relics of S. Marta's church, desecrated at the end of the 18th century and partially demolished, of which also the reuse as a cultural centre and museum has been designed; the Trotti Palace, currently venue of the Municipality, built in the first half of 18th century but never finished, which served for a deep understanding of traditional building techniques; the abandoned buildings of the former Hospital, which have been assessed in sight of an overall adaptive reuse in the frame of an urban regeneration process; the facades of the most interesting buildings in the historic centre, investigated for their conservation issues but also thinking at the theme of urban landscape [11, 12].

The limited dimensions of the historic centre enabled also on one hand to link the students' work to real problems and tasks, on the other hand to exploit their colorful presence as a pretext to call attention on the local heritage by the resident community, which did not care so much for it in the last decades, especially through the municipal town planning. This relationship between the didactic activity and the real processes has been supported by the continuity with some ongoing projects [13–16].

The pure technical aspects of conservation of the structures and the surfaces of historic buildings have been placed in such a frame, highlighting the multidisciplinary features involved in the conservation process: cultural approaches, regulations about protected and not protected buildings, urban planning, urban quality and regeneration processes, energy efficiency and fiscal incentives, and cost-benefit evaluation of a knowledge based approach. This latter aspect is the most important in order to pinpoint the impacts of conservation activities. It seems important to strengthen the students' awareness about the role of historic buildings, protected or not, and of their conservation in the development of the set of cultural values that a community can share, and therefore can give sound foundations to the management of urban spaces.

2 Description of the Teaching Methodology

2.1 Geometrical Survey

The geometrical survey, which is part of the official syllabus of the course, is understood as the basis of any knowledge approach, and as the preparation of the tools to control the transformation processes of existing buildings.

The proposed experience is based on the implementation of both direct and indirect techniques, i.e. the most advanced acquisition techniques (laser scanner and terrestrial photogrammetry) as well as the direct measurement (Fig. 1), usually for details and for the sake of integration on less accessible areas, so that students can deeply understand the complementarity of the techniques and the themes of the tolerances and the propagation of error.

The exercise includes also the organization of information in digital 2D and 3D files and the introduction to GIS and HBIM, implementing the outcomes of the researches carried out by the teaching team [e.g. 17].

Fig. 1. Example of geometric survey activity based on laser scanner and common measurements, implemented with image rectification.

2.2 Stratigraphic and Archaeometric Analysis

The development by students of an attitude to the stratigraphic approach to historic buildings is an important target of the Authors' courses, therefore being given quite a long time in classes and on the site, supplying often unpublished documentary stuff, training students to deal with the different kinds of information. Several selected case studies featured not plastered facades, enabling a direct exam of the building materials and their arrangement in the walls. As in Vimercate a widespread use of bricks mixed to pebbles has been detected (Fig. 2), a special research has been carried out to test whether the dimensions of the bricks through time followed the trends of the Milan city market [18], or the local production had a fuzzy logic due to the local production uses. As a first result, it was concluded that the examined samples tend to follow the Milan trends, at least for the 15th–18th centuries: this statement confirms a strict dependency of the local patronage and building organization from the city, and several other historic issues. On the other hand, a building technique was studied, which exploited the potentialities of locally available pebbles as building materials. Relevant data from literature have been used to demonstrate the effectiveness of brick strips in confining and strengthening masonry made up of pebbles with a wide use of very weak mortar. This technique was widespread in Italy, but Vimercate area provided a rich set of case studies, enabling to highlight the long lasting evolution of the technique for several centuries, during which the characteristics of the typology evolved towards an optimal ratio between height and thickness.

The detected data therefore enabled observations on the production and supply of bricks in this typical subalpine area, and on construction processes at large. This knowledge provides also useful references for the conservation of historic buildings in Vimercate and surroundings, making understandable the distinctive characteristics of the buildings, thus also enhancing their significance [19].

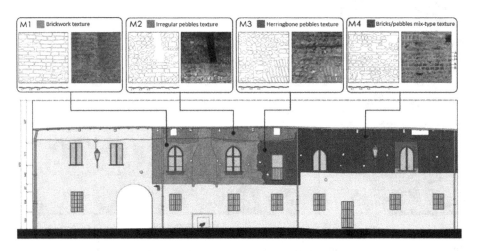

Fig. 2. Stratigraphic analysis through the masonry texture classification.

2.3 Thermography

Among the different diagnostic techniques, some on-site applications of thermographic tests were shown to the class. This investigation technique, being totally non-destructive, provides qualitative results concerning different aspects: masonry textures (for plastered surfaces), hidden structural elements (like technological elements realized in different periods) and also decay phenomena afflicting the building. The distribution of temperatures recorded by the thermo-camera can drive to different interpretations and since the acquisition phase to the final elaborations of the radiometric images, the students were guided to get in touch with the common issues of a diagnostic campaign, improving the role of the knowledge path into the complex of operations forming the conservation process.

The application of thermographic tests offered a fruitful support to the survey analysis, allowing the identification of the component materials, when covered by external coatings, and providing indications on the moisture presence or on diffused detachments, showing the causes of some decays. For the present work, the acquired thermograms were useful for confirming the use of well codified building techniques, like masonry textures composed by pebbles stones or mix typologies combining pebbles and regular brick courses (Fig. 3).

Fig. 3. Examples of thermographic analysis for the masonry texture detection and the integration with the historical sources for interpreting the evolution of the buildings

2.4   Georadar

The roadshow of georadar technique enabled to point out how the recognition of probable underground structures can give important inputs to correct reuse and valorization projects of buildings in historic areas. Meaningful hints of pre-existent structures have been detected both near the former hospital, maybe pointing out the position of the ancient urban walls, and in Palazzo Trotti courtyard, where the walls of the medieval house had been supposed to be located [20]. The most interesting data concern the relics of the former S. Marta church, where the tomb has been detected, which was used for the death executed comforted by S. Marta's confraternity, and beneath it, at a deeper level, a wall similar by depth and orientation to the archaeological relics of the Roman age discovered in the nearby S. Maria's church [21].

Furthermore, the investigation aimed at recognizing exactly the positions of the seven urban doorways demolished by the end of 19th c. is still ongoing; it will be based on the crosscheck of georadar data and the georeferencing of eighteenth-century cartography. The first encouraging results consist in the localization of S. Damiano doorway, nearby the former hospital (Fig. 4).

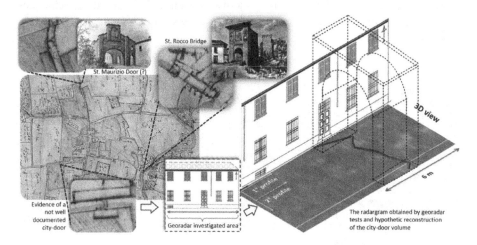

Fig. 4. Implementation of the evidences found through the historical analysis based on the door-system traces reported in the 1721 cadaster map of Vimercate by applying georadar for detecting the archaeological remains of an ancient city-door.

3   The Dialogue with the Local Community

Among the outputs of this ongoing process, the most important achievements may spring out from the discussion on the urban regeneration project concerning the area of the former hospital, changing decisions and attitudes. However, for the time being the processes involving the recognition of the values of the urban facades produced the most mature results.

In terms of Economics, the façades of historic buildings in the urban environment have the characteristic of not-excludability (the passing-by citizen cannot be excluded from seeing the facades on the public scene) and not-rivalry (among the facades no competition exists). Therefore, the facades of historic buildings can be described as a pure public cultural good, even if they keep belonging to private owners: they are on the threshold between private and public property, in physical not less than legal and symbolic sense.

Reasoning in terms of heritage protection, today we would be keen to think the topic in terms of the required proper maintenance that means also prevention of the risks caused by ageing, and limits to the transformation of the façade. In the past the façade was instead conceived as the place where the social status and the civic pride had to be exhibited. A long ribbon unravels from the medieval statutes, to the renaissance norms on "laute aedificare", to the nineteenth-century discipline of public decorum: in the evolution of administration, the consistency among the owner's culture and the governmental culture used to produce a common understanding of magnificence and good state of repair as a converging interest of the public and the private as well.

In the last years this concept has been developed in terms of Historic Urban Landscape, on which Unesco Recommendations have been issued since 2011 [11], while ICOMOS developed the principles approved at La Valletta again in 2011.

This approach is based on the idea that culture, in its diverse forms, should play a role in facilitating and carrying on sustainable development [22, 23]. In other words, the emphasis in no longer on visual integrity and protection, instead it's on the involvement of local communities and stakeholders.

The reference of scientific research and international documents are often those historic urban centres, which outstanding value is widely acknowledged. In these cases, the recognition of cultural values usually coincides with the touristic exploitation: touristification and gentrification, adaptations and bombastic restorations threaten authenticity, as values related just to an external and aestheticizing approach often produce conflicts with the willingness to develop on behalf of local communities, as it has been detected and discussed [24].

Vimercate is not crossed by waves of tourists, but is located in the rich Brianza region, at the heart of a high-tech district that had better times some years ago, but anyway with a productive environment that could give many opportunities. Its ancient town core, with outstanding features, faces nowadays a crisis in its traditional nature of retail area, suffering the competition of other commercial locations, which look more consistent with contemporary lifestyle. Even if in a contradictory way, local dealers are aware of this problem, and therefore pay attention to the opportunities, which could be produced by the rediscovery of the urban quality of the historic neighborhood. In this climate, the work done by Politecnico students for the conservation of the urban historic landscape has been highly appreciated by the local stakeholders, even inviting the students to exhibit their work in the streets.

The Vimercate experience became thus an example of the inseparable theoretical relationship among conservation and valorisation, giving the latter the utmost political value, because of its aims of promoting knowledge and public enjoyment. Valorisation should be understood and put into practice with reference to the themes heralded by the Faro Convention [25]. Particularly meaningful is the definition proposed by Pietro Petraroia, speaking of "valorisation as the relational dimension of protection" [26].

Even the simplest didactic implementation of a knowledge-based approach (historic investigation, 3D survey, multispectral analyses…) triggered curiosity and several inspiring reflections, even concerning facades, which used to tell little stories. Then the students, mostly foreigners, imagined the presentation of their outputs as a mean to propose an enhancement of the urban quality and of the relationships in the city core, with some interesting proposals.

At the same time, the interventions executed in the recent past on the historic buildings started to be seen by local people as lost opportunities, and even as misdeeds in some cases. It's still unknown when this awareness will be transferred into actual planning: the point here is the capability of this knowledge-based approach to rise interest (valorisation) and to start participation practices, which are the premise to make protection grow up from being authoritarian to becoming proactive and relational [27].

From university teaching a cultural transfer has been produced, already implemented in the process, through which the new municipal plan is developing, with the maximum commitment to put into action the innovative directives of regional norms about participated planning [28]. If in the past the planning document of Vimercate Municipality had been pretty rough on the provisions for the preservation and management of the historic centre [15], thanks to this new sensitivity the future plan is going to pay better attention in defining targets and tools.

On the other hand, students had the opportunity to understand what could be the role and the potential of heritage preservation in local processes. They had the opportunity to develop their exercise as a real task, similar to service-learning experiences, which are particularly proficient as implemented in heritage field [29].

4   Conclusions

In a vision of integrated conservation, the theme of urban environment requires awareness of the many analysis levels involved.

The advanced technical-methodological progresses are not useful, it they are implemented within a top-down logic separate from the mediation with stakeholders; on the contrary, by means of a knowledge-based approach the coordination among conservation activities, investigation being the first one, and social inclusion projects can build the conditions for a proactive administrative action, which in turn will require decision making systems adequate to the challenge [12].

On the other hand, the concreteness of the applied approach on case studies framed in real, and often problematic processes of protection and valorization is an invaluable opportunity to let the students learn the fundamentals of techniques but also become aware of the implementation contexts.

References

1. Sacchi, F.: Vimercate in età romana. In: Marchesi, A., Pesenti, M. (eds.) MUST Museo del territorio, pp. 49–69. Electa, Milano (2011)
2. Jokilehto, J.: An international perspective to conservation education. Built Environ. 33(3), 275–286 (2007)
3. Musso, S.F., De Marco, L. (eds.): Teaching Conservation/Restoration of the Architectural Heritage, Goals, Contents and Methods. EAAE (2008)

4. Lobovikov-Katz, A.: Heritage education for heritage conservation. A teaching approach (Contribution of educational codes to study of deterioration of natural building stone in historic monuments). Strain **45**(5), 480–484 (2009)
5. Lobovikov-Katz, A., Konstanti, A., Labropoulos, K., Moropoulou, A., Cassar, J., De Angelis, R.: The EUROMED 4 project "ELAICH": e-Tools for a teaching environment on EU Mediterranean cultural heritage. In: Ioannides, M., Fritsch, D., Leissner, J., Davies, R., Remondino, F., Caffo, R. (eds.) EuroMed 2012. LNCS, vol. 7616, pp. 710–719. Springer, Heidelberg (2012). https://doi.org/10.1007/978-3-642-34234-9_75
6. Embaby, M.E.: Heritage conservation and architectural education: "an educational methodology for design studios". HBRC J. **10**, 339–350 (2014)
7. Lobovikov-Katz, A., et al.: Tangible versus intangible in e-Learning on cultural heritage: from online learning to on-site study of historic sites. In: Ioannides, M., Magnenat-Thalmann, N., Fink, E., Žarnić, R., Yen, A.-Y., Quak, E. (eds.) EuroMed 2014. LNCS, vol. 8740, pp. 819–828. Springer, Cham (2014). https://doi.org/10.1007/978-3-319-13695-0_84
8. Merillas, O.F., Terradellas, R.J.: La educación patrimonial: una disciplina útil y rentable en el ámbito de la gestión del patrimonio cultural. CADMO **1**, 9–25 (2015)
9. Tedeschi, C., Bortolotto, S., Cucchi, M., Tonna, S.: Laboratori di diagnostica: attività didattica e di ricerca, all'interno dei corsi di restauro in Italia e all'estero. In: Fiorani, D. (ed.) I° Convegno Nazionale SIRA "RICerca/REStauro", pp. 978–986. Quasar, Roma (2017)
10. Bairati, E.: Due episodi di architettura "minore". In: Vergani, G.A. (ed.) Mirabilia Vicomercati. Itinerario in un patrimonio d'arte: il Medioevo, pp. 223–225. Marsilio, Venezia (1994)
11. UNESCO: Recommendation on the Historic Urban Landscape. UNESCO, Paris (2011)
12. Veldpaus, L.: Historic urban landscapes, framing the integration of urban and heritage planning in multilevel governance. Eindhoven University of Technology, Eindhoven (2015)
13. Moioli, R.: Preventive and planned conservation and economies of scale. Conservation process for 12 churches. In: Van Balen, K., Vandesande, A. (eds.) Community Involvement in Heritage, pp. 107–120. Garant Publishers, Antwerp – Apeldoorn (2015)
14. CHCfE Consortium: Cultural Heritage counts for Europe full report (2015). http://blogs.encatc.org/culturalheritagecountsforeurope/outcomes/. Accessed 9 Aug 2018
15. Moioli, R.: Centro storico? Solo se brand new. Il centro storico di Vimercate tra tutela e pianificazione. In: Biscontin, G., Driussi, G. (eds.) Le nuove frontiere del restauro. Trasferimenti, contaminazioni, ibridazioni. Arcadia Ricerche, Venezia, pp. 155–165 (2017)
16. Moioli, R.: La conservazione delle facciate nei centri storici: perché, come e per chi. In: Biscontin, G., Driussi, G. (eds.) Intervenire sulle superfici dell'architettura tra bilanci e prospettive, pp. 23–32. Arcadia Ricerche, Venezia (2018)
17. Oreni, D., Brumana, R., Della Torre, S., Banfi, F., Previtali, M.: Survey turned into HBIM: the restoration and the work involved concerning the Basilica di Collemaggio after the earthquake (L'Aquila). ISPRS Ann. Photogram. Remote Sens. Spat. Inf. Sci. **2**(5), 267–273 (2014)
18. Casolo Ginelli, L.: Indagini mensiocronologiche in area milanese. Archeologia dell'architettura **III**, 53–60 (1998)
19. Della Torre, S., Cantini, L., Moioli, R.: Stone masonry with brick stripe courses: study on a historical building technique diffused in Brianza district. In: Aguilar, R., Torrealva, D., Moreira, S., Pando, M.A., Ramos, L.F. (eds.) Structural Analysis of Historical Constructions. RB, vol. 18, pp. 275–284. Springer, Cham (2019). https://doi.org/10.1007/978-3-319-99441-3_29
20. Ferruzzi, A.: Il Palazzo Secco Borella e la sua documentazione. In: Il Palazzo Trotti di Vimercate, pp. 33–44. Comune di Vimercate, Vimercate (1990)

21. De Angelis d'Ossat, M.: Le ricerche archeologiche. In: Corbetta, M., Venturelli, P. (eds.) Luogo di meraviglie. Il Santuario della Beata Vergine in Vimercate, pp. 109–112. Il Polifilo, Milano (1995)

22. Bandarin, F., Van Oers, R.: The Historic Urban Landscape: Managing Heritage in an Urban Century. Wiley-Blackwell, Oxford (2012)

23. Bandarin, F., Van Oers, R. (eds.): Reconnecting the City: The Historic Urban Landscape Approach and the Future of Urban Heritage. Wiley-Blackwell, Oxford (2014)

24. Pendlebury, J., Short, M., While, A.: Urban World Heritage Sites and the problem of authenticity. Cities **26**, 349–358 (2009)

25. Thérond, D. (ed.): Heritage and Beyond. Council of Europe, Strasbourg (2009)

26. Petraroia, P.: La valorizzazione come dimensione relazionale della tutela. In: Negri-Clementi, G., Stabile, S. (eds.) Il diritto dell'arte. La protezione del patrimonio artistico, pp. 41–49. Skira, Milano (2014)

27. Moioli, R.: Valuing the impact: reflections and examples on single buildings and their contexts. In: Van Balen, K., Vandesande, A. (eds.) Heritage Counts, pp. 197–210. Garant Publishers, Antwerp – Apeldoorn (2015)

28. Balducci, A., Calvaresi, C.: Participation and leadership in planning theory and practices. In: Haus, M., Heinelt, H., Steward, M. (eds.) Democratic Choices for Cities. Routledge, London (2005)

29. Dewoolkar, M.M., Porter, D., Hayden, N.J.: Service-learning in engineering education and heritage preservation. Int. J. Archit. Herit. Conserv. Anal. Restor. **5**, 613–628 (2011)

Connections with the Cultural Heritage in Formal and Informal Learning: The Case of an Interactive Visual Game of School Life Museum in Chania

Maria A. Drakaki[1,2,3](✉)

[1] Friends Association of School Life Museum, 7 I. Lekanidis Str.,
731 34 Chania, Greece
mdrakaki65@gmail.com
[2] Hellenic Open University, Patras, Greece
[3] Department of Cultural Technology and Communication,
University of the Aegean, Mytilene, Greece
http://www.school-life-museum.gr

Abstract. The game frames the collection of School Life Museum of Chania since September 2018 and is an output of a European partnership between Greece and Turkey under the ERASMUS program on digital story telling and visual teaching and learning in primary education. The game was attempted to support links with open digital resources and national repositories, highlighting cultural heritage as a pedagogical tool and inspiration for formal and informal education.

In the interactive game meet and talk through the canvas of an original digital narrative points of the cultural heritage of Crete and Ephesus. Apart from Photodentro and cultural metadata, links are also encouraged with the Center for Greek Language with a digital scenario of the writer focusing on educational cultural heritage. In addition, the game leads users to the digital library for European culture, the Europeana.

Keywords: Digital storytelling · Cultural heritage · Visual game

Digital teaching and learning is not just a method of teaching but a practice that can creatively link education to culture, showing common elements in the cultural heritage of the peoples but more widely in their culture, contributing decisively to the respect of the different and intercultural education (Tzovara 2013).

Digital storytelling can be implemented in all levels of education, including adult education and lifelong learning, in all fields of science and can be combined with various learning strategies such as experiential learning discourse, the reflective approach of an artwork (Perkins model), interactive learning (Meimaris 2013).

Within the framework of the European program ERASMUS KA2 "Visual Teaching and Digital Storytelling as an Educational Tool", the Friends Association of the School Life Museum, that was founded in 2005 with the mission to promote and strengthen the vision and aspirations of the Museum[1], undertook the obligation to

---

[1] silogosfilonmsl.wixsite.com/school-life-museum.

A. Moropoulou et al. (Eds.): TMM_CH 2018, CCIS 961, pp. 329–336, 2019.
https://doi.org/10.1007/978-3-030-12957-6_23

produce an interactive application for the Museum of School Life. The conditions for the creation of the digital game were to highlight through its identity the rare evidence of the library of the School Life Museum in connection with the library of Celsus in Ephesus, Minoan culture and digital libraries all around Europe.

The game has framed the collection of School Life Museum of Chania[2] since September 2018 and is an output of a European partnership between Greece and Turkey under the ERASMUS program on digital story telling, visual teaching and learning in primary education. The game was intented to support multiple connections and promote links with open digital resources and national repositories, highlighting cultural heritage as a pedagogical tool and inspiration for formal and informal education.

In the interactive visual game there are given opportunities to meet and talk through the canvas of an original digital narrative with elements of the cultural heritage both of Crete and Ephesus. Apart from Photodentro and cultural metadata, links are also encouraged with the Center for Greek Language with a digital scenario of the writer focusing on educational cultural heritage. In addition, the game leads users to the digital library for European culture, the Europeana.

The purpose of the game is to prepare the participants for a visit to the School Life Museum with the scope for an interesting and critical approach of the exhibits that function as heroes of the digital storytelling. In addition, they are emotionally identified with the scenario hero by developing empathy skills and forming attitudes of respect for the different and common European cultural heritage.

The goal of the game is to overcome the sterilized technocratic process of a digital application in a museum environment that focuses mainly on information and attractive graphics, and the creators set two challenges:

Interaction and links. Links to open national repositories and open digital resources that highlight European cultural pluralism.

An attempt to create a framework that could be used in interactive exhibition planning or in evaluating the effectiveness of interactive exhibits is the five areas proposed by Adams and Moussouri (2002, 11–17), based on data from visitors' surveys on what constitutes a successful museum interactive experience:

1. Multi-sensory dialogue, exploration and discovery. Sensitization of many senses responds to the so-called multimodal learning and communication. Exploration involves the element of problem provocation and problem solving that enhances creativity.
2. Cultural links. It has been noticed that successful interactive experiences also provide a wider conceptual cultural context that helps the visitors group (especially families) to feel members of a wider community and to gain a sense of the social, political, cultural context of their everyday life.
3. Embracing. Emancipation refers to the range of choices and controls that visitors can have in their visit to the museum, and includes issues of intellectual orientation and accessibility for all ages.

---

[2] www.school-life-museum.gr.

4. Uniqueness. Unique, unusual or authentic is usually what fascinates visitors and makes museums so distinctive. Thus, an interactive application based solely on its aesthetic or technological value does not prove popular to the public.
5. Structure of meaning. It is important the design facilitates and enriches the process of understanding, both at a personal and social level. Let's not forget that visitors belong to interpreting communities and build their personal interpretation, often regardless of the intentions of the curators.

By exploring a combination of the above-mentioned elements both at the level of learning value and at the level of interpretative principles, some of the key features of an interactive game can be summarized as following:

- Immediate exploration, physical contact and handling of objects (authentic, copy and/ or special educational material).
- Multi-media interactive approach on behalf of the museum to its visitors: different types of involvement in the physical, mental, emotional and social level.
- Empowering the visitor: developing an initiative and possibilities for control and selection within the exhibition.
- Employability for visitors of different ages and abilities.
- Emphasis on building the meaning and the personal context of the visit
- Promote self-directed learning.
- Encouragement of social learning.
- "Meaningful play" elements combining educational and learning purposes into creative activities.
- Organic linking of interactive applications with museum collections and museum mission.
- Link to the wider cultural and social context in which they form part
- Uniqueness of experience, i.e. the ability to make the visitor into the museum things he can not do elsewhere.

These data are, by no means, an exhaustive list of design criteria, neither are they all necessary for any interactive application. These are features encountered in the large and heterogeneous family of interactive museum experiences across the spectrum of museum practice (Halinidou 2013).

The multiple links attempted to be supported through the game are as follows:

- **Connections between formal and informal education.** The digital narrative of the game connects the subjects of formal learning, e.g. history of the City Hall with the museum environment (School Life Museum). The bees that star in the script come from the exhibit of the Malia Jewelry hosted at the Archaeological Museum of Heraklion, but also in an old reading book of the Museum's collection.
- **Connections with cultural reference sites.** The scenario connects the Museum of School Life with the Archaeological Museum of Heraklion and the Library of Celsus with the archaeological site of Ephesus. The forms of sculptures on the front of the Celsus monument are inherently linked to the theoretical background of Edgar Dale's experience cone of the School Life Museum of Chania. Virtue, Science and Concept come alive and through the puzzles given to the users of the application, encourage activities of all types of learning resources proposed in the

cone of experience. Wisdom is the one that illustrates giving the necessary help. The cone presents a classification of learning resources. According to this, greater emphasis has to be placed on the basis presenting the resources used for learning by action. While the top is simply the verbal reason, that is, learning by explanation. The intermediate level is learning by observation. Dale Edgar's experience cone expresses the core of the theoretical background of the Third Periodic Report of the MAS titled: "LEARNING INSTRUMENTS". Edgar Dale's Experience Cone is the most well-known classification of teaching tools (Dale 1946).

- **Connections between different cultures.** The game was produced in the framework of a bilateral Erasmus KA2 partnership between Greece and Turkey. The shadow theater is undoubtedly dynamic in the folk tradition of both Greece and Turkey. In one scenario version, the heroes were designed within the eShadow workshop for the partners of the program implemented with the collaboration of the Technical University of Crete and in particular the Laboratory of Distributed Information Systems and Multimedia Applications of the Hellenic Chamber of Commerce and Industry.

EShadow launches from the rich tradition of shadow theater and exploits modern internet technologies and interactive graphics to form a creative digital platform in harmony with modern cross-thematic learning trends. EShadow provides a digital narrative environment with distinctive and unique features. The most important of these is the realistic simulation of the movement of shadow theater figures.

- **Connection with rare tokens of the collection of the School Life Museum.** On the occasion of the game, the visitor identifies relevant pages from the Museum's reading book and studies the bee's life and behavior within the bee community.
- **Connecting Open National Digital Resources.** The scenario is linked to the Digital School and to cultural learning objects hosted in it through the riddles the protagonists are to decode. A Learning Object Repository (LOR) is a repository that provides the infrastructure for storing, managing, retrieving and delivering digital resources. The open education resources give the user free access to educational resources with the permission to revise, reuse, remix and redistribute their content. A learning object is the source that can be reused for free. PHOTODENTRO CULTURE is also used by the Institute for Educational Policy (IEP) for the selection and extra pedagogical transfer of cultural objects from collections of the Europeana Portal. The result of the process includes some thousands of cultural learning objects and is available at the Repository Philodendron Culture. Personally, I have participated in this project as one of the writers of the culture learning objects (from Museums, Libraries, Research Institutes etc.) and as a special scientist in Action Δ.2.2, implemented by the Institute for Educational Policy (IEP), entitled "Selection and Enrichment with Educational Metadata of Existing Digital Content by collections of cultural institutions (Museums, Libraries, Audiovisual Archives, etc.) "of Sub-project 12 for Digital School (PHOTODENDRO)[3].

---

[3] dschool.edu.gr.

- **Link to digital libraries and digital museum collections.** The link here once again occurs through the riddles that connect other museums that have granted material to the digital school with learning objects and metadata from Europeana.
- **Linking of different age groups.** At the climax of the script, the protagonists are strongly encouraged to follow the steps of a web exploration that constitutes an organic part of the writer's cross-curricular script, posted on the PROTEAS repository of the Center for Greek Language[4] with focus on the culture of school life. The visual game is also connected with Drakaki's script about school life all over the world[5] through the platform PROTEAS of the Greek Language Center, a research institute based in Thessaloniki. It is funded and supervised by the Ministry of Education, Research and Religious Affairs and it is administrated by the Ministry of Foreign Affairs and the Ministry of Culture. The identity of this link is a script that is based in visual course methodology and supports 14 + hours for teaching and learning through ICT focused on the culture of school life all over the world, for example in the webquest: «School life in space and time in the world»[6]. It also uses creatively:
- Many free digital web02 tools
- Blogs
- E books
- Educational national TV
- Videos from Youtube

In order to achieve these links and the interactive criteria we set out, it was necessary that the script of our digital narrative be inventive and precede a very systematic inquiry to collect the sources that will document the links. According to the 5-phase model, digital narrative dictates the following stages:

The process begins with the creation of the script or the choice of a ready-made script and possibly its adaptation. This is followed by a pre-production phase that includes the selection and/or creation of figurines and sceneries, the composition of the individual scenes of the story and their detailed design (optional) through storyboards. Following is the production itself (shooting scenes). The post-production phase involves the processing of reportings and recordings (where required) and ends with editing to produce the final digital story. The fifth and final phase concerns the distribution (sharing) of digital history.

The game was varied in two script scenarios and thus, unfolds two different stories that inherently support its purpose and goals. Our hero, Sifis who receives bullying at his school due to the special sign on his face, realises during visit at the Museum that he looks like the Phaistos disc. There, he has an unexpected encounter with Sophy Owl, prompting him to enter the museum and discover himself.

In the other version of the scenario the heroes are two children playing in the courtyard of the Museum escaping the educational visit to the Museum that is boring and get involved with the Celsus Library on a fantastic journey of adventurous quest

---

[4] proteas.greek-language.gr.

[5] http://proteas.greek-language.gr/scenario.html?sid=1539.

[6] http://zunal.com/process.php?w=161325.

for knowledge. The adventure is uninterrupted because they have to manage various puzzles and difficulties related to the Minoan civilization, Ephesus and the history of the library of Celsus,

Both versions of the game serve the following principles and values:

- Promotion of interculturalism in a given multicultural society centered on art, history and education.
- Exploitation of new technologies in the educational process, such as digital narrative tools.
- Raising awareness and highlighting the role of the "self" as a unique and integral social element, as well as an active part of society.
- Promotion of values such as respect, diversity, solidarity.
- Exploitation of art as a timeless instrument of culture for the mental empowerment and improvement of the social and cultural parameters that govern every community.
- Adoption of a humanistic vision of modern reality with empathy.
- Alternative exploitation of the Phaistos Disk as a pedagogical tool for strengthening our mental resilience and interactive learning.

The game was connected with the tour of all the Museum's premises in order to activate the interest and curiosity of the participants to visit the Museum and encounter everything concerning the narrative heroes.

The game is about to be uploaded to the platform of the ERASMUS Bilateral Partnership and will be accessible not only to visitors of the Museum but also to any user visiting the site. The game will be available in three languages, Greek, Turkish and English[7].

For these reasons, the association's pedagogical team attempted to support as many connections with open cultural resources and national repositories as possible through its scenario. More precisely, connections with all key digital open source projects that the writer was fortunate enough to become involved as a special scientist of culture, namely Photodentro and the Center for the Greek Language.[8]

It is noteworthy that the inspiration for creating the links in this game was an open platform for collective interactive digital narrative titled "THE MILIA". Milia is an open interactive digital storytelling platform with the scientific support of Professor Michael Meimaris, which supports the digital recording, presentation and co-creation of narratives of all kinds and forms and has been honored with the Euromedia Seal of Approval of the Erasmus Euromedia Awards 2011. Its applications include digital storytelling, education, publishing, and more generally the publication and collective creation of digital material and digital creation. The concept is either alone or in group cant create and plant a story and see it grow in the form of a tree. Several stories are listed on the site on a chronological order list. In the "How to Prepare" section, general

---

[7] http://www.visi-teaching.com/haber/hakkinda/37.

[8] http://photodentro.edu.gr/cultural/.

instructions are given for creating digital narratives. This is a page with general tips for the narration economy and the best use of multimedia image, vide The idea is either alone or in group to plant a story and see it grow in the form of a tree. Several stories are listed on the site on a chronological order list. In the "How to Prepare" section, general instructions are given for creating digital narratives. This is a page with general tips for the narration economy and the best use of multimedia image, video, etc., etc.

The game was completed in August 2018 while there is a thought for the game to be installed not simply on a computer but in the form of a three-dimensional art installation that will introduce visitors to the adventurous journey to the library of Celsus of Ephesus within the Museum's activities hall.

Finally, the members of the pedagogical group Manolakis Konstantinos and Drakaki Maria worked for the script while a visual artistic installation is expected to be the contribution of an architect-visual artist, a member of the Association of Friends of the Museum of School Life. There will procede actions of diffusion of the produced bilateral European Partnership, while emphasis will be placed on experiential workshops for the animation of the interactive game for teachers, families and museum educators.

Conclusions

Storytelling and learning are inextricably intertwined because the process of composing a story is also a process of meaning-making. Integrating opportunities for "storytelling" into coursework strengthens course participant learning. Through storytelling, students are asked to reflect on what they know, to examine their (often unquestioned) assumptions, and – through a cyclical process of revision – to record their "cognitive development process." Because the stories provide a record of students' thinking, teachers can use them in assessing student progress toward learning goals. (Matthews-DeNatale 2008).

Next goal will be the School Life Museum to use this interactive game and the experience of its process as an effective educational tool in visual teaching training courses for various aged groups.

References

English

Dale, E.: Audio Visual Methods in Teaching. Dryden Press, New York (1946)

Robin, B.R., Pierson, M.E.: A multilevel approach to using digital storytelling in the classroom. Paper presented at the Annual Meeting of the Society for Information Technology Teacher Education. Phoenix, AZ (2005)

Matthews-DeNatale, G.: How is Digital Storytelling Relevant to Teaching & Learning, Academic Technology Simmons College, Boston, MA (2008)

Greek

Μεϊμάρης, Μ.: Χτίζοντας γέφυρες, δημιουργώντας νοήματα: η χρήση της ψηφιακής αφήγησης (digital storytelling) στον ευρύτερο χώρο της εκπαιδευτικής μονάδας. επιστημονική ημερίδα Κοινωνικά Δίκτυα και Σχολική Μονάδα: Γέφυρες και Νοήματα, Πανεπιστήμιο Αιγαίου, 1718/5/2013, Ρόδος (2013)

Τζοβάρα, Ν.: Digital Storytelling in the Classroom. New Media Pathways to LITERACY, LEARNING & CREATIVITY του Jason Ohler, εργασία για το μάθημα « Δημιουργία Ψηφιακών Εφαρμογών », Πανεπιστήμιο Αθηνών ΔΔΜΠΣ « Τεχνολογίες της Πληροφορίας & της Επικοινωνίας για την Εκπαίδευση » (2013)

Χαλινίδου, Β.Ζ.: Διαδραστικότητα, Μάθηση και Εμπειρία, Μελέτη περίπτωσης: Η διαδραστική έκθεση Επιστήμης και Τεχνολογίας του Ιδρύματος Ευγενίδου , Διπλωματική εργασία Πάντειο Πανεπιστήμιο Τμήμα Επικοινωνίας Μέσων και Πολιτισμού, Μεταπτυχιακό Πρόγραμμα Πολιτιστική Διαχείριση, , Ιούνιος 2013 , Αθήνα (2013)

Networking (Last access on 3/07/2018)

Adams, M., Moussouri, Th.: The interactive Experience: Linking Research and Practice. Πρακτικά Συνεδρίου Interactive in Museums of Art, Victoria & Albert Museum (2002). online στο http://media.vam.ac.uk/media/documents/legacy_documents/file_upload/5748_file.pdf

http://proteas.greek-language.gr/

http://proteas.greek-language.gr/scenario.html?sid=1539

http://zunal.com/process.php?w=161325

https://economu.wordpress.com/εκπαιδευτικό-υλικό/αφήγηση-storytelling/

National Storytelling Network στο. https://storynet.org/gr

http://www.visi-teaching.com/haber/hakkinda/37

http://photodentro.edu.gr/aggregator/lo/photodentro-aggregatedcontent-8526-6091

http://photodentro.edu.gr/aggregator/lo/photodentro-aggregatedcontent-8526-6093

http://photodentro.edu.gr/aggregator/lo/photodentro-aggregatedcontent-8526-6148

https://economu.wordpress.com/%CE%B5%CE%BA%CF%80%CE%B1%CE%B9%CE%B4%CE%B5%CF%85%CF%84%CE%B9%CE%BA%CF%8C-%CF%85%CE%BB%CE%B9%CE%BA%CF%8C/%CE%B1%CF%86%CE%AE%CE%B3%CE%B7%CF%83%CE%B7-storytelling/

http://photodentro.edu.gr/cultural/

www.school-life-museum.gr

www.digitalschool.gr

silogosfilonmsl.wixsite.com/school-life-museum

From Discovery to Exhibition - Recomposing History: Digitizing a Cultural Educational Program Using 3D Modeling and Gamification

Maria Xipnitou[1], Sofia Soile[2(✉)], Ioannis Tziranis[3],
Michail Skourtis[1], Alcestis Papadimitriou[1], Athanasios Voulodimos[3],
Georgios Miaoulis[3], and Charalabos Ioannidis[2]

[1] Ephorate of Antiquities of Argolida, Ministry of Culture, Nafplio, Greece
mxipnitou@gmail.com, mmskourtis@yahoo.gr,
alcpapadimitriou@gmail.com
[2] Laboratory of Photogrammetry, School of Rural and Surveying Engineering,
National Technical University of Athens, Athens, Greece
{ssoile, cioannid}@survey.ntua.gr
[3] Department of Informatics and Computer Engineering,
University of West Attica, Egaleo, Greece
y.tziranis@gmail.com, thanosv@mail.ntua.gr

Abstract. The main objective of this paper is to show the need for redefining the role of digitalization in the field of interpreting archaeological sites and museum exhibitions. Digitalization of archaeological fragments provides an excellent research platform, useful to distant researchers. New experiential activities will be provided in the process of knowledge development, communication and sharing. All ages and levels of students make practical, effective use of the produced platform, where, using 3D models, they will study and recompose the past. The aim is to enrich the educational methods, so that they respond fully to sustainable economic and social development. The research presented is developed in two levels. The first creates digital laboratories, where students follow findings from revealing till exhibition. The second investigates the use of this platform as a management tool by professional users (archaeologists, historians, conservators) in order to elaborate fragments of objects or even the whole findings, without real contact. The project enables distant research and mainly the possibility to examine findings that may be lost in future.

To meet the above objectives, high-accuracy digital copies and 3D models are of significant importance. The project investigates the potential of using a structured light scanner and image-based modeling. The multi-image techniques using images of findings taken from different angles, distances, level of illumination, create a dense point cloud in 3D space and develop an interface, which enables users to process it using simple tools. The produced 'game' offers a playful, interactive and engaging experience. The application developed encompasses important functionality (kinematics, collision detection, point intersection, movement validation, lightmapping, skinning, animation state machines). In addition, the environment of the tool enables the students to provide feedback about the application and the entire experience, which can be used for further study and improvement planning.

© Springer Nature Switzerland AG 2019
A. Moropoulou et al. (Eds.): TMM_CH 2018, CCIS 961, pp. 337–349, 2019.
https://doi.org/10.1007/978-3-030-12957-6_24

Keywords: Educational programs · Virtual laboratory · Artifacts ·
3D textured model · MVS · Visualization · Educational games

1 Introduction

The current paper focuses on the possibility of transferring an educational programme, currently taking place in a museum, specifically Argos Archaeological Museum, in a digital platform. The purpose is twofold: The first is to make the museum approachable from remote visitors and the other is to promote the digital education as its main target. ICOM, at the 21st General Conference in Vienna (2007), elaborated the role of the museum as "an institution which acquires, conserves, researches, communicates and exhibits for purpose of study, educational and enjoyment, material and evidence of people and their environment". Its role is not only to preserve and store artifacts of national treasures, but also to provide a source of knowledge for a holistic view towards all ways of life.

Many researchers had recognized that museums are an important path of broader learning and knowledge institution for the society. Recently, museums have developed a strong interest in technology. However, there are contradicting opinions arising between using traditional approach and interactive technology exhibition techniques for visitor's learning. That is why, until now, the use of communication technologies for learning purposes is relatively limited. Furthermore, being to a degree accustomed to digital environments, students are more likely expected to respond to forms of coaching and practices being implemented in distance -digital- learning. The developments in technology offer a variety home entertainment, additionally to the emergence of digital toys and technology media, that leads children and adults not to be always interested in what many museums offer. The new challenges demand museums to be more creative and innovative. An exhibition with the latest techniques, such as interactive tools or 3D, is proven more effective in captivating younger people's attention. Additionally, studies have shown that museum visitors sometimes have difficulty in identifying the relevance between a static textual information and the exhibits, when they learn more with interactive actions.

The main objective of this paper is to show the need for redefining the role of digitalization in this field. Digitalization of fragments of archaeological objects may provide an excellent educational and research platform, which may be useful to students and specialists that do distant research. The aim is to show that through digitalization of exhibits, new experientially activities will be provided in the procedure of knowledge development and sharing. The digital exhibits become then the tools for any user, even a distant one, to experiment and communicate his/her own personalized interpretation.

The development and visualization of 3D models of artifacts is a remarkable tool for studying and recomposing the past. In its initial form, the developed tools in this project aim to reform and enrich the educational methods, in order to respond fully to the current challenges, as well as the sustainable economic and social development within a knowledge society. It is our intention to make all students sensitive to matters related to the protection of cultural heritage through simple ways, such as interactive participatory procedures and the use of technology of data processing and communication. All ages and levels of students would be able to make practical and effective use of the produced platform.

2 Museum Education

With the term "cultural educational programme" we mean certain activities that take place in a museum from a small team of students. A functional target is to avoid an exhausting visit. The programmes are organized around particular distinct findings, that attract student's eyes and minds. Through these, they develop critical capacity to understand abstract meanings, like past, time, historical changes, social conformations and cultural differences. The aim is to be integrated in the school programme. It usually has three stages: a. the preparation by teachers at school, b. the visit to the museum, where students observe and work in groups with specific task sheets (additionally, a theatrical activity or treasure hunt can take place), c. activities back in the classroom.

2.1 "From Discovery to Exhibition"

In a museum's exhibition, findings often are presented as work of arts, detached from their initial position. Visitors see them in showcases and get to know them through their captions. In other ways, they obtain certain, fragmentary information's, concerning the position they found, their dating, materials and use. Correspondingly, the educational programmes are planned based on the same aspect. It is undeniable that visiting a museum, with family or classmates, is a great experience. You get acquainted with history and past, while obtaining knowledge. An educational programme widens all these concepts. It transforms the images to an insight trip, because it encourages experiment and gaming, motivates the environmental exploration and encourages creative education and expression.

The programme "From discovery to exhibition" (see Fig. 1) is planned in a way that teachers can practice it by themselves. Through narration, game, observation and decorative creations introduces children to the exciting world of discovery and collective memory.

Fig. 1. The educational programme's poster.

In the first stage, they realize that archaeological research studies the physical remains of human activity in order to understand both the historical as well as the social conditions that produced them. They learn about the excavation and the specialties they need. They evaluate different materials and methods of construction and learn their relationship with each historical age. They deal with the peculiarities and difficulties from the moment of finding, through the eyes of the experts involved (archaeologists, conservationists, architects, topographers, geologists, paleontologists) until the moment of the exhibition.

In the second stage, they become themselves members of an excavation research, with all the works that precede and follow it. They excavate, collect, record, evaluate, study, date, preserve, study and plan.

In the third stage they create their own museum exhibition. They learn what collective effort and work means. They transform from archaeologists and conservators into museographers, architects and musicologists, transferring their own experience and interpretation to their 'audience'.

All stages come to life with real and creative workshops:

- the excavation workshop, where children turn into small archaeologists. They excavate in a dip that 'hides' an ancient residence, specially prepared for the purpose. They evaluate and date the findings, create logs, and keep calendars.
- the conservation workshop (see Fig. 2). The small conservators fix, weld and reconstruct the findings.
- the exhibitors workshop. Small museologist undertake to highlight the history of their findings and their values equally and over time.

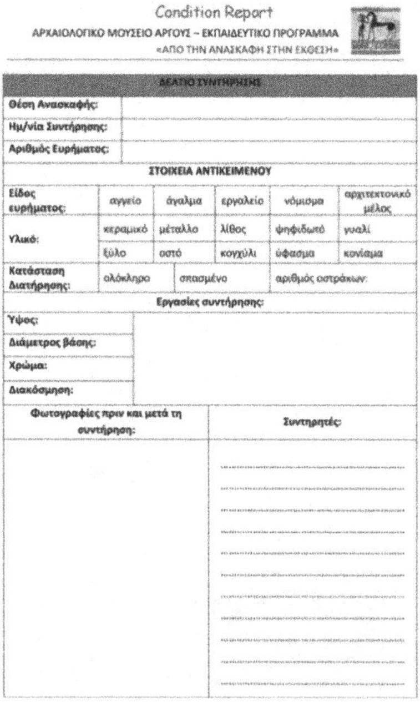

Condition Report

ΑΡΧΑΙΟΛΟΓΙΚΟ ΜΟΥΣΕΙΟ ΑΡΓΟΥΣ – ΕΚΠΑΙΔΕΥΤΙΚΟ ΠΡΟΓΡΑΜΜΑ

«ΑΠΟ ΤΗΝ ΑΝΑΣΚΑΦΗ ΣΤΗΝ ΕΚΘΕΣΗ»

ΔΕΛΤΙΟ ΣΥΝΤΗΡΗΣΗΣ					
Θέση Ανασκαφής:					
Ημ/νία Συντήρησης:					
Αριθμός Ευρήματος:					
ΣΤΟΙΧΕΙΑ ΑΝΤΙΚΕΙΜΕΝΟΥ					
Είδος ευρήματος:	αγγείο	άγαλμα	εργαλείο	νόμισμα	αρχιτεκτονικό μέλος
Υλικό:	κεραμικό	μέταλλο	λίθος	ψηφιδωτό	γυαλί
	ξύλο	οστό	κογχύλι	ύφασμα	κονίαμα
Κατάσταση Διατήρησης:	ολόκληρο	σπασμένο		αριθμός οστράκων:	
Εργασίες συντήρησης:					
Ύψος:					
Διάμετρος βάσης:					
Χρώμα:					
Διακόσμηση:					
Φωτογραφίες πριν και μετά τη συντήρηση:			Συντηρητές:		

Fig. 2. The conservation's workshop report

2.2 Visualization of the Programme

A new digital, interactive multimedia museum highlights the historical and archeological evidence using the latest technologies. The integration of digital technologies challenges museum learning. This answers the following questions. Are the existing learning frameworks and assumptions still valid? How do they influence different audiences? Which is the role of museums in our century?

Responding to a growing demand, digital heritage content has become available online on museum websites and cultural platforms, on social media and crowd sourced platforms, on mobile devices, making knowledge accessible for free-choice learning (see Fig. 3). In order to increase cultural participation and learning through co-creation, museums open up their online collections to social tagging (see Fig. 4). Furthermore, learning becomes accessible to disabled people through digital applications tailored to their needs [1].

Fig. 3. Time Machine, a digital storytelling exhibit, Museum of Prehistory and Early History, Neues Museum, Berlin.

Fig. 4. Marker scanning during the session "Passport to the afterlife", a mobile augmented reality trail at the British Museum.

Educational programmes, on their turn, gradually participate in the digital era. Online learning opportunities range from educational resources to prepare for a museum visit or to memorize it afterwards [2]. Moreover, online environments with digital resources and tools allow teachers and students to personalize, annotate and share, or to create their own tools [3]. Digital learning centers offer visitors, with different levels of digital literacy, an in-depth experience of the museum's collections, or their participation in digital crafting workshops [4]. As recent studies show, young generations are increasingly becoming co-creators of digital content, using interactive technologies towards a participatory culture [5].

Until a few years ago, archaeological objects in exhibitions, were simply treated in a flat manner by type and typology. Today findings owe to treat as information carriers for the time and the society that produced them. In this way, the exhibitions are educational and entertaining, and their reach for the general public is greater.

Given the great impact of the educational program "From Discovery to Exhibition" in schools of the wider region of Argos Archaeological's Museum, and delighted by the demands of remote areas for implementation and monitoring of the program, we decided to exploit the new technologies that have already entered the museum practice. We focused on creating an interactive online program which will give the visitor the opportunity to choose the kind, quality and quantity of information he wants.

The program will be divided into two levels. In the first the children-users aim to update and familiarize themselves with history, excavation and museum functions. Each child or school group will be able to replicate all the laboratories that have already been analyzed, with different levels of difficulty. The second level will be addressed to the remote researcher, who, through the platform, will be able to study and reconstruct findings and thus develop his research faster. At the same time, he will be able to engage with other scholars, utilizing in the most flexible way the information society, which is now transformed into the society of group knowledge.

At the moment, we pay attention to the task of conservation. Thus, through the platform we create, the user participates in the identification, design, visualization and reconstructing the fragments of a finding, providing a whole and proper composition (see Figs. 5 and 6). After grouping the fragments, it is possible to move them around on the computer screen and try to find pieces that fit together. Hypotheses of the patterns could be made and compared to the literature. The main advantage of the computer puzzling system is that it limits the manual handling of the fragile fragments.

In later stages, the user will be able to choose ways to examine and clean the findings, through a series of suggested procedures. He will also be able to keep a record of photos and attempts, where he can see the exhibits before and after their preservation. In summary, the weight of the study is based on the ability of an interactive digital exhibit to display and transmit "hidden" information to the visitor, where a real object, exposed in a showcase, cannot.

Fig. 5. Vase fragment's.

Fig. 6. Reconstructed vase (krater)

3 Digital Platform

The research presented in this paper is developed in two levels. The first part applies to the students of primary and secondary education. The usual educational programs in museums or exhibitions are organized around the archaeological findings, that immediately provoke the interest, because they excite the natural curiosity and the inquiring mood of the children (see Fig. 7). Our objective is to further develop this procedure and to create digital laboratories, which may operate either close to their natural field or in a distant school classroom. The digitalization of exhibits and archaeological sites enables the students to participate to the excavation procedures and practice the various ways of observing things and interpreting the history of signs and finally understand the multidimensional meaning of an exhibition and its offer to the

Fig. 7. The educational programs in museums or exhibitions are organized around the archaeological findings.

community memory and education. The students may follow the findings from the time of their revealing in the earth till the moment they are placed in an exhibition.

The second part of the project investigates the use of this digital platform as a management tool by professional users, such as archaeologists, historians and conservators. Using the same tools but different branches, professionals may elaborate fragments of objects and/or the whole findings. The program provides with the ability to redefine, recompose, investigate and finally create new data and new signs of history without contact with the real finding. This project enables distant research, and the possibility to examine findings that may be lost or destroyed sometime in future. With the developed tools, a wide working environment is created, which converse with an open, digital and continuously 'supplied' museum. We hereby present the process in a "proof of concept" sense, focusing on the creation of an application that permits students to examine various fragments of an "artifact" and put them together to reconstruct it. The different steps of the process from data capture, to 3D modeling and application design and development are described.

3.1 Data Capture

For the data capture process, we used the Canon EOS 6D digital SLR (Single Lens Reflex) camera. Some of its main features are: (i) CMOS sensor, full frame, dimension 35.8 mm × 23.9 mm; (ii) 20.2 active megapixels; (iii) Canon EF 50 mm Lens 1:1.2 L USM. We used the smallest aperture of the lens, i.e. f16, in order to achieve the widest possible depth of field, which ensures high quality clarity in the final images produced. The corresponding speeds that occurred after the metering were very slow (i.e. t 1/2 s). This was treated by using a tripod to support and thereby stabilize the machine.

The *Foldio360,* a smart turntable that allows you to create high quality 360° photos easily, was used for rotating and lighting the model. The model was placed in the turntable, in a floor plane with recognizable patterns, which helped in scaling the model (Fig. 8). A photo was taken every 10°, the rotation controlled by Foldio360 application for smartphones.

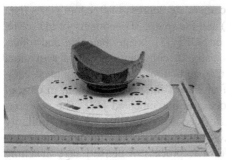 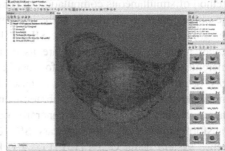

Fig. 8. Left: A fragment being photographed in Foldio360. Right: The Point Cloud, 112 photos aligned, 63.418 tie points found by Photoscan.

 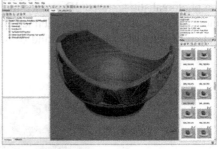

Fig. 9. Left: The position of the cameras. Right: The Dense Cloud, calculated in High Quality.

The model used was a vase, approximately 20 cm tall. It is a replica of an ancient Greek amphorae style and it bears the representations of two figures (one of Goddess Demeter and one of God Ares of the Greek Olympian pantheon). The model was broken into 8 pieces and photographed in Foldio360, with approximately 80 to 100 photos taken for every single piece. Both sides of every piece were photographed for better quality result. Where needed, i.e. in the small fragments, were also photographed in different angles.

3.2 Data Processing

To meet the above objectives, high-accuracy 3D models of parts and pieces of archaeological artifacts, as discovered, are of significant importance, both when professional users are involved, and when these pieces are to be used as teaching material for students. The geometry of the object may be accurately represented including the damages and weaknesses of the artifact.

The automatic creation of 3D models using image sequences relies on photogrammetric and computer vision algorithms. The first step of the image-based modelling pipeline is the Structure from Motion (SfM) process [6], which refers to the method of simultaneous computation of the camera six degrees-of-freedom poses (i.e., the camera exterior orientation of the images) and the sparse 3D geometry of the scene. Feature extraction algorithms, like SIFT [7] or SURF [8], image matching and robust outlier rejection techniques [9] are used, usually in combination with a sequential (incremental) algorithm [10, 11] for metric reconstruction, in case of calibrated cameras, or projective reconstruction, in case of unknown interior orientation; in case of uncalibrated cameras an auto-calibration process is implemented [12]. The georeferencing of the SfM outputs is generally performed by estimating the 3D similarity transformation between the arbitrary SfM coordinate system and the coordinate reference system using GCPs [13] (Fig. 9 Left). The process of dense image matching [14] is applied for the creation of a dense 3D point cloud using the camera exterior and interior orientation of the imagery, estimated by the SfM process (Fig. 9 Right). Then, a 3D mesh model can be created using the generated dense point cloud [15] via a surface reconstruction algorithm. Finally, texture mapping may be applied to the mesh model using the oriented images. The result of the image-based 3D modelling process is a photorealistic textured 3D surface model of the scene depicted in the imagery that refers to the coordinate system defined by the GCPs or the arbitrary SfM coordinate system, in case of lack of such kind of information (Fig. 10).

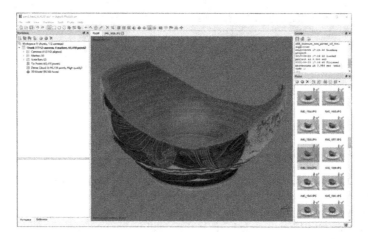

Fig. 10. The final goal of photographs processing with PhotoScan is to build a textured 3D model. The 3D model can be exported for further usage in external programs and in our case, Unity. 90.000 faces were connected.

The Agisoft PhotoScan software was used for the creation of dense point clouds using images taken by the Canon EOS-1Ds Mark III camera. The processing of each Vase fragment's was performed separately. The alignment of the images was the first step performed by PhotoScan; it is a SfM process that generates a sparse point cloud of the scene and computes the camera interior and exterior orientation, using a modification of the SIFT algorithm for the extraction of feature points. Dense image matching was the next step applied through PhotoScan; the software calculates depth information for each camera and combines it into a single dense point cloud.

The reconstructed of the surfaces was made into the PhotoScan environment creating 3D model without texture. The final step are 3D textured models for every internal and external part of the Vase fragments. The multi-image (MVS) techniques use images of a finding taken from multiple different angles or unordered collections of images taken from multiple distances, viewing angles and level of illumination to create a dense point cloud in 3D space. Accurate models can be built by matching unique features in each image and determining from which position and direction the images were taken. The well-known procedure consists of 3 steps: calculation of interior and relative orientation of the images and creation of sparse point cloud by applying Structure from Motion method (SfM); dense image matching; production of the meshes and 3D textured model.

3.3   Game Design and Development in Unity Game Engine

The application is being developed in Unity game engine. A game engine, is a software that provides game creators with the necessary set of features to build games quickly and efficiently. The core functionality typically provided by a game engine includes a rendering engine ("renderer") for 2D or 3D graphics, a physics engine or collision detection (and collision response), sound, scripting, animation, artificial intelligence, networking, streaming, memory management, threading, localization support, scene graph, and may include video support for cinematics. The actual game logic has to be implemented by some algorithms.

Unity is a cross-platform game engine with a built-in Integrated Development Environment (IDE). It is used to develop video games for web plugins, desktop platforms, consoles and mobile devices (it supports 27 platforms). Unity gives users the ability to create games in both 2D and 3D, and the engine offers a primary scripting application programming interface (API) in C# programming language.

The reason we chose Unity is because as a development tool it is completely free and royalty free and it gives a set of tools that is fairly easy to use. Unity is a complete package that allows the user to simultaneously play your game, edit it, and test it as well, create environments, add physics and lighting, manage audio and video, handle animation, and more. As such, it is being used for the development of a variety of games and applications, including cultural heritage related ones [16, 17, 18].

The final goal of photographs processing with PhotoScan is to build a textured 3D model. The 3D model can be exported for further usage in external programs and in our case, Unity. PhotoScan supports model export in numerous formats, like Wavefront OBJ, Collada DAE, 3DS file format, Autodesk FBX or DXF and Google Earth KMZ to name a few, with most of them being able to imported in Unity (Fig. 11).

The introduction of the 3D models into Unity is simple, and usually dragging and dropping the exported files will add/import them in the editor's Assets panel. In our case the export model is in OBJ format, composed by three files; the model's file, a material library file and the texture's image, all must be imported. The 3D models are then available for usage.

The basic elements of our application, are eight 3D models (a model for every piece of the amphorae – vase). In the application, the user has to move and rotate every piece,

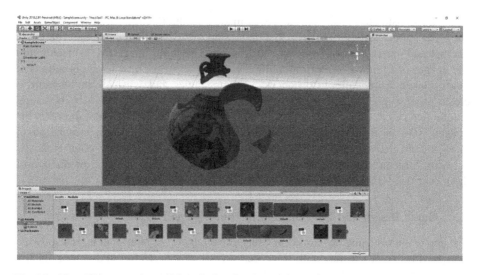

Fig. 11. The 3D textured models of the fragments in Unity Game Engine during the development of the application.

so to assemble the vase in its original form. The challenge is to solve the 3D puzzle, finding the correct place and angle of every piece.

The application presents the vase's fragments randomly put in an excavation and will enable the user to process the 3D models of the fragments of each artifact, using the mouse and the keyboard. The 3D models of all the fragments are of the same scale but are randomly placed and oriented in space.

The user (student) is asked to recognize through the specific characteristics and find the right position of each 3D model of the fragmented parts of the artifact so that the artifact will be digitally reconstructed. The user may zoom in, zoom out, move and 3D rotate the models in order to reconstruct the artifact and give to it its real shape. The environment of the tool enables the students to be critical on the procedure used and give their comments about the artifact under study.

4   Conclusion

The use of multimedia applications and interactive exhibits is not a privilege and right only for children's museums, science museums, natural history or theme parks, but is applicable, with very good results, in more traditional cultural spaces and organizations such as art galleries, archaeological and historical museums. Museums as sites of informal education can, now, emphasize the social and recreational nature of a visit, disperse information and transmit knowledge by engaging the visitor in interactive and creative processes, enabling him to follow his own "path" in conquering education.

Museums need to enrich the form of static display with interactive technology and entertainment in order to give a potential tool in best-practice for future exhibitions. Moreover, museums need to acquire a high position in education and the only way to do it is by incorporating similar programmes, that minimize distances and differences.

As regards the technical aspect of the work, future directions include the consideration of augmented reality technologies for "filling in" missing fragments that were not retrieved during an excavation, as well as the creation of an online repository of artifacts, including a detailed description of their characteristics, usage and historical time that can provide a useful reference base for expert users, such as archaeologists, museum curators, etc.

References

1. Prado Museum. www.openculture.com/2015/03/prado-creates-first-art-exhibition-for-visually-impaired.html
2. LeMO: The Living Museum Online. www.dhm.de/lemo
3. Sayre, S., Wetterlund, K.: The social life of technology for museum visitors. Vis. Arts Res. **34**, 85–94 (2008)
4. The Taylor Digital Centre at the Tate. www.tate.org.uk/whats-on/tate-britain/daily-activities/taylordigital-studio-public-access-drop
5. Jenkins, H., et al.: Confronting the Challenges of Participatory Culture: Media Education for the 21st Century. MIT Press, Cambridge (2009)
6. Hartley, R., Zisserman, A.: Multiple View Geometry in Computer Vision. Cambridge University Press, Cambridge (2003)
7. Lowe, D.G.: Distinctive image features from scale-invariant keypoints. Int. J. Comput. Vis. **60**(2), 91–110 (2004)
8. Bay, H., Tuytelaars, T., Van Gool, L.: SURF: Speeded up robust features. In: Leonardis, A., Bischof, H., Pinz, A. (eds.) ECCV 2006. LNCS, vol. 3951, pp. 404–417. Springer, Heidelberg (2006). https://doi.org/10.1007/11744023_32
9. Fischler, M., Bolles, R.: Random sample consensus: a paradigm for model fitting with applications to image analysis and automated cartography. Commun. ACM **24**(6), 381–395 (1981)
10. Snavely, N., Seitz, S., Szeliski, R.: Photo tourism: exploring photo collections in 3D. In: SIGGRAPH 2006, pp. 835–846 (2006)
11. Klopschitz, M., Irschara, A., Reitmayr, G., Schmalstieg, D.: Robust incremental structure from motion. In: 3DPVT, vol. 2, pp. 1–8 (2010)
12. Hemayed, E.E.: A survey of camera self-calibration. In: IEEE Conference on Advanced Video and Signal Based Surveillance, pp. 351–357. IEEE Press (2003)
13. Nilosek, D., Walvoord, D.J., Salvaggio, C.: Assessing geoaccuracy of structure from motion point clouds from long-range image collections. Opt. Eng. **53**(11), 113112 (2004)
14. Remondino, F., Spera, M.G., Nocerino, E., Menna, F., Nex, F., Gonizzi-Barsanti, S.: Dense image matching: comparisons and analyses. In: Digital Heritage International Congress, vol. 1, pp. 47–54 (2013)
15. Remondino, F.: From point cloud to surface: the modeling and visualization problem. Int. Arch. Photogram. Remote Sens. Spat. Inf. Sci. 34(5) (2003)
16. Mortara, M., Catalano, C.E., Bellotti, F., Fiucci, G., Houry-Panchetti, M., Petridis, P.: Learning cultural heritage by serious games. J. Cult. Heritage **15**(3), 318–325 (2014)
17. Kiourt, C., Koutsoudis, A., Pavlidis, G.: DynaMus: a fully dynamic 3D virtual museum framework. J. Cult. Heritage **22**, 984–991 (2016)
18. Vourvopoulos, A., Liarokapis, F., Petridis, P.: Brain-controlled serious games for cultural heritage. In: 2012 18th International Conference on Virtual Systems and Multimedia (VSMM), pp. 291–298. IEEE, September 2012

Resilience to Climate Change and Natural Hazards

Climate Information for the Preservation of Cultural Heritage: Needs and Challenges

Lola Kotova[1](⊠) ⓘ, Daniela Jacob[1], Johanna Leissner[2],
Moritz Mathis[3], and Uwe Mikolajewicz[3]

[1] Climate Service Center Germany (GERICS), Helmholtz-Zentrum Geesthacht,
Hamburg, Germany
lola.kotova@hzg.de
http://www.gerics.de
[2] Fraunhofer Gesellschaft zur Foerderung der angewandten Forschung eV,
Munich, Germany
[3] Max Planck Institute for Meteorology, Hamburg, Germany

Abstract. Continued preservation of cultural heritage requires reliable climate information as input for an accurate projection of possible impacts of climate change. Future climate-induced outdoor risks for cultural heritage can in general be estimated from the information provided by Earth System Models (ESMs). In this paper we present the results of the project Climate for Culture. The project focused on damage risk assessment, economic impact and mitigation strategies for sustainable preservation of cultural heritage in times of anthropogenic climate change. We utilized advanced climate modelling techniques with dynamical downscaling and novel analysis methodology allowing better integration of climate information to impact studies on cultural heritage. Challenges of bridging supply and demand of climate information relevant for adaptation measures for the preservation of cultural heritage are also addressed.

Keywords: Climate change · Climate modelling · Sea level rise ·
Cultural heritage

1 Introduction

Reliable climate information is required for a sustainable management of cultural heritage. Future climate-induced outdoor risks for historic buildings can in general be estimated from the information provided by Earth System Models (ESMs). ESMs describe the global climate system and its development in time by a combination of coupled physical and biogeochemical model compartments representing the different components of the Earth System, e.g. the atmosphere, ocean, soil, vegetation, etc. [1].

In the last years two large-scale integrated European research projects were conducted to investigate the impacts of climate change on cultural heritage. These are Noah's Ark (www.noahsark.isac.cnr.it) and Climate for Culture (www.climateforculture.eu). Noah's Ark studied the impact of the climate change of outdoor environments on typical materials, structures and infrastructures of built cultural heritage. The General Circulation Climate Model HadCM3 and the European set up of the Regional Climate Model of the

Hadley Centre (UK) have been used to generate the Climate Risk Maps for entire Europe [2]. The results presented in the atlas indicate clearly that serious impacts of anthropogenic climate change on cultural heritage are likely to take place, especially towards the end of this century.

In this paper we present the results of the project Climate for Culture. The project focused on damage risk assessment, economic impact and mitigation strategies for sustainable preservation of cultural heritage in times of anthropogenic climate change [3]. We utilized advanced climate modelling techniques with dynamical downscaling and novel analysis methodology allowing better integration of climate information to impact studies on cultural heritage. Challenges of bridging supply and demand of climate information relevant for adaptation measures for the preservation of cultural heritage are also addressed.

2   Climate Information for Cultural Heritage: *Needs*

2.1   Regional Atmospheric Modelling

In the project, we applied the regional atmosphere model REMO in its most recent hydrostatic version (REMO 2009) [4]. The regional simulations downscaled not only the climate projections of the Coupled Model Intercomparison Project 3 (CMIP3) [5] under the assumptions of the Special Report on Emissions Scenarios (SRES) A1B scenario [6], but also the new CMIP5 global climate projections [7] for the representative concentration pathway RCP4.5 [8].

REMO has been run with a horizontal resolution of 12.5 km (EUR-11) at the EURO-CORDEX domain [www.euro-cordex.net] and 27 vertical levels [9]. In addition to the scenario simulations for near (2021 to 2050) and far future (2071 to 2100) conditions for scenarios A1B and RCP4.5, a control simulation for the recent past (1961 to 1990) forced with observed greenhouse gases concentration was analyzed.

Special emphasis has been placed on assessment of the robustness of projected climate change patterns. Main sources of uncertainty are human external forcing (e.g. future emission scenarios) and climate models. The internal climate variability is well known to be the largest source of uncertainty in the projections with temporal scale less than 10 years. In this context we analyzed periods of 30 years to avoid a substantial impact of natural variability on the estimated climate signal. The climate change signal expressed a relative change between the atmosphere time-mean states in near/far future and recent past. To estimate statistical significance, we applied a two-sided student t-test [10]. The climate signal is called statistically significant if the level of significance reaches 95% or more.

Two global models ECHAM5-MPIOM [11] and MPI-ESM [12] and two emissions scenarios A1B and RCP4.5 were used in the project to create a multi-model-multi-emission ensemble and to estimate the anthropogenic climate change signal in case of moderate emission control.

The climate simulation based on the A1B scenario provides a good mid-line scenario for carbon dioxide output and economic growth. RCP4.5 is the scenario of the long-term, global emissions of greenhouse gases, short-lived species and land-use-land-cover which stabilizes radiative forcing at 4.5 W/m^2 (approximately 650 ppm CO_2-equivalents) in the year 2100 without ever exceeding that value [8].

The results of the simulations with the regional model REMO show that the annual mean of near-surface air temperature increases statistically significant for the entire model domain in all simulations for near and far future. For the far future (2071–2100), the RCP4.5 simulations showed that the temperature might rise between 1 and 3 °C. The A1B scenario simulations showed projected future warming of 2 to 4.5 °C for the same period. (see Fig. 1).

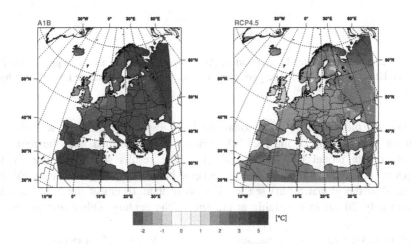

Fig. 1. Changes in near-surface air temperature, [°C] (2071–2100 relative to 1961–1990) for different emission scenarios

The projected spatial patterns are very similar in all scenarios with stronger annual mean warming in Southern Europe and in the northeastern areas. Whereas near-surface air temperature is rising everywhere, the REMO model does not simulate a consistent tendency in precipitation for entire Europe. The results show that the general tendency is enhanced precipitation for most regions in central and northern Europe and decreased precipitation in the Mediterranean region (up to 40% over the Iberian Peninsula for A1B) (see Fig. 2).

2.2 Sea Level Rise

For coastal regions relative sea level change is an important feature especially for shelf areas. This quantity, however cannot easily be derived from climate models, several components with different regional distributions contribute to it. Changes in ocean circulation, which cannot be reliably estimated from global climate models, are

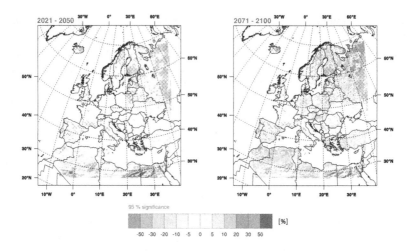

Fig. 2. Relative change in total precipitation, [%]. 30-year seasonal mean changes for the A1B scenario relative to 1961–1990. The pink lines indicate the 95% significance level (Color figure online)

important for the evolution of regional sea level, especially on shelf seas. Here, a dynamical downscaling of the RCP4.5 scenario was performed with the regionally coupled climate model REMO/MPIOM [13]. The results show (see Fig. 3), that changes in the ocean circulation and the inverse barometer effect lead to an elevated (compared to the global mean) sea level of more than 10 cm in the North Sea and approximately 20 cm in the Baltic at the end of the century. Other components are

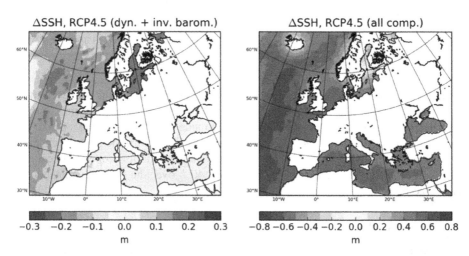

Fig. 3. Relative sea level change according to emission scenario RCP4.5 (2081–2100 minus 1986–2005). Left: sea level change due to the ocean circulation and the inverse barometer effect (simulated by MPIOM/REMO). Right: added components due to global thermosteric expansion (simulated by MPIOM/REMO), dynamic ice and surface mass balance of Greenland and Antarctica, deglaciation, land water storage and glacial isostatic adjustment (from [14]).

important as well. This includes the thermal expansion of sea water, and the contribution from cryospheric components (melt of ice sheets and glaciers). For those ones (except for thermal expansion) the estimates from the IPCC 5th Assessment Report have been used [14]. Together this yields typical sea level rises of 40 cm. An additional component is the relative movement of the continental crust in response to changes in ice loading. This leads de facto to a decrease of relative sea level around the northern Baltic, which was covered by a huge ice sheet during the last ice age. In contrast it enhances sea level rise at the Dutch coast.

2.3 Tailoring Climate Information

The research community of Climate for Culture was very diverse ranging from climate modelers to conservation scientists, architects and heritage managers. Furthermore, the required meteorological variables for the models used for different impact studies and risk assessments varied strongly. In this context, easily understandable and accessible climate information was required.

At the first step, a number of climate variables has been selected to estimate how different types of buildings will respond to the outdoor climate change and how heritage artifacts inside the building might be affected.

While future climate-induced outdoor risks on the pan-European scale can be accessed with the information provided directly by climate models, a more complete understanding of risks for indoor interiors of historic buildings requires the application of whole building simulation tools [15]. A number of case studies representing historical buildings all over Europe were therefore selected.

For each of the selected historical buildings the results of the climate simulations were provided and served as an input for hydrothermal building simulations. In addition, these simulated data were compared with the observed meteorological parameters form several stations of the global meteorological database METEONORM [https:// meteonorm.com]. A novel methodology has been developed within the project for simulating and mapping indoor climate and indoor risks. For this, additional 474 locations have been selected. They are equally distributed over entire Europe and are located in the center of each 12th grid box of the REMO model. For each location the climate information has been provided and served as an input for the State Space Model. This inverse modeling approach was applied for compiling a transfer function and enabled to simulate the thermo-hygrometric performance of a generic building [16].

3   Climate Information for Cultural Heritage: *Challenges*

Our experiences show that climate projections can be successfully applied for investigating the impacts of climate change on cultural heritage.

Nevertheless, climate information cannot be treated as an isolated topic and should be effectively embedded into the real world of cultural heritage. The output of climate models needs to be tailored to practical needs. In Climate for Culture we used a novel approach of offline coupling of climate modelling with building simulation. By doing so, future indoor climates and energy demands were calculated not only for selected buildings but also on a pan-European level and thus suitable mitigation strategies were developed for a number of case studies.

It is also important to mention, that climate models should fit to the required applications, e.g. the simulations of the local characteristics of climate. Despite the skills in simulating the Earth's climate, global climate models can provide information only at a relative coarse spatial scale, which is not sufficient for assessment of climate change impact on cultural heritage items. In the project we applied high-resolution (EUR-11) projections of future climate. This information has been used to visualize the effects of climate change on cultural heritage across Europe. Nevertheless, a much finer spatial scale is required to calculate outdoor/indoor transfer functions for a selected type of movable or immovable cultural heritage.

In the project we provided such information on climatic time scales only. Different types of buildings, however, respond differently to climate change. Furthermore, changing climate impacts on indoor interiors and collections that requires conservation measurements. In this context shorter time-scale predictions, e.g. seasonal and sub-seasonal, can play a significant role.

Evaluation and quality control of available climate data and derived climate information are one of the most important issues when developing mitigation strategies to prevent damages of cultural heritage. Climate information from a single model is not sufficient. To estimate the range of possible scenarios, we therefore used a multi-model-multi-scenario ensemble.

4   Conclusion

Although the project Climate for Culture has made a significant contribution to understanding the effects of anthropogenic climate change on cultural heritage across Europe, there is still a demand for translating climate information into practical applications. Thus, the development of climate services for cultural heritage is of high importance.

The European Road Map of Climate Services [17] facilitates "the transformation of climate-related data – together with other relevant information – into customized products such as projections, forecasts, information, trends, economic analysis, assessments (including technology assessment), counselling on best practices, development and evaluation of solutions and any other service in relation to climate that may be of use for society at large".

In this context climate services not only provide climate information, they also respond to user needs. Our experiences underline the need for a dialogue between the research communities involved and users of climate information in the cultural heritage sector. Bringing together climate scientists, conservation scientists and heritage managers as well as SMEs working in the field of conservation helps to exchange existing knowledge and to translate all available information into adaptation and precautionary measures to protect in a sustainable way cultural heritage from future climate change impacts.

References

1. Collins, M., et al.: Long-term climate change: projections, commitments and irreversibility. In: Stocker, T.F., Qin, D., Plattner, G.-K., Tignor, M., Allen, S.K., Boschung, J., et al. (Eds.) Climate Change 2013: The Physical Science Basis. Contribution of Working Group I to the Fifth Assessment Report of the Intergovernmental Panel on Climate Change, Cambridge (2013)
2. Sabbioni, C., Brimblecombe, P., Cassar, M. (eds.): Atlas of Climate Change Impact on European Cultural Heritage. Scientific Analysis and Managements Strategies, The Anthem-European Union Series,160 p. (2010). ISBN 9781843313953
3. Leissner, J., et al.: Climate for culture: assessing the impact of climate change on the future indoor climate in historic buildings using simulations. Herit. Sci. **3**, 38 (2015). https://doi.org/10.1186/s40494-015-0067-9
4. Jacob, D., Petersen, J., Eggert, B., Haensler, A., et al.: EURO-CORDEX: new high-resolution climate change projections for European impact research. Reg. Environ. Change **14**, 563 (2014). https://doi.org/10.1007/s10113-013-0499-2
5. van der Linden, P., Mitchell, J.F.B. (eds.): EMSEMBLES: Climate Change and Its Impacts: Summary of Research and Results from EMSEMBLES Project. Met Office Hadley Center, Exeter (2009)
6. Nakićenović, N., et al.: Special Report on Emissions Scenarios: A Special Report of Working Group III of the Intergovernmental Panel on Climate Change. Cambridge University Press, Cambridge (2000)
7. Taylor, K., Stouffer, R.J., Meehl, G.A.: An overview of CMIP5 and the experiment design. Bull. Am. Meteorol. Soc. **93**, 485–498 (2012). https://doi.org/10.1175/BAMS-D-11-0094.1
8. van Vuuren, D.P., et al.: Representative concentration pathways: an overview. Clim. Change **109**, 5–31 (2011). https://doi.org/10.1007/s10584-011-0148-z
9. Kotova, L., Mikolajewicz, U., Jacob, D.: Climate modelling. In: Leissner, J., Kaiser, U., Kilian, R. (eds.) Climate for Culture. Fraunhofer-Center for Central and Eastern Europe MoEZ (2014). ISBN 978-3-00-048328-8
10. von Storch, H., Zwiers, F.: Testing ensembles of climate change scenarios for "statistical significance". Clim. Change **117**(1–2), 1–9 (2013)
11. Roeckner, E., et al.: The atmospheric general circulation model ECHAM 5. PART I: model description. MPI-Report no. 349 (2003)
12. Giorgetta, M.A., et al.: Climate and carbon-cycle changes from 1850 to 2100 in MPI-ESM simulations for the coupled model intercomparision project phase 5. J. Adv. Model Earth Syst. **5**(3), 572–597 (2013)
13. Mathis, M., Elizalde, A., Mikolajewicz, U.: The future regime of Atlantic nutrient supply to the Northwest European shelf. J. Mar. Syst. **189**, 98–115 (2019)
14. Church, J.A., et al.: Sea level change. In: Stocker, T.F., Qin, D., Plattner, G.-K., Tignor, M., Allen, S.K., Boschung, J., et al. (Eds.) Climate Change 2013: The Physical Science Basis. Contribution of Working Group I to the Fifth Assessment Report of the Intergovernmental Panel on Climate Change, Cambridge (2013)
15. Bertolin, C., et al.: Results of the EU project climate for culture: future climate-induced risks to historic buildings and their interiors. In: Proceedings of the Second Annual Conference (SICS) on Climate Change Scenarios, Impacts and Policy, Venice, 29–30 September (2014)
16. Leissner, J., Kaiser, U., Kilian, R. (eds.): Climate for Culture. Fraunhofer-Center for Central and Eastern Europe MoEZ (2014). ISBN 978-3-00-048328-8
17. A European Research and Innovation Roadmap for Climate Services, European Commission, Directorate-General for Research and Innovation, European Union (2015). https://doi.org/10.2777/702151. ISBN 978-92-79-44341-1

Heritage Resilience Against Climate Events on Site - HERACLES Project: Mission and Vision

Giuseppina Padeletti[(✉)] and HERACLES Consortium Staff

CNR-ISMN, Via Salaria Km 29,300,
00015 Monterotondo Stazione (Rome), Italy
giuseppina.padeletti@cnr.it, gpadeletti@gmail.com

Abstract. Europe has a significant cultural diversity together with exceptional ancient architectures and artefact collections that attract millions of tourists every year. This incalculable value and global assets have to be preserved for future generations. The effects of floods, extreme wind storms or rains on these assets are clearly identifiable but it should be worth to note that all these effects are seriously amplified on ancient and fragile assets where advanced techniques, commonly used for modern buildings and structures, cannot be applied to preserve their originality. Moreover, ancient paintings as well as ancient structures, are very sensitive to environmental parameters (e.g. level of moisture in the air, temperature, etc.) and their relative quick leaps undermine the asset from the inside (mould, thermal stress, etc.). Again, in order to preserve ancient buildings and artefacts, dedicated technique, material and methodologies have to be applied to these important and valuable heritages. In order to address all the above challenges, the concept underpinning HERACLES project is to propose a holistic, multidisciplinary and multi-sectorial approach with the aim to provide an operative system and eco-solutions to innovate and to promote a strategy and vision of the future of the CH resilience. The HERACLES vision will be tested in Greece (Koules Fortress in Heraklion and Knossos Palace) and in Italy (historical town of Gubbio - Town Walls and Consoli Palace).

Keywords: Cultural heritage · Climate change · Resilience

1 CH and CC: HERACLES Overall Approach

In the last years the world is necessarily facing to the effects of the climate change (CC), which require interventions in many different fields, from the environment, agriculture, to land protection.

The cultural heritage (CH) too, in particular that one in Europe and in the Mediterranean basin, where many important and prestigious monuments and sites are located, must face this emergency: in many situations, the presence of meteorological extreme events can severely damage historical buildings and works of art.

HERACLES Consortium Staff—Details on HERACLES Website: www.heracles-project.eu.

A. Moropoulou et al. (Eds.): TMM_CH 2018, CCIS 961, pp. 360–375, 2019.
https://doi.org/10.1007/978-3-030-12957-6_26

Effects on cultural heritage assets deriving from natural and environmental hazards related to climate change are many and complex. Environmental and natural hazards can cause damaging or destruction of cultural heritage assets through various natural catastrophic processes; this entails the necessity to consider preservation and protection issues. Different types of natural hazards may have different impacts on cultural heritage assets.

Sites of cultural significance can be affected by catastrophic events of both endogenous (earthquakes, volcanic eruptions, tsunamis) and exogenous origin (landslides, floods, ground collapses, wildfires, cyclones), for which little or no warning has been received. However, these sites may also suffer from processes, which are not catastrophic in the conventional sense but their cumulative effects and aging in the long term may have a highly adverse impact. These include ground subsidence, especially in coastal settings, accelerated weathering of building stone, sandstorms, and recession of coastal cliffs.

Climate Change impact is functioning as a risk multiplier to already existing problems and increases and accelerates them. Climate stressors can directly affect cultural heritage buildings, monuments, and settlements. Sea level rise threatens coastal assets with increased erosion and salt water intrusion. More frequent and intense storms and flood events can damage structures, which were not designed to withstand prolonged structural pressure, erosion, and immersion. Changing precipitation patterns can quickly erode assets built for a different climate. Also, stability issues can arise since during extreme rainfall events, the increased ground soil moisture can reduce the physical stability and trigger landslides. Warmer temperatures and increased humidity can damage building materials and structures by encouraging rot, pest infestations (e.g. wood materials), and erosion.

Several key indicators are used in the scientific literature to describe CC among which: greenhouse gas composition (in particular CO_2), outdoor surface temperature, precipitation (rain, snow, hail), snow cover, sea and river ice, glaciers, sea level, climate variability, extreme weather events over time. In the complex world of interaction between CC and CH one of the first step is to define proper indicators to assess the overall impact of climate on cultural World Heritage [1].

The most significant global CC risks and impacts on CH are well known and an example is reported in the table *Principal Climate Change risks and impacts on cultural heritage of Working Document 30* [2]. In the last years, many European projects have addressed these problems [3–15].

In this framework, HERACLES proposes a novel systematic approach to ensure the sustainable management and protection of the different cultural heritage typologies in Europe and worldwide territory with respect to the CC impacts. The approach benefits of a multidisciplinary methodology that bridges the gap between the two different worlds: the CH stakeholders and the scientific/technological experts, both involved in the project.

To find solutions to this problem is the aim of the HERACLES project, funded by the EU within the Horizon 2020 research and innovation programme under grant agreement n 700395, and coordinated by CNR (National Council of Research of Italy). The aim of this project is to increase the resilience of cultural heritage assets against the effects deriving by CC. To accomplish this, a partnership made up of 7 different

Countries, integrating research institutions, universities, small, medium and large international enterprises, as well as international organizations, was set up. The end-users, such as the Ephorate of Antiquities of Heraklion and the Municipality of Gubbio, are active parts of the HERACLES Consortium [16].

The multidisciplinary HERACLES approach aims to:

- deal with widespread of worldwide CH (historical towns, different ages sites/elements, new architecture, etc.) reducing fragmentation in this sector through increased collaboration and cooperation and fostering of an integrated and interdisciplinary approach;
- face risks factors related with CC also in combination with other environmental risks;
- achieve a multi-temporal and spatial situational awareness of the site itself and as element of a more general context including also the territories in which the CH is located (wide and local monitoring, material diagnosis);
- assure sustainable long-term preservation and maintenance coupled with capabilities of risk reduction and crisis management;
- respect the historic and cultural integrity and valorise the social-economic value of the CH:
- strengthen and improve research and industrial excellence in CH preservation.

The key elements of the HERACLES systematic and interdisciplinary approach starts from the identified needs. Both are summarized as follows (Fig. 1):

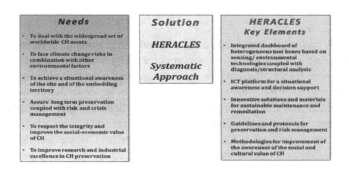

Fig. 1. Elements of the HERACLES systematic approach

- An integrated dashboard populated by heterogeneous tool boxes based on sensing/environmental technologies (at wide and local scale) coupled with diagnosis/structural analysis methodologies for a multi-temporal/multi-spatial monitoring of the CH asset;
- An ICT platform able to provide a situational awareness about the CH status and support the short and long term decisions of the stakeholder involved in CH site management for risk reduction;

- Innovative solutions and materials, for the economically sustainable maintenance and remediation, preserving the integrity and improving the social value of the CH;
- Guidelines and protocols not only for the preservation of the CH site but also able to manage the overall risk cycle management;
- Methodologies and strategies aiming at improving the awareness of the social and cultural value of CH from the different communities.

Specifically, the integrated observation and monitoring of the CH assets was planned and organized at two scale levels:

(I) a macro-level to gain a wide vision of the site and to detect and predict the long term CC impacts and
(II) a micro (local) level through monitoring and diagnosis of buildings and artefacts present on the sites.

HERACLES will allow to constantly monitor these CH assets and their surroundings combining data collected by the different observational platforms (satellite, aerial, traditional as well as innovative in-situ sensing technologies).

An innovative flexible and scalable ICT (Information Communication Technology) platform is being designed and developed in order to integrate, correlate and manage the data collected over a long period of time from external sources and intrinsically related to CH object behaviour and to its physico-chemical status in terms of:

- integration with historical information of the site (including past critical events);
- integration of different monitoring data to obtain an on-line updated situation of the site and its surroundings;
- vulnerability and risk evaluation by means of advanced modelling (geomorphological site modelling, climate change and extreme weather condition modelling, anthropogenic pressure modelling);
- integration of information about the structural and physico-chemical status of materials and site;
- visualization (3D) of information through maps and 3D models;
- situational assessment, information for awareness building and decision support system (including warning and alert messages).

Such a structured information is crucial for the development of effective solutions for the mitigation and remediation of the CC effects.

In particular, by taking into account the complete set of information derived from the platform, HERACLES includes also activities related to:

- New environmentally sustainable materials and solutions for innovative, fast and effective maintenance and restoration of damaged sites;
- New procedures to respond to operational requirements for heritage sites suffering CC risks and damaging effects.

2 HERACLES Methodology

A flow chart of the HERACLES methodological approach is depicted in Fig. 2, with the analysis to be carried on. As previously mentioned, it basically consists of the integration of innovative and complementary elements aimed at monitoring, preserving and valorising CH sites affected by CC events and progressively increasing risk. The multidisciplinary method and composition of the partnership allow to be focused on each component of the project and, at the same time, to carry out a strong and effective integration and systematic effort, combining the progressive steps starting from the site analysis and monitoring, up to the development of new operative solutions for restoring and minimizing CH damage risk.

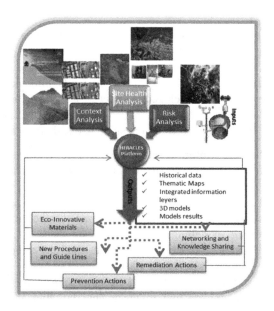

Fig. 2. HERACLES methodology

In details:

Context Analysis
The geographical location and geomorphological features affect the safety and security of the site itself. Moreover, human presence close to the site can represent a criticality: the construction of new structures can compromise the stability of the site and highly urbanized areas can generate high level of pollutant impacting on the site structures/artefacts. Under these premises, the analysis of the overall site context is

necessary; the use of remote sensing analysis integrated with local information can effectively satisfy this need thanks to its "wide eye" on the earth. It consists in:

- *Weather Monitoring and climate modelling*
  Weather monitoring and climate modelling is a key step of context analysis since it allows the forecast/prediction of the CC and weather conditions at different spatial and temporal scales.
  Global models are exploited in order to simulate the climate changes in the regional areas surrounding the sites for a long term prediction. These global models are then downscaled to simulate/predict meteorological fields over a smaller grid size (3 km × 3 km) comparable with the extent of a CH site. The improvement of the accuracy of these downscaled models is ensured by incorporation of local information about topography, coastline and land uses (achieved also thanks to Earth Observation data).
- *Change Detection/Wide Area Surveillance*
  Context analysis will benefit from the adoption of multi-temporal analysis techniques for change detection, based not only on the comparison of layers extracted from images collected at different date, but also detecting selective changes directly processing the multiresolution image data. The project is assessing a general methodology for the analysis of the evolution of a site starting from multisource information, to be adopted not only for the cultural heritage but also in other contexts. Actual and evolving situation will be described starting from the information collected by satellite, airborne and Remotely Piloted Aircraft Systems (RPAS) platforms. All information layers derived by the remote sensing data and having different resolutions and spectral features, are seamless integrated within a single geographic data base. Thus, multiple georeferenced layers, with different level of detail, are made available for the detection and analysis of sites evolution and changes throughout a wide temporal range by means of modelling tools developed in the project techniques.
- *Anthropic Pressure Analysis*
  Changes to CH caused by climate factors cannot be considered separately from changes in society, demographics, and land use planning. In addition, socio-economic activities (e.g. tourism) have a more and more increasing effect. In this context, the effects of anthropogenic pressures on the areas of interest, as air pollution, local increase of air temperature, humidity, and energy need from the grid, are modelled with the final aim to predict the long term impact of these factors. This modeling benefits from the historical datasets and from the outcome of the sensors deployed for the site monitoring.

Risk Analysis

- *Assess climate threats and their impacts*
  The first step of the risk analysis regards the assessment of the climate threats (characterized in terms of relative frequency and intensity) in terms of adverse impacts on CH assets. In particular, geomorphological dynamics resulting from climate change natural hazards (e.g. landslides, floods, and coastal erosion) is examined. In this frame, HERACLES solution is based on the identification of the

environmental characteristics by using geomorphological and structural site modelling and remote and in-situ sensing technologies.

– *Evaluate climate-related risks in light of all existing risks to cultural heritage*
Risk is defined by the European Commission (ISO/IEC5, 2009) [17] as "the probability of harmful consequences, or expected losses (deaths, injuries to property, livelihoods, disruption to economic activities or environment), resulting from interaction between vulnerability and exposure". Therefore, in order to estimate the risk it is important to specify the vulnerability of cultural heritage to each hazard (e.g. floods, landslides, slope stability, storm impact, intensity and frequency of storms) and to determine its exposure to the factors of CC (e.g. probability of flooding), and to socioeconomic processes (e.g. tourism revenues). A standard approach for risk evaluation is used, based on a ranking matrix for exposure and vulnerability. Risk, vulnerability and exposure variables are evaluated on a scale of 1 to 5, corresponding to the characterization of Very Low (1), Low (2), Medium (3), High (4) and Very High (5) risk. Risk maps are proposed in order to provide a geographical reference to risk evaluation and thus achieve additional information to enable better prediction.

Site Health Analysis

– *Ground/Structure Stability*
Remote sensing technologies such as Synthetic Aperture Radar (SAR) persistent scatterer interferometry (PSI) and SAR tomography are exploited in order to monitor at different scales the sites of interest, allowing the measurement of differential deformation of the ground and each single structure, with millimetre/centimetre accuracy. This information about the deformation is integrated with data obtained by traditional in-situ continuous monitoring systems. In this way, it is possible to improve the existing models used in the field of Structural Health Monitoring (SHM) of historical and monumental constructions. In fact, this integration allows to monitor continuously and with a very high resolution the condition of each single element of the site.

– *Site Diagnostic analysis*
The remote sensing technologies are complemented by in-situ (ground based) technologies in order to provide the user with the detailed high-resolution information on the single CH elements. HERACLES uses minimally or non-invasive diagnostic technologies. Ground penetrating radar, electrical resistivity tomography and Low frequency electromagnetic methods are combined to obtain information about the inner of the underground and of the structures (masonries, foundations, pillars, columns, etc.). Traditional contact monitoring techniques, using off-the-shelf sensors such as accelerometers and displacement transducers, are also applied on the most relevant structures and data acquired from such sensing technologies are processed for health monitoring purposes. All these techniques are integrated by high frequency methods providing information about the surface and very shallow layers of the structures. In this way, it is possible to enable a long-term multi-temporal multi-spatial monitoring conveying information of different quantities regarding structural defects and deterioration. In this frame, HERACLES also benefits from the

fusion and integration of these monitoring data as inputs to structural engineering models for vulnerability assessment of the single elements of the site.

- *In-situ, ex-situ material diagnosis*
  The characterization of the original materials (lime stones, sand stones and binders) and the identification of their degradation processes are carried out by non-destructive in-situ analysis and in some cases, with the minimally invasive ex-situ (laboratory) analysis. Their structural, physico-chemical, morphological, mechanical and rheological features are assessed by the use of XRD, XRF, FT-IR, XPS, SEM/FEG EDS, OM, PLM, porosimetry, optic and mechanical tests. Furthermore, state-of-the-art analysis based on coherent light metrology and non-linear microscopy are implemented for diagnostic purposes. This information is extremely important since in conservation science it is mandatory to use materials with physico-chemical characteristics that are similar as much as possible to the original ones.

- *Micro-climate and thermal energy analysis and modelling (Local Scale)*
  The analysis and the identification of the relationships between meteo climate parameters and environmental risk for CH, and for the development of a model for monitoring and mitigation of this risk is carried out thanks to a micro-climate data processing chain. The first steps of this processing chain is the analysis of a series of meteo-climate parameters (temperature, moisture, particulate matter etc.), the characterization of local scale meteorological conditions in the studied areas, the CC and pollution effects on historical structures, the study and the development of a series of risk analysis. Different thermal-energy analysis is developed for the different analyzed parameters. Therefore, starting from these results, a risk analysis is carried out by identifying the most influential parameters affecting the environmental performance of historic buildings in terms of indoor-outdoor comfort conditions, CO_{2eq} emissions, and energy need. All the collected meteo-climate data are used in order to develop a series of numerical models that allows to evaluate the effects that meteo climate conditions have on historical buildings. The development of the existing models for the study of meteo-climate effects of historical sites are carried on by the analysis and the collection of a series of different data, such as satellite data, in-situ measurements, existing modelling on meteo-climate parameters, geomorphological and structural analysis of study sites, analysis of anthropogenic pressure. All these data are analyzed by the collection of time series of the meteo-climate parameters, in order to provide a series of robust CC indicators and extreme weather forecast. Starting from the meteo-climate data, two different models are used in order to generate meteo-climate parameters and particulate matter data: the WRF (Weather Research and Forecast) model (at regional scale), and the micro- SWIFT model [18], which works at a local scale (in this case, the provided satellite data are requiring a downgrade of their resolution). The micro-climate chain outputs is providing thematic maps of meteorological and climatic parameters (medium and high resolution maps from optical satellite data), to be used for the estimation of climate impacts on the monuments over the two analysed areas. HERACLES provides meteoclimatic data which will be included in the risk analysis and in the development of models for effects of meteo-climate parameters

on historical buildings. Data provided for the processing chain are of two different types: satellite data and in-situ measured data.

– *Air Pollution Monitoring*

The regional scale modelling tools are generally built around nested versions of the WRF (Weather Research and Forecast) mesoscale meteorological model and of the CHIMERE or FARM [18] reactive transport and dispersion models, which take into account all atmospheric emissions and their interactions. The solutions can then be nested down to metric resolution (3 m) with the PMSS model (Parallel Micro SWIFT SPRAY) [18], readily applicable to the detailed description of air pollution at the local scale, and its effects on buildings facades, monuments and statues. The combination of regional and local scale tools allows the combination of the contribution to air pollution coming from distant sources, and of the contribution from local sources (traffic, airports, industry). In order to understand the potential evolution of the pressure on CH linked to CC, a collection of numerical results of IPCC (Intergovernmental Panel on Climate Change) climatic simulations for Horizons 2050/2070 will be gathered. This allows to compare the simulation of a current yearly evolution of air quality in the test cities, with the simulation of the air quality evolution for a typical year after 2050. Such complex and referenced model is today available just in 5 Research Centre worldwide.

Interesting project achievements can be found in the HERACLES deliverable D3.7 [19].

3   HERACLES Platform

HERACLES ICT platform will allow:

- to acquire, process and integrate heterogeneous data from both external causes (climate changes) and intrinsic CH objects behaviour/status;
- the handling of algorithms to correlate/fuse the data of different information sources (sensors and other kind of information), assisted by experts of the CH sector who know physico-chemical and structural values affecting mainly CH assets;
- to complement the data with semantic information;
- to provide a workflow engine to implement specific methodologies for a proper event detection and classification, but also promoting additional investigations due to anomalous values for refining forecasts and updating risk evaluation;
- to make available a multi-user interface, providing end users with CH awareness and decision support dashboards.

Platform requirements are:

– *Data Acquisition*

The platform will collect data and info from heterogeneous sources such as:
- Different sensor data (satellite, airborne, ground based platforms, local sensing technologies, raw or processed).
- Public meteo information: climate changes trends and specific extreme weather events affecting the site/artefact.

- Historical data: information about the architectural and historical value of the artefact.
- Structural and material data: information about the materials and the structure, including historical damage events and subsequent restoration interventions.
- *Information Storage (DB- Data Base)*
  A database that can be queried by the different stakeholders as scientists, structural engineers, responsible and managers of the CH, will be provided. The information will be stored for the processing with different aims.
  - Data Processing and Fusion
  - Workflow Engine, Awareness Building & Event Detector
  - Presentation Layer
  - Decision Support System.

3.1 HERACLES Platform Front-End

The front-end will be a web interface designed by exploiting innovative technologies in the frame of data visualization and will be accessible by means of different devices currently adopted for internet connection, i.e., desktop, laptop, tablet, smartphone.

The HERACLES portal will permit to visualize different levels of representation of the same scenario (from a global view to a detailed vision).

The **Context Analysis section** will deliver information regarding geomorphological and structural data; climate changes and extreme weather conditions; anthropogenic pressure phenomena.

The **Health State section** will provide information regarding ground and elements of the CH site.

The **Prevention section** will provide a map of the surveyed site classifying the areas based on their vulnerability level and for each critical area and will give a synthetic summary on its health state.

The **Remediation section** will describe the results concerning new materials and techniques describing best practices designed during the project to face the deterioration phenomenon.

4 Eco-Innovative Materials

The partner background and the experience in the field of CH in studying ancient materials such as ceramics, pigments, metallic decorations (lustre), clays, stones, protective materials for stone monuments and metallic artefacts, are addressed in the assessment of degradation phenomena affecting the CH asset and in studying and designing innovative protection solutions aiming at site proper preservation; also new materials to safeguard the site from atmospheric agents are developed and implemented in HERACLES. These innovative materials and accompanying solutions will embrace the concepts of smart design, eco-friendliness, and multi-functionality, always with respect to the guidelines and ethics of CH restoration as indicated by the articles 9 and 10 of the Venice Charter [20]. These experimentations will generate sufficient information related to the definition of sustainable best practices and policies to guide restorers, conservators and people that actively operate on site.

5 Prevention and Remediation Actions

HERACLES will suggest maintenance, remediation and restoration actions in view of a reliable, cost-effective conservation, respecting the integrity of CH. Maintenance will be performed by means of preventive and routinely actions based on periodic inspections and cyclical actions for mitigating the slow deterioration of the artefacts. Condition-based maintenance (CBM) will also be undertaken for relevant CH structures, as a cost-effective strategy improving sustainability of heritage preservation. CBM will be triggered by warnings provided by Structural Health Monitoring systems using both traditional and innovative sensing technologies. A reliable preventive maintenance coupled with the remediation represents a keystrategy to prevent decay/damage and mitigate the necessity of large-scale conservation-restoration treatments. Restoration actions will be also activated with the aim to preserve and reveal the aesthetic and historic value of the monument by respecting the integrity in terms of original materials and authentic documents. All the actions will benefit from the holistic vision of the site and of the single elements provided by the HERACLES platform philosophy.

Defining **best practices** (in terms of protocols and guidelines) is an important aspect and should account for the cooperation/interaction of different Authorities responsible for the site in case of ordinary situations and of critical events (as floods, high humidity, heat waves, storms, earthquake, etc.).

5.1 New Procedures and Guidelines

HERACLES will exploit operational procedures for disaster prevention and risk management by exploiting the outcomes of the previous phases regarding the risk analysis, the holistic vision of the site and its elements and solutions and materials for conservation. In fact, conservation principles should be integrated in all phases of disaster planning, response and recovery, as it is envisaged in the project. Operational procedures are being developed and included in the HERACLES platform also for crisis management, where the first key step is the quick damage assessment, which also enables CBM. This step will be activated through a proper observation/sensing chain defined on the basis of the event and complying with requirements of fastness and economic sustainability. Quick damage assessment opens the way to the other phases, where the first effort should be made [20] for a prioritization of the intervention to be done, in order to ensure sustainable restoration and repair of the damaged structures.

5.2 Networking and Knowledge Sharing

Natural and social systems of different sites/regions present different characteristics, and consequently will be subjected to different pressures (including CC), generating differences in adaptive capacities. A strong point of the HERACLES approach is to take advice of the real knowledge acquired on specific cases, by networking and sharing the different experiences among different communities (researchers, operators, governmental authorities, decision makers, industries). All the operational CH chain will benefit from this networking and knowledge sharing approach.

6 HERACLES Sites and Test-Beds

Italy and Greece are among the Countries more rich in cultural emergencies and in this context the HERACLES choice was not to focus on very famous locations, already object of attention, but on minor historical centers/areas since they represent the essence of the European Countries, often not greatly taken into account, even if characterising European Countries, Culture, Identity, Economy, where people lives, and works. The Sea Fortress of "Koules" is located in the port of Heraklion and symbolises all the monuments and sites facing the risk of hazards from climate change, such as significant impact from the sea (sea level rising, increasing intensity of extreme weather phenomena that combined with the air and land associated hazards, increased salinity are accelerating corrosion and deterioration of materials and structures, etc.) (Fig. 3a).

Gubbio instead, wants to represent all the historical monumental towns in Italy and in Europe, which were conceived and built in the past following criteria when the climate conditions were very different from nowadays and that suffers at present the effects of climate changes, which would endanger their safeguard, particularly the hydrogeological instability (heavy rains, flood, landslides), worsened by the seismic risk (Fig. 4a and b). The Minoan Palace of Knossos, is a spectacular Bronze-Age archaeological citadel representing the ceremonial, economical, social and political centre of the first European civilization of the Mediterranean basin, namely the Minoan civilization (Fig. 3b), suffer from the sea-linked effects.

a)

b)

Fig. 3. (a) Koules Fortress; (b) Knossos Palace

a)   b)

Fig. 4. Gubbio Town: (a) View with Consoli Palace (b) Detail of Town Walls

7   Protocols Definitions

Activities carried out in the HERACLES framework were directed towards the definition of protocols for each monument of interest in HERACLES test-bed on the basis of its structural and material preservation state. To this end, the available (satellite, airborne and in situ) sensors together with a number of laboratory-based material characterization instruments and techniques, able to give relevant information to assess the weathering state and the degradation processes of the investigated materials, are considered.

The defined protocols will be verified, during the demonstration activities of project, to assess their efficiency and validity. The final scope is their generalization for a wider applicability.

In Fig. 5 is shown a general HERACLES protocol, proposed to study a general CH asset, and assuring the Context, the Site and the Risk assessments.

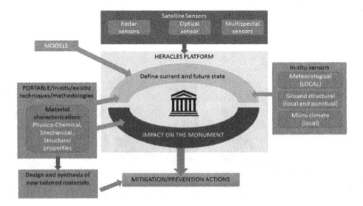

Fig. 5. Schematic representation of the HERACLES general protocol approach as regard the sensing, diagnostic and analytical strategies.

The most important phase is the definition of the user's need for each test-bed to study. It was done in terms of the different risks affecting the different sites and and in terms of precisely assess the state of the CH asset and its surroundings. Details of these activities can be found in the HERACLES documents: precisely, an accurate description of the end users' needs in deliverable D1.2 [21] and the evaluation of climate change impact and risks on resulting vulnerability in the deliverable D1.3 [22].

To investigate the risk context, the structural state and the environmental factors affecting the test-beds, several sensors (satellite, airborne and in-situ) are employed. These sensors acquire information on the conditions around the monument, which are input to the platform. Assimilation of the sensor measurements in the models is fundamental step for assessing the risk factor and the current and future status of the monuments.

Each test-bed is taking advantage of a multi-scalar strategy though sensors and techniques that examine/evaluate the CH assets, the surrounding areas, as well as the building and/or restoring materials. As already said, this is carried out by means of satellite and airborne sensors [23], in-situ sensors [24] and material characterization methods [25, 26].

In fact, specific information on the physico-chemical properties of the involved materials are also crucial. They are characterized through a number of portable (in-situ at the monument) and laboratory (ex-situ in the laboratory, by means of samples) analytical techniques. The protocol for the use of portable sensor and in-situ analytical techniques is described in more detail in the deliverables D3.2 [25] and the ex-situ in D3.5 [26].

8   Conclusions

In this paper a brief and general overview on the HERACLES project aim and activities have been provided. Summarising, HERACLES project has the ambition to design, implement, validate and promote a full innovative systematic strategy to the CH management and protection in order to increase CH resilience with respect to the climate change impacts. The HERACLES key point is the multidisciplinary and collaborative approach based on a large set of expertise present in the Consortium and Advisory Board, such as CH Domain Experts and Technological and Problem Solving Expertise.

HERACLES presents an innovative perspective, with many strengths and advancements beyond the-state-of-the-art:

1. An innovative context analysis based on an approach integrating climate and weather conditions, anthropogenic pressure and socio-impact activities modelling.
2. Multi-dimensional observation capability based on the integration of sensors (mostly non-invasive) acting from different platforms (satellite, airborne, RPAS; ground based) in order to enable a multi-scale (spatial and temporal), multi-sensing, multi-depth monitoring and diagnosis.

3. Multi-scale modelling capabilities to support the different requirements of HERA-CLES based on data processing and fusion from multiple and heterogeneous sources.
4. Integration of the results of multi-source information in the structural models for an improved vulnerability assessment of the single element of a CH site.
5. Imaging and mapping techniques for material characterization at micro and macro scale, with the final aim to assess the preservation state and the material functionality.
6. Novel solutions and materials for CH preservation and protection.
7. Advanced scalable ICT platform for an efficient use and integration of multiple information sources as well as provision of the HERACLES services to the stakeholders.
8. Guidelines and protocols for a sustainable long-term maintenance as key element for adaptation strategy.
9. Operational procedures for risk management and crisis mitigation. Further details on the HERACLES activities and achievements can be found in the technical reports and at: www.heracles-project.eu.

References

1. Reyes-García, V., et al.: Local indicators of climate change: the potential contribution of local knowledge to climate research. Wiley Interdiscip. Rev. Clim. Change 7(1), 109–124 (2016)
2. Working Document 30, Com 7.1 prepared for the 30th. Session on the World Heritage Committee, July 2006
3. NOAH'S ARK eu project. https://cordis.europa.eu/result/rcn/47770_en.html
4. RISC-KIT eu project. https://cordis.europa.eu/project/rcn/110483_it.html
5. ITACA eu project. https://cordis.europa.eu/project/rcn/188817_it.html
6. REDMONEST jpi project. http://www.jpi-culturalheritage.eu/wp-content/uploads/REDMONEST1.pdf
7. CLIMATE for CULTURE eu project. https://cordis.europa.eu/result/rcn/165682_en.html
8. CHANGES jpi project. http://www.jpi-culturalheritage.eu/wp-content/uploads/CHANGES1.pdf
9. PROTHEGO jpi project. https://www.era-learn.eu/network-information/networks/heritage-plus/jpi-cultural-heritage-and-global-change/protection-of-european-cultural-heritage-from-geo-hazards
10. CLIMA jpi project. http://www.jpi-culturalheritage.eu/wp-content/uploads/CLIMA1.pdf
11. ChT2 jpi project. http://www.jpi-culturalheritage.eu/wp-content/uploads/CHT21.pdf
12. STORM eu project. https://cordis.europa.eu/project/rcn/202681_en.html
13. RESIN eu project. https://cordis.europa.eu/project/rcn/196890_en.html
14. RESCCUE eu project. https://cordis.europa.eu/project/rcn/202678_en.html
15. URBANFLUXES eu project. https://cordis.europa.eu/project/rcn/193502_it.html
16. HERACLES. www.heracles-project.eu
17. ISO/IEC ISO Guide 73 – Risk Management – Vocabulary (2009). http://www.iso.org
18. HERACLES D2.3: Approaches for correlation/integration of the sensing technologies. www.heracles-project.eu

19. HERACLES D3.7: Determination of the degradation state of the studied structures, structural materials and proposed solutions. www.heracles-project.eu
20. Venice Charter. http://www.icomos.org/charters/venice_e.pdf
21. HERACLES D1.2: Definition of the end-users requirements with emphasis on HERACLES test beds. www.heracles-project.eu
22. HERACLES D1.3: Definition of methodologies for climate change impact evaluation and risk and vulnerability analysis. www.heracles-project.eu
23. HERACLES D3.3: Intermediate analysis of the experimental and theoretical aspects underlying the state-of-the-arts application of the satellite and airborne sensing technologies. www.heracles-project.eu
24. HERACLES D3.4: Intermediate analysis of the experimental and theoretical aspects underlying the state-of-the-arts application of the in-situ sensing technologies. www.heracles-project.eu
25. HERACLES D3.2: Development of an in-situ diagnostic protocol for quick assessment and monitoring of the weathering state and its progress on the areas of interest for the studied test beds. www.heracles-project.eu
26. HERACLES D3.5: Development of a detailed analysis protocol for laboratory analysis for the determination of materials and their alteration for the studied test beds using high resolution sophisticated state-of-the-art analysis and microscopies. www.heracles-project.eu

Resilient Eco-Smart Strategies and Innovative Technologies to Protect Cultural Heritage

Anastasios Doulamis[(✉)], Kyriakos Lambropoulos,
Dimosthenis Kyriazis, and Antonia Moropoulou

National Technical University of Athens,
9th, Heroon Polytechniou str., 15773 Zografou, Athens, Greece
adoulam@cs.ntua.gr

Abstract. In all parts of Europe Cultural Heritage represents a significant economic sector on which many communities depend. However, built CH artefacts are under continuous degradation from environmental and anthropological effects, and face significant risk of catastrophic damage from events such as flooding, earthquakes, storm and fires, which are themselves exacerbated by the impact of climate change. Action must be taken to manage these risks and mitigate their effect, however traditional approaches have been fragmented between different areas of expertise and disciplines, without regard to the combined effect of such uncoordinated interventions, leading to ineffective and often detrimental impact on the very assets they seek to protect.

Therefore, in this paper we propose a new holistic approach to the effective safeguarding and management of built CH artifacts through the provision of a decision support platform based upon interdisciplinary resilience modelling of current and future risks and interventions, monitoring and analysis of CH assets and natural hazards, creation of a semantic knowledge base and vulnerability, risk and cost modelling for planning and implementing intervention strategies.

The main goal of this research is to deliver an architecture that takes account of and supports mitigation of the inter-related impact of environmental, climatic and anthropogenic factors on such significant cultural heritage assets as are represented by the world heritage sites acting as primary use cases for the project. The economic impact of such interventions is also addressed, since the financial importance of tourism at these sites must be balanced with the cost of intervention to protect them whilst recognising the socio-economic benefits to be gained.

1 Introduction

In the recent years, it has been recognized that environmental neglect can have severe economic and social impacts which outweigh the cost of protection, therefore, environmental considerations are often mainstreamed into policy and are an integral part of the overall economic model [1–3]. Cultural Heritage (CH) is nowadays increasingly regarded as a positive contributor to European GDP and an essential part of Europe's underlying socioeconomic, cultural and natural capital. CH is also a major contributor to social cohesion and engagement [4].

© Springer Nature Switzerland AG 2019
A. Moropoulou et al. (Eds.): TMM_CH 2018, CCIS 961, pp. 376–384, 2019.
https://doi.org/10.1007/978-3-030-12957-6_27

In an era with declining acidic pollutants in European atmosphere [5, 6], climate change may play a more significant role in the decay of monuments and buildings. A robust and consistent picture of climate change over the Mediterranean emerges: consisting of a pronounced decrease in precipitation in the warm season and of an intensification of the meteoric phenomena (precipitation, storms) in the winter period, in particular in the northern Mediterranean areas (e.g. the Alps) [7]. Strictly connected to climate change, the rising of temperature, the intensification of rain precipitation, the rise in sea level and the possibility of flooding in particular for costal area and cities is also getting more attention and importance [8]. All the Mediterranean basin has been identified as a climate change "hotspot", expected to undergo environmental impacts considerably greater than in other places around the world [7]. Until now, however, the analysis of the impact that climate change might produce on historic buildings has been quite limited, despite the many researches dealing with the damages caused by atmospheric pollution, global warming and sea level rise [9]. While there is a generally good knowledge and awareness regarding degradation factors in respect to cultural heritage, so far specific studies and data are few and limited: on the interactions between building materials and the surrounding environments, on the most important degradation factors related to climate change, and on the possible solutions for reducing the impact of future climate change on the architectural surfaces [10, 11]. An integrated and interdisciplinary approach should start from a holistic evaluation of the existing and fragmentary data with new ones, such in order to individuate meaningful parameters and relative threshold values, efficient mitigation strategies to be adapted and validated for Cultural Heritage assets in the specific situation [12].

Nowadays, the future relation between CH materials and the environment is difficult to predict and to understand as heavily influenced. New materials and technological solutions are produced according to market-industrial logic and applied in the restoration intervention, even if their long-term behavior is unknown. The EU has funded and supported the development of systems, centres and services such as the European Flood Awareness System (EFAS) [13], the EC's Emergency Response Coordination Centre (ERCC) [14], the COPERNICUS Emergency Management Service [15].

In this paper, we describe a new architecture that supports a novel decision support platform to increase the efficiency and effectiveness of conservation and management strategies through advanced processes, tools, models and services for identification, assessment and mitigation of the effects of environmental changes, natural hazards and the impact of climate change on cultural heritage assets (i.e. sites, structures and artefacts). The architecture can increase the resilience and sustainability of European cultural heritage through a paradigm shift that transforms conservation policies and intervention strategies.

Figure 1 shows the way that our architecture integrates cross-disciplinary knowledge regarding both the built structure and the hazard environment to provide the depth of knowledge necessary to develop approaches spanning the complete intervention lifecycle. The Decision Support Platform (DSP) integrates the related functions of prediction, planning, implementation, documentation, monitoring anod diagnosis. Expert knowledge of degradation and environment is combined with asset knowledge and monitoring data to predict the developing risks; the planning process evaluates

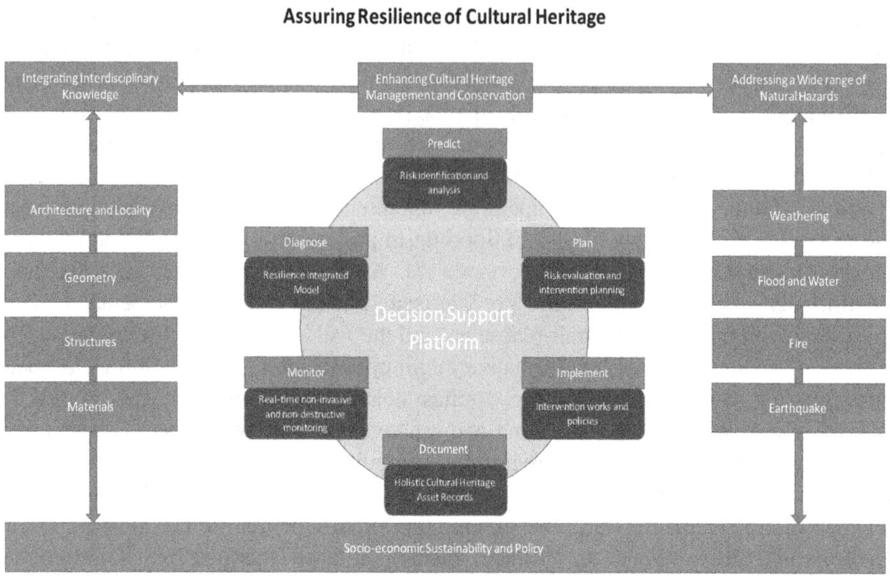

Assuring Resilience of Cultural Heritage

Fig. 1. Our proposed model for assuring resilience of CH assets.

those risks and supports the development of an intervention strategy. As the strategy is implemented, so real-time monitoring and diagnosis augments and maintains the knowledge base, which feeds back into the ongoing risk prediction process and ensures the planning and intervention strategies remain current and effective.

2   Overall Approach

The overall concept of our approach is illustrated in Fig. 2. Our method can be applicable to a wide range of outdoor cultural sites, undergone different environmental and climate change phenomena (weathering, earthquakes, fires, floods/water). The first core research module of our architecture is the "situation awareness subsystem" responsible for monitoring of defects over CH sites, assessing their current status, collecting geospatial data and applying a semantic interoperability analysis on them. Then, our architecture triggers an inter/multi-disciplinary research group covering numerous domains including structural analysis, material decay characteristics, geometric properties, architectural assessment, survey of climate changes and the effect of natural hazards, and urban properties and characteristics. These inputs are combined together to create a set of Resilience Features which are incorporated into the Resilience Integrated Model (RIM) alongside risks and threats aspects (risk assessment and vulnerability functions) and socio-economical strategies. A horizontal knowledge base toolkit is deployed to support these functionalities, adopting the concepts of holistic records through ontologies, dynamic linkage of assets to environmental stress as well as ontology evolution properties [16, 17]. The final outcome is a Decision Support

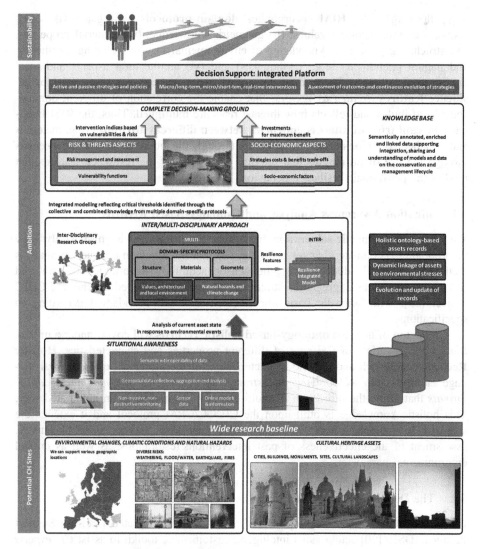

Fig. 2. A diagram of the proposed approach.

Integrated Platform that recommends to experts active/passive strategies/policies, long/short-term interventions at micro, meso and macro scale while assessing outcomes to set guidelines for policy makers and the updating of existing strategies.

3 Components for a Resilience Integrated Model (RIM)

Through **a holistic approach,** we propose a **Resilience Integrated Model (RIM)** to manage cultural heritage sites and conservation solutions, while mitigating the effects of climate change, natural phenomena (e.g., earthquakes) and weathering factors (e.g., flooding). The **RIM** incorporates **domain-protocols** as regards (i) socio-economical and historical values, (ii) geometric knowledge, (iii) material properties, (iv) structuring issues, (v) knowledge of environmental factors including weathering and natural phenomena (e.g., earthquakes) and (vi) architectural aspects and local environmental factors. The combination of all these aspects comprises a set of **resilience features** *(treated as vectors)* that enable the correlation of different domain-specific attributes and reflects how threats/ risks are multiplied. Thus, the RIM facilitates the **capturing of interdependencies between different attributes** (e.g. materials and weather conditions). This is achieved through the building of **holistic records** through the use of **ontologies** that provide a snapshot of the resilience features that modify the per-domain resilience behavior [18].

3.1 Situation Awareness Analysis and Ontology-Based Records

We will build **diagnostic schemes** in the form of **Situational Awareness Analysis** by incorporating (i) climate changes profiling, (ii) non-invasive/non-destructive analysis methods while (iii) documenting the sensor acquired spatio-temporal data and (iv) supporting data aggregation strategies. Protocols developed in the EU-CHIC project [19], will form the basis our research and will be adapted according to specifications.

On the top of that, the **ontology-based holistic records** will cover, under a unified framework, a variety of CH sites of different properties and characteristics. **Holistic Records** enable rich and generic characterisation of the aforementioned cultural heritage elements into *well-defined and structured components* and *machine-readable formats* that allow the utilization of multi-domain information in an automated way. This holistic knowledge is built upon the CIDOC/CRM protocol and it extends the standard to include historic documentation, geometric survey, material survey, assessment of the effectiveness of past interventions/restorations, as well as critical environmental parameters.

3.2 The Decision Support Platform

Our main outcome is the development of an innovative integrated Decision Support Platform (DSP) [20] acting as an intelligent, customizable toolkit to assist *CH experts and authorities* to take crucial decisions, covering both short-term and long-term aspects regarding the current status of a CH monument, providing risk assessment actions, intervention strategies and conservation policies. Core elements of this DSP are the aforementioned holistic records and the Resilience Integrated Model of the cultural heritage asset, without which any decisions regarding mitigation measures and adaptation strategies would be ineffective. The DSP will provide sustainable multi-modal and multi-faceted management and conservation measures by enabling the development of various "templates" of adaptation and mitigation strategies that can be applied in different temporal scales. Long (short)-term strategies refer to decisions that may affect the asset with respect to global (asse-level)-level information (e.g. climate change). Real-time strategies are applicable to the asset and are driven by real-time monitoring data that highlight an emerging risk. Both short-term and long-term strategies will be triggered at macro, meso and micro scale level of a monument analysis [21].

3.3 Resilience Integrated Model and Invention Necessity Indices

Central to the our approach (as depicted in the following figure), is the **Resilience Integrated Model (RIM)** that incorporates multi-disciplinary knowledge and assessments in a combined way, leading to the identification and definition of **intervention necessity indexes** *based on this collective and synthesized knowledge of different domains.* The intervention necessity indexes developed will be multi-disciplinary, encompassing the holistic approach adopted throughout the project. Intervention necessity indexes are not, however, the sole tool with which decisions on whether to intervene or not will be taken by stakeholders. Instead, a complete risk identification and assessment framework is developed to assess different threats and raise warning signals which, according to the corresponding intervention necessity indexes, trigger the employment of active and passive adaptation strategies and policies. In this process, socio-economic aspects will be considered through a trade-off analysis framework and a business simulation engine, incorporated as modules in the integrated platform. Our architecture proposes a continuous feedback approach in which many workflows can be executed simultaneously, while the final outcome of the Decision Support Integrated Platform feeds back into the Resilience Model to get more reliable analysis for the CH site protection and preservation.

For example, in the materials domain case, the time evolution of an intervention necessity can be seen as a curve of a positive slope (see Fig. 3) indicating that as materials deteriorate the necessity for intervention increases and this accumulative deterioration becomes worse with time. A similar behaviour can be seen in other disciplines, i.e., with time the necessity to intervene generally increases (linear, exponentially etc). Based on the traditional per-domain approach, stakeholders/scientists could then define intervention necessity thresholds, that when reached, would trigger the process of initiating the implementation of interventions.

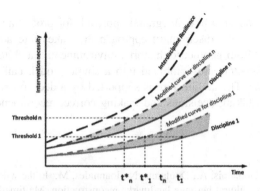

Fig. 3. Example of intervention necessity indices.

4   Strategic Impact

Monitoring, assessing, digitising and surveying CH assets is one of the most salient research actions in the cultural community. This assists conservators to check how restoration methods affect an object or how materials decay through time; archaeologists to better document an item; curators to properly display them and the creative industries to disseminate cultural knowledge in a digestible way worldwide. Europe's cultural treasure is also laying onto outdoor cultural sites, on historic cities, and on buildings and monuments being visited by thousands of users. These CH objects undergo material degradation mainly due to environmental effects, climate change, and spatial changes due to restoration actions applied. Thus, there is an imminent need for CH actors to spatial-temporally monitor critical aspects on the monuments and extract concrete conclusions on the impact different restoration strategies can have on them, how its materials behave through time, how the behaviour of material decay is accelerating/decelerating with respect to the micro-climate conditions, and what parts of an object needs more precise reconstructions than other parts. On the other hand, the involvement of many and different players in the CH community, each having different preferences and needs, makes the delivery of the same scale of the digital counterpart insufficient let alone non-affordable and time consuming. Last but not least, exploiting the CH model across different devices and platforms requires the implementation of a multi-level scalable scheme. Our approach addresses these issues by proposing an innovative holistic framework over which critical decisions and resilience actions are taken to prevent material degradation of historic monuments. This is achieved using a series of actions at different research fields (a) including climate models, (b) exploiting monuments' integrated protocols, (c) supporting time evolved 3D geometry digital technologies (4D modelling) and the creation of change history maps to accelerate reconstruction time at forthcoming time intervals, (d) material decay properties, (e) historical factors, (f) spatial and geometric characteristics of a monument [22].

5   Conclusions

In this paper, we propose a new integrated approach for protecting CH outdoor sites. We introduce a new inter-disciplinary approach that takes into account a variety of parameters ranging from geometric, historic, environmental, architectural documentation. All these parameters are combined into a single model, called Resilience Integrated Model (RIM). The architecture is supported by a decision support mechanism, assisting experts in taking decisions and making correct recommendations.

References

1. Makantasis, K., Doulamis, A., Doulamis, N., Ioannides, M.: In the wild image retrieval and clustering for 3D cultural heritage landmarks reconstruction. Multimed. Tools Appl. **75**(7), 3593–3629 (2016)
2. Kyriakaki, G., et al.: 4D reconstruction of tangible cultural heritage objects from web-retrieved images. Int. J. Heritage Digit. Era **3**(2), 431–451 (2014)

3. Ioannides, M., et al.: Online 4D reconstruction using multi-images available under open access. In: ISPRS Annals of the Photogrammetry, Remote Sensing and Saptial Information Sciences, II-5 W 1, pp. 169–174 (2013)
4. Doulamis, A., et al.: 5D modelling: an efficient approach for creating spatiotemporal predictive 3D maps of large-scale cultural resources. In: ISPRS Annals of Photogrammetry, Remote Sensing & Spatial Information Sciences (2015)
5. Bonazza, A., Messina, P., Sabbioni, C., Grossi, C.M., Brimblecombe, P.: Mapping the impact of climate change on surface recession of carbonate buildings in Europe. Sci. Total Environ. **407**, 2039–2050 (2009)
6. Sabbioni, C., et al.: NOAH'S ARK (Global Climate Change Impact on Built Heritage and Cultural Landscape). Pr. ref: 501837 (2006)
7. Barbi, A., Monai, M., Racca, R., Rossa, A.M.: Recurring features of extreme autumnal rainfall events on the Veneto coastal area. Nat. Hazards Earth Syst. Sci. **12**, 2463–2477 (2012)
8. Santoro, F., Tonino, M., Torresan, S., Critto, A., Marcomini, A.: Involve to improve: a participatory approach for a decision support system for coastal climate change impacts assessment. North Adriatic Case Ocean Coastal Manage. **78**, 101–111 (2013)
9. Sabbioni, C., Brimblecombe, P., Cassar, M.: The Atlas of Climate Change Impact on European Cultural. Scientific Analysis and management strategies. Anthem Press, London (2010)
10. Bernardi, A., et al.: A methodology to monitor the pollution impact on historic buildings surfaces: the TeACH project. In: Ioannides, M., Fritsch, D., Leissner, J., Davies, R., Remondino, F., Caffo, R. (eds.) EuroMed 2012. LNCS, vol. 7616, pp. 765–775. Springer, Heidelberg (2012). https://doi.org/10.1007/978-3-642-34234-9_81
11. Brimblecombe, P.: Mapping heritage climatology. In: Bunnik, T., De Clercq, H., Van Hees, R., Schellen, H., Schueremans, L. (eds.) Effect of Climate Change on Cultural Heritage, pp. 18–30. WTA Publications, Pfaffenhofen (2010)
12. Zendri, E., et al.: The monitoring of architectural stone surface treatments: choice of the parameters and their threshold limit values in Southeast University of Nanjing, UNESCO, RLICC Leuven, Monumentenwacht Belgium. In: International Conference on Preventive Conservation Of Architectural Heritage, Nanjing, China, pp. 123–130. UNESCO Asia-Pacific World Heritage Training and Research (2011)
13. European Flood Awareness System (EFAS). https://www.efas.eu/
14. EC's Emergency Response Coordination Centre (ERCC)
15. COPERNICUS Emergency Management Service. http://emergency.copernicus.eu/
16. Ntalianis, K., Doulamis, N.: An automatic event-complementing human life summarization scheme based on a social computing method over social media content. Multimed. Tools Appl. **75**(22), 15123–15149 (2016)
17. Doulamis, N., Ntalianis, K.: On the fly semantic annotation and modelling of multimedia. In: 16th IEEE International Conference on Systems, Signals and Image Processing (2009)
18. Spala, P., Malamos, A.G., Doulamis, A., Mamakis, G.: Extending MPEG-7 for efficient annotation of complex web 3D scenes. Multimed. Tools Appl. **59**(2), 463–504 (2012)
19. EU CHIC (European Cultural Heritage Identity Card). http://www.eu-chic.eu/
20. Sabeur, Z., Doulamis, N., Middleton, L., Arbab-Zavar, B., Correndo, G., Amditis, A.: Multimodal computer vision for the detection of multi-scale crowd physical motions and behavior in confined spaces. In: Bebis, G., et al. (eds.) ISVC 2015. LNCS, vol. 9474, pp. 162–173. Springer, Cham (2015). https://doi.org/10.1007/978-3-319-27857-5_15

21. Kioussi, A., Labropoulos, K., Karoglou, M., Moropoulou, A., Zarnic, R.: Recommendations and strategies for the establishment of a guideline for monument documentation harmonized with the existing European standards and codes. J. Geoinf. FCE CTU **6**, 178–184 (2011)
22. Verykokou, S., Ioannidis, C., Athanasiou, G., Doulamis, N., Amditis, A.: 3D reconstruction of disaster scenes for urban search and rescue. Multimed. Tools Appl. **77**(8), 9691–9717 (2018)

Interventions on Coastal Monuments Against Climatic Change

George Alexandrakis[✉], Georgios V. Kozyrakis,
and Nikolaos Kampanis

Coastal and Marine Research Lab, Institute of Applied and Computational
Mathematics, Foundation for Research & Technology-Hellas, Heraklion, Greece
alexandrakis@iacm.forth.gr

Abstract. Climate change impacts are functioning as risk multipliers to problems which are already apparent and affect cultural heritage sites. Sea Level Rise and increased storm events can damage structures that were not designed to withstand prolonged structural pressure, erosion, and immersion. Risks affecting coastal cultural heritage may stem from exposure to one or more hazards and it is important to facilitate a holistic understanding of factors driving them. Wave energy and overtopping of coastal structures represents a potential hazard for people, property and infrastructure. Especially when the coastal structure is a monument or landmark, mitigation measures and monitoring are needed. Depending on the level of acceptable risk and required degree of certainty related to wave overtopping, coastal engineers rely on predictions from semi-empirical desktop methods and numerical models for answers. Moreover, the anticipated increase in extreme events due to climatic change make protection and prevention action even more necessary. In this work the combination of risk assessment analysis related to increasing sea level and storm frequency, wave numerical modelling, breakwater design and economic sustainability is presented. As a case study, the Venetian Coastal walls of the city of Heraklion are considered. Numerical modelling results were generally found to be consistent with overtopping wave measurements. For the analysis of the wind regime in the near and far future, climatic modelling has been used. Climatic modelling results indicate that for the coastal area of Heraklion the wind speed and directions are expected to change in the near and far future, with an increase in wind speeds but also an increase in the frequency of the wind directions that effect the monuments the most. Based on the results of the measurements and numerical modelling, mitigation actions were proposed that include, increasing the submerged armouring of the Venetian City walls and the use of natural based solutions for low slope areas in order to reduce wave energy, run up and overtopping, reconstruction of the natural environment, so that the monument can be made accessible for longer periods of time.

Keywords: Wave overtopping · Breakwaters · Storms · Heraklion

© Springer Nature Switzerland AG 2019
A. Moropoulou et al. (Eds.): TMM_CH 2018, CCIS 961, pp. 385–401, 2019.
https://doi.org/10.1007/978-3-030-12957-6_28

1 Introduction

Built heritage is a cultural asset inherited through generations which defines the origin and identity of a place (ICOMOS 2002). The present concern is to preserve important structures and sites that promote identity and continuity of place, without compromising on development that is essential for the present times. Evolution of surrounding locations and natural aging of the historical structures is inevitable but it is essential to check both for sustained development. UNESCO (1972) defined that built Cultural Heritage comprises Architectural works which are of outstanding universal value. It was found that the built heritage across the globe is increasingly threatened with defects being caused not only by the natural causes of decay like gradual weathering and biochemical factors, but also by the variations in Climatic, social and economic conditions. Climate change is a substantial and inevitable threat to the built heritage of our coasts and the way of life which co-exist with these environments, and the overall wellbeing. The major indicators of climate change anticipated are mainly the increased mean annual temperatures; increased mean annual rainfall, but with the likely drier summers; More extreme weather including heavy downpours and more intense storms and Higher sea levels. The direct impacts of these climate change phenomena would be higher evaporation leading to regular summer drought conditions, combined with heavy downpours; Alluvial flooding and sudden flash floods; Increased coastal erosion and coastal flooding due to the combined effect of sea level rise and occurrences of storms. These impacts present serious consequences on heritage structures and socioeconomic activity that is directly or indirectly associated with it, including tourism (Kelly and Stack 2009). Study by Intergovernmental Panel on Climate Change (IPCC), links vulnerability with climatic change, and point out that the vulnerability of a region depends to a great extent on its wealth, and that poverty limits adaptive capabilities (IPCC 2000).

Protection of Built heritage from climate change effects were the subject of discussions on the effect of (GCC) Global Climate Change (ICOMOS 2007, 2008) which produced recommendations for materials or components which withstand prevailing environmental conditions are often likely to fail when those conditions changed. Extreme weather and rising sea levels are more likely to cause catastrophic damage and destruction to cultural heritage. As losses occur in the physical environment, intangible heritage values associated with the environment will also be lost, and hard choices will need to be made about what to try to preserve and what to compromise on, based on context specific issues. Adjusting to Global Climate Change will require improved monitoring so that changes can be identified in time for immediate response, and improved maintenance to make cultural heritage more resilient to changing environments and disasters.

The coastal zone is an area of high interest, characterized by increased population density, hosting important commercial activities and constituting habitats of high socioeconomic value (Costanza 1999). Sea level rise (SLR) in view of climate change poses a serious threat to coastal areas and therefore, much research effort has focused on this aspect of coastal hazard (Church and White 2011; Hinkel et al. 2014; Hogarth 2014; Hoggart et al. 2014; Jevrejeva et al. 2014; Losada et al. 2013; Tol 2009).

Extreme events, however, determine an additional hazard component. Some studies report an increased intensity and frequency of extreme water levels along several coastal regions in the world (Izaguirre et al. 2013; Ullmann and Monbaliu 2010; Wang et al. 2014; Weisse et al. 2014).

Wave overtopping has always been of prime concern for coastal structures constructed to defend against flooding: often termed sea defences. Similar structures may also be used to provide protection against coastal erosion: sometimes termed coast protection. Other structures may be built to protect ship navigation or mooring within ports, harbours or marinas formed by breakwaters. Within harbours, or along shorelines, reclaimed areas must be defended against both erosion and flooding. Some structures may be detached from the shoreline, often termed offshore, nearshore or detached, but most structures used for sea defence or similar function form a part of the shoreline.

2   Case Study

Heraklion is the largest urban centre in Crete, the capital of the region; it represents the most important place for the cultural, social and economic development of Crete. The population of the municipality of Heraklion is approximately 150.000 people. It is a very dynamic town, due to the natural and cultural touristic attractions and the main occupations of the inhabitants are tourism, agriculture and commerce. In fact, the presence of very attractive sea locations and the importance of its museums, monuments and archaeological sites make Heraklion as a place where culture, economy, and daily life are strongly related. The particularity of Heraklion is that it includes areas characterised by different levels of anthropogenic pressures. Coastal areas are the most exploited and inhabited areas, due to their rich resources (Kiousopoulos 2008). These pressures, enhanced by the effects of climate change and natural hazards, are endangering the viability and conservation of coastal resources and increase the risks that coastal population face (Tsilimigkas et al. 2016). Even though threats to the coastal environment arising either from natural hazards, the main triggering factor is the anthropogenic innervations.

The quality of the heritage site and the visiting experience can be determined by a number of factors, with the level of sites preservation among them. Site maintenance and its level of restoration/preservation are included in the site management activities. The level of restoration and how it is associated with the actual and perceived authenticity of the site are of great significance. The above social and economic factors have a strong influence on the site and on the management, decision making context. To place the heritage sites in their own context, helps in identifying what impacts should be evaluated. CH sites have a greater potential to influence and to have an impact on both a micro and a macro context. In particular, Knossos and Koules are contributing to the local economy through increased number of visitors, capital expenditures as well as the development of the collateral activities. The Sea Fortress of "Koules", located in the port of Heraklion, constitutes a characteristic type of the Venetian military architecture. Similar fortifications can be found in all major cities in Crete (e.g. Rethymnon, Chania) as well as in other locations in the Mediterranean basin (e.g. Cyprus). In fact, an important sector of CH in Greece and Europe is located on coastal areas throughout the Mediterranean (cities, ports, lighthouses, fortresses and other monuments). These monuments face hazard risks due to climate change (sea level rising, increasing intensity of extreme weather phenomena), which can be combined with other air and land associated hazards, increased salinity accelerating corrosion and deterioration of materials and structures, etc. (Fig. 1).

Fig. 1. Study area

During the first decade of 2000, the Greek Ministry of Culture, anticipating the problems that the monument was facing, decided to take new measures for its protection and safeguarding and a restoration and conservation project concerning the of the Venetian Fortress (Koules), took place in 2011–2016, which focused on the building. In the conservation program the main concern was related to the static and reinforcement aspects of the monument. The continuous exposure of the fortress to marine aerosol has produced severe weathering of the building stone, which is a porous material susceptible to the action of soluble salts and environmental conditions.

2.1   Risks/Hazards and Technical Aspects

The immediate contact of Koules with the sea makes the fortress vulnerable to salty northern winds, which are often very severe, reaching 9, 10 or even 11 units in the Beaufort scale. Especially during the winter season high waves are often literally covering the monument (Fig. 2).

Fig. 2. Wave impact on Koules during winter

The fortress of Koules is affected by weather conditions coupled with pollution which can initiate and accelerate the deterioration mechanisms for both original and restoration materials. Crusts of salts and black hard encrustations are observed on the walls in several rooms at the interior of the monument. Similar black crusts have also been observed at the areas around the joints both internally and externally to the monument. These crusts are rough and inhomogeneous and aggressive to the materials. The deterioration of the stone, along with the detachment of the grain aggregates, proceeds to selective pitting, resulting to the formation of deep interconnected cavities. The stone appears to have suffered an irregular loss of material, which follows the alveolar weathering pattern. Furthermore, a number of cracks have been detected around the monument and their restoration is important.

3   Methodology

Wind parameters were measured by a meteorological station deployed in Koules, while sea wave parameters are measured by two wave gauges offshore at depths of 6 m and 10 m respectively. Moreover, hourly time series of wind direction and wind speed observed during the period 1953–2015, were used. The meteorological data statistical analysis focuses on two different time-scales (long and short-term analysis).

3.1 Long-Term Analysis

First, the data from the long-term analysis are evaluated on the base of the results from the EURO-CORDEX project providing regional climate projections for Europe at 12.5 km resolution (the reader can refer to relevant report for more information). The provided data for the two future periods (near future: 2036–2065, far future: 2071–2100) are used to extract the near and far future wind time-series data, which are used as input data to the wave model analysis software. In this way, the significant wave height and period can be estimated for the coastal front of Heraklion and the impact on Koules monument can be addressed for the near and far future.

3.2 Coastal Hydrodynamic Regime

The significant wave height, H_s, is defined as the mean of the highest one-third of the waves present in the sea. The maximum wave height, Hs, is the maximum vertical distance between the highest crest to the lowest trough; T_p is the wave period corresponding to H_s, and T_z is the mean zero-up crossing period of the wave field. The dimensionless wave parameters, namely, the wave steepness (H_s/L) and the wave age ($C/U10$), where C is the phase speed, are often used to determine the nature of the sea state. Wave steepness is usually expressed as the ratio between the significant wave height and the wave length, of the peak period. For the calculation of the wave climate off the area, the significant wave height H_s, the peak energy period T_s, and the mean period T_z, are estimated by JONSWAP model. Wave estimations were made for wind waves from 5 directions in origin (N, NW, NE, E, and W). The wind blowing length (fetch) used in the above forecasting equations is determined in a straight line along the wind direction. For stretches limited by irregular coastlines, it is recommended to measure the length of the stretch at nine points (along the direction of the wind direction and four points per 2° on each side) and use the average of the measurements for the waveform calculations. The spectral significant wave height and spectral period calculated are the maximum values that can be derived for the maximum wind speed. The minimum required duration for the winds affecting the area is within the range observed in the wider area and therefore the assumption is that the maximum wave growth is only limited by wind blowing. To study the distribution of wave energy along the coast in the area, wave refraction diagrams were constructed using numerical models. For each wave direction affecting the coastal zone, refraction diagrams for waves with higher wavelengths were constructed, as well as the usual peak values for the area calculated on a weighted average basis for the incidence. Wave run up height is given by $R_{u2\%}$. This is the wave run-up level, measured vertically from the still water line, which is exceeded by 2% of the number of incident waves. The number of waves exceeding this level is hereby related to the number of incoming waves and not to the number that run-up. The wave run-up (R_W) is given from Maze's (1989) equations for:
breaking waves:

$$\frac{R_{u2\%}}{H_S} = 2.32 \zeta_0^{0.17},$$

and non-breaking waves:

$$\frac{R_{u2\%}}{H_s} = \sqrt{2\pi} \cdot \left(\frac{\pi}{2B}\right)^{1/4},$$

where, H_s is the offshore significant wave height; B is the berm height, and ξ: the Irribaren number (Irribaren and Nogales 1949) $\xi_o = \tan\beta\sqrt{\frac{L_0}{H_s}}$, where β is the beach slope and Lo the wave length.

The wave run-up level on smooth slopes is determined by the level at which the water tongue becomes less than 2 cm thick. Thin layers blown onto the slope are not seen as wave run-up. Run-up is relevant for smooth slopes and embankments and sometimes for rough slopes armoured with rock or concrete armour. The percentage or number of overtopping waves, however, is relevant for each type of structure.

Wave overtopping is the mean discharge per linear meter of width, q, for example in m3/s/m or in l/s/m. The methods described here calculate all overtopping discharges in m3/s/m unless otherwise stated; it is, however, often more convenient to multiply by 1000 and quote the discharge in l/s/m.

The process of wave overtopping is very random in time and volume. The highest waves will push a large amount of water over the crest in a short period of time, less than a wave period. Lower waves will not produce any overtopping. Still, a mean overtopping discharge is widely used as it can easily be measured and also classified as (a) insignificant with respect to strength of crest and rear of structure $(q < 0.1 \text{ l/s/m}), q = 1 \text{ l/s per m}$: On crest and inner slopes grass and/or clay may start to erode $(0.1 < q < 1 \text{ l/s/m})$, (b) Significant overtopping for dikes and embankments. Some overtopping for rubble mound breakwaters. $(1 < q < 10 \text{ l/s/m})$, and crest and inner slopes of dikes have to be protected by asphalt or concrete; for rubble mound breakwaters transmitted waves may be generated $(10 < q < 100 \text{ l/s/m})$. The mean overtopping discharge, q, is the main parameter in the overtopping process. The overtopping discharge is given in m3/s per meter width and in practical applications often in litres/s per meter width. Although it is given as a discharge, the actual process of wave overtopping is much more dynamic. Only large waves will reach the crest of the structure and will overtop with a lot of water in a few seconds. This wave by wave overtopping is more difficult to measure in a laboratory, compared to the mean over-topping discharge. As the mean overtopping discharge is quite easy to measure many physical model tests have been performed all over the world, both for scientific (ide-alised) structures and real applications or designs. The structures considered for the Koules Fortress in Heraklion are rubble mound structures such as breakwaters and rock slopes. The principal formula used for wave overtopping is:

$$\frac{q}{\sqrt{gH_{m0}^3}} = a\,exp\left(-\frac{bR_c}{H_{m0}}\right)$$

It is an exponential function with the dimensionless overtopping discharge $q/(gH^3m0)1/2$ and the relative crest freeboard Rc/Hm0.

Damage to armour layers is characterized either by counting the number of displaced units or by measuring the eroded surface profile of the armoured slope. In both cases, damage is related to a specific sea state of a given duration. When damage is characterized in terms of displaced units, it is usually given as the relative displacement, D (the ratio of displaced units to the total number of units, or preferably, to the number of units within a specific zone around the still water level. In fact, structures with the same geometry and armour-layered units that are subjected to the same incident sea states (significant wave height and peak period or wavelength) are expected to have a similar number of displaced units, but they would have different relative displacements if they have different lengths of the respective armoured slopes. Damage can be related to any definition of armoured layer movements, including rocking. The relative number of moving units can also be obtained with the total number of units within a strip of width Dn stretching along the slope from the bottom to the top of the armoured layer. Because almost all armour unit movements take place within the strip comprised between still water level ± Hs, the number of armour units within this reference area is adopted to calculate D. Since the width of such a strip (and consequently the number of armoured units there) changes with Hs, it is recommended to use the number of units within the region defined by the still water level ± n x Dn, where: Dn is the nominal diameter; n is chosen such that almost all movements take place within these levels (Shore Protection Manual (1984). Damage characterization based on the eroded cross-section area are around the still water level was used by Iribarren (1948) and Hudson (1959). Hudson defined D as the percent erosion of original volume. Iribarren defined the limit of severe damage that could occur when the erosion depth (de) in the main armoured layer reached the nominal diameter of the armor layer elements (Dn). Broderick and Ahrens [8] and Van der Meer [9] defined a dimensionless damage parameter, given by:

$$S = \frac{A_e}{D_n^2}$$

where Ae is the eroded cross-section area around the still water level and Dn the nominal diameter of the armored units. Thus, S is a dimensionless damage parameter, independent of slope length. Another quantity that may be used to characterize the damage in a cross-section of the armour layer of a rubble-mound breakwater, is the eroded length Le, which is the distance measured along the idealized design slope between the extremities of the eroded area.

The conducted seafloor survey includes sediment samples and Starfish 450F Side Scan Sonar images (Fig. 3). Side scan sonar images were used to investigate the composition of the see bottom surface of the wider area and underwater breakwater state.

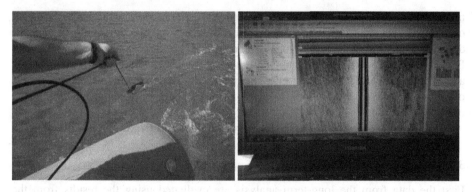

Fig. 3. Side scan sonar starfish 450F (Left) and log screen (right)

3.3 Results

The analysis of the wind data for the period 1956–2015 shows that in the wider area of the coastline of Heraklion, including the location of the Koules fortress, the winds N, NW, W and NE prevail with an annual incidence of more than 65%. Within the 7.23% of the year, wind speed velocities are very low. The area is mainly affected by NW winds with an annual frequency of 28.85%. The most commonly occurring wind speed, regardless of direction, is 4Bf (11–16 kn) with an annual incidence of 25.08%. At 90.85% of the time the wind speed does not exceed 5 Bf (<22 kn), and 99.75% of the time the wind speed does not exceed 8 Bf (<40 kn). The maximum wind speed is 9 Bf (41–47 kn). It has an annual occurrence rate of 0.25% and is mainly due to N winds. The frequency of prevailing winds in the area is shown in the rose diagram of Fig. 4. From the two rose diagrams it can be observed that, there is a change in wind frequencies with an increase of winds of southern origin. The coastal zone of the area is

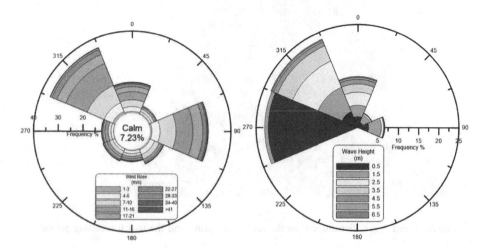

Fig. 4. Wind and wave rose diagram for the period 1956–2015.

mainly affected by wind waves of N, NE and NW origin, while wind waves W and E originally affect the coastal zone after diffraction and intense refraction. N and NE direction wind waves have an annual frequency of 12.43% and 3.74% respectively. East origin wind waves have an annual frequency of 6.79%, while W and NW of origin are 28.85% and 23.62% respectively. The maximum expected wind waves are of NW origin, with a wave height of 7.57 m and a period of 11.03 s, with an annual frequency of 0.11% (Fig. 4).

The meteorological data statistical analysis focuses on two different time-scales.

3.4 Long-Term Analysis

First the data from the long-term analysis are evaluated using the results from the EURO-CORDEX project that provides regional climate projections for Europe at 12.5 km resolution. The provided data for the two future periods (near future: 2036–2065, far future: 2071–2100) are used to extract the near and far future wind time-series data to introduce to the wave model analysis software. This way the significant wave height and period can be estimated for the coastal front of Heraklion and the impact on Koules monument can be addressed for the near and far future. Examining Figs. 5 and 6 for the Heraklion coastal area, there is an obvious wind speed shift to higher values over the near and far future distributions. The wind-rose diagram and Table 1 shows that the prevailing wind direction remains the NW (315°) throughout the two forecasting periods (near and far future periods) with the mean wind speed increasing by 0.15 m/s in the near future and by 0.24 m/s for the far future with respect to the reference period. The secondary wind direction is the Western direction (270°).

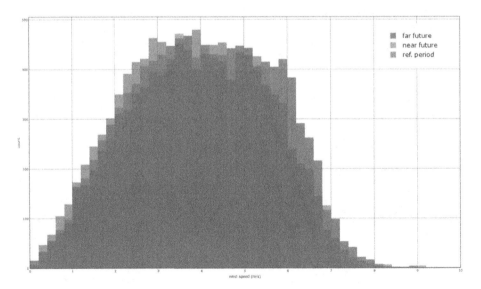

Fig. 5. Wind speed distribution for the reference period and the two forecasting periods

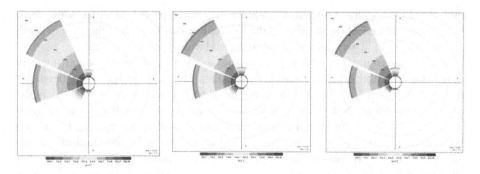

Fig. 6. Heraklion coastal area windrose distribution for the reference period (left) the near future (middle) and the far future (right).

Table 1. Occurrence (%) of wind direction sector for the reference period and the near and far future (Bold: Prevailing Wind Direction, Italic: Secondary Wind Direction)

%	45°	90°	135°	180°	225°	270°	315°	360°
Ref.	0,49	0,24	0,37	2,53	8,33	*36,39*	**46,10**	5,55
Near	0,50	0,26	0,30	1,85	6,77	*34,75*	**49,07**	6,50
Far	0,54	0,26	0,38	1,68	5,18	*33,77*	**51,16**	7,03

3.5   Wave Propagation

To study the distribution of wave energy along the coast in the area, wave refraction diagrams were constructed using numerical models. For each wave direction affecting the coastal zone, refraction diagrams for waves with higher wavelengths were constructed, as well as the usual peak values for the area were calculated on a weighted average basis for the incidence. The data of these wave's characteristics, (Table 2) include the relation of

Table 2. Wave characteristics

		U_a (m/sec)	H_s (m)	T_s (sec)	h_c (m)	L_o (m)	H_b (m)	d_b (m)
N	Mean	21,27	4,09	8,69	7,77	117,81	4,53	5,24
	Max	33,91	6,80	10,28	12,45	164,84	7,27	8,72
NE	Mean	18.00	2.64	4.69	1.73	76,92	3,18	3,75
	Max	33.91	4.98	8.37	8.88	109,24	5,22	6,39
NW	Mean	20.62	4.6	9.36	14.6	136,62	5,12	5,90
	Max	33.91	7,57	11.03	13.97	189,76	8,15	9,70
E	Mean	21,26	3.15	7.22	5.86	81,26	3,41	4,04
	Max	33.91	5.03	8.42	8.97	110,58	5,27	6,45
W	Mean	19.94	0.81	3.01	1.34	20,06	1,33	1,77
	Max	33.91	1.34	3.59	2.11	14,13	0,81	1,04

key: U_a wind speed, T_p wave period, H_s significant wave height h_c closure depth; L_o wave length; H_b breaking wave heightand d_b breaking wave depth.

wind speed (U_a) with significant wave height H_s, wave period T_s. Also, some wave characteristic that effect the stability of the Koules breakwater such as the closure depth (h_c), wave length L_o; wave breaking height H_b and wave breaking depth d_b.

The waves originating from the north and northeast and the maximum intensity waves approach the coastline almost in parallel, with an average wave height of between 4.5 m and 5 m. For normal northern origin waves, the wave height is in the range of 3 m–3.5 m. Near the coastline, the wave height is smaller, up to approximately 2.5 m, due to shallow bathymetry. This is because the breaking zone begins at a distance about 500 m for the north waves, and about 450 m for the Northeast waves (see Figs. 7 and 8). Waves coming from northwest are approaching the coastline after refraction at the Cape Panagia. The average wave height of normal waves is 2–2.5 m, while for normal maximum observed waves is between 2.5 m and 3 m. The breaking zone is even wider, and about 600 m from the coast (Fig. 9). The eastern origin waves approach the coastline with average wave height of ranging between 1.5 m and 2.5 m, while for the corresponding average maximum waves observed between 3 m and 3.5 m. For the waves coming from the west, the wave height appears to be in the range of 0.5–1 m for normal waves, and 1–2 m for the maximum (Fig. 11). From the analysis of the refraction diagrams, it appears that the waves affecting the wider area are those originating from N, NE, NW and E. West origins waves arrive at the shoreline after

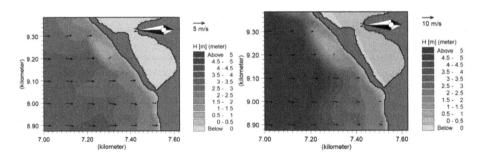

Fig. 7. Wave propagation of Northern waves Hs: 4,09 m - Ts: 8,69 s and Hs: 6,80 m - Ts: 10,28 s

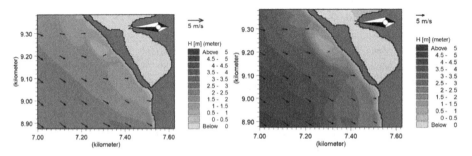

Fig. 8. Wave propagation of Northeastern waves Hs: 2,64 m - Ts: 4,69 s and Hs: 4,98 m - Ts: 8,37 s

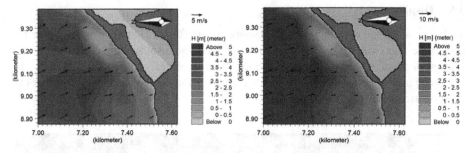

Fig. 9. Wave propagation of Northwestern waves Hs: 4,60 m - Ts: 9,36 s and Hs: 7,57 m - Ts: 11,03 s

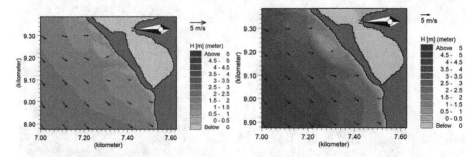

Fig. 10. Wave propagation of eastern waves Hs: 3,30 m - Ts: 7,57 s and Hs: 5,03 m - Ts: 8,42 s

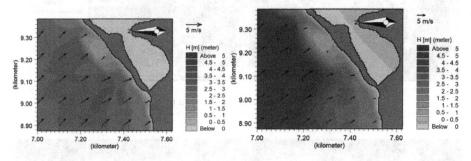

Fig. 11. Wave propagation of western waves Hs: 0,81 m - Ts: 3,01 s and Hs: 1,34 m - Ts: 3,59 s

refraction and diffraction, causing greatly weakening of the wave energy in the near-shore. The breaking depth (d_b) and the breaking wave height (H_b) in the coastal zone were determined. From the calculation of the maximum breaking depth for the maximum expected wave characteristics, the width of surf zone was determined as the area enclosed between the maximum breaking depth and the coastline (Table 5). The maximum expected wave height at break in the studied area is derived from NW waves and is 8.15 m and waves breaking is expected to start at a depth of 9.70 m (Fig. 10).

3.6   Displacements

With the use of the correlation of displacement data (Fig. 12) and wave height for the area of the breakwater (Fig. 13) it can be seen that large displacements occur in periods after strong wave events. Thus, the correlation analyses capabilities will help the interested user to better investigate the displacements in the Koules breakwater that can be supported by the wave sensor data.

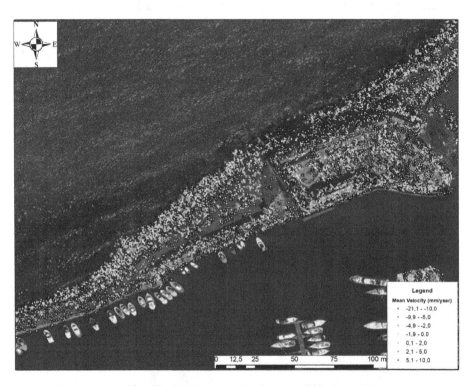

Fig. 12. Displacements in the area of Koules

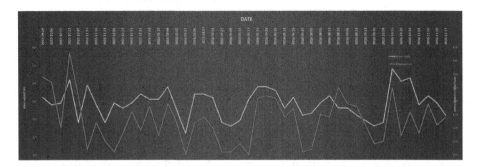

Fig. 13. Correlation of displacement with wave height.

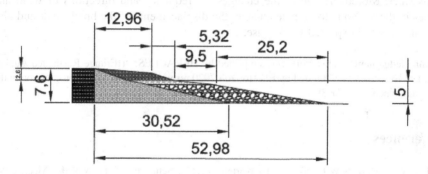

Fig. 14. Proposed solution

4   Conclusions

The aim of this work is to estimate the problems caused by the wave impact in the area of the Koules fortress and to propose a solution for the degradation of the monument. The results have shown that the main impacts are from the N, NE, E and NW origin waves, which influence the area and impact mechanical stresses to the breakwater and have significant overtopping events. The problem of the degradation of the breakwater armour in the area of the Venetian Fortress of Koules is proposed to be addressed by strengthening the existing shielding of the area to limit the wave run-up. The proposed solution involves the completion of the existing armour along its entire length in the area of the Venetian Fortress of Koules, aiming to reduce the wave energy impacting the walls (through induced wave breakage) at greater distances compared to current impacts. The materials are composed of natural boulders in the brim weighting 150 kg. each, while in the reef section of artificial boulders weighing 4 tn. each. The future constructions are environmentally compatible, but also comply with the monument's requirements, so that the new works will harmonize with the existing historical constructions. The extent of the project is shown in Fig. 14 with an indicative section of armour reinforcement. Moreover, for the safety of the public and passers-by, a small concrete vertical wall with a seating area will be built, to limit sea water from high energy events to reach the breakwater deck.

Moreover, salts and crusts accumulation due to sea proximity, are proposed to be periodically monitored to control and check the stone erosion process. Moreover, seasonal monitoring of wind and wave effects and the observed structural cracks on the Venetian Fortress of Koules might be either related to vertical dead loads, or to scouring of foundations on the sea due to sea waves. At present, periodic analysis of satellite data aiming at understanding the stability of the scouring process is recommended as a solution for preventive conservation of the Fortress.

With the use of the correlation of displacement data and wave height for the area of the breakwater it can be seen that large displacements occur in periods after strong wave events. Thus, the correlation analyses capabilities, supported by the wave sensor data, will help the interested party to better investigate the displacements in the Koules breakwater. Moreover, due to the changes in frequency and directions of wind and waves in the area due to climate change, the displacements of the breakwater and also the salt stray are expected to increase.

Acknowledgements. This work was supported by HERACLES: "HEritage Resilience Against CLimate Events on Site" funded by EU Horizon 2020 research and innovation programme under grant agreement No. 700395.

References

Pullen, T., Allsop, N.W.H., Bruce, T., Kortenhaus, A., Schüttrumpf, H., van der Meer, J.W.: Assessment Manual. Places of Cultural and Heritage Significance. ICOMOS International Cultural Tourism Committee (2002)

Costanza, R., et al.: The value of the world's ecosystem services and natural capital. Nature **387**, 253–260 (1997). https://doi.org/10.1038/387253a0

EurOtop Overtopping Manual (2007). Wave Overtopping of Sea Defences and Related Structures –

Hinkel, J., et al.: Coastal flood damage and adaptation costs under 21st century sea-level rise. Proc. Nat. Acad. Sci. **111**, 3292–3297 (2014)

Hogarth, P.: Preliminary analysis of acceleration of sea level rise through the twentieth century using extended tide gauge data sets. J. Geophys. Res. Oceans **119**, 7645–7659 (2014)

Hoggart, S.P.G., et al.: The consequences of doing nothing: the effects of seawater flooding on coastal zones. Coastal Eng. **87**, 169–182 (2014)

Hudson, R.Y.: Laboratory investigation of rubble-mound breakwaters. A.S.C.E. Waterways & Harbours Div., September 1959

ICOMOS: International Workshop on Impact of Climate Change on Cultural Heritage, New Delhi.2007. Issued: May 2008

ICOMOS: Thematic Workshop on Cultural Heritage and Climate Change Report. 16th General Assembly and Scientific Symposium Quebec, Canada, 2008, E News No. 18, Issued, March 2009

ICOMOS: International Cultural Tourism Charter. Principles and Guidelines for Managing Tourism

IPCC: Special report on Emissions Scenarios. Intergovernmental Panel on Climate Change (IPCC) (2000)

Iribarren, R.: Una formula para el oalculo de los cliques de escollera. July 1938. Translated Fluid Mechanics Laboratory, University of California, Berkeley, Technical Report HE-116–295 (1948)

Izaguirre, C., Méndez, F.J., Espejo, A., Losada, I.J., Reguero, B.G.: Extreme wave climate changes in Central-South America. Clim. Change **119**, 277–290 (2013)

Church, J.A., White, N.J.: Sea-level rise from the late 19th to the early 21st century. Surv. Geophys. **32**(4–5), 585–602 (2011)

Jevrejeva, S., Moore, J.C., Grinsted, A., Matthews, A.P., Spada, G.: Trends and acceleration in global and regional sea levels since 1807. Global Planet. Change **113**, 11–22 (2014)

Kiousopoulos, J.: Methodological approach of coastal areas concerning typology and spatial indicators, in the context of integrated management and environmental assessment. J. Coast. Conserv. **12**(1), 19–25 (2008)

Losada, I.J., Reguero, B.G., Méndez, F.J., Castanedo, S., Abascal, A.J., Mínguez, R.: Long-term changes in sea-level components in Latin America and the Caribbean. Global Planet. Change **104**, 34–50 (2013)

Shore Protection Manual. US Army Corps of Engrs, CERC, US Govt. Printing Office, Washington, DC (1984)

Tol, R.S.J.: Economics of sea level rise. In: John, K.T.K., Steve, A.T. (eds.) Encyclopedia of Ocean Sciences, 2nd edn, pp. 197–200. Academic Press, Oxford (2009)

Tsilimigkas, G., Deligianni, M., Zerbopoulos, T.: Spatial typologies of Greek coastal zones and unregulated Urban growth. J. Coast. Conserv. **20**(5), 397–408 (2016)

Ullmann, A., Monbaliu, J.: Changes in atmospheric circulation over the North Atlantic and sea-surge variations along the Belgian coast during the twentieth century. Int. J. Climatol. **30**, 558–568 (2010)

van der Meer. www.overtopping-manual.com

Wang, X.L., Feng, Y., Swail, V.R.: Changes in global ocean wave heights as projected using multimodel CMIP5 simulations. Geophys. Res. Lett. **41**, 1026–1034 (2014)

Weisse, R., Bellafiore, D., Menéndez, M., Méndez, F., Nicholls, R.J., Umgiesser, G., Willems, P.: Changing extreme sea levels along European coasts. Coastal. Eng. **87**, 4–14 (2014)

World Heritage report 22. 2006, Climate change and Heritage. World Heritage centre, UNESCO (2006)

Increasing the Resilience of Cultural Heritage to Climate Change Through the Application of a Learning Strategy

Elena Sesana[1] , Chiara Bertolin[2] , Arian Loli[2] ,
Alexandre S. Gagnon[3] , John Hughes[1(✉)] ,
and Johanna Leissner[4]

[1] School of Computing, Engineering and Physical Sciences,
University of the West of Scotland, Paisley PA1 2BE, UK
john.hughes@uws.ac.uk
[2] Department of Architecture and Technology, Norwegian University of Science
and Technology (NTNU), 7491 Trondheim, Norway
[3] School of Natural Sciences and Psychology, Liverpool John Moores
University, James Parsons Building, Byrom Street, Liverpool L3 3AF, UK
[4] Fraunhofer Gesellschaft zur Förderung der angewandten Forschung,
Munich, Germany

Abstract. There is growing concern about the threat posed by climate change to cultural heritage, notably to World Heritage properties. Climate change is triggering changes in rainfall patterns, humidity and temperature, as well as increasing exposure to severe weather events that can negatively impact on cultural heritage materials and structures by enhancing the mechanical, chemical and biological processes causing degradation. In response to this climate change challenge, the Climate for Culture (CfC) project, funded by the European Commission, investigated the impacts of climate change on the European cultural heritage through the use of a high-resolution regional climate model that projected future changes in climatic conditions, and simulated the future conditions of the interiors of historical buildings and their impacts on the collections they hold using building simulation tools. This paper compares the climate change impacts on cultural heritage identified by the CfC project with semi-structured interviews with experts working on cultural heritage preservation in Norway, Italy and the UK. Hence, the perceptions of the cultural heritage community on the impacts of climate change on heritage assets are first explored, which are then compared with the risk matrices produced by the CfC project as a decision-support tool to inform managers involved in the preservation of cultural heritage. In addition, the learning strategy underpinning examples of climate change adaptive measures applied to cultural heritage is discussed. Through the identification of the current learning strategy in the case study sites, this research highlights the lack of dissemination of the outcomes of scientific research to managers of cultural heritage in the context of adaptation to climate change impacts.

Keywords: Resilience · Adaptation · Climate change · Cultural heritage · Learning strategies

A. Moropoulou et al. (Eds.): TMM_CH 2018, CCIS 961, pp. 402–423, 2019.
https://doi.org/10.1007/978-3-030-12957-6_29

1 Introduction

Cultural heritage is susceptible to fluctuations in climatic conditions and extreme weather events. Changes in temperature, precipitation and relative humidity can impact on the historical materials and structures that comprise cultural heritage assets, through variation in the mechanical, chemical and biological mechanisms of material and structural degradation. Cultural heritage is also affected by extreme sea level rise and flooding, for example during storm surges, causing coastal impacts and landslides; the intensity and occurrence of which are predicted to increase with climate change. Coastal erosion is also seen as a particular risk for heritage, potentially resulting in the complete loss of sites (Sabbioni et al. 2010; Brimblecombe et al. 2011).

To assess the risk that changes in climatic conditions pose for cultural heritage, two projects were funded by the European Union: the Noah's Ark project (2004–2007) and the Climate for Culture (CfC) project (2009–2014). Both projects developed predictive models of the impacts of climate change on cultural heritage in Europe. This was done mainly through the development of maps projecting variations of climatic conditions into the future, which were related in turn to the mechanical, chemical and biological degradation mechanisms that affect cultural heritage (Sabbioni et al. 2010; Leissner et al. 2015; Bertolin and Camuffo 2014; Sabbioni et al. 2008). These two projects attempted to overcome a barrier to climate change adaptation in the field of cultural heritage by introducing climate modelling and building simulation tools, an approach not commonly used within the heritage sector.

Scenarios developed by projects of this sort can be used to inform stakeholders about the possible risks and impacts that are predicted to affect cultural heritage in the near and far future. There is an associated urgency to protect threatened heritage sites from climate change impacts, for which these new tools may be expected to find valuable application. However, to date it has not been possible to evaluate the impact of such projects' outputs and predictions on conservation awareness and practice. Have these scenarios been used in cultural heritage preservation or have they remained a mere scientific exercise?

This paper aims to compare the results from the scientific community, focusing specifically on the more recent CfC project, with the perceptions and awareness of the cultural heritage preservation stakeholders' community in selected locations in Europe. The paper also examines whether there are connections between awareness of climate change risks and the propensity to take adaptive actions as well as identifying the learning process behind the adopted adaptation strategies.

2 Cultural Heritage and Climate Change

Most research published to date in the field of climate change and cultural heritage has focused on assessing the risks and impacts of climate change on cultural heritage in Europe (Sabbioni et al. 2010; Bertolin and Camuffo 2014; Leissner et al. 2015; Cassar 2005; Brimblecombe et al. 2011; Cassar and Pender 2005). The CfC project developed a methodology to assess the climatic risks for the indoor European cultural heritage. Maps, at a 12.5 km resolution, projecting potential scenarios of change for a number of climatic variables affecting cultural heritage, were developed using the Regional Climate Model (RCM) REMO for the baseline (1961–1990), near future (2021–2050) and the far future (2071–2100) time-periods. The project used two moderate greenhouse gas (GHG) emission and concentration scenarios: the A1B and the Representative Concentration Pathway (RCP) 4.5 of the Fourth and Fifth Assessment Report of Intergovernmental Panel on Climate Change (IPCC), respectively (IPCC 2000; IPCC 2014; Thomson et al. 2011). Potential variations of the mechanical, chemical and biological indoor degradation of light-weight (i.e. wooden) and heavy-weight (i.e. masonry) buildings into the future were estimated on the basis of those climate change projections (Leissner et al. 2015).

Although there is increasing research on climate change impacts on cultural heritage, there remains a paucity of studies reporting on the awareness of the cultural heritage community on those impacts and the use of the outcomes from such research in adaptation decision-making. The way decision-makers perceive the risks and impacts of climate change is likely to influence the choice and development of adaptation strategies (Gray et al. 2014), but this has yet to be examined in the field of cultural heritage.

Research accomplished to date on adapting our cultural heritage to climate change centres on the dissemination of guidelines and recommendations to implement adaptation measures (Sabbioni et al. 2010; Sabbioni et al. 2008; Heathcote et al. 2017; Haugen and Mattsson 2011; Cassar 2016; Pollard-Belsheim et al. 2014; Carmichael et al. 2017a; Fatorić and Seekamp 2017a; Grøntoft 2011; Hall et al. 2015; Hall 2015) and on the identification of opportunities and barriers to adaptation (Phillips 2014; Fatorić and Seekamp 2017b; Carmichael et al. 2017b; Casey 2018; Sesana et al. 2018). Preserving cultural heritage from the impacts of climate change requires a shift from reactive to proactive adaptation (Sesana et al. 2018). However, the process of deciding when and how proactive adaptation is appropriate, and its connection to the knowledge base amongst stakeholders is unclear. How do decision makers react to climate change impacts? Where do they get the knowledge required to inform the adaptation process? What approach do they follow to gather and apply that knowledge? Building on the knowledge requirement for adaptation reported in the literature (Sesana et al. 2018), we argue that an increased understanding of the learning process underpinning the adaptation measures that have occurred in cultural heritage preservation would better inform the management of cultural heritage (McDonald-Madden et al. 2010; Williams 2011).

3 Methodology

In this paper, a comparison between qualitative data obtained from semi-structured interviews with the cultural heritage management community and a quantitative risk assessment developed as part of the CfC project using regional climate change projections is made. Information from semi-structured interviews conducted as part of a larger study (also presented elsewhere, e.g. Sesana et al. 2018) were extrapolated in order to understand the perceptions of climate change impacts on cultural heritage amongst selected experts working in the field of cultural heritage in three European countries: Norway, Italy and the UK. The data collected during those interviews were then compared with the results from the CfC project, which estimated the impacts of future climate change on cultural heritage in sites or in the region where the interviewed stakeholders are located. In addition, the data from the interviews were used to identify examples of climate change adaptation measures adopted and matching the adaptation approach to two learning strategies.

3.1   Qualitative Data

Forty-five semi-structured interviews were conducted (Table 1). The interviews focused on three main questions, namely (1) Are you aware of any changes in the climate that are affecting cultural heritage? (2) Are you aware of climate change projections for your country? (3) Are you aware of any examples of adaptation measures or case studies where adaptation measures have been adopted to preserve cultural heritage from climate change impacts?

Table 1. Professional affiliations of the interviewees.

Interviewees	Number of interviews
Academics and researchers	19
Managers of cultural heritage	26
Total	45

The cultural heritage experts interviewed and falling within the 'academics and researchers' category are specialists in anthropology, archaeology, architecture, biology, conservation science, climate science and geology, while the other category comprises heritage site managers and coordinators, sustainability officers and urban planners, and architects and conservators working within heritage sites. The interviews were audio recorded, transcribed and analysed using the NVivo software. Ethical approval was sought and obtained through the University of the West of Scotland procedure.

3.2   Quantitative Data

The CfC project assessed the risks of climate change on cultural heritage for locations distributed over a regular grid across Europe. Hence, on the basis of the outcomes of the CfC project, matrices compiling the risks of climate change were developed by obtaining data for the grid points in the three countries were the interviews took place, i.e. Norway, Italy and the UK, as depicted in Figs. 1, 2 and 3, with the coordinates of the grid points shown in Tables 2, 3 and 4. The information from the CfC project has

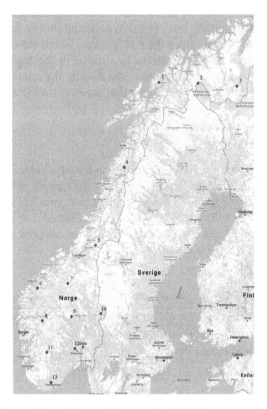

Fig. 1. Locations where climate change risks on cultural heritage were estimated in Norway as part of the CfC project. Map data ©2018 Google.

Table 2. Locations of variables (Latitude, Longitude and ID) simulated in the CfC project for Norway.

Country	ID	Lat.	Long.
Norway	1	69.2898	17.5711
	2	69.2659	21.1079
	3	69.1791	24.6277
	4	66.6144	14.4121
	5	63.8928	11.7972
	6	62.2671	6.5031
	7	62.4467	9.2598
	8	60.9630	6.9321
	9	61.1359	9.5886
	10	61.2618	12.269
	11	59.6575	7.3273
	12	59.8244	9.8912
	13	58.3510	7.6928

Fig. 2. Locations where climate change risks on cultural heritage were estimated in Italy as part of the CfC project. Map data ©2018 Google.

Table 3. Locations of variables (Latitude, Longitude and ID) simulated in the CfC project for Italy.

Country	ID	Lat.	Long.
Italy	1	46.5594	10.0936
	2	46.6832	12.0026
	3	45.0934	8.4419
	4	45.2468	10.2938
	5	45.3674	12.155
	6	43.9338	10.4849
	7	44.0515	12.3004
	8	42.7353	12.4395
	9	41.5003	14.3083
	10	39.8539	9.32880
	11	40.2311	16.0938
	12	37.5451	14.5604

Fig. 3. Locations where climate change risks on cultural heritage were estimated in the UK as part of the CfC project. Map data ©2018 Google.

Table 4. Locations of variables (Latitude, Longitude and ID) simulated in the CfC project for the UK.

Country	ID	Lat.	Long.
United Kingdom	1	58.1739	−4.9767
	2	56.9151	−4.2360
	3	55.6523	−3.5437
	4	54.3858	−2.8948
	5	53.1160	−2.2848
	6	53.4525	−0.1554
	7	51.8433	−1.7098
	8	52.1706	0.3652

been re-analysed and reinterpreted for the purpose of producing the matrices for the current paper. These locations were selected by the CfC project team so as to represent a range of different geographical contexts in each country considered. Projections for a number of climatic variables for the grid cells centred on the points with the coordinates shown in Tables 2, 3 and 4 were obtained from the RCM, which were then used to determine the risks of those climatic factors on cultural heritage. For this purpose, the climate change projections were coupled with buildings simulation tools to estimate mechanical, chemical and biological degradation damage variables. The matrices provided in this paper (Figs. 4, 5, 6 and 7) show the cumulative mechanical, chemical and biological risks for the indoor cultural heritage depending on structure typology, location of the site and the time-scale of the projections. Loli and Bertolin (2018) provide more details on the development of the matrices.

4 Results and Discussion

4.1 Perception and Awareness of Climate Change Impacts on Cultural Heritage

The interviewees' answers to the question on whether they are aware of climate change impacts on cultural heritage assets are summarised in Table 5. The responses are divided according to the three countries in which the interviews took place and according to the background of the interviewees (i.e. whether they are academics or cultural heritage professionals, for example professionals involved in the management of cultural heritage). All interviewees in Norway and Italy noted impacts of changes in climate on cultural heritage, but this figure was found to be lower in Italy, particularly for the managers of heritage sites.

Table 5. Interviewees' answer to the question: "Are you aware of any changes in the climate that are affecting cultural heritage?"

		Academia	Managers of cultural heritage
Norway	Yes	100%	100%
	No	0%	0%
	Some	0%	0%
Italy	Yes	66.6%	14.3%
	No	16.7%	71.4%
	Some	16.7%	14.3%
UK	Yes	100%	100%
	No	0%	0%
	Some	0%	0%

Table 6 provides examples of direct quotations from a selected number of interviews to the first question mentioned in Table 5. These quotes provide examples of specific impacts perceived to be occurring by the interviewees.

Table 6. Examples of interviewees' answer to: "Are you aware of any changes in the climate that are affecting cultural heritage"?

Interviewee	Country	Quotation
Academic	Norway	*"The weather is wetter; there is much more rain, a lot of river flooding. According to climate change this is more dangerous, more rain, strong rain which makes the river flood. You have also the rise of sea level. There are some areas of cultural heritage that are exposed to that for example the houses exposed to the coast. (...) The rise of humidity is a danger to older building construction."*
WHS coordinator	Norway	*"We have a general knowledge about prediction of climate change. (...) rainfall (...) flood (...) erosion (...) heavy rain and landslides."*
Academic	Italy	*"The change of precipitation pattern, the seasonality of the precipitation, drier summers, wetter winters, an increase in the frequency of heavy rain events. It is not the quantity of precipitation that is changing, what is projected to change is the number of extreme events and their seasonality. The changes in the hygro-thermal parameters such as temperature and relative humidity will influence the cohesion and cracking (of historical materials) due to salt crystallization. Or (they will influence) the biological growth."*
Member of governmental institution	Italy	*"I am aware of climate change, but I am not aware that climate change is influencing cultural heritage."*
Academic	UK	*"Rainfall has increased, and temperature has increased by half a degree, at least on average. (...) Increase in intensity of rainfall. The guttering is not good enough and we probably look at an increase in soil moisture, so probably the largest problem is rising damp in buildings."*
Member of governmental institution	UK	*"Increased precipitation, rising temperature, higher humidity and that's important because insects like humidity. We are getting less frost so the freeze and thaw is getting better. Frost does not matter if the building is dry, but if the building is wet it is a problem. (...) we have quite a lot of data here, which essentially show wetter winter, drier summers and intense rainfall in summer. So it is not saying that it going to be drier, but that summer is drier."*

The interviewees' answers in Table 7 highlight where the interviewees have consulted or have seen climate change projections for their country. The answers are divided, as in Table 6, according to the three different countries investigated and the field that the interviewees work in. In all three countries, the percentage of interviewees aware of climate change projections for their country is higher for the academics than the managers of the heritage sites, but the difference between the two groups of stakeholders is much smaller in the UK. Managers of heritage sites in Italy appear to be unaware of climate change projections for their country. This is in line with the results depicted in Table 5 as one could argue that as the changes in climate that have occurred to date are not perceived to have impacted cultural heritage assets, the stakeholders are less likely to consult projections of future climatic changes.

Table 7. Interviewees' answer to the question: "Are you aware of climate change projections for your country?"

		Academia	Cultural heritage management and preservation
Norway	Yes	75%	40%
	No	0%	0%
	Some	25%	60%
	No answer	0%	0%
Italy	Yes	80%	0%
	No	0%	14.3%
	Some	0%	0%
	No answer	20%	85.7%
UK	Yes	70%	61.5%
	No	0%	7.7%
	Some	20%	23.1%
	No answer	10%	7.7%

Table 8 summarises the main consequences of climate change on cultural heritage as identified by the interviewees in the three selected countries.

A comparison between the risk matrices shown in the subsequent section with the interviewees' answers reveals that, overall, the interviewees are aware of climate change impacts on cultural heritage. However, in Italy, there is a contrast between the recognition of climate change impacts affecting cultural heritage amongst heritage managers in comparison to the perceptions of academics, with the latter appearing more knowledgeable on this issue. In all three countries, heritage site managers showed less awareness of climate change projections than the academics. All interviewees were aware of climate change, but some of them were not aware of the possible consequences for cultural heritage. In fact, some interviewees appeared more knowledgeable about climate change mitigation (i.e., reducing the carbon footprint of historical buildings) rather than on the risks and impacts of climate change on the heritage sites; the latter was particularly the case at the management level in Italy. One reason might be that northern European countries may already be experiencing more negative

Table 8. Consequences of climate change on cultural heritage as identified by the interviewees.

Norway	Italy	UK
Increase in biological growth	Decline in freeze-thaw weathering	Stone erosion is a risk
Increase in insect growth	Decrease of decohesion of porous materials due to a decrease of freeze-thaw cycles	Increase in biological growth
Increase of biological patina due to increased humidity and increased concentration of nitrogen in the air	Change in decohesion and in cracking of materials due to a change in salt-crystallization phenomenon under a variation of temperature and relative humidity	Increase in insect growth
Increased humidity	Gutters may not cope with extreme rainfall	Increase of biological patina due to increased humidity and increased concentration of nitrogen in air
Warmer winters increases decay of wooden structures	Accelerated decay of roofs under increased rainfall	Water ingress into buildings
Warmer and more humid climate is not good for timber structures	Sudden changes in temperature crossing the zero degrees with subsequent serious condensation	Dampness penetration into buildings
Fungal decay	Increased condensation in summer	Water saturation
Blackening of wooden panelling	Condensation over 70–75% of humidity can lead to biological growth. In Italy there is an alternation of wet and dry periods with sun	Blockage of gutters
	Anomalous condensation on frescoes	General increased in decay and particularly for sandstone buildings
		Climate change is a risk multiplier
		Increased decay on unroofed castles and monuments
		Decrease of freeze-thaw cycles

impacts from climate change on cultural heritage in comparison to the Mediterranean region. This could be investigated further by consulting a larger sample of stakeholders over a larger geographical area.

It was also found that the European projects mentioned above, i.e. Noah's Ark and CfC, were known mostly in academia and not amongst managers of cultural heritage. Moreover, the majority of interviewees that indicated considering climate change impacts in their decision-making did so using readily available and nationally produced climate change projections (for example, those produced by the United Kingdom Climate Impacts Programme (UKCIP) in the UK and the report Climate in Norway 2100 (Hanssen-Bauer et al. 2017)) rather than using the maps developed in projects such as CfC that estimated potential changes in the degradation of historical materials of cultural heritage resources on the basis of regional climate change projections. With regards to the impact of such project outputs, and to emphasise the need for improved communication, an interviewee mentioned the following:

"The problem is that these instruments should be known by the people that need to use them. Instead, a lot of times, those operating in the heritage sector do not know what has been developed and produced (at a research level). Who's guilty? The one who developed the product or the one who needs to use it? There is a link that is missing. We need to understand if those in charge of cultural heritage have this sensibility. Climate change is not considered and this is the major problem. There is a need for a connection between local realities and research centres, and this connection should be made by governmental institutions."

4.2   Projections of Climate Change Risks on Cultural Heritage

Figures 4, 5, 6 and 7 summarise the mechanical, chemical and biological risks projected by the CfC project in different locations in the three studied countries. The risks of climate change were estimated for two different building typologies: lightweight (i.e. wooden) and heavyweight (i.e. masonry) buildings for the baseline ((a) 1961–1990), the near future ((b) 2021–2050) and the far future ((c) 2071–2100). The ID numbers refer to the locations where climate change risks on cultural heritage are estimated in Norway (Fig. 1), Italy (Fig. 2) and the UK (Fig. 3). Six levels of decay were depicted using different colours to display a combination of the likelihood and the impact of the decay: very low (green), low (light-green), medium (yellow), medium-high (orange), high (red) and very high (dark red). The boundary value for each decay level was established according to the variables identified in the CfC project as described in Loli and Bertolin (2018, p. 6).

If we compare the risks projected by the CfC project with the issues of concern identified by the interviewees we can see that:

- The issues of concern identified by the stakeholders agree with the projected climate change risks depicted in the matrices, however, the matrices are more detailed in terms of information on the specific decay mechanisms and types, and linking this to specific locations. Accordingly, the matrices confirm the increase in biological degradation on Norwegian cultural heritage perceived by the interviewees. The matrices also show a decrease in mechanical risk due to a decrease in the number of freeze-thaw cycles and a change in the mechanical risk for salt crystallization for some parts of Norway. In Italy, the matrices classify an increase in decay for the wooden cultural heritage, showing a medium-high risk in some locations. Few changes were identified for masonry buildings as a result of a decrease in decay due to a reduction in the freeze-thaw cycles in the northern regions of Italy. In the UK, the interviewees recognized the increase in biological decay as indicated in the matrices with a high risk for wooden buildings and a high or medium-high risk for masonry buildings mainly by mould degradation. An increase in the chemical risk on cultural heritage was also identified for the UK. Moreover, the interviewees pinpointed an increase in mechanical degradation, that is confirmed by the matrices which show that UK cultural heritage is currently at risk of mechanical degradation, and that this risk will remain high in future.
- Some of the interviewees identified climate change risks for the more common building materials in their country, i.e. wooden buildings in Norway and stone buildings in the UK. For instance, Norwegian interviewees expressed awareness of the possible increase in decay of wooden built heritage, but they did not show

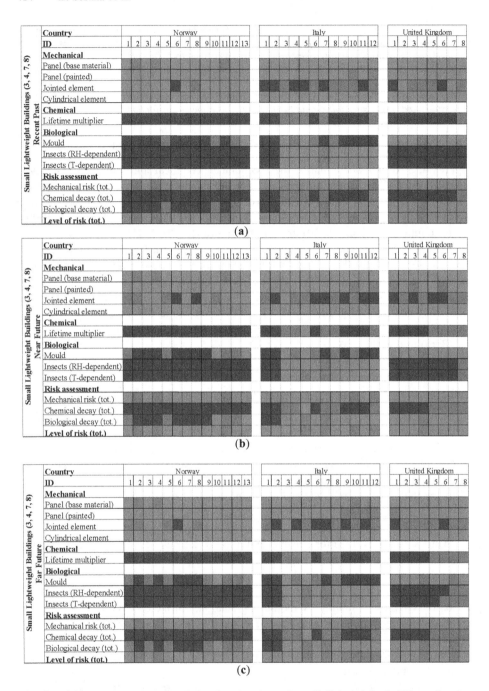

Fig. 4. Risk assessment matrix of the deterioration of small lightweight buildings for the: (**a**) Baseline (1961–1990); (**b**) Near Future (2021–2050); (**c**) Far Future (2071–2100). (Color figure online)

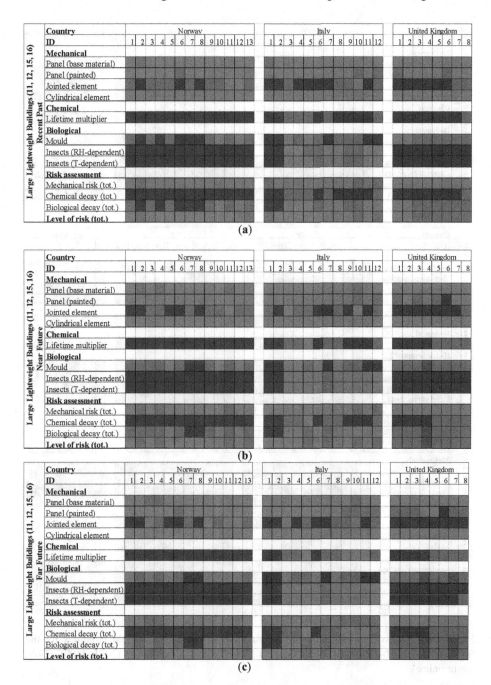

Fig. 5. Risk assessment matrix of the deterioration of large lightweight buildings for the: (**a**) Baseline (1961–1990); (**b**) Near Future (2021–2050); (**c**) Far Future (2071–2100). (Color figure online)

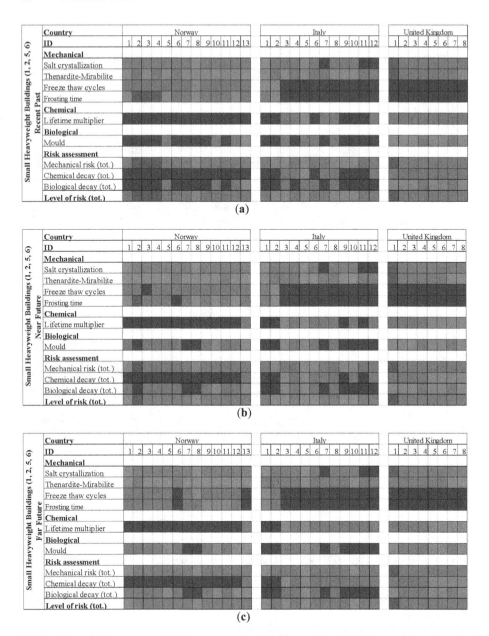

Fig. 6. Risk assessment matrix of the deterioration of small heavyweight buildings for the: (a) Baseline (1961–1990); (b) Near Future (2021–2050); (c) Far Future (2071–2100). (Color figure online)

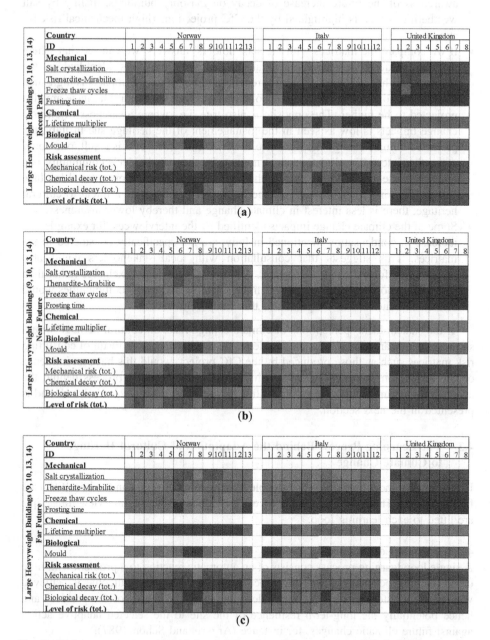

Fig. 7. Risk assessment matrix of the deterioration of large heavyweight buildings for the: (**a**) Baseline (1961–1990); (**b**) Near Future (2021–2050); (**c**) Far Future (2071–2100). (Color figure online)

awareness of the future increase of decay on masonry buildings, mainly by salt weathering, which is highlighted by the CfC projections (high mechanical risk for stone).

- If we compare the matrices with the stakeholders' answers, we can see that where the stakeholders' perceptions generally agree with the risks identified in the matrices and hence stakeholders' awareness of climate change impacts is high such as observed for the UK site, the risks of climate change on cultural heritage are also projected to be higher. The opposite is also true, i.e., where awareness of climate change impacts is low as seen in the management of the heritage site in Italy, the projected climate change risks are also lower. On the basis of this result, one could argue that interest in climate change impacts by the stakeholder's community is a reactive consequence to the current threats of climate change on cultural heritage and in locations where climate change is not yet perceived as a threat to cultural heritage, there is less interest in climate change and thereby lower awareness.
- Some of the climate change impacts identified by the interviewees, for example, the increase in condensation mentioned in Italy, are not specifically considered in the matrices. This suggests that consultation with local stakeholders could also potentially inform the risk assessment. A two-way knowledge exchange rather than a one-way knowledge transfer between the scientific and users' communities would clearly be beneficial as the results of this study demonstrate.

If the stakeholders' perceptions of climate change risks on cultural heritage would show greater awareness of what the problems are, for instance, as a result of consulting risk matrices such as those produced by the CfC project, would this have an influence on the adaptation process? To answer this question the learning process behind the adaptation measures and strategies identified by the interviewees was analysed, and is presented in the next section.

4.3   The Learning Process Behind the Adaptation of Cultural Heritage to Climate Change

In this section, models of single and double loop learning (Argyris and Schön 1987) are first presented. Then, the adaptation process as deduced from the interview transcripts are fitted to each learning loop.

4.3.1   Single Loop Learning Process

A single loop learning process consists of an automatic reaction to a problem with little or no learning occurring during the process (Fig. 8). The final outcome is achieved without taking steps to improve the understanding of the causes of the problem and hence potentially the long-term resilience of the site to the selected adaptive action against future climatic changes, for instance (Argyris and Schön 1987).

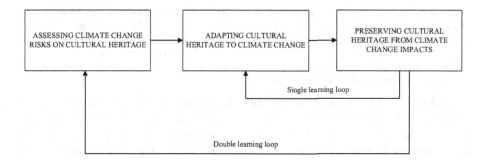

Fig. 8. Single and double learning loop applied to the preservation of cultural heritage from climate change impacts.

The existence of a single loop learning in the experts' responses to climate change impacts on heritage and the identification of adaptation solutions can be seen in the following quotes from selected interviewees:

"Here in Norway, to adapt cultural heritage located near the coast from sea level rising they moved the small groups of wooden houses to the internal land." (Interviewee, Academic)

"In Cesenatico the house of Marino Moretti on Porto Canale can be flooded and on the ground floor you cannot put any of the museum collections." (Interviewee, researcher)

"I am aware that in some cases they increased the capacity of gutters and downpipes, because they can be overwhelmed by the volume of water and then it overflows and (the water) can come into the building." (Interviewee, Member of governmental institution)

These are three examples of reactive adaptation measures adopted after hazardous events affected cultural heritage sites. The action adopted is a reaction to a specific impact, but the response does not involve a deeper consideration or research into the causes for the occurrence of the impact, and a longer-term planned response. In other words, a single loop learning process focuses on the management of the change rather than on the implementation of a long-term strategy.

4.3.2   Double Loop Learning Process

A double loop learning process refers to a rethinking of the current norm, rules and procedures (Fig. 8). This type of learning thus requires a certain degree of critical thinking in the identification of the best solution to a problem or to accomplish an objective. Argyris and Schön (1987) considers double loop learning as the best learning approach for addressing problems that can evolve with a change of circumstances.

The double learning loop involves the evolution of the operational schemes and theories behind the action. For example, this is illustrated in the following quote that expresses concerns with regards to planning for the impacts of climate change on a coastal archaeological site in Scotland:

"In the World Heritage Site of Skara Brae the effects of climate change are very well known: sea level rise, increased storminess. (...) There are a lot of issues there, trying to understand what is happening with the coastal erosion. (...) There is a hard seawall that did his job (of

protection) so far, but there are other questions. At the moment (...) we are studying what happens using laser scanning (...) trying to find out if the sea wall deflects the waves (...) and trying to pull all this information for a better understanding of what is happening so that we can make better plans to mitigate [the risks]. And that has been combined with an annual photographic survey with fixed points for a visual (record) as well as 3D modelling. " (Interviewee, Member of governmental institution).

Within the governmental organization in charge of the preservation of this archaeological site, Skara Brae, a risk assessment for understanding the specific impacts of natural hazards on the site has been developed. For example, the results of this assessment highlight that the site is at risk of groundwater flooding and of slope instability (HES 2018).

In this example, the adaptation measures to be adopted at the site are informed by projections of climate change risks and by monitoring the site through laser scanning and photographic surveys. This is a longer term adaptation planning process that can be correlated with a double loop learning process. The custodians of the site are not waiting for a disaster to befall the site. It indicates that they have prior knowledge of the likely outcomes of, in particular, extreme events. The learning process behind the collection of this information will be able to inform the adaptation process. The double loop learning process shows that new understanding on the possible climate change risks and vulnerabilities should inform future adaptation interventions. This learning approach may help those involved in cultural heritage preservation in planning preventive adaptation interventions. It might also be used as a re-thinking reactive approach after the occurrence of hazardous events.

5   Discussion and Conclusion

This paper investigated the perceptions and awareness of the cultural heritage community of the risks and impacts of climate change on cultural heritage. The learning process behind the identification and implementation of the adaptation measures adopted by cultural heritage managers to mitigate against specific climate change impacts was also examined, as a way, to assess the potential of scientific information and tools that project climate change risks on cultural heritage to support adaptation decision-making.

A number of interviewees showed awareness of some of the impacts of climate change on cultural heritage. The CfC matrices of risks on cultural heritage, which were developed on the basis of climate change projections from a RCM, or any other tools previously developed can be useful to inform cultural heritage managers or those in charge of cultural heritage preservation on the possible future changes in decay on cultural heritage resources under climate change, but, further efforts are required by the scientific community to disseminate those tools so that site managers can integrate them into cultural heritage management. Simulations such as those produced as part of the CfC project can support and improve effectiveness of adaptation practices. For example, giving heritage site managers quantitative data about the future rate of decay on groups of cultural heritage resources with similar characteristics (e.g. materials, dimensions) according to their locations. However, it should be emphasized that the risks identified and depicted in the figures presented in this paper are based on moderate GHG emission scenarios (and pathways). Given the current population growth projections and our continued reliance on fossil fuels as our main source of energy, the RCP8.5 worst-case emission scenario is becoming a more realistic trajectory, which would result in more drastic climatic changes and consequently a stronger magnitude of the risks and ensuing decay on cultural heritage sites.

Disseminating the outcomes of scientific research on the identification of the risks of climate change on cultural heritage (e.g. Noah's Ark and CfC projects or other developed risks assessments) can increase decision-makers' awareness on those issues and help in moving forward climate change adaptation. However, this paper identified a lack of communication between the academic and management sector. We believe that there is scope for better designing effective adaptation measures and strategies to preserve cultural heritage against climate change impacts by the application of double loop learning as described in the case of a heritage site in Orkney, Scotland. A double loop learning process can be used to implement preventive measures for the conservation of cultural heritage against the impacts of climate change. The use of this learning mechanism as a preventive measure by cultural heritage site managers (i.e. before a hazardous event occurs) can contribute to increasing the resilience of cultural heritage sites.

References

Argyris, C., Schön, D.: Organizational Learning: A Theory of Action Perspective. Addison Wesley, Reading (1987)

Bertolin, C., Camuffo, D.: Climate change impact on movable and immovable cultural heritage throughout Europe, Climate for Culture, Deliverable 5.2 (2014)

Brimblecombe, P., Grossi, C.M., Harris, I.: Climate change critical to cultural heritage. In: Gökçekus, H., Türker, U., Lamoreaux, J.W. (eds.) Survival and Sustainability: Environmental Concerns in the 21st Century, pp. 195–205. Springer, Heidelberg (2011). https://doi.org/10.1007/978-3-540-95991-5_20

Carmichael, B., et al.: Local and Indigenous management of climate change risks to archaeological sites. Mitig. Adapt. Strateg. Glob. Change 23, 231–255 (2017a)

Carmichael, B., Wilson, G., Namarnyilk, I., Nadji, S., Cahill, J., Bird, D.: Testing the scoping phase of a bottom-up planning guide designed to support Australian Indigenous rangers manage the impacts of climate change on cultural heritage sites. Local Environ. 22, 1197–1216 (2017b)

Casey, A.: Climate Change and Coastal Cultural Heritage: Insights from Three National Parks. Open Access Dissertations, University of Rhode Island (2018)

Cassar, J.: Climate change and archaeological sites: adaptation strategies. In: Lefèvre, R.-A., Sabbioni, C. (eds.) Cultural Heritage from Pollution to Climate Change. Edipuglia srl., Barri (2016)

Cassar, M.: Climate Change and the Historic Environment. Centre for Sustainable Heritage, University College London, London (2005)

Cassar, M., Pender, R.: The impact of climate change on cultural heritage: evidence and response. In: ICOM Committee for Conservation: 14th Triennial Meeting the Hague Preprints (2005)

Fatorić, S., Seekamp, E.: Evaluating a decision analytic approach to climate change adaptation of cultural resources along the Atlantic Coast of the United States. Land Use Policy **68**, 254–263 (2017a)

Fatorić, S., Seekamp, E.: Securing the future of cultural heritage by identifying barriers to and strategizing solutions for preservation under changing climate conditions. Sustainability **9**, 2143 (2017b)

Gray, S.R.J., et al.: Are coastal managers detecting the problem? Assessing stakeholder perception of climate vulnerability using fuzzy cognitive mapping. Ocean Coastal Manag. **94**, 74–89 (2014)

Grøntoft, T.: Climate change impact on building surfaces and façades. Int. J. Clim. Change Strateg. Manag. **3**, 374–385 (2011)

Hall, C.M.: Heritage, heritage tourism and climate change. J. Herit. Tour. **11**, 1–9 (2015)

Hall, C.M., Baird, T., James, M., Ram, Y.: Climate change and cultural heritage: conservation and heritage tourism in the Anthropocene. J. Herit. Tour. **11**, 10–24 (2015)

Hanssen-Bauer, I., et al.: Climate in Norway 2100 NCCS report no. 1/2017 Lead authors – a knowledge base for climate adaptation. Miljødirektoratet (2017). http://www.miljodirektor atet.no/M741/. Accessed 25 Sep 2018

Haugen, A., Mattsson, J.: Preparations for climate change's influences on cultural heritage. Int. J. Clim. Change Strateg. Manag. **3**, 386–401 (2011)

Heathcote, J., Fluck, H., Wiggins, M.: Predicting and adapting to climate change: challenges for the historic environment. Hist. Environ. Policy Pract. **8**, 89–100 (2017)

HES: A climate change risk assessment. screening for natural hazards to inform of the properties in care of historic environment Scotland. Historic Environment Scotland (2018)

IPCC: Summary for Policymakers, Emissions Scenarios, A Special Report of IPCC Working Group III, Intergovernmental Panel on Climate Change (2000). https://ipcc.ch/pdf/special-reports/spm/sres-en.pdf. Accessed 25 Sep 2018

IPCC: Climate Change 2014: Synthesis Report. Contribution of Working Groups I, II and III to the Fifth Assessment Report of the Intergovernmental Panel on Climate Change [Core Writing Team, Pachauri, R.K., Meyer, L.A. (eds.)]. IPCC, Geneva, Switzerland, 151 pp. (2014)

Leissner, J., et al.: Climate for Culture: assessing the impact of climate change on the future indoor climate in historic buildings using simulations. Herit. Sci. **3**, 38 (2015)

Loli, A., Bertolin, C.: Indoor multi-risk scenarios of climate change effects on building materials in Scandinavian countries. Geosciences **8**, 347 (2018)

McDonald-Madden, E., et al.: Active adaptive conservation of threatened species in the face of uncertainty. Ecol. Appl. **20**, 1476–1489 (2010)

Phillips, H.: Adaptation to climate change at UK world heritage sites: progress and challenges. Hist. Environ. Policy Pract. **5**, 288–299 (2014)

Pollard-Belsheim, A., Storey, M., Robinson, C., Bell, T.: The CARRA project: developing tools to help managers identify and respond to coastal hazard impacts on archaeological resources. IEEE (2014)

Sabbioni, C., Brimblecombe, P., Cassar, M.: The Atlas of Climate Change Impact on European Cultural Heritage. Scientific Analysis and Management Strategies. Anthem Press, London (2010)

Sabbioni, C., Cassar, M., Brimblecombe, P., Lefevre, R.A.: Vulnerability of cultural heritage to climate change. European and Mediterranean Major Hazards Agreement (EUR-OPA), Council of Europe, Strasbourg (2008)

Sesana, E., Gagnon, A., Bertolin, C., Hughes, J.: Adapting cultural heritage to climate change risks: perspectives of cultural heritage experts in Europe. Geosciences **8**, 305 (2018)

Tanner-Mcallister, S.L., Rhodes, J., Hockings, M.: Managing for climate change on protected areas: an adaptive management decision making framework. J. Environ. Manag. **204**, 510–518 (2017)

Thomson, A.M., et al.: RCP4.5: a pathway for stabilization of radiative forcing by 2100. Clim. Change **109**, 77 (2011). https://doi.org/10.1007/s10584-011-0151-4

Williams, B.K.: Passive and active adaptive management: approaches and an example. J. Environ. Manag. **92**, 1371–1378 (2011)

Conserving Sustainably the Materiality of Structures and Architectural Authenticity

Taxonomy of Architectural Styles and Movements Worldwide Since 8500 BC

Christos Floros[✉]

Architect NTUA, Athens, Greece
cfloros@windtools.gr

Abstract. An appropriate taxonomy of architectural styles and movements worldwide, since 8500 BC, is provided to be utilized as a basis for the digital documentation of buildings and other architectural constructions.

Architecture is classified in unities, so that each one of them is relatively autonomous. Each architectural style or movement corresponds to a particular culture (tribe, religion, ideology), to a particular geographic region (landscape, climate, available materials), to particular functional needs and demands, and to a particular technology.

The number of proposed unities is 119. The same classification criteria are applied upon all geographical areas and periods, so that the same scale is provided.

Keywords: Taxonomy · Architectural styles · Architectural movements · Historic buildings · Historic constructions

1 Introduction

I tried to find out the components of all aspects of architecture that always fascinated me, and their space-time and cultural interrelationships.

I am engaged in architecture for forty-five years, but at the beginning, I was centered upon the architecture of the western world. Thus, I had to proceed further, all over the globe, in the depth of eleven millenniums. However, if I spread in this vast space and time without the proper compass, I would have been lost.

I had to operate mathematically; to get away from details and to classify the essentials, retaining though equal scale on my approach to all the civilizations. I had to simplify the problems to be able to fully understand them, avoiding, though any simplifications that might distort their essence.

2 Research Aim

Books on the world history of architecture do not include all architectural unities. Some significant architectural creations, in remote places, are not popular and are presented only as case studies in specialised papers. Also, architecture is often classified according to dynasties and historical periods, without following architectural criteria.

© Springer Nature Switzerland AG 2019
A. Moropoulou et al. (Eds.): TMM_CH 2018, CCIS 961, pp. 427–448, 2019.
https://doi.org/10.1007/978-3-030-12957-6_30

We aim to organize through taxonomy the basic knowledge on all architectural creations worldwide, using architectural criteria, to provide digital documentation and assist anyone interested in the past of architecture.

A "periodic table of architecture", such as the periodic table of the chemical elements, is required, for the taxonomy of architectural creations in space, time and civilization, within a framework of Comparative Architecture. I tried to respond to this need, as I believe that there is no future without a past.

3   Methodology

Only surviving architectural creations are considered. Exceptionally, in the case of ancient civilizations where complete buildings did not survive, architectural ruins are taken into account, provided that they are sufficient to help us realize how the original building looked like.

3.1   Process in Successive Steps

Step 1: Investigating bibliography and every other source of information I could find, including personal visits, I recorded all architectural styles and movements that have existed all over the world.

Step 2: I classified them according to the major cultures where they belong.

Step 3: I investigated the various architectural characteristics of the buildings in each particular major culture. These are: function, layout, form, structural system, materials and decoration.

Step 4: Whenever the architectural characteristics of the buildings of the same culture, are significantly different, they are classified in subcategories (architectural unities). Differentiations might exist due to alternative functions, religion, ideology, landscape, climate, available materials and technological evolution from era to era.

3.2   Result Processing

The results of the above process led to 119 distinct architectural unities. In any case, the same classification criteria are applied upon all geographical areas and periods, so that the same scale is provided.

Considering that the architectural revolution has its own way, which does not lose its continuity according to the succession of dynasties, I did not follow the classification by dynasty, which is the most common classification of east Asia's art history and even in some European countries such as the United Kingdom. For example, Chinese architecture has not been divided into the 16 dynasties that lasted till 1911 but has been divided into 6 crucial sections (Pre-Imperial, Confucian-Taoist, Secular, Buddhist, Garden, Euro-Chinese).

The chronological order of the unities is based on their initiation time. The chronology of the beginning and the end of a time period refers to the architectural creation and does not necessarily coincide with the particular years, which mark the beginning and the end of the historical period.

3.3 Timeline

The text is accompanied by a timeline, where we can see all the architectural styles (or movements) that existed all over the world at any particular time.

The Timeline is organized in five horizontal rows for the continents: Europe, Asia, America, Africa, Oceania and islands of the Pacific Ocean.

4 Presentation of the Taxonomy

The 119 architectural unities are listed in chronological order, in nine successive periods, from 8500 BC till now, according to their initiation time.

After the name of each unity, the time of its beginning and end is shown in brackets, followed by the region where it was developed. The Characteristic Building Types of each unity are shown following the abbreviation CBT.

4.1 Period 8500 BC – 1700 BC

It was eleven millenniums ago, when our ancestors, who were food gatherers and hunters, started to become farmers, settling permanently in the same place. That was the terming point for the beginning of architecture. The first settlements were created in Asia, Europe and Africa in the era 8500 BC-4000 BC (see A1).

The construction of megalithic monuments in Europe (see A2) and in Asia (see A6) followed.

The first properly designed, complete, free-standing, megalithic buildings started to be constructed on the Islands of Malta and Gozo, about 3500 BC (see A3) and it was not till seven centuries later that the famous, magnificent, Egyptian architecture began to flourish (see A5).

In the 4th millennium BC in Mesopotamia (see A4) and in the 3rd millennium BC in Hindus Valley (see A8), the first cities were built, following large scale agricultural development that created food surplus.

In the 3rd millennium BC, on Crete and on the Cycladic Islands, the first naval civilization in the world, Cycladic-Minoan, developed and created an elaborated, functional architecture on a human scale (see A7).

In the 3rd millennium BC also, in the area of today's Peru, the earliest architecture in the American continent was created (see A9).

At the beginning of the 2nd millennium BC, following the Neolithic settlements in China, Chinese architecture started to develop its own character (see A10).

A1 Neolithic Settlements (8500–2000 BC) in Asia, Europe & Africa,
CBT: settlement, walled enclosure
A2 European Megalithic (4000–1500 BC) in Europe,
CBT: menhir, dolmen, stone circle, burial chamber
A3 Megalithic Malta Temples (3500–2500 BC) in Malta,
CBT: megalithic temple, rock-hewn hypogeum temple-tomb
A4 Mesopotamian (3500–540 BC) in Iraq & Iran,
CBT: ziggurat, temple, palace, fortification, settlement, house

A5 Egyptian (3100–30 BC) in Egypt,
CBT: pyramid, temple, tomb, rock-hewn tomb, obelisk, walled enclosure
A6 Asian Megalithic (3000–1000 BC) in the Korean Peninsula, China, Mongolia, Japan, Vietnam, Laos, India, Indonesia & Jordan,
CBT: menhir, stone circle, dolmen, burial shaft
A7 Cycladic and Minoan (2800–1450 BC) in Greece,
CBT: palace, administration center, settlement, villa, tomb, store, workshop
A8 Indus Valley (2750–1550 BC) in Pakistan & India,
CBT: settlement, communal, fortification, wheat store, bath
A9 Early Peruvian (2600–300 BC) in Peru,
CBT: artificial platform mound (sometimes pyramidal), ceremonial center, temple, altar, tomb
A10 Pre – Imperial Chinese (2000–221BC) in China,
CBT: walled enclosure, palace, workshop, ritual complex, tomb

4.2   Period 1700 BC – 650 BC

For one thousand years, from the 17th to the 7th century AD, the megalithic architecture of the Mycenaeans (see A11), the Hittites (see A13) and the Nuraghi and other Mediterranean people (see A14) developed in the area of the Mediterranean Sea, while the Egyptian architecture was still flourishing.

The architecture of the maritime civilization of the Phoenicians-Carthaginians (seeA16) developed on the Mediterranean coasts but left trivial architectural remains.

The architecture of the temples of the Jewish religion dates since the erection of the first Temple of Solomon (953 BC) and continuous up today, but it has not developed a particular architectural style (see A18).

A remarkable rock-hewn architecture developed mainly in west Asia, but also in south Europe; it lasted for centuries and created functional dwellings and religious spaces (see A20). In west Asia, the Mesopotamian architecture was still flourishing.

Early Japanese architecture (see A17) appeared in East Asia, while the Pre-Imperial architecture was lasting still in China.

Early Peruvian architecture was still lasting in South America. In Central America, Olmec architecture (see A15) was created around 1500 BC and Pre-Classic Mayan architecture (see A19) was created around 800 BC; they both founded the pre-Columbian architecture in Central America.

In this period, the earliest architecture surviving in North America, appeared in the Mississippi valley (see A12).

A11 Mycenaean (1700–1100 BC) in Greece, Cyprus & Syria,
CBT: fortification, megaron, settlement, funeral tholos, temple
A12 Early Mississippi (1700–700 BC) in USA,
CBT: artificial mound
A13 Hittite (1600–1180 BC) in Turkey & Syria,
CBT: fortification, palace, temple, sanctuary

A14 Nuraghe and similar structures in west Mediterranean Sea (1500–600 BC) in Italy, Spain & France,
CBT: tower, settlement, house, workshop, sacred well, tomb
A15 Olmec (1500–400 BC) in Mexico,
CBT: artificial mound, pyramid, ball court, ritual site, tomb
A16 Phoenician – Carthaginian (1500–146 BC) in Lebanon, Tunisia, Cyprus, Morocco, Algeria, Libya, Egypt, Israel, Syria, Turkey, Italy, Spain & Malta,
CBT: settlement, sanctuary, Tophet (children necropolis), cistern
A17 Early Japanese (1500 BC – 350 AD) in Japan,
CBT: house, shrine, fenced enclosure, store, kofun (burial mound)
A18 Synagogue (953 BC – today) all over the world,
CBT: synagogue
A19 Pre-Classic Mayan (800 BC– 300 AD) in Guatemala, Belize, El Salvador & Mexico,
CBT: artificial mound, temple, pyramid, tomb
A20 West Asia & South Europe Rock – Hewn (800 BC – 1350 AD) mainly in Jordan, Lebanon, Turkey, Armenia, Iran, Saudi Arabia, Bulgaria, Greece, Cyprus, Italy & France,
CBT: temple, church, monastery, house, tomb

4.3   Period 650 BC – 1 AD

This period is marked in the Western world by birth and development of the Greco-Roman civilization.

Greek architecture developed in three successive phases, the Archaic Greek (see A21), the Classical Greek (see A24) and the Hellenistic (see A26).

Etruscan architecture (see A22) was influenced by the Greek architecture and influenced the Roman architecture.

Roman architecture (see A33) emerged within the Hellenistic world and acquired a dominant role since 146 AD, when the Romans conquered and destroyed Carthage and subjugated Greece.

Achaemenids' civilization (see A23) and Parthian civilization (see A31) developed in west Asia; they were both influenced by the Greek architecture, but they developed their own identity.

Buddhism was born in India and the Early Indian Buddhist architecture was created (see A27).

The unification of China into a single empire was realized in 221 BC and so started the secular architecture of the Chinese Empire (see A32), which lasted up to the 20th century. Confucianism and Taoism and their architecture (see A25) developed also in China. The Korean secular architecture (see A35) started in the Korean Peninsula and lasted up to the 20th century. In Japan, the Early Japanese architecture was lasting still.

In Africa, the Meroitic architectural style developed in Sudan (see A28), under the influence of the Egyptian architecture. The African Megalithic architecture (see A30) appeared in the same period.

In Peru, Nazca & Moche architecture (see A29) succeeded the Early Peruvian architecture.

In Mexico, Teotihuacan was created, which was the greatest city in America until modern times. Very significant town planning and architecture developed there and evolved later in Xochicalco (see A34). In parallel, the Pre-Classic Mayan architecture was lasting still in Central America.

A21 Archaic Greek (650–480 BC) in Greece, Italy, Turkey & Libya,
CBT: temple, sanctuary, open theatre, stadium
A22 Etruscan (650–300 BC) in Italy,
CBT: tomb, temple, necropolis, house
A23 Achaemenids (550–330 BC) West Asia up to Indus River and Egypt and Libya,
CBT: palace, fire temple, fortress, tomb
A24 Classical Greek (480–330 BC) in Greece, Italy, Turkey & Libya,
CBT: Temple, Sanctuary, Treasury, Propelaea, Agora, Stoa, Open Theatre, Arsenal, Stadium, Fortificatio
A25 Confucian – Taoist (479 BC – 1911 AD) in China, S. & N. Korea & Vietnam,
CBT: Confucius Commemoration Temple, Monastery, Academy, Palace for oblation & sacrifice, Altar, Cubby for religious service, Sacred Garden, Hut, Citadel
A26 Hellenistic (330–30 BC) in Greece, Albania, F.Y.R.O.M., Bulgaria, Italy, north Africa and West Asia up to the Indus River,
CBT: temple, sanctuary, mausoleum, altar, agora, commercial building, stoa, open theatre, asclepeion, arsenal, stadium, library, fortification, house, necropolis
A27 Early Indian Buddhist (322 BC – 320 AD) in India, Pakistan, Nepal, Sri Lanka, Afghanistan & Bangladesh,
CBT: stupa, chaitya, vihara
A28 Meroitic (300 BC – 350 AD) in Sudan,
CBT: temple, pyramid
A29 Moche, Lima & Nazca (300 BC – 800 AD) in Peru,
CBT: pyramidal artificial mound, aqueduct
A30 African Megalithic (250 BC – 1550 AD) in Senegal, The Gambia, Ethiopia & Central African Republic,
CBT: menhir, obelisk, stone circle, burial chamber
A31 Parthian – Sassanid (240 BC – 651 AD) in Turkmenistan, Iraq, Iran and parts of its neighboring countries,
CBT: fortress, settlement, palace, fire temple, watermill
A32 Secular in the Chinese Empire (221 BC – 1911 AD) in China,
CBT: palace, mausoleum, tomb, commercial, observatory, housing complex, city walls
A33 Roman (146 BC – 565 AD) in the countries included in the Roman Empire at some time,
CBT: temple, palace, basilica, amphitheatre, forum, library, villa, fortification, triumph arch, bath, olive oil manufactory
A34 Teotihuacan & Xochicalco (50 BC – 900 AD) in Mexico,
CBT: temple, pyramid, tomb, palace, ball court

A35 Secular Korean (50 BC – 1910 AD) in S. & N. Korea,
CBT: palace, estate building, seowon (academy), observatory, house, fortress, rural
pavilion, garden, tomb

4.4   Period 1 AD – 450 AD

This period begins with the birth of Christ, the year 1 AD corresponding to the year
4239 of the Egyptian calendar, the year 3763 of the Jewish calendar, the year 3144 of
the Olmec calendar, the year 3100 of the Maya Long Count Calendar, the year 756 of
the Roman calendar, the year 752 of the Babylonian calendar and the year 547 of the
Buddhist calendar. The year 622 AD of the Christian calendar corresponds to year 1 of
the Islamic calendar.

Christianity took its first steps into the late Roman Empire, eastern Europe, western
Asia and northeastern Africa, despite successive persecutions. The Edict of Milan (313
AD) ensured religious freedom for Christians throughout Roman Empire and Christian
civilization started to flourish (see A36).

In India, Late Indian Buddhist architecture (see A40) succeeded the Early Indian
Buddhist architecture and Early Hindu architecture took its first steps (see A39).

Cham Architecture (see A41) started to develop in Southeast Asia, in the Hindu
kingdom of Champa, in nowadays southern and central Vietnam.

In Japan, the Main Secular Japanese architecture (see A43) succeeded the Early one
and Shinto architecture was created (see A42).

In Central America, the Classic Mayan architecture (see A38) succeeded their Pre-
Classic one.

Notable Megalithic architecture appeared on several islands of the Pacific Ocean
(see A37).

A36 Early Christian (50–527 AD) in Europe, West Asia & Northeast Africa,
CBT: church, martyrion, baptistery, monastery, catacomb
A37 Pacific Megalithic (200–1500 AD) in the Islands of the Pacific Ocean,
CBT: stone platform, elevated house, monolithic statue, fortification, step pyramid,
religious centre, tomb
A38 Classic Mayan (300–900 AD) in Mexico, Guatemala, Honduras, Belize & El
Salvador,
CBT: pyramid, temple, altar, palace, ball court, observatory, fort, house, tomb
A39 Early Hindu (320–700 AD) in India & Nepal,
CBT: temple, ratha
A40 Late Indian Buddhist (320–1450 AD) in India, Nepal, Bangladesh & Sri
Lanka,
CBT: stupa (buddhist funerary mound), chaitya (buddhist prayer hall), vihara
(monastery), tower-shaped temple
A41 Cham (350–1300 AD) in Vietnam,
CBT: temple, monastery
A42 Shinto (350–1912 AD) in Japan,
CBT: shrine, monastery

A43 Main Secular Japanese (350–1868 AD) in Japan,
CBT: house, farmhouse, palace, mausoleum, imperial villa, castle, tower, theatre, ageya (pleasure house), garden

4.5   Period 450 AD – 900 AD

This period is signified by the spread of Christianism, the birth and spread of Islamism and several crucial geopolitical events that created a great variety of architectural styles all over the world.

Christianity

Following the division of the Roman Empire into the Eastern and the Western (395), the Eastern evolved into the Byzantine Empire, which during the era of Justinian the Great (527–565) reached its maximum extent and created architectural masterpieces, such as Hagia Sophia (see A50).

The fall of the Western Roman Empire (476) created a total collapse, which marked the beginning of the Middle Ages. In the Middle Ages, a variety of vernacular architectural styles was created by the various local cultures (see A45).

At the end of the 6th century, in some western Europe territories, few Pre-Romanesque architectural styles started to emerge, resulting from the reminiscence of the glamour of the Roman Empire (see A53).

The Christianization of Ireland and Scotland created an affecting, austere monastic architecture (see A52).

Vikings developed their architecture in northern Europe, in the 9th century (see A62). Vikings were Christianized in 960.

Ethiopia started to be Christianized in the 6th century and it was then that the foundations for the creation of a unique architecture of rock-hewn churches were laid (see A54).

Islamism

Islam was an immense power that changed the course of history since the 7th century AD.

At this period, the Muslims dominated north Africa and most of western Asia and created the Islamic architecture (see A55).

In 711, the Muslims invaded from Africa, began conquering the Iberian Peninsula and developed the marvelous Moorish architecture that lasted till the "Reconquista", the reconquest by the Spaniards, which was accomplished in 1492 (see A59).

Asia

Additionally to the Islamic and to the Christian architecture, several other architectural styles were created in Asia.

Since the middle of the 1st millennium AD in the Arabian Peninsula, High-rise Mud Buildings were constructed (see A48).

In India, the Main Indian Hindu architecture (see A57) succeeded the Early one, while the Late Indian Buddhist architecture (see A40) was still lasting. Indian Tantrism, Jainism & Sikhism architecture emerged also in India (see A65).

Tibetan-Bhutanese Buddhist architecture (see A61) emerged in Tibet, the highest plateau on earth, and on the Himalayas.

In east Asia, Buddhism was introduced to China from India, and Buddhist architecture started to develop in China (see A44), Korea (see A49) and Japan (see A51).

In southeast Asia, Hindu and Buddhist architecture emerged in the Shailendra (see A60) and Khmer (see A64) kingdoms, creating marvelous monuments.

The idiosyncratic architecture of buildings on pilotis in Indonesia (see A63) is of a great interest.

Africa

Apart from the Islamic architecture that developed in North Africa, there was a notable architectural creation developed in sub-Saharan Africa.

Indigenous tribes and ethnicities created the Bantu architecture (see A58) that has a wide variety regarding function, form and building materials. The Berbers, those considered as "barbarians" by the Romans because they were living out of the borders of the Roman Empire, created both a notable free-standing architecture (see A58) and a rock-hewn architecture (see A54).

America

The Pueblo civilization, developed in North America, created its own notable architecture (see A56).

In Central America, the Classic Mayan architecture was still lasting (see A38).

In South America, the Tiwanaku empire emerged and created an important megalithic architecture (see A47), which is considered as a forerunner of Inca architecture, and on the Andes, in Colombia, an elaborated Tomb architecture developed (see A46).

A44 Chinese Buddhist (464 – 1911 AD) in China & Mongolia,
CBT: temple, pagoda, chaityas (buddhist prayer hall), monastery
A45 Medieval (476 – 1500 AD) in West and Central Europe,
CBT: fortification, settlement, castle, church, mansion, urban house, peasant house, town hall, hospital, workshop, tannery, windmill, watermill, store
A46 Andean Tombs (500–900 AD) in Colombia,
CBT: funeral dolmen, hypogea (underground tomb)
A47 Tiwanaku (500–1050 AD) in Bolivia, Chile, & Peru,
CBT: temple, sacred space, pyramidal artificial mound, sunken plaza, fortification
A48 Arabian Peninsula High-rise Mud Buildings (527– 1910 AD) in Yemen, Oman & Saudi Arabia,
CBT: palace, multi-storey house, fort
A49 Korean Buddhist (527– 1910 AD) in S. & N. Korea,
CBT: temple, pagoda, monastery
A50 Byzantine (527– 1453 AD) in today's countries that belonged to the Byzantine Empire and countries in Caucasus,
CBT: church, monastery, castle, megaron
A51 Japanese Buddhist (538 – 1868 AD) in Japan,
CBT: temple, pagoda, monastery

A52 Irish Celtic (540–1168 AD) in Ireland & Scotland,
CBT: monastery, round tower, ringfort
A53 Pre – Romanesque (570–914 AD) in Germany, France, Italy, Croatia, north Spain and their neighbouring areas,
CBT: church, monastery, castle
A54 Africa Rock-Hewn (600–1250 AD) in Ethiopia, Tunisia, Libya, Morocco and some countries in Sub-Saharan Africa,
CBT: bet (church), tomb, dwelling
A55 Early West Asia & North Africa Islamic (630–1260 AD) in the countries of West Asia, except West Turkey, up to the Persian area and all coastal Mediterranean countries of North Africa,
CBT: mosque, madrassa, palace, harem, mausoleum, castle, ribat (small fortification) ksour (fortified city), souk (market), mâristân (hospital), caravansary (inn for caravans), bath
A56 Pueblo (650–1525 AD) in USA & Mexico,
CBT: pueblo (settlement), castle, kiva (communal hall), house
A57 Main Indian Hindu (700–1858 AD) in India & Nepal,
CBT: temple, ratha, step well, palace
A58 Indigenous African tribes architecture – Bantu & Berber Architecture (700–1900 AD) Bantu in Zimbabwe, Nigeria, Ghana, Burkina Faso, Kenya, Uganda, Madagascar, Benin and other Sub-Saharan countries. Berber in Morocco, Algeria, Tunisia & Libya,
CBT: fortress, tower, settlement, house, ksar or gasr (fortified granary)
A59 Moorish (756 – 1492 AD) in Spain & Portugal,
CBT: mosque, madrasa, palace, castle, funduq (caravanserai), hammâm (bath), dâral-imâra (palace of a governor or ruler), mâristân (hospital), munya (villa), qaysâriyya (commercial building), ribât (small fortification), sûq (market)
A60 Shailendra (760–900 AD) in Indonesia,
CBT: temple, palace
A61 Tibetan – Bhutanese Buddhist (770 AD – today) in China (including Tibet) & Bhutan,
CBT: monastery, dzong, chorten, kumbum, palace
A62 Viking (800–1070 AD) in Norway, Sweden, Denmark, Scotland, Germany & Canada (Newfoundland),
CBT: tomb, settlement, house, boathouse, ring fort or Trelleborg
A63 Indonesian Buildings on Pilotis (800–1900 AD) in Indonesia,
CBT: dwelling, workspace, storehouse, granary, meeting space
A64 Khmer (802 – 1431 AD) in Cambodia, Thailand, Laos, Vietnam, Myanmar & Malaysia,
CBT: temple, wat (monastery with temple)
A65 Indian Tantrism, Jainism & Sikhism (850–1858 AD) in India & Pakistan,
CBT: temple, stambha

4.6   Period 900 AD – 1250 AD

The dominance of the Byzantine Empire was still lasting in eastern Europe and in a part of the western Asia, while the Russian Orthodox architecture was born in the state of Rus (see A71).

Romanesque architecture (see A72) developed in western Europe; Gothic architecture (see A75) emerged at the end of this period and its cathedrals continue to be city landmarks, till our days.

The architecture of the Scandinavian Stave Churches emerged in Norway and its neighboring areas (see A73).

The architecture of the Knights (see A74) was born in the Holy Lands and extended all over the Mediterranean area. The Venetian Colonial architecture emerged, on the Mediterranean coasts, at the end of this period (see A78).

Moorish architecture was still lasting in the Iberian peninsula (see A59).

Islam was flourishing and created the Islamic Persian architecture (see A68), which still adorns Isfahan and other cities of that area. At the end of this period, Islam reached to the Deccan Sultanates in India, contributing to the evolution of its architectural styles (see A76).

Late Buddhist and Main Hindu architecture were still lasting in India (see A40, A57) and Chinese Buddhist architecture (see A44) was still lasting in East Asia.

In Southeast Asia, Pagan Buddhist architecture developed in nowadays Myanmar (see A69), while the architectural creation in the Shailendra (see A60) and Khmer (see A64) kingdoms were still lasting.

A unique architecture of rock-hewn churches developed in Ethiopia (see A54).

In the Sudano-Sahelian zone of Africa, a marvelous mud architecture emerged, which created buildings that look like sculptures, molded by hand (see A70).

In North America, the architecture of the pre-Columbian Puebloans was still lasting (see A56).

In Central America, the latest period of Mayan civilization (see A66) succeeded its classic one.

In South America, in Peru, the Chimú sculptural architecture of adobe buildings was born (see A67), while at the end of this period, the austere, megalithic Inca architecture emerged (see A77).

A66 Post Classic Mayan (900–1524 AD) in Mexico & Guatemala,
CBT: hypostyle hall, fortification, temple, pyramid, observatory, castillo
A67 Chimú (900–1470 AD) in Peru,
CBT: ciudadela (walled town), settlement, palace, mausoleum, temple
A68 Islamic Persian (900–1760 AD) in Uzbekistan, Iran, Kazakhstan, Afghanistan & Pakistan,
CBT: mosque, madrassa, palace, mausoleum, castle, Tomb, souk (market), caravansary, bath, khurshu (bazaar complex)
A69 Pagan Buddhist (1000–1300 AD) in Myanmar,
CBT: temple, stupa, pagoda, monastery, library

A70 Sudano – Sahelian (1000–1900 AD) in Mali, Mauritania, Burkina Faso, Togo, Niger, & Ivory Coast,
CBT: mosque, house, store, granary, library, tomb
A71 Russian Orthodox (1010–1917 AD) in Russia, Ukraine, Belarus, Finland & Greece,
CBT: church, monastery
A72 Romanesque (1020–1200 AD) in West and Central Europe and in Italy,
CBT: church, baptistery, monastery, castle
A73 Scandinavian Stave Church (1070–1300 AD) in Norway,
CBT: stavkirke (wooden pillars church)
A74 Knights (1080–1700 AD) in Malta, Greece, Syria, Israel & Cyprus,
CBT: crac (castle), hospital, palace, auberge (residence of knights), church
A75 Gothic (1140–1500 AD) in West and Central Europe,
CBT: church, monastery, government, castle, palazzo, town hall
A76 Deccan Sultanates (1192–1690 AD) in India & Bangladesh,
CBT: mosque, mausoleum, palace
A77 Inca (1200–1533 AD) in Peru, Bolivia, Ecuador and areas of Chile and Argentina,
CBT: settlement, sanctuary, temple, palace, house, fortification
A78 Venetian Colonial (1204 – 1670 AD) in Greece, Slovenia, Croatia & Cyprus,
CBT: castle, government, arsenal, barracks, urban mansion, country villa, church

4.7   Period 1250 AD – 1550 AD

In eastern Europe, the Byzantine Empire came to its end, while the Orthodoxarchitecture was still lasting in Russia (see A71).

In western Europe, the stunning Gothic architecture was still lasting (see A75).

Yet, the most important evolution of the European civilization took place at the beginning of the 15th century, in Florence, where the Renaissance was born or, literally, the classical civilization, whose roots were still alive in Italy, was "reborn" (see A86).

In UK, at the end of this period, following the end of the War of the Roses, the Tudor Dynasty initiated a period of political stability and prosperity, within which the homonymous architectural style developed (see A87).

In the Iberian Peninsula, the Moorish architecture was still lasting (see A59).

In the Mediterranean area, the architecture of the Knights (see A74) and the Venetian Colonial architecture (see A78) were still lasting.

In west Asia and north Africa, the Islamic Mamluk Sultanate emerged and developed a functional robust architecture (see A80). Subsequently, the Ottoman Empire prevailed over this region and over the Balkans, until the early 20th century, and created functional buildings for all uses (see A84).

In Asia, the Islamic Persian architecture, the architecture of the Deccan Sultanates, the Late Indian Buddhist, the Main Hindu architecture, the Chinese Buddhist architecture, the Tibetan-Bhutanese Buddhist architecture and the Khmer architecture were still lasting.

In eastern Asia, Thai architecture (see A79), East Asia Islamic architecture (see A82) and Chinese Garden architecture (see A85) emerged.

In Africa, the Sudano-Sahelian architecture was still lasting (see A70), while in Maghreb, Late Islamic architecture (see A81) succeeded the Early one.

In North America, the Pueblo architecture was still lasting (see A56).

In Central America, Aztec architecture emerged (see A83), while Post Classic Mayan architecture was still lasting (see A66).

In South America, in Peru, Chimú architecture (see A67) and Inca architecture (see A77) were still lasting.

A79 Thai (1250–1900 AD) in Thailand,
CBT: wat (temple complex) consisting of the phutthawat and of the sanghawat, palace

A80 Mamluk (1260–1517 AD) in Egypt, Syria, Lebanon, Israel, Jordan and parts of Saudi Arabia and Libya,
CBT: mosque, madrassa, palace, mausoleum, castle, ribat (small fortification), wikala (market complex), hospital, bath

A81 Late Maghreb Islamic (1260–1900 AD) in Morocco, Algeria, Tunisia & Libya,
CBT: mosque, madrassa, palace, mausoleum, castle, ksar (fortified settlement) souk (market), caravansary (inn for caravans), workshop, barn, stable, bath, dar al-magana (clockhouse)

A82 East Asia Islamic (1300–1900 AD) in Islamic countries and countries with considerable Islamic population in East Asia,
CBT: mosque, madrassa

A83 Aztec (1325 – 1520 AD) in Mexico,
CBT: pyramid-temple, shrine-altar, palace, house, steam bath, ball court, market

A84 Ottoman (1326 – 1923 AD) in the countries in West Asia, up to the Persian area, the countries of the Balkan peninsula and the countries of North Africa excluding Morocco,
CBT: mosque, madrasa, palace, mausoleum, castle, barracks, souk (market), hospital, caravansary (inn for caravans), hammâm (bath), house

A85 Chinese Garden (1368– 1911 AD) in China,
CBT: garden, summer palace, temple, pavilion

A86 Renaissance (1420–1620 AD) in taly and in countries of west and central Europe and colonies,
CBT: church, monastery, palazzo, chateau, castle, hospital, villa, town hall, government, library, theatre, opera

A87 Tudor (1485 – 1558 AD) in UK,
CBT: castle, palace, government, church, monastery, college, market hall, hospital, mansion, house

4.8   Period 1550 AD – 1800 AD

Following the discovery and conquest of the American continent and the establishment of colonies on all the continents, by European powers, an early globalization began, centered on Europe.

At the end of this period, the English Renaissance architectural style (see A91) developed in Europe, in parallel with the final period of Renaissance architecture, till the beginning of the 19th century.

In the 17th century, Baroque architecture (see A92) and the Palladian architectural style (see A93) developed in Europe and spread abroad. In the 18th century, the Gothic architectural style revived (see A95) and both Rococo architectural style (see A96) and Neoclassical architecture (see A97) were created. Neoclassical architecture became international architecture and lasted till the beginning of the 20th century.

In the Ottoman Empire, its architecture was still lasting.

Spanish Colonial architecture (see A89) was created in this period and lasted till the end of the 19th century.

Mughal architecture (see A88) developed in India and created masterpieces, such as Taj Mahal.

Euro-Chinese architecture (see A90) was created in East Asia, as a product of cultural exchange between European and Chinese civilizations.

The idiosyncratic colonial Cape Dutch architectural style (see A94) developed in South Africa, while Late Islamic architecture was still lasting in Maghreb.

Thai architecture, East Asia Islamic architecture and Chinese Garden architecture were still lasting.

Revival of Islamic architectural styles (see A98) emerged in Islamic states and in several other states having Islamic population.

A88 Mughal (1550–1857 AD) in India, Pakistan, Afghanistan & Bangladesh,
CBT: mausoleum, temple, palace, fort
A89 Spanish Colonial (1550–1898 AD) in America,
CBT: church, castle, palacio (mansion), hacienda (farm), casa (house)
A90 Euro – Chinese (1550–1949 AD) in China, Vietnam, Philippines & Laos,
CBT: market, theatre, opera, museum, library, hotel, government, mansion, villa, house, church, temple, fortress
A91 English Renaissance (1558–1625 AD) in UK & Ireland,
CBT: mansion, palace, castle, government, church, monastery, college, market hall, hospital, house
A92 Baroque (1600–1800 AD) in Europe and America and in few countries in Asia,
CBT: church, palace, palazzo, chateau, castle, villa, town hall, government, library, theatre, opera, guildhall, hotel, monastery
A93 Palladian (1614–1800 AD) in west Europe and in North America,
CBT: government, college, market hall, hospital, mansion, house
A94 Cape Dutch (1650–1850 AD) in the Republic of South Africa,
CBT: urban house, farmhouse, winery, church
A95 Gothic Revival (1700–1900 AD) in West Europe,
CBT: château, government, church, college, theatre, opera, house, hotel
A96 Rococo (1710–1800 AD) in Europe,
CBT: church, schloss, palazzo (palace), chateau, castle, villa, town hall, government, library, theatre, opera, hotel

A97 Neoclassical (1750–1920 AD) in countries in Europe and in North America and Australia,
CBT: all building types
A98 Islamic Revival (1760 AD – today) in Islamic countries and countries with considerable Islamic population in Asia and Africa,
CBT: mosque, madrassa, palace, mausoleum, souk (market), caravansary (inn for caravans), bath

4.9 Period 1800 AD – Today

In the 19th and 20th centuries, architecture stops looking backwards to copy the past and looks forward, with only few exceptions, the Eclectic (A99), Beaux-Arts (A100), Shingle (A102) and Edwardian (A109) and probably the Post-Modern (A118).

Many architectural movements, all born in Western Europe and USA, tried to express the identity of the new west civilization. These movements started in the 19th and at the beginning of the 20th centuries (see A101, A103, A104, A105, A106, A107, A108, A111, A112).

However, the new identity was expressed more successfully in that era, by the Early Modern Movement (see A110).

Following the Early Modern Movement, the Interwar Modern Movement (see A113) dominated over all the other architectural movements, which emerged till the Second World War and founded the architecture of the second half of the 20th century that is still lasting. The International Modern Architecture (see A114), the architecture of globalization, is its own legitimate child. The Regional Modern (A115) and the Eco Architecture (A117) are its children, who revolted towards right directions. High Tech Architecture (A116) developed within the framework of International Modern Architecture.

Post-Modern Architecture (A118) made a revolution that did not last long.

Deconstructivist architecture (A119) is a revolution, which applies solely to prestige buildings, till our days.

A99 Eclectic (1820–1920 AD) all over the world,
CBT: all building types
A100 Beaux – Arts (1840–1920 AD) in Europe, America & Australia,
CBT: opera, theatre, mansion, government, railway station, office building
A101 Industrial Style (1840–1920 AD) mainly in the most technologically advanced countries in Europe,
CBT: exhibition hall, greenhouse, railway station, multifunctional hall, shopping gallery, factory
A102 Shingle (1870–1910 AD) in USA,
CBT: house, communal building

A103 Chicago School (1879–1900 AD) in USA & Canada,
CBT: office, commercial, hotel
A104 Arts & Crafts (1880–1920 AD) in UK, Ireland, USA, Canada, in west and north Europe and in English colonies,
CBT: house, college, culture building, church
A105 Prairie (1893–1920 AD) in USA,
CBT: house, welfare building
A106 Art Nouveau (1893–1920 AD) mainly in Europe, but also in America and Australia,
CBT: all building types
A107 Sezession (1897–1920 AD) in Austria, Belgium & Germany,
CBT: house, public, railway station
A108 Catalan Art Nouveau (1900–1920 AD) in Spain,
CBT: church, villa, apartment building, college, hospital, culture building, park
A109 Edwardian (1900–1920 AD) in countries within the former British empire,
CBT: mansion, house, government, town hall, public
A110 Early Modern (1900–1920 AD) in West Europe and USA,
CBT: house, office, commercial, industrial
A111 Russian Constructivism (1920–1931 AD) in Russia (Soviet Union at that time),
CBT: public, house
A112 Art Deco (1920–1939 AD) all over the world, but mainly in West Europe and USA,
CBT: all building types
A113 Interwar Modern (1920–1939 AD) mainly in West Europe and USA,
CBT: all building types
A114 International Modern (1945 AD – today) all over the world,
CBT: all building types
A115 Regional Modern (1945 AD - today) all over the world,
CBT: all building types
A116 High Tech (1965 AD – today) all over the world,
CBT: all building types
A117 Eco (1968 AD – today) all over the world,
CBT: all building types
A118 Post – Modern (1975 – 2000 AD) in Europe, America, Asia & Oceania,
CBT: all building types
A119 Deconstructivist (1975 – today) in Europe, America, Asia & Oceania,
CBT: all building types

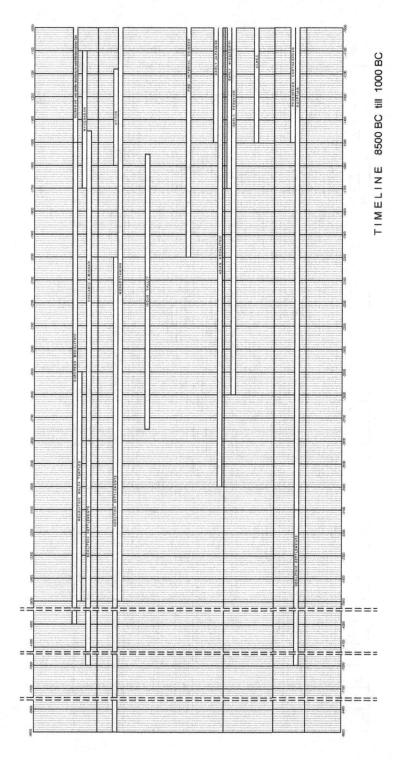

TIMELINE 8500 BC till 1000 BC

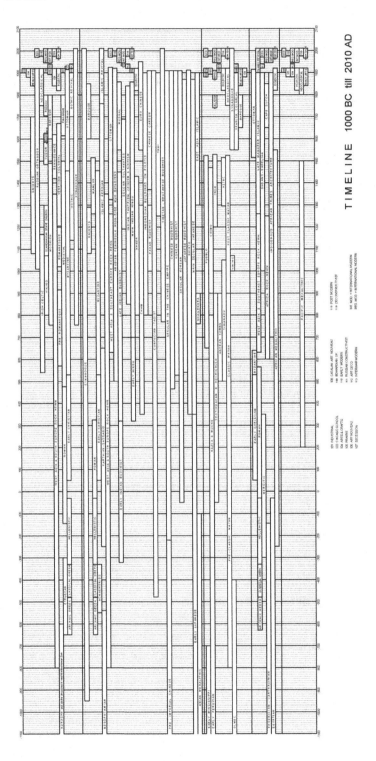

T I M E L I N E   1000 BC till 2010 AD

5 Conclusions – Achievements and Limitations

This classification can be utilized as a basis for the digital documentation of buildings and other architectural constructions. Architectural monuments included on the UNESCO World Heritage List are the first to be classified in the proposed unities.

It can be used for organising lessons for students of architecture and history of cultures and for providing organised information for those interested to visit foreign countries.

Investigating the attached timeline, several findings may arise; e.g. noticing that today exist only 8 unities worldwide, while in 1100 BC existed 28 unities, since in our globalization era pluralism is denied.

This research can develop to subdivide each one of the 119 unities into sub-unities, wherever a more detailed classification is required. Future research could add unities for any architectural creations that might have been omitted.

Indicative Bibliography

Alfieri, B.M.: Islamic Architecture of the Indian Subcontinent. Laurence King Publishing, London (2000)

Auboyer, J., Goepper, R.: The Oriental World. Paul Hamlyn Ltd., London (1967)

Baines, J., Malek, J.: Atlas of Ancient Egypt. Time-Life Books, Amsterdam (1987)

Bandinelli, B.R., Giuliano, A.: Etruschi e Italici prima del dominio di Roma. BUR Biblioteca Univ. Rizzoli, Milano (2000)

Bandinelli, B.R.: Rome, the late empire: Roman art AD 200–400, G. Braziller (1971)

Bandinelli, B.R.: Rome: The Center of Power, 500 B.C. to A.D. 200, G. Braziller (1970)

Banham, R.: Age of the Masters. Architectural Press, England (1975)

Barral i Altet, X.: The Early Middle Ages, From Late Antiquity to A.D. 1000, Taschen, Köln (1997)

Barral i Altet, X.: The Romanesque - Towns, Cathedrals and Monasteries, Taschen, Köln (1998)

Barrucand, M., Bednorz, A.: Moorish Architecture in Andalusia. Taschen, Köln (2002)

Baudez, C.-F., Becquelin, P., Mayas, L.: Collection L'Univers des Formes, Gallimard, Paris (1984)

Bernal, I., Simoni-Abbat, M., Le Monde précolombien: Le Mexique des origines aux Aztèques, Gallimard, Paris (1986)

Binding, G.: High Gothic - The Age of the Great Cathedrals. Taschen, Köln (1999)

Biris, M., Kardamitsi-Adami, M.: Neoclassical Architecture in Greece. Melissa Publishing House, Athens (2004)

Blake, S.: Half the World: The Social Architecture of Safavid Isfahan, 1590–1722. Mazda Pub, Costa Mesa (1999)

Bloom, J., Blair, S.: The Art and Architecture of Islam: 1250–1800. Yale University Press, New Haven and London (1994)

Bonanno, A., Malta: An Archaeological Paradise, OTS, Malta (1997)

Calloway, S., Cromley, E., Powers, A. (eds.): The Elements of Style: An Encyclopedia of Domestic Architectural Detail. M. Beazley Publishers (1991)

Casson, L.: Ancient Egypt. Time Inc., USA (1965)

Charbonneaux, J., Martin, R., Villard, Fr.: Greek Art, 3 volumes (Archaic Greek Art, Classical Greek Art, Hellenistic Art), London, Thames and Hudson (1971–1973)

Chéhab, M.H., Moscati, S., Parrot, A.: Les Phéniciens - L'expansion Phénicienne Carthage. Collection L'Univers des Formes, Gallimard, Paris (2007)

Chihara, D.: Hindu-Buddhist Architecture in Southeast Asia, E. J. Brill (1996)

Ching, F., Jarzombek, M., Prakash, V.: A Global History of Architecture. Wiley, New Jersey (2007)

Coe, M.D., Koontz, R.: Mexico - From the Olmecs to the Aztecs. Thames & Hudson, London (2002)

Copplestone, T. (ed.): World Architecture. The Hamlyn Publishing Group Ltd., Verona (1968)

Crouch, D.P., Johnson, J.G.: Traditions in Architecture - Africa, America, Asia and Oceania. Oxford University Press, New York (2001)

Demargne, P.: Aegean Art - The Origins of Greek Art. Thames & Hudson, London (1964)

Duval, P.-M.: Les Celtes, Collection L'Univers des Formes. Gallimard, Paris (1977)

Evers, B., et al.: Architectural Theory from the Renaissance to the present, Taschen (2006)

Garbini, G.: The Ancient World. Paul Hamlyn Ltd., London (1966)

Gombrich, E.H.: A Little History of the World. Yale University Press (2005)

Gombrich, E.H.: The Story of Art. Phaidon Press Ltd., London (1995)

Gössel, P., Leuthäuser, G.: Architecture in the Twentieth Century. Taschen, Köln (1991)

Grube, E.J.: The World of Islam. Paul Hamlyn Ltd., London (1966)

Gympel, J.: The Story of Architecture from Antiquity to the Present. Könemann, Köln (1996)

Hagen, V., Wolfgang, V.: The Desert Kingdoms of Peru. New American Library, New York (1968)

Hardoy, J.: Urban Planning in Pre-Columbian America. George Braziller, Netherlands (1968)

Jencks, Ch.: Modern Movements in Architecture. Penguin Books, London (1973)

Jencks, Ch.: Late - Modern Architecture. Academy Editions, London (1980)

Jencks, Ch.: The Language of Post-Modern Architecture. Rizioli, New York (1991)

Kanellopoulos, P.: The History of the European Spirit (in Greek), Athens (1976)

Kidson, P.: The Medieval World. Paul Hamlyn Ltd., London (1967)

Kitson, M.: The Age of Baroque. Paul Hamlyn Ltd., London (1966)

Krautheimer, R.: Early Christian and Byzantine Architecture. Penguin Books Ltd., England (1979)

Kubler, G.: The Art and Architecture of Ancient America - The Mexican, Maya, and Andean Peoples. Yale University Press, New Haven (1990)

Lassus, J.: The Early Christian and Byzantine World. Paul Hamlyn Ltd., London (1967)

Lavallee, D., Lumbreras, L.G.: Les Andes. De la Préhistoire aux Incas, L'univers des formes, Paris, Gallimard (1985)

Leclant, J. (ed.): Le temps des pyramides, L'univers des formes. Gallimard, Paris (1974)

Leroux-Dhuys, J.F.: Cistercian Abbeys- History and Architecture. Könemann, Cologne (1998)

Lommel, A.: Prehistoric & Primitive Man. Paul Hamlyn Ltd., London (1966)

Lynton, N.: The Modern World. Paul Hamlyn Ltd., London (1965)

Martindale, A.: Man and the Renaissance. Paul Hamlyn Ltd., London (1966)

Michell, G.: Architecture of the Islamic World - Its History and Social Meaning. Thames & Hudson, New York (1984)

Mignot, C.: Architecture of the 19th Century. Taschen, Köln (1994)

Norwich, J.J. (ed.): The Great Cities in History. Thames & Hudson, London (2009)

Onians, J. (ed.): The Art Atlas. Laurence King Publishing, Singapore (2008)

Parrot, A.: The Arts of Assyria, The Arts of Mankind. Gallimard, Paris (1961a)

Parrot, A.: Sumer: The Dawn of Art, The Arts of Mankind. Gallimard, Paris (1961b)

Pevsner, N.: A History of Building Types. Thames & Hudson, London (1979)

Ponting, C.: A New Green History of the World. Penguin Paperbacks, New York (2007)

Robinson, F.: Atlas of the Islamic World since 1500. Time-Life Books, Amsterdam (1991)

Scoufopoulos, N.C.: Mycenaean Citadels, Gothenburg. P. Astrom, Sweden (1971)

Schiltz, V.: Les Scythes et les nomades des steppes, L'univers des formes. Gallimard, Paris (1994)

Spawforth, T.: The Complete Greek Temples. Thames & Hudson, London (2006)

Steinhardt-Shatzman, N.: Chinese Architecture. Yale University Press, New World Press, Beijing, New Haven (2002)

Stierlin, H.: Islam, Volume I - Early Architecture from Baghdad to Cordoba, Taschen, Köln (1996)

Stierlin, H.: The Roman Empire, vol. I. Taschen, Köln (1996)

Stierlin, H.: Greece - From Mycenae to the Parthenon. Taschen, Köln (1997)

Stierlin, H.: The Maya – Palaces and Pyramids of the Rainforest Taschen, Köln (1997)

Stierlin, H.: Turkey - From the Seljuks to the Ottomans. Taschen, Köln (1998a)

Stierlin, H.: Hindu India - From Khajuraho to the Temple City of Madurai. Taschen, Köln (1998b)

Strong, D.: The Classical World. Paul Hamlyn Ltd., London (1965)

Tadgell, C.: The History of Architecture in India - From the Dawn of Civilization to the End of the Raj. Architecture Design and Technology Press, London (1990)

Taylour, W.L.: The Mycenaeans. Thames & Hudson, London (1964)

Thapar, R.: Early India – From the Origins to AD 1300. University of California Press, Berkeley and Los Angeles (2002)

The Architectural Review, No1191, May 1996, The First 100 Years (1896–1996)

Tietz, J.: The Story of Architecture of the 20th Century. Könemann, Cologne (1999)

Tournikiotis, P. (ed.): The Parthenon: And Its Impact in Modern Times. Melissa Publishing House, Athens (1994)

Trachtenberg, M., Hyman, I.: Architecture - From Prehistory to Postmodernity. Harry N. Abrams, New York (2002)

Trump, D.H.: Malta - Prehistory and Temples. Midsea Books, Italy (2002)

UNESCO, History of Humanity, Scientific and Cultural Development, UNESCO Publishing (1994)

Vellinga, M., Oliver, P., Bridge, A.: Atlas of Vernacular Architecture of the World, Abingdon (2007)

Vialou, D., La Préhistoire, L'univers des formes, Gallimard, Paris (2006)

Ten Vitruvius, The: Books on Architecture. Dover Publications, New York (1960)

Wildung, D.: Egypt - From Prehistory to the Romans. Taschen, Köln (1997)

Willetts, R.F.: The Civilization of Ancient Crete. University of California Press, Berkeley and Los Angeles (1976)

Wines, J.: Green Architecture. Taschen, Köln (2000)

Online Resources

http://architecture.about.com/od/periodsstyles/Periods_and_Styles.htm
http://library.thinkquest.org/16545/data/low/time.htm
http://odysseus.culture.gr/index_gr.html
http://whc.unesco.org/en/list
http://www.archnet.org
http://www.brynmawr.edu/Acads/Cities/wld/wcapts1.html#etr
http://www.culturopedia.com/Architecture/architecture.html
http://www.eie.gr/archaeologia/gr/index.aspx
http://www.essential-architecture.com/MISC/MISC-hist.htm

http://www.greatbuildings.com
http://www.historyworld.net/wrldhis/PlainTextHistories.asp?historyid=ab27
http://www.indiapicks.com/annapurna/B_Buddhist.htm
http://www.metmuseum.org/toah
http://www.mcah.columbia.edu/ma/htm/ma_home.htm
http://www.orientalarchitecture.com/

Monastery of Kimisis Theotokou, Valtessiniko, Arcadia, Greece - Restoration of the Temple and Integration of New Structures

Vobiri Julia[✉]

NTUA, Athens, Greece
vobirijulia@yahoo.gr

The presentation concerns the restoration of the temple of Kimisis Theotokou and the older cells of the monastery, as well as the integration of the new buildings into the surrounding area.

Today the Monastery consists of the temple, the old cells and the cells built in 1956. The purpose of this project has been to bring out the historical value of the Monastery as well as the search for elements that will help with the integration of new structures.

Some historical information about the Monument

The exact period during which the monastery was founded is unknown. According to T. Gartsopoulos, it was built during the mid 17th century. The monastery underwent many renovations, so we can't be sure of the original condition of the buildings it consisted of. Nowadays, we only have information about the temple and the buildings that we see at the entrance of the monastery, which were the old cells. As far as this building is concerned, only the ground floor still remains, from the first period of construction. The upper floor has been replaced by a contemporary incompatible construction.

The monastery has been listed as a historic monument since 1962.

Description of the monastery as a whole

The monastery has a perimeter wall which initially consisted of drystone walls, that have since been reinforced. The compound has two entrances, the main one, which is situated next to the old cells as well as a secondary one. The floor of the court consists of compressed soil.

Main temple

The main temple of the monastery, probably built in mid 17th century, is dedicated to the Dormition of the Mother of God. It is a building whose internal dimensions are approximately 11,30 × 7,75 m, and the entrance, with a height of 1,20 m, is situated on the south side of the temple. Today, the temple has the form of a three-nave basilica, but we can't reach safe conclusions as to the original shape of the temple. The area of the sanctum is not connected internally with the area of the sacristy. The main nave of the temple has a width of 3,0 m, the left one is 1,25 m wide and the right one is 1,15. The naves are connected with pairs of arches, which are supported by pairs of columns with approximate dimensions of 0,45 × 0,40 m. The arches on the east are based on the separating walls of the sanctum and the western ones are based on fairly long pillars

© Springer Nature Switzerland AG 2019
A. Moropoulou et al. (Eds.): TMM_CH 2018, CCIS 961, pp. 449–455, 2019.
https://doi.org/10.1007/978-3-030-12957-6_31

(0,50 × 1,70 m). The separation of the naves and the arrangement of the sanctum are the elements that make the temple unusual. Most of the interior of the temple is decorated with murals by an anonymous folk artist and a large portion of the murals has been destroyed.

Old cells
To the west of the entrance and south of the main temple, there is a building which today houses the basic functions of the monastery. There used to be a two story structure that housed the cells, but the floor was demolished when it suffered extensive damage because of an earthquake. The new structure consists of two floors and the top floor is constructed with reinforced concrete.

The cells of 1965
The 1965 earthquake created cracks in the original stone cells, which were deemed unsafe. The building is abandoned and the new stone cells are built to the south of the original ones. Originally, they consisted of two elongated buildings, with their long sidesadjoined. During the writing of this paper, the cells were under repair.

History of Reconstructions
Through the analysis and documentation process, the following information has emerged about the interventions that took place in the monastery:
   Early 18th century: Completion of the murals of the temple
   1736: Renovation of the monastery
   1773: New renovation of the monastery due to damage caused during the Orlof revolution.
   1823: Another renovation of the monastery
   1826: Ibrahim burns the monastery. Due to the fire, news repairs took place.
   1833: Testimony concerning the coated exterior of the temple
   1939: Testimony concerning the poor state of the murals. Murals partially destroyed.
   1950s: After the war, the temple had several problems which were repaired by the two nuns of the monastery. In photographs of the time, the temple seems to be coated externally with cement mortar.
   1955: The Restoration Department of the Ministry approved the repairs and completion of the murals by Ag. Karitzi, under the supervision of F. Kontoglou.
   1956: An earthquake caused extensive damage to the cells. The building is abandoned and new stone cells were built at the west side of the original ones.
   1957: The roof is repaired, using the original tiles. During the repair, the roof is placed 60 cm higher and the wall is also raised. At the same time, a cement roof was built on the south side of the monument.
   1964: Construction of bell tower. Later it was demolished as it put pressure on the masonry and was the main cause of water coming into the temple, from the roof.
   1986: The floor of the old cells, which was deemed inappropriate, was demolished, the ground floor was repaired and a new concrete floor was constructed. This is the current condition of the building.
   1990: Maintenance and cleaning of murals

1990s: The outer layer of plaster is removed from the walls, the unstable wall on the south side is replaced and the walls undergo mild repairs with exterior grouting, using cement and sea sand.

Pathology and Diagnosis

Main Temple

The main temple's problems can be separated in four main categories:

Structural problems:

The building as a whole did not face serious structural problems. But localized damages and material failures, although not posing an immediate danger, needed careful monitoring, explanation and future treatment, so that they would not deteriorate even more. The most serious structural problems consisted of cracks which were due to a combination of seismic activity. In the northeastern corner of the sanctum's alcove there were indications that there might be some differential collapse. This suspicion is caused by a crack with an upwards direction. Also, in the outer side of the sanctum there is a well that leads us to assume that the ground in that side might not be sufficiently stable. After surveying with the traditional method, this indication was not confirmed.

Construction problems:

Construction problems were located in the building elements of the monument, but they were not an immediate threat to its stability. Some of these are the following:

(a) An extensive use of grouting with cement, which played an important role in the appearance of humidity issues.
(b) Extensive upward humidity.

Aesthetic problems:

The monument suffered aesthetically because of the newer interventions, such as the use of contemporary cement grouting, which was the main reason for altering the original walls, but it is also due to the new roof. The alteration is so apparent that by looking at the monument externally, its age cannot by assessed. The more recent interior coating reduces the monument's aesthetic value. Also, the overwhelming presence of electrical equipment and wires spoil the monument's aesthetic value. Finally, the northern face's cement buttress complete a picture of lower aesthetics caused by the newer interventions.

Problems of highlighting the monument's historical value:

The above problems contributed to decreasing the artistic and historical value of the monument and of the murals, which face an immediate danger of being completely destroyed.

Old cells

The old cells mainly had problems with very serious aesthetic alterations. No part of the existing building represented the original phase of the building. The top floor was completely new and the old masonry wall of the ground floor was covered by newer coating and cement. Also, the windows seemed to be imitations of the old ones, but the result was not satisfactory. Apart from that, since the building underwent recent repairs and changes, there were no structural and construction problems.

Cells of 1956

The newer cells had extensive structural and construction problems. Specifically, only the original outer stone wall remained, which had to be reinforced, since the connecting cement was for the most part broken. Also, there was no roof and sills, resulting in the walls being even more exposed to the elements.

All the cells, old and newer ones, did not have any particular historical or scientific value. They were deemed to be humble secondary buildings which form a whole, along with the main temple.

Proposal

Goals of the restoration:

- To face structural and construction problems, to remove the hazards caused by their presence, and to stop further damages.
- To highlight the historical and aesthetic value of the monument.
- To continue to utilize the monument as a place of worship.

The principles used to achieve the above goals are based on the Map of Venice and the other internationally recognized treaties for the restoration of monuments, adapted to the particularities and conditions of the specific monument.

Main principles of the restoration:

- To maintain the authenticity of the monument through the maintaining, restoring and preserving of the elements that have been unaltered, as well as to highlight those that have been destroyed, but their restoration is feasible.
- The use of traditional materials and building methods, with the commitment that if they are deemed inadequate, there will be a parallel use of modern methods and materials, that have been tested in other restorations.
- The reversibility of the interventions, where it is feasible.
- The discreet differentiation between new materials and original parts of the monument, which is deemed necessary for functional or other reasons. In this particular building, this is the case with new floors etc.
- The guaranteed continuous protection of the monument.

General description of the Restoration

Based on the above goals and principles, it was suggested that the structural and construction problems of the monument be dealt with, as well as the removal of most of the previous interventions, since they altered extensively the aesthetic value of the monument. Specifically, the following was suggested: the walls should return to their original height, a new roof should be built, the newer and badly built cement with deep grouting would be carefully removed and new grouting with a mild indentation would be applied, with material that is compatible with the wall materials.

Application of localized injection grouting in the presence of a restoration specialist and the construction of tension rods.

The humidity will be dealt with by insulating the foundation and creating a draining moat.

Replacement of the interior flooring of the temple. Specifically, it was suggested that the floor of cement tiles be carefully removed and that it should be replaced with new ones, of the same quality and color, apart from the decorative motif, that will be conserved.

The overall highlighting and aesthetical improvement of the temple, with the removal of electrical equipment. In order to restore the appearance of the temple, it was suggested that exterior cables be removed and interior electrical equipment be rearranged. Also, fixed radiators will be replaced by movable ones.

Replacement of metal door and windows.

Integration of new buildings

Concept

Until 2003, the monastery of Kimisis Theotokou was partially closed, although it is an important site of pilgrimage for the broader region. Because of that, the Archdiocese wanted to improve the monastery's infrastructure, so that it could fully operate.

Following this rationale and in the context of writing this paper, apart from completing the technical part, which consists of the surveying, the diagnosis-pathology and the restoration proposal, an effort was made to investigate the inclusion of new uses.

The new architectural interventions were aimed at improving the functional needs of the whole and its modernization, as well as to improve it aesthetically. Changes had to be distinct, to express their time and, if possible, to be reversible. Such an approach is a declaration of respect from our society towards future generations, which have the right to reinterpret and evaluate cultural heritage with other criteria and methods.

Due to this need, I felt the desire, apart from restoring the main temple, to also study the means by which the existing buildings (old and new) could be highlighted, as well as to integrate and arrange functions and building structures.

To complete this goal, the proposed intervention moved along the following axes:

- The restoration, highlighting and reuse of existing buildings.
- The study of monastery architecture and the particular parameters that it presents, through relevant bibliography, visits to other monasteries and discussions with monks, in order to understand the functional layout of monasteries.
- The study of the greater region's morphology, in order to discover traditional styles and materials, that can be subsequently interpreted and expressed in the buildings, through a new spirit.
- Detailed cataloguing of options to include new forms & uses

The layout of functions and forms:

The whole:

The allocation of functions and forms was affected by the geomorphology and the existence of buildings, as well as the tradition of monasteries, in order to highlight the main temple.

The tradition of monastery architecture calls for a circumferential linear layout of the buildings, leaving a single free space in the center, the monastery's court. The main temple is placed in the courtyard, independent of other buildings, and with its central position it becomes the main element of the whole, as well as its ideological center.

In the specific case that we are studying, this enclosed shape that characterizes a large portion of the monasteries was missing, leaving the main temple unsupported, without a context of reference. This parameter created the need to place structures around it. Since in the restoration it is proposed to bring the main temple to its original height, it was decided to place around it single story buildings, at a distance that would create around it a backdrop and not a stifling environment.

Across the entrance of the main temple, as it is common in most monasteries of Agion Oros, is the place of the sanctum, so that the monks can easily proceed to it after Mass. The movement of monks and visitors played a defining role in the arrangement of functions. The functions that were directly connected to the monks were close to the main temple, so that it would always be a reference point for them. So, the cells, the priory, the library, the storage rooms and the chapel are placed in such a way that they are interconnected and their access by visitors is controlled, creating a conceivable border, since this is part of the charter of many monasteries.

On the contrary, the reception area for the guests (Archontariki) and the shop-exhibition are placed in positions where they will be immediately be noticed by the visitors and therefore are easily accessible. As a matter of fact, these activities take place in the main courtyard, which functions as the intermediate space where the main users of the monastery, pilgrims and monks, appear and mingle. The presence and contact between monks and visitors, which could be perceived as a functional weakness, actually makes the space more interesting and gives it a broader and more complex character, allowing the pilgrim to choose by himself the route that he will follow. So, the monastery space has a more open, social character, despite its isolation. Besides, Eastern monasticism never avoided coming into contact with the outside world.

Existing buildings:
The choice of uses for the existing buildings was made according to what better served the whole and especially to distribute the movement of visitors in the monastery. As it was explained above, the surrounding space of the monastery has been separated into two areas. The area where only monks can go and the area used by pilgrims and monks. Since the structure of the old cells is near the entrance and inside the zone where pilgrims move, it was decided to make it a part of the exhibition-shop and the reception area, as they are both directly connected to the pilgrimage. The reception area was placed on the floor, to be more visible and have a distinguished character. The cells of 1956 are the most remote building of the compound. That explains the position of the guest houses, so that during their stay, the monastery's guests would make their presence as discreet as possible. The guest houses can be reached by passing in front of the great reception area. In the meantime, the monks can have easy access to the guest houses through the elevated wing of the cells, without having to cover a long distance.

Morphology
Existing buildings:
Old cells:
In the old cells we found photographs that provided sufficient information about their original form. Since this is the second oldest building of the compound, the choice was made to restore it to its original state. To achieve this goal, it is proposed to remove the reinforced concrete floor and the roof, as well as all the newer windows, that were imitations of the original ones. Only the stone walls of the ground will remain, which will be restored after the newer cement grouting is removed and replaced by a compatible grout. The material and method to be used is similar to that which will be used for the main temple. The floor will be made of stone, with openings similar to those seen in the photographs. The internal partitioning will be made with plaster boards. On the east side, there will be a balcony whose shape will be a modern version of the traditional balconies of the time.

The cells of 1956:
The western cells, as mentioned above, will include the guest houses. Since the building is relatively recent, a more liberal restoration is preferred, compared to the one proposed for the old cells. We propose the restoration of the walls, by removing unstable cement and reinforcing them with deep grouting. The openings will remain as they are, with some modifications made to the west side, which are necessary to better serve new uses. The windows will be made of wood, but they will follow the shape of the windows of the new buildings. Since the building already exists, the roof will be hipped, to distinguish itself from the new buildings. We propose the use of plaster boards for internal partitioning.

New buildings:
The shape of the new buildings was dictated by their function as well as the long standing tradition of monasteries. Still, every change must express the time in which it took place. Therefore, the solution that was chosen moves along the lines of interpreting traditional styles based on a modern approach. The cells retain their traditional placement, one next to the other, giving them an elongated form. Although the corridor that passes in front of the cells is still present as an element, the expression of the openings and the façade follows a more modern approach, avoiding traditional forms. Also, the structure of the refectory as well as secondary spaces placed around the main temple follow traditional motifs, mainly as to their place in space. Traditional materials will be used, but the interplay between stone and adobe exterior walls presents a more modern style. Reinforced concrete was deemed unsuitable for use in a space which is by definition created to last for a long time. It is proposed that all of the additional structures (floors, roofs, beams, windows and doors) should be made using lumber. Nonetheless, the form of the roofs is modern, since they will be pitched or hipped, depending on the case, but they are constructed with scotia moldings above the walls.

Deterioration of Monument Building Materials: Mechanistic Models as Guides for Conservation Strategies

Dimitra G. Kanellopoulou[1] , Aikaterini I. Vavouraki[2] ,
and Petros G. Koutsoukos[1(✉)]

[1] University of Patras and FORTH-ICEHT, Patras, Greece
pgk@chemeng.upatras.gr
[2] School of Mineral Resources Engineering, Technical University of Crete,
73100 Chania, Greece

Abstract. Chemical dissolution and salt crystallization are very important factors contributing to the deterioration and damage of the built cultural heritage consisting of marble and limestone. Understanding of the underlying mechanisms of the calcitic materials damage is necessary for the design of efficient strategy for the prevention of deterioration can be done only through the appropriate kinetics measurements. Kinetics of dissolution of calcitic materials and of the growth of salts of interest should be measured, precisely and reproducibly. In the present work, the rates of dissolution of calcitic marbles (>98% calcite) and sandstone were measured at constant undersaturation and were correlated with the respective solution undersaturation. The dissolution kinetics measurements showed that calcitic limestone at pH 8.25 had a lower dissolution rate constant in comparison with the respective value for Pentelic marble, a calcitic material (ca. 98% calcite). The mechanism was surface diffusion controlled at alkaline pH values. In more acid pH values mass transport became more significant. Reduction of the rates of dissolution was achieved by the addition of substances with functional groups, which may interact with the surface of the calcite crystals. Several inorganic ions (including orthophosphates, pyrophosphates, phosphonates, fluoride and sulfate) and one organic environmentally friendly compound, polycarboxymethyl inulin (CMI) (MW 15000) were tested and their effect on the rates of dissolution was discussed. Phosphonates, were found to have a beneficial effect on the control not only of marble dissolution, but also on the crystallization of damaging soluble salts like NaCl and $Na_2SO_4 \cdot 10H_2O$ (mirabilite).

Keywords: Dissolution kinetics · Constant undersaturation · Inhibition

1 Introduction

The preservation of the built cultural heritage consists to a large extent on the resistance of the built elements towards dissolution and to the formation inside their pore structure of soluble salts, the volume of which increases upon crystallization [1, 2]. Dissolution processes may be regarded as the opposite of crystal growth, i.e. they may be described as a series of events leading to the disintegration of basic building blocks of the minerals making up the bulk materials [3–6].

© Springer Nature Switzerland AG 2019
A. Moropoulou et al. (Eds.): TMM_CH 2018, CCIS 961, pp. 456–469, 2019.
https://doi.org/10.1007/978-3-030-12957-6_32

Carbonate stone dissolution has been the focus of research over the past decades and several books have been written on the topic [7–9]. However, there is still considerable doubt concerning the mechanisms underlying the dissolution of the carbonate materials an issue which is very important in the design of the damage remediation strategies.

Knowledge of the mechanism of dissolution is expected to contribute in finding processes and methods to retard dissolution, through the appropriate interaction of chemicals with the surface of the calcitic building materials. Moreover, the measurement of the kinetics parameters is also very important to decide for the prevalent mechanism over a range of conditions of interest. The development of the constant composition methodology and its adaptation to the dissolution of calcium carbonates provided the ability to measure rates of dissolution at constant thermodynamic driving force [10, 11]. The maintenance of constant driving force ensures not only accurate kinetics measurements and most important events taking place at the first stages of dissolution, which are otherwise obscured by the rapid changes of the solution saturation with respect to the dissolving mineral phase. This is most important at the acid pH range where dissolution is quite fast. Generally, at conditions close to equilibrium the mechanism is expected to be surface diffusion controlled, whereas at large deviations from equilibrium, as in the case of the acid pH domain, the mechanism controlling dissolution is bulk diffusion. In the former case, processes that modify the surface of the minerals do affect the kinetics of dissolution while in the latter the process is controlled mostly by fluid dynamics.

Based on this fact and considering that calcitic materials deterioration takes place under close to equilibrium conditions, except for the case of heavily polluted environments in which pH of the solution in contact with the stone is very low, surfaces may be protected by adsorption of compounds which through the interaction with ions of the surface. It is for this reason that compounds like phosphonates which have been shown to be efficient inhibitors for the crystal growth of calcium carbonate [12–14]. At the same time phosphonates have also been shown to be efficient inhibitors of the dissolution of calcite and calcitic materials [15].

Salt damage is another serious mechanism of calcitic stone deterioration and is due to the formation of soluble salts in the pore structure of the building materials [16]. Crystal growth of soluble salts is often controlled by surface diffusion [17].

In the present work, the kinetics of dissolution of a calcitic marble used extensively for the construction of historical monuments and artifacts, Pentelic Marble (PM) and Granada Sandstone (GS) rich in calcium carbonate were investigated. The kinetics of dissolution, were studied at constant undersaturation and the thermodynamic conditions differed from equilibrium by little. Since in both cases surface diffusion is prevalent, the effect on the kinetics of dissolution of small inorganic ions like orthophosphate, pyrophosphate, F^-, $C_2O_4^{2-}$ and SO_4^{2-}, and organic compounds like hydroxy ethylideno, 1,1, hydroxy phosphonic acid (HEDP) and carboxymethylinulin (CMI) which are expected to interact with the positively charged calcium ions of the calcitic materials surface, was investigated at constant thermodynamic driving force. Recently CMI has been shown to behave as a green inhibitor of the crystal growth process of calcium carbonate and silicates [18, 19].

2 Experimental Part

2.1 Materials

Calcium chloride stock solutions were prepared from the respective crystalline solid ($CaCl_2 \cdot 2H_2O$, Merck, Puriss.), filtered through membrane filetrs (Sartorius 0.22 μm) and was standardized by titrimetry with standard EDTA solutions and with atomic absorption spectrometry (AAS, Perkin Elmer AAnalyst 300). Sodium chloride stock solutions were prepared from the respective solid (Merck, Puriss.) dried overnight at 70 °C. Sodium bicarbonate solutions were prepared fresh for each experiments from the crystalline solid (Merck Puriss.). The test inhibitor stock solutions for orthophosphate, pyrophosphate, F, SO_4^{2-}, and $C_2O_4^{2-}$ were prepared from the respective crystalline sodium salts (Merck, Puriss.). Carboxymethylinulin, MW 15,000 (CMI-15, Solutia, Belgium) was obtained as a 20% solution in water (w/w). PM (98% calcite), pure synthetic calcite and GS (>95% calcite) have been used as test materials in the dissolution experiments. HEDP was prepared form 60% solution (Solutia, Belgium). The Prague sandstone was a calcitic stone, for which chemical analysis showed that it consisted (w/w) of: Calcium (95.6%), Magnesium (0.15%), Manganese (0.22%), Iron (1.65%) and silicon (0.65%). Analyses were done by fusing the materials with $LiBO_2$, followed by dissolution in 2N hydrochloric acid solution. X-ray diffraction analysis confirmed the results.

2.2 Method

2.2.1 Dissolution Experiments

The powder dissolution experiments were done in undersaturated solutions prepared from standardized stock calcium chloride and sodium bicarbonate solutions as described in detail elsewhere. All experiments were done at 25 °C, pH 8.25, and ionic strength 0.1 M adjusted by NaCl. The master variable used for monitoring the dissolution process was H^+ activity, measured by a combination glass-Ag/AgCl electrode. The dissolution was initiated with the introduction of carefully weighed amounts of the solids. pH changes as small as 0.005 units in the solution triggered the addition of titrant solutions from two mechanically coupled burets to the working solution. The titrants consisted of $CaCl_2$, $NaHCO_3$, NaCl as inert electrolyte, and HCl to adjust the solution pH. The concentrations of the titrants were calculated from the mass balance for each component so that the constancy of the activities of all species was maintained throughout the course of dissolution.

A schematic outline of the reactor simulating the water-solids interaction and data recording is shown in Fig. 1. All experiments were done at 25 °C. The undersaturated solutions were prepared directly inside the reactor, volume totaling 200 mL and pH was adjusted by the addition of standard sodium hydroxide solution (0.1N Merck, Titrisol] to a value of 8.25 (equilibrium with atmospheric CO_2). The test conditions allowed the treatment of the system as closed.

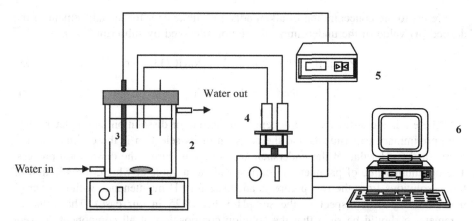

Fig. 1. Schematic representation of the experimental set-up for the measurement of the kinetics of dissolution processes at sustained saturation. 1: Magnetic stirrer; 2: Double walled Pyrex glass reactor thermostated by circulating constant temperature water; 3: Sensor to monitor process (combination glass/reference electrode); 4: Electric pumps, mechanically coupled to deliver titrant solutions; 5: Control unit (pH set point and pumps trigger); 6: PC with A/D converter for the control and data acquisition.

Following the preparation of the undersaturated solutions (saturation with respect to calcite was one of the most important experimental variables other than the concentration of the additives), an accurately weighted quantity of the test solid was introduced in the reactor in the form of powder. Dissolution started immediately without any appreciable induction time and was monitored on the basis of OH- released during dissolution:

$$CaCO_3(s) + H_2O(l) \rightarrow Ca^{2+}(aq) + CO_3^{2-}(aq) \tag{1}$$

$$CO_3^{2-}(aq) + H_2O(l) \rightarrow HCO_3^{-}(aq) + OH^{-} \tag{2}$$

Changes in the pH of the solution as little as 0.005 pH units, triggered the addition of solutions with composition appropriately calculated to achieve dilution of the released ions according to the stoichiometry of Eq. (1), taking into consideration the mass balances for the process. The solutions added from the two pumps and their respective concentration was as follows:

$$\text{Pump A:} \quad [CaCl_2] = 2 \times [CaCl_2]_{us} - c \tag{3}$$

$$[HCl] = -2 \times [NaOH]_{us}^{*} + 2c \tag{4}$$

*refers to the concentration of alkali added in the reactor for the adjustment of the desired pH value of the undersaturated solution (denoted by subscript "us").

$$\text{Pump B:} \quad [\text{NaHCO}_3] = 2 \times [\text{NaHCO3}]_{us} - c \qquad (5)$$

$$[\text{NaCl}] = 2 \times [\text{NaCl}]_{us} \qquad (6)$$

Where c is a constant, which depends on the rate of dissolution and is determined from preliminary experiments and the criterion of its selection is the optimal maintenance of the pH value of the undersaturated solutions during the dissolution process. The optimum value of parameter c, corresponds to a value at which, the addition of titrant solutions from the two pumps is adequate for pH maintenance without over or undershooting with respect to the set pH value (8.25 in our case). The value of parameter c should be such that the solution composition of all components is kept constant to within ±0.5%. During dissolution, samples were withdrawn, filtered through membrane filters and the filtrates were analyzed for calcium by AAS. The solids on the filter were characterized by powder x-ray diffraction (SIEMENS D-5000) and scanning electron microscopy (SEM, Leo Supra, 35 VP, Bruker).

2.2.2   Salt Crystallization Experiments

Accelerated tests were conducted in a salt spray chamber in which Granada limestone (size of $1 \times 1 \times 10$ cm) specimens were exposed to salt spray consisting of a sodium sulfate solution dispersed in small droplets with the aid of a nebulizer (Fig. 2). More specifically the specimens were exposed at constant temperature (30 °C) to a mist of 0.1 M Na2SO4 solution for duration of six days (one cycle). The exposed specimens included both untreated and treated calcarenite specimens by prior impregnation in HEDP solutions of concentration 0.2–1% (w/w), pH = 5 and 10. The exposure duration of calcarenite samples in the sodium sulfate chamber was three cycles (18 days in total).

(a)   **(b)**

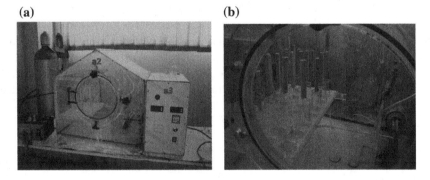

Fig. 2. (a) The salt spray chamber: a1: Peristaltic pump for salt supply to nozzle; a2: Specimen chamber; a3: Temperature controller unit and (b) Interior of the specimens exposure area with suspended limestone rods, conical drainage flasks and the glass nozzle spraying sodium sulfate solution at constant temperature (30 °C).

3 Results and Discussion

The thermodynamic driving force for the dissolution of the calcitic marbles tested is expressed in terms of the relative undersaturation, σ, with respect to the principal chemical component, calcite:

$$\sigma = 1 - \left\{ \frac{(Ca^{2+})(CO_3^{2-})}{K_{s,calcite}^0} \right\}^{\frac{1}{2}} \tag{7}$$

In Eq. (1) parentheses denote activities and K_s^0 is the thermodynamic solubility of calcite. The calculations of the relative undersaturation was done from the mass balances for calcium and carbonate considering all possible equilibria using PHREEQC software [20]. As may be seen from Eq. 7, $0 \leq \sigma < 1$. The rate of dissolution, R_{diss}, is a function of the relative undersaturation and may be expressed by the power law, as in Eq. (8):

$$R_{diss} = k_{diss} \sigma^n \tag{8}$$

In Eq. 8, k_{diss} is the dissolution rate constant and n the apparent order for the dissolution, which is indicative of the dissolution mechanism. Plotting the experimentally measured rates of dissolution as a function of the relative undersaturation with respect to calcite for the calcitic marble and the rest of the materials tested, the kinetics plot, shown in Fig. 3 was obtained.

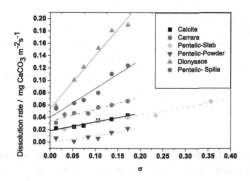

Fig. 3. Plot of the rates of dissolution of: powdered Pentelic marble and of various building materials (shown in the inset) as a function of the relative solution under-saturation with respect to calcite; pH 8.25, 25 °C, 0.1M NaCl

As may be seen from Fig. 3, in all cases the fit of the kinetics data yielded a first order dependence of the rate of dissolution on the relative undersaturation with respect to calcite. The possibility of bulk diffusion-controlled mechanism was ruled out, because if valid it would yield values for the diffusion coefficient of the order of 10^{-5} m^2s^{-1} which are unrealistically high. The first order dependence of the rates on the relative solution undersaturation with respect to calcite was interpreted therefore as a proof that at the conditions studied, dissolution was controlled by surface diffusion. It is interesting to note that the rates of dissolution measured for sandstone were comparable with the respective on synthetic calcite, although lower, while marble gave higher dissolution rates. This could perhaps be ascribed to the presence of relatively high iron levels which are known to retard crystal growth of calcite even at $1 \mu M$ concentration level (and most likely acting as inhibitors of dissolution as well) [21–23]. The rates of dissolution measured were 9.2×10^{-3}, 3.4×10^{-2} and 7.0×10^{-3} $mg \cdot m^{-2} \cdot s^{-1}$ for the reference material (calcite), a dense material like Pentelic marble and Prague sandstone (porous material).

It is interesting to note that mass loss rate was slower in the more porous material. It is obvious that other factors, related with the active sites for dissolution are prevalent. Moreover, the high concentration of silicates in the sandstone, may contribute to the release in solution of soluble silicate species which most likely act as inhibitors of calcium carbonate dissolution, in a similar way as they retard its crystal growth. The mechanism however was the same, as may be seen from the dependence of the rate of dissolution on the relative supersaturation with respect to calcite, shown in Fig. 4.

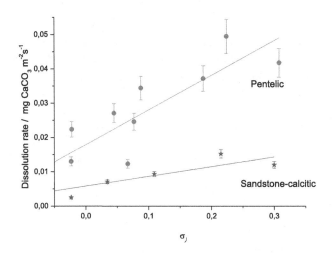

Fig. 4. Plot of the rates of dissolution of: powdered Pentelic marble and sandstone as a function of the relative solution under-saturation with respect to calcite; pH 8.25, 25 °C, 0.1M NaCl

For the same material, higher porosity, normally leads to higher dissolution rates. Thus e.g. calcitic marble (Carrara) thermally treated (550 °C for 2, 5 h) with significantly higher porosity, resulted in higher rates of dissolution, as may be seen in Fig. 5.

This result, suggested that the lower rates of the porous sandstone material should be attributed to foreign ions, present as "impurities" with silicates perhaps acting as inhibitors of calcite dissolution.

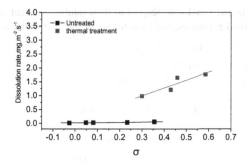

Fig. 5. Plot of the rates of dissolution of Carrara marble powder before and after thermal treatment as a function of the relative solution under-saturation with respect to calcite; pH 8.25, 25 °C, 0.1M NaCl.

It is interesting to note that this dependence of the rates on the relative supersaturation was found to be valid over a range of pH values of the aqueous environment, as may be seen in the kinetics plot shown in Fig. 6.

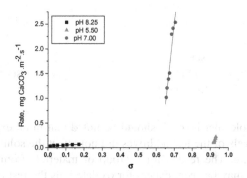

Fig. 6. Plot of the rates of dissolution of: powdered Carrara marble (99.8% calcite) as a function of the relative solution under-saturation with respect to calcite; pH 8.25, 25 °C, 0.1M NaCl

At pH 7.0, the mechanism was mixed (surface and bulk diffusion controlled) and at pH 5.5 bulk diffusion control had a major contribution in the overall dissolution mechanism.

The compounds tested for the investigation of their possible inhibitory activity on the kinetics of calcitic materials dissolution are summarized in Table 1. All compounds are water soluble with ionized functional groups.

The test additives, were selected on account of the anionic sites they possess, which give them the ability to bind with the active sites of the prevalent 001 crystal faces of the calcitic material and thus block the active sites for dissolution. The investigation was done mainly at alkaline pH, where the mechanism is purely surface diffusion controlled and the respective experiments could serve as a proof of the concept that materials conservation design can be based on understanding the dissolution

Table 1. Compounds investigated with respect to their inhibitory effect on marble, selected on the basis of the anticipated mechanism of dissolution, dominated by surface diffusion.

Compound	Chemical formula	# of functional groups
Sodium orthophosphate	Na_2HPO_4	1
Sodium pyrophosphate	$NaO-\overset{\overset{\displaystyle O}{\parallel}}{\underset{\underset{\displaystyle ONa}{\mid}}{P}}-O-\overset{\overset{\displaystyle O}{\parallel}}{\underset{\underset{\displaystyle ONa}{\mid}}{P}}-ONa$	2
Sodium tripolyphosphate	$NaO-\overset{\overset{\displaystyle O}{\parallel}}{\underset{\underset{\displaystyle ONa}{\mid}}{P}}-O-\overset{\overset{\displaystyle O}{\parallel}}{\underset{\underset{\displaystyle ONa}{\mid}}{P}}-O-\overset{\overset{\displaystyle O}{\parallel}}{\underset{\underset{\displaystyle ONa}{\mid}}{P}}-ONa$	3
HEDP	$\begin{array}{c} HO \quad\ PO_3H_2 \\ \diagdown\ / \\ C \\ \diagup\ \diagdown \\ H_3C \quad\ PO_3H_2 \end{array}$	2
Carboxymethyl Inulin (CMI), MW 15000		2 carboxyl groups per monomer unit
Sodium fluoride	NaF	F^-
Sodium sulfate	Na_2SO_4	SO_4^{2-}
Sodium oxalate	$Na_2C_2O_4$	2

mechanism at the molecular level. It should be noted that in all experiments involved the concentration levels of the test additives did not affect the solution undersaturation with respect to calcite. The results for a group of materials from Table 1 tested are shown in Fig. 7. As may be seen, except for oxalate ions the rest of the species tested showed considerable reduction of the rates of dissolution. CMI as perhaps expected because of the large number of anionic sites gave the best reduction of the rates of dissolution, while, the retardation effect of SO_4^{2-} is interesting. It is reported that sulphate ions inhibit vaterite transformation and enhance the calcite formation [21–23].

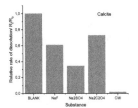
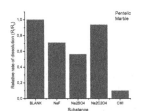
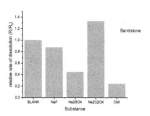

Fig. 7. Effect of the presence of additives on the rate of dissolution of calcite (reference) and natural building stones (Pentelic marble and Granada sandstone (calcitic)); 25 °C, pH 8.25, $\sigma = 0.14$.

The phosphate containing additives, showed an impressive effect on the dissolution of Carrara marble as may be seen in Fig. 8.

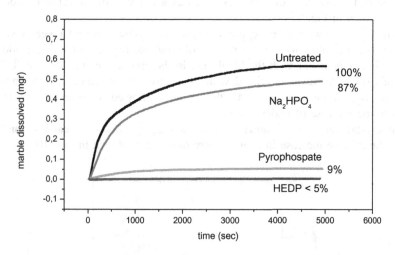

Fig. 8. Comparison of kinetics of marble dissolution Carrara, without and past the treatment with phosphate containing compounds. 25 °C, pH 8.25, 0.1M NaCl

It should be noted that all compounds tested were adsorbed strongly on the marble surface and their activity was attributed to the blockage of the active dissolution sites.

The effect of HEDP was most pronounced, and as may be seen in Fig. 8, it showed the same excellent inhibitory activity, when marble specimens were immersed in undersaturated solutions.

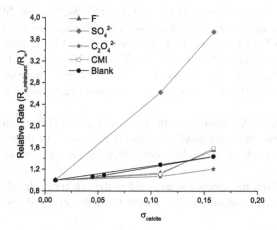

Fig. 9. Plot of the ratio of the rates of dissolution at the undersaturation tested over the respective rate at the least undersaturation with respect to calcite as a function of the relative undersaturation. pH 8.25, 25 °C (0.5 mM NaF, 0.5 mM Na_2SO_4, 0.1 mM $Na_2C_2O_4$ and 10 ppm CMI).

It should be noted that at all undersaturations tested, the efficiency of rate decrease was reduced with increasing undersaturation (i.e. higher deviation from equilibrium). The trend of the presence of additives tested as a function of the relative undersaturation is shown in Fig. 9.

As may be seen with the exception of sulfate ions which showed a weak resistance of the respective inhibition of the rate of dissolution, the rest of the test additives showed similar behavior with the blank, i.e. in the absence of the additive. The low effect of oxalate however is related with its relative ineffectiveness as an inhibitor. However, as may be seen for CMI, this additive showed a higher resistance with respect to the increase of undersaturation.

Salt crystallization was shown to have damaging effects on calcitic sandstone tested. The relative increase in average pore diameter is shown in Fig. 10.

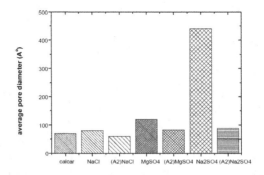

Fig. 10. Average pore diameter (APD) values of immersed limestone obtained from BET analysis.

The APD for untreated limestone is ~ 70 Å; NaCl, $MgSO_4$, Na_2SO_4 correspond to APD for limestone specimens after immersions in salt solutions and (A2) NaCl, (A2) $MgSO_4$, (A2) Na2SO4 correspond to specimens with A2 treatment (immersion of materials first in salt solution followed by 0.1% (w/w) HEDP for sulfate salts and 10^{-3} M potassium ferricyanide, $K_3Fe(CN)_6$ for NaCl). All BET measurements were for limestone specimens following the seventh immersion cycle (except for Na_2SO_4 - fifth cycle).

As may be seen the treatment with the test inhibitor (HEDP) resulted in the efficient reduction in the pore size, apparently because it prevented the respective salt crystallization, which in its absence develops crystallization pressure, which is damaging to the growing crystals.

Calcium loss from the specimens exposed in the salt spray chamber was measured by analyzing for total calcium in the runoff from the exposed specimens by AAS (Fig. 11).

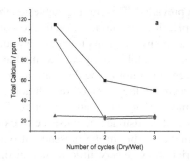
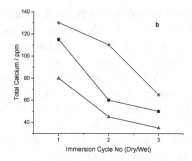

Fig. 11. Results of dissolution of sandstone exposed in spray chamber at 30 °C. Calcium loss (dissolution) as a function of the number of dry/wet cycles for sandstone specimens with and without treatment with HEDP. a: pH 5.0 b: pH 10. (■) untreated sandstone; (●) sandstone treated with 0.5% HEDP; (▲) treatment of sandstone slabs with 1% w/w HEDP.

For untreated limestone specimens exposed for six-days to 0.1 M Na_2SO_4 spraying (one cycle), the total calcium was ~ 115 mg/L. For the following cycles, the calcium loss was reduced (56 and 46 mg/L, for second and last third cycle, respectively). Immersion of the limestone specimens in HEDP solutions at pH = 5 resulted in the decrease of calcium release in comparison with the untreated specimens (Fig. 11a). The total calcium concentration in the runoff from limestone specimens pre-treated with HEDP solutions (pH = 5) after the first cycle of exposure was ~ 95 mg/L, lower than the corresponding to untreated specimens (~ 115 mg/L). The successive cycles showed the same trend of decrease of the calcium concentration in the runoff. Limestone pre-treated with HEDP solutions 1% w/w, suppressed total calcium concentration in the runoff to ~ 25 mg/L and remained constant during the next series of exposure cycles. The presence of HEDP solutions (1% w/w and pH = 5) in the spray solutions resulted in significant protection. The minimal mass loss found in the presence of this compound was attributed to the limited formation of Ca-HEDP salts. At higher pH, only treatment with 1% w/w HEDP of the sandstone specimens proved to be efficiently protective, although to a lower extent in comparison with the respective results at acid pH. Again this finding is attributed to the formation of Ca-HEDP salts which are relatively soluble but at higher concentrations they seem to form a more protective layer.

4   Conclusions

In conclusion, the investigation of the kinetics of dissolution of several of the most commonly employed building materials: different types of calcitic marbles, not only in powder but also in the form of slabs, Granada sandstone and synthetic calcite as reference material were measured at conditions of sustained undersaturation. The measurements showed that although there are quantitative differences in the measurement of the rate constants, mechanistically, the process was controlled by the same mechanism: surface diffusion. At neutral pH the mechanism was mixed while at acid pH values bulk diffusion controlled mechanism prevailed. This finding suggested that it is possible, at least at the neutral and alkaline pH, through surface modification of the materials to ameliorate their behavior in aggressive environments towards dissolution. Phosphonate ions, were found to be highly efficient inhibitors of calcitic marble dissolution. CMI was shown to be an efficient "green" inhibitor of the dissolution process. The presence of sulfate ions was found to decrease the rates of dissolution more significantly than the presence of fluoride while oxalate gave rather poor results in terms of inhibition of the dissolution. In the case of sandstone, the presence of oxalate accelerated dissolution. The dissolution of Pentelic marble in the presence of oxalate was hardly reduced while the reference material showed very little reduction of the respective rate of dissolution in its presence. Possible alterations of the solubility of the minerals tested in the presence of the additives investigated, need to be assessed quantitatively, as this parameter affects the effective thermodynamic driving force for dissolution. Finally accelerated dissolution measurements in a salt-spray chamber, showed that treatment of the specimens with 1% w/w solution of HEDP resulted in the efficient protection of sandstone dissolution both in acid (pH 5) and in alkaline (pH 10.0) environment.

Acknowledgment. The authors wish to acknowledge support of the work from KRHPIS programme- POLITEIA II (Region of Western Greece» Code # (MIS) 5002478.

References

1. Keller, W.D.: Progress and problems in rock weathering related to stone decay. In: Geological Society of America, Engineering Geology Case Histories, no. 11, pp. 37–46 (1978)
2. Likens, G.E.: Acid Precipitation, Chemical and Engineering News, 29–44, 22 November 1976
3. Winkler, E.M.: Stone: Properties, Durability in Man's Environment, 2nd edn. Springer, New York (1975)
4. Schaffer, R.J.: The Weathering of Natural Building Stone, Building Research Special Report No. 18 (Garston, 1931); Reprinted 1972 by Building Research Station, Garston, England
5. Baedecker, P.A., Reddy, M.M.: The erosion of carbonate stone by acid rain. Laboratory and field investigations. J. Chem. Educ. **70**, 104–108 (1993)
6. Graedel, T.E.: Mechanisms for the atmospheric corrosion of carbonate stone. J. Electrochem. Soc. **147**, 1006–1009 (2000)
7. Carrels, R.M., Christ, C.L.: Solutions, Minerals and Equilibrium. Harper and Row, New York (1966)
8. Krauskopf, K.B.: Introduction to Geochemistry. McGraw-Hill, New York (1961)
9. Stumm, W., Morgan, J.: Aquatic Chemistry, 2nd edn. Wiley, New York (1981)
10. Wang, L., Nancollas, G.H.: Calcium orthophosphates: crystallization and dissolution. Chem. Rev. **108**, 4628–4669 (2008)

11. Kanellopoulou, D.G., Koutsoukos, P.G.: The calcite marble/water interface: kinetics of dissolution and inhibition with potential implications in stone conservation. Langmuir **19**, 5691–5699 (2003)

12. Xyla, A.G., Mikroyannidis, J., Koutsoukos, P.G.: The inhibition of calcium carbonate precipitation in aqueous solutions by organophosphorus compounds. J. Colloid Interface Sci. **153**, 537–551 (1989)

13. Knepper, T.P.: Synthetic chelating agents and compounds exhibiting complexing properties in the aquatic environment. Trends Anal. Chem. **22**, 708–724 (2003)

14. Guo, J., Severtson, S.J.: Inhibition of calcium carbonate nucleation with aminophosphonates at high temperature, pH and ionic strength. Ind. Eng. Chem. Res. **43**, 5411–5417 (2004)

15. Spanos, N., Kanellopoulou, D.G., Koutsoukos, P.G.: The interaction of diphosphonates with calcitic surfaces: understanding the inhibition activity in marble dissolution. Langmuir **22**, 2074–2081 (2006)

16. Rodriguez-Navarro, C., Doehne, E.: Salt weathering: influence of evaporation rate, supersaturation and crystallization pattern. Earth Surf. Proc. Land. **24**, 191–209 (1999)

17. Vavouraki, A.I., Koutsoukos, P.G.: Kinetics of crystal growth of mirabilite in aqueous supersaturated solutions. J. Cryst. Growth **338**, 189–195 (2012)

18. Boels, L., Witkamp, G.J.: Carboxymethyl Inulin biopolymers: a green alternative for phosphonate calcium carbonate growth inhibitors. Cryst. Growth Des. **11**, 4155–4165 (2011)

19. Demadis, K.D., Preari, M.: Green scale inhibitors in water treatment processes: the case of silica scale inhibition. In: Lekkas, D.F., Belgiorno, V., Voulvoulis, N. (eds.) Proceedings of the 13th International Conference on Environmental Science and Technology, CEST2013, Athens, Greece, 5–7 September 2013, Paper No. 0280 (2013)

20. Parkhurst, D.L., Appelo, C.A.J.: Description of input and examples for PHREEQC version 3–a computer program for speciation, batch-reaction, one-dimensional transport, and inverse geochemical calculations: U.S. Geological Survey Techniques and Methods, Book 6, Chap. A43, 497 p. (2013) http://pubs.usgs.gov/tm/06/a43

21. Cailleau, P., Jaquin, C., Dragone, D., Girou, A., Roques, H., Humbert, L.: Influence of foreign ions and of organic matter on the crystallization of calcium carbonates. Oil Gas Sci. Technol. Rev. IFP **34**, 83–112 (1979)

22. Tlili, M.M., Ben, A.M., Gabrielli, C., Joiret, S., Maurin, G.: On the initial stages of calcium carbonate precipitation. Eur. J. Water Qual. **37**, 89–108 (1979)

23. Paquette, J., Vali, H., Mucci, A.: TEM study of Pt-C replicas of calcite overgrowths precipitated from electrolyte solutions. Geochim. Cosmochim. Acta **60**(23), 4689–4701 (1996)

Interdisciplinary Preservation and Management of Cultural Heritage

Technological Innovations in Architecture During Antiquity. The Case of Cyprus

Maria Philokyprou[(✉)]

Department of Architecture, University of Cyprus, Nicosia, Cyprus
mphiloky@ucy.ac.cy

Abstract. Stone, adobe and mortars have constituted the primary building materials throughout antiquity in Cyprus and many other countries in the Mediterranean area. What is impressive is that many structural innovations took place during the first periods of antiquity such as the first use of adobes, the investigation of gypsum and lime manufacture technology as well as the appearance of ashlar stone. In this paper the results of a systematic research regarding the different building materials used during the earliest periods of antiquity in Cyprus are presented. The investigation of prehistoric mortars demonstrated that the discovery of lime and gypsum technology occurred on the island during the Neolithic period while lime mortars were widely disseminated during the Chalcolithic period. Although the production of adobes in Cyprus seems to have been known since the Neolithic period, the use of moulds during their preparation was identified during the Late Bronze Age. The appearance of ashlar stone occurred during the Late Bronze Age, while in the earlier periods only rubble stone was used. It is interesting that each innovation in building technology is associated with other social, economic and political factors of each period. Thus, the first appearance of lime and gypsum mortars as well as the first use of adobe coincided with the first permanent habitation on the island. In parallel, the discovery of crushed-brick lime mortars, the use of moulds in the manufacture of adobes as well as the use of ashlar during the Late Bronze Age can be connected with the development of the first urban settlements with monumental public buildings.

Keywords: Adobes · Ashlar stone · Mortars and plasters

1 Introduction

Stone, adobe and mortars have constituted the primary building materials throughout antiquity in Cyprus, as well as in many other countries especially in the Mediterranean area. What is impressive is that many technological innovations in construction specifically took place during the prehistoric period such as the first use of adobes, the investigation of gypsum, lime and hydraulic lime manufacture technology as well as the appearance of ashlar stone. In this paper the results of a systematic research regarding the different building materials and techniques used during the earliest periods of antiquity in Cyprus are presented. The information described herein derives from the first all-encompassing research involving microscopic and laboratory analyses of a large number of samples collected from various prehistoric sites of the island.

© Springer Nature Switzerland AG 2019
A. Moropoulou et al. (Eds.): TMM_CH 2018, CCIS 961, pp. 473–486, 2019.
https://doi.org/10.1007/978-3-030-12957-6_33

This investigation was carried out in the framework of a PhD thesis on the prehistoric building materials [1] as well as in the context of two research programs funded by the Research Promotion Foundation of Cyprus. Interestingly, the findings of this research indicate that the various innovations in building technology seem to be associated with other social, economic and political factors of each period (first permanent habitation on the island, prosperity of the era, development of the first urban settlements with monumental public buildings etc.).

Throughout antiquity the geographical position of the island, its close relations with important prehistoric civilizations of neighbouring countries, as well as its climatic conditions, geology and geomorphology have played a very important role in the development of its civilization. The island can be separated in three geomorphological zones: the mountainous area of Pentadaktylos in the north, the Troodos mountainous area in the south, and the plain of Mesaoria in between. The Mesaoria area consists mainly of sedimentary rocks of various formations that were used for building purposes as these rocks were ideal as ashlar stones. During antiquity the island was almost completely covered with trees.

The prehistoric period in Cyprus shows a very impressive development in many aspects of civilization and it is divided in three main periods: the Neolithic period (7500–3900 B.C.), the Chalcolithic (3900–2500 B.C.) and the Bronze Age (2300–1050 B.C.). The latter is further divided in three periods – The Early, Middle and Late Bronze Age. The first permanent architectural evidences in the island dated to the Neolithic period are small or large settlements consisting of circular dwellings that replaced some more primitive and temporary timber structures that already existed. The Neolithic civilization seems to be a very well-developed civilization in matters of technology and culture. The circular dwellings that were widely used in the island throughout the Neolithic and Chalcolithic period were later replaced by linear structures. The exploration of copper mines, which constitute some of the main sources of wealth during the period, led to a more rapid development of the civilization and more innovations in architecture. During the Late Bronze Age rectangular dwellings appeared following some common typologies as well as elaborate houses with central hearths and bathtubs. At the same time, urban centres with parallel and intersecting roads [2] and architectural structures with different uses (industrial, religious, administrative etc.) as well as monumental, impressive public buildings appear. Additionally, new types of construction appeared during this period such as adobes made with the use of moulds, ashlar masonry, *cyclopean* walls, hydraulic plasters etc. [1].

2   Building Materials and Techniques

Stone, earth and timber, always readily available in nature and within the vicinity of the various settlements, were the basic building materials for the construction of circular as well as rectangular dwellings during the prehistoric period [3]. Metal and ceramics (in the form of bricks and tiles) were also sparsely used during this period. Stone either as rubble or in a dressed form (ashlar) was used for the construction of the base of the dwellings, earth in the form of adobes constituted the upper part, whereas timber was used mostly as a roofing material. The walls as well as the floors of the dwellings were often covered with plasters made of mud or lime.

The results presented herein derived from a systematic research regarding all the different building materials and techniques used during the earliest periods of antiquity in Cyprus. More specifically a large number of samples of stone, adobes and mortars (mud, lime and gypsum) were selected from Neolithic, Chalcolithic and Bronze Age settlements such as Khirokitia, Kalavasos-*Tenta*, Kissonerga-*Mosphilia*, Maroni-*Vournes*, Kalavasos-*Ayios Dhimitrios*, Kition-*Kathari*, Kouklia-*Palepaphos*, Alassa-*Paliotaverna*. Samples of rocks were also selected from nearby quarries in order to verify the provenance of ashlar stone. Stone and adobe samples were examined mainly under a petrographic microscope. Samples of plasters were put under many laboratory tests such as chemical, mineral (XRD), petrographic and thermal analyses as well as examination under a scanning electron microscope, in order to investigate the microstructure of the samples and the method of their preparation [1].

2.1   Stone. Rubble and the First Use of Ashlar

In Cyprus stone was the main building material, since the most primitive periods of antiquity. Initially stone was used only as rubble but during the Late Bronze Age ashlar stone appeared for the first time. In-situ observations in many settlements has shown that during the first periods of antiquity various types (sedimentary as well as igneous) and sizes of stone were used as rubble in circular and rectangular structures. The two external faces of the walls were constructed using stones of large size whereas the space in between was filled with smaller stones and mud mortar. The first use of ashlar in the island is dated to the Late Bronze Age and it is related to the widespread use of bronze tools during that period [4]. The use of ashlar demanded the existence of appropriate type of rock, tools, wealth, social organization and available workforce. Ashlar stone was mainly used in public buildings and administrative complexes (Alassa-*Paliotaverna*, Kalavasos-*Ayios Dhimitrios*, Maroni-*Vounres*), as well as places of worship (Kition-*Kathari*, Kouklia-*Palepaphos)* and less in residential complexes (Hala-Sultan Tekke) [5].

The examination of a large number of samples of ashlar stones selected from Late Bronze Age settlements and rocks derived from surrounding quarries has shown that the use of stone depended on the geology of each area (availability of suitable material to be quarried). Ashlars were mainly derived from local sedimentary rocks of various geological formations, which could be easily quarried. The petrographic examination has shown that the most widely used stone was the fine-grained calcareous sandstone that belongs to the Pachna geological formation, as this stone is especially durable and has a certain aesthetic appearance. A highly porous yellowish-ochre calcareous sandstone that belongs to the Nicosia-Athalassa geological formation, had a rather extensive use mainly in secondary structures of important buildings. The whitish reef limestone that belongs to the Koronia geological formation, was used in significant public buildings such as the Kition temples (Kition-*Kathari*). The petrographic examination of samples derived from the nearby quarries of Potamos Liopetriou and Xylofagou has shown that the ashlar stone used in Kition most probably derived from these quarries. The fine-grain chalk that belongs to the Lefkara geological formation had a limited use that was restricted to certain architectural components, which demanded fine carving. The research has shown that prehistoric people transported large pieces of ashlar stone from a very long distance on land or via the sea in order to obtain the most appropriate and durable rock for the construction of impressive buildings. The stone to be quarried was removed from the parent rock by the opening of circumferential trenches and undercutting of the rock. Most of the ancient quarries were on hill slopes or near the sea but there were also cases of underground quarries or quarries in caves.

The in-situ observation on ashlar stone of the Bronze Age has shown that ashlars are not perfectly worked to shape. The visible faces are rectangular whereas the faces that were not visible are mostly unwrought (Fig. 1). A characteristic feature of the ashlar masonry is the dressing and smoothing of only its faces which were meant to be visible (free visible surfaces) and partially of the surfaces set against other stones, whereas the back of the stone almost always remained "unworked" [5]. The sizes of the blocks vary considerably from 0.50 × 0.30 × 0.30 m, to very large sizes with a length of 1.00–5.00 m, a height of 0.50–1.50 m, and a thickness of 0.50–0.90 m.

Fig. 1. Ashlar wall masonry of shell type from Late Bronze Age settlements (Hala Sultan Tekke, Kition-*Kathari*)

In many cases there is a smooth, carefully-worked peripheral band – the *drafted margins* – on the external visible surfaces of ashlars surrounding a central panel – the *apergon* – that protrudes slightly (Fig. 2). The margins served as a guide for the placing of the blocks. The central surface was often carved after placing the stone. Sometimes the smoothing work was never completed and thus the margins are still visible today. Some interesting structural features of the ashlar stones are the small protrusions (bosses) that are observed in the middle of their visible surfaces that were left on the stone surface at the beginning of its working so as to ease its handling with the use of levers [6]. During the final working of the stone surfaces the bosses were cut off. There are, however, examples where these remained, probably for economic reasons, thus forming a decorative element showing the initial size of the stone which had been quarried (Fig. 2).

Fig. 2. Ashlar masonry showing drafting margins and bosses on the external face of walls from the Late Bronze Age settlements of Kalavasos-*Ayios Dhimitrios* and Alassa-*Paliotaverna*

On the upper surface of some *orthostats* (upright placed stones) and on the upper surface of the lower course of the stones of the walls on which the *orthostats* stood, small-sized shallow mortises (blind holes) were cut. The irregular size and shape of these holes at the top of the lower course of stones lead to their possible relation to the use of a lever for the placement of the stones above them. On the other hand, the uniform regular form and size of the mortises and the frequency of their appearance on the top of the *orthostats* indicates their possible use for fastening the timber elements of the superstructure [7].

During the Late Bronze Age, ashlar masonry was not used for building entire structures, mainly due to economic reasons. Its use was often limited to the lower part of the walls and frequently in the construction of only certain parts of the structures. Most of the public buildings have the lower part of their walls (*socle*) constructed entirely of ashlar for better stability and for the protection of the masonry against rising damp. In many cases, the use of ashlar was restricted to the visible and thus more important parts of the buildings. The use of ashlar can also be observed in wall corners and frames of openings. This selective use aiming to reinforce the structures create at the same time an interesting aesthetic effect.

Ashlar stones, were usually used in such a way as to create a shell-type construction, laid lengthways along both the interior and exterior faces of a wall (Fig. 1). A small number of stones with a width equal or almost equal to the thickness of the wall was often placed transversely at regular intervals as headers connecting the two faces to reinforce the masonry. The intermediate space between the two parallel stone faces in the shell-type walls was usually filled with rubble, mud and earth (Fig. 1). Ashlar, in most cases, was used in combination with other materials (rubble stone, mudbricks), both along the length of the wall as well as along its height.

Two different systems of construction using ashlar were observed in the Late Bronze Age – the method of setting ashlars in successive horizontal courses and the method of *orthostats* (Fig. 3). In the few examples of coursed masonry available, each successive course was of a different height and the stones did not have a uniform length. The second building technique, that was more common during this period, involved the use of *orthostats* and their setting on a horizontal course of larger, slightly protruding stones called *plinths*. In the construction of walls with this method, several variations of placing the *orhtostats* and forming their base can be observed.

Fig. 3. Different systems of construction using ashlars: coursed masonry (Hala Sultan Tekke) and method of *orthtostats* (Kouklia-*Palepaphos*)

Ashlar, apart from its use for walls, was used for the construction of floors and as lining for walls in bathrooms as well as for structures with industrial use in order to secure water-tightness. Dressed masonry was also used in the construction of channels, tanks and olive presses due to the impermeability of ashlar stone [5] as well as in the construction of staircases and pillars to acquire a greater stability and to achieve a carefully prepared final appearance of certain architectural elements. In addition, dressed stone was used on the bases of timber posts in order to protect the timber elements from moisture.

Only in a limited number of cases incised decoration was observed during the Late Bronze Age in Cyprus. Such decoration can be found mainly in some stepped bases and in some other members of the stone masonry walls, especially of temples. The decoration of the bases includes geometric shapes (rosettes, circles), while the decoration of the masonry incorporates illustrations mainly of boats that may have a symbolic significance (sacred offerings) [7].

The appearance of ashlar stone in Cyprus during the Late Bronze Age can be associated with the great prosperity of the area during this period and the establishment of well-organized urban centres [8]. The use of ashlar stone was the result of the general prosperity of Cyprus linked directly to the intensification of the exploitation of copper and the active participation in commerce with neighbouring countries [9]. The use of ashlar in Cyprus involved high costs depending on the availability of the appropriate tools and labour necessary for the quarrying, transportation, processing and installation, making ashlar a symbol of prestige demonstrating wealth and power. The use of ashlar shows an excellent construction development and a high level of technology during the early periods of antiquity. The choice of the stones used as ashlars was always related to the geology of each area. Since prehistoric time, builders acquired the knowledge of selecting and transporting the best quality of rock for each particular use. The transportation of stone over long distances reveals key facts about the organization of workers responsible for the erection of the monumental Late Bronze Age public buildings.

2.2   Adobes. Initial Use, Methods of Preparation and Construction

The investigation of the construction methods of the prehistoric masonries has shown that adobe was one of the primary building materials used in the island [10]. The earliest use of adobes can be dated in the Neolithic period, which coincides with the first permanent habitation on the island whereas the initial use of moulds in their preparation was identified at later periods (Late Bronze Age). Wall construction made of amorphous clay in successive layers, a technique similar to rammed earth, is very limited and appears in some dwellings of the Neolithic period (Khirokitia, Kalavasos-*Tenta*) parallel to the use of adobes. This method was more widely used during the Chalcolithic period as the predominant method of earth masonry construction [11] but was abandoned in the Bronze Age.

The process of soil selection and preparation for the manufacture of adobe bricks is very simple. Initially, the large stones contained in the soil were removed and the soil was placed in a natural trench. Then the required water was added as well as some organic material (straw, reeds etc.). The mixture was left in the pit for some time to mature and ferment. The paste after fermentation was shaped either by hand or by means of suitable moulds. During the prehistoric period, two methods were used for the manufacture of adobes with moulds. The most common method was the one that involved the placing of the mould on the ground, which was then filled with clay, smoothing the upper surface, and leaving the mixture to dry. The second method involved the pressing of a rectangular mould on top of an amount of wet mixture placed by hand on the ground [12]. The latter method was faster but resulted in the creation of adobes with straight sides with curved upper surfaces (plano-convex) [13, 14]. After the completion of the first step the moulds were removed, and the adobes remained in place for two to three days to dry and become solid, then adobes were turned over so that the lower surface could dry and at the end of this procedure they were placed upright on one side.

The earliest adobes recorded on the island had an irregular shape (Fig. 4), similar to rubble stones, in the form of hand-shaped loaves. A large variety in the dimensions of adobes in the various Neolithic constructions, as well as in each masonry, has been identified [15, 16]. Their length ranges between 0.21 and 0.35 m, their width ranges between 0.12 and 0.26 m and their thickness between 0.06 and 0.08 m. A layer of adobes on the upper part of a foundation of a rough stone construction that came to light in Cape Andreas-*Kastros* had a hole on the top of 0.04 m in diameter and 0.08 m in depth. It is possible that these holes supported a lightweight sloping frame of branches [10].

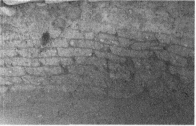

Fig. 4. Neolithic adobes of irregular shape from Khirokitia

Adobes of regular rectangular shape of the Bronze Age were manufactured using rectangular timber moulds and had fixed dimensions (Fig. 5). Adobes of this type initially appeared on the island during the Early and Middle Bronze Age and coexisted with adobes of irregular shape. These were widely used during the Late Bronze Age. In Dali-*Ambeleri* adobe bricks of particular interest were found. They can be divided into three types of such sizes that when combined together, they create a compact construction. The three types of adobe had equal thicknesses. The width of type 3 is equal to the length of type 1 and 2, while the length of type 3 is equal to the sum of the width of type 1 and 2. Adobes of rectangular shape are the predominant kind in most of the Bronze Age settlements. Adobe bricks generally have a substantial thickness of 0.10–0.15 m. Some fixed ratios of 4:3:1, 4:2:1, 2:3:1, 5:4:1, 5:6:1 can be observed, with a relative preference to the dimensions of 0.40 m, 0.50 m, 0.60 m.

Fig. 5. Late Bronze Age masonries using rectangular adobes from the settlement of Maroni-*Vournes*

Adobes have diachronically been combined with other materials such as stone (rubble and ashlar) and were used for the construction of the upper part of load bearing walls. Adobes were arranged in successive layers in such a way that the vertical joints of each layer bisected the underlying earth bricks, thus avoiding continuous joints (Fig. 5). Neolithic masonry, constructed with varying combinations of rubble stones and adobes, both at height and width had an overall thickness of 0.70–1.70 m. During the Late Bronze Age period, adobes were placed either lengthwise (as stretchers) creating thin walls or as headers *(diatoni)* to achieve thicker walls (Fig. 6). Adobes were often placed in two, three or even four rows in combination with headers and stretchers resulting into more complex arrangements.

Fig. 6. Different methods of construction using adobes of rectangular shape (sketches prepared by the author)

The chemical and mineralogical analyses of various prehistoric adobes have shown that manufacture of adobes involved the use of soil with high content of calcite (30–40%), a significantly high content of silicon oxide (10% and 20% or higher) and relative significant proportion of aluminum oxide (2% to 5% or higher 7.45% to 10%). Aluminum and silicon oxides seem to have come from clay minerals (montmorillonite, kaolinite, nontronite, etc.), which assign the clay its cohesive properties. The laboratory analysis of the adobes derived from Kalavasos-*Tenta* has shown a rather high percentage of gypsum. The existence of gypsum in adobes is quite unusual and this was attributed to the use of soil from the immediate surroundings of the settlement consisting of gypsum and covered by secondary deposits of lime and gypsum compositions [16]. Another peculiarity revealed in the adobes of the Neolithic settlement of Khirokitia is the existence of a high rate of strontium (sulfate celestite). This mineral is quite rare on the island and is mainly found in the geological strata of the area near the Neolithic settlement. It is obvious that the manufacture of adobes was achieved using local soils (or a combination of two in order to achieve the proper amount of clay) readily available in the immediate environment of the settlements. The detailed investigation of samples has shown various plant residues added in the mixture such as straw and broken stalks and cobs (Khirokitia, Kalavasos-*Tenta*), traces of grass and other plant residues, reeds and sea weed in coastal settlements (Maa-*Palaiokastro*).

In conclusion, it can be noted that the earliest use of adobes in Cyprus coincides with the first permanent habitation on the island. An evolution in the construction of earthen walls can be observed during the prehistoric period on the island. The hand-shaped adobes that constitute the main method of earth wall construction in the Neolithic period, were succeeded during the Bronze Age by the rectangular adobes made with the use of moulds. The latter seems to coincide with the introduction of rectangular buildings with linear elements in Cypriot architecture, while their wider use 1000 years later is associated with the urbanization of the island, as well as with the appearance of public and administrative buildings. The widespread use of earthen structures in Cyprus can be associated with the fact that adobes are more economic, due to the abundance of raw material on the island suitable for this purpose, as well as to the arid climate of the island which allows quick drying and easy maintenance.

2.3   Plasters. The Use of Lime and Gypsum

Lime and gypsum mortars have been used extensively in various structures in Cyprus since the prehistoric period employed mainly as plasters. Although the production of lime and gypsum seems to have been known since the Neolithic period, coinciding with the first permanent habitation on the island, it was widely disseminated during the Chalcolithic period. On the other hand, mud was mainly used as mortar in most of the structures during the whole of the prehistoric period. The discovery of lime and gypsum plasters introduced a revolutionary pyrochemical industry, in which the natural rocks (limestone and gypsum rocks) underwent a chemical change when heated, and when crushed to powder and mixed with water created a paste that could be easily worked [17]. The technology in the production of pozzolanic or crushed brick-lime mortars is more complex [18]. These hydraulic mortars can be manufactured by mixing lime with pulverized clay materials called pozzolans (natural or artificial, mainly in the form of ceramics) [19].

The main type of mortar used between stones and adobes during the prehistoric period was mud due to the economic procedure of preparation (Fig. 7). The raw material (soil with some clay content) collected from the neighbouring areas around the settlements was mixed with water and in some cases with an organic material such as straw. This material was mainly used as a binder between stones filling also the gap between the two shells of the walls.

Fig. 7. Rubble wall and lime plaster floor from the Chalcolithic settlement of Kissonerga-*Mosphilia*

On the other hand, for plastering the external and internal faces of adobes as well as stone walls and floors, lime plasters were mainly used while the use of mud and gypsum was very limited. The coating plaster of the walls was very often set in successive layers of small thicknesses for structural reasons in order to avoid the collapse under their own weight. The final thin coating of lime, often applied over a mud base layer, contributed to the better adhesion of the coating to the structure of the wall especially in the cases of adobes. In the lower part of the vertical walls, where they met the horizontal floors, series of small rounded stones (pebbles) were sometimes laid for easing the curvature of the wall coating [1] which continued as floor coatings. The floor plaster often constituted of two or three successive layers of considerable thickness. A fine-grained coating was often laid over one or two coarser substrates and sometimes this was placed as a base layer for flat stones [1].

During the Neolithic period, lime and gypsum either separately or in combination, were used as plaster coatings for walls and floors. The chemical and mineralogical analyses have shown that plasters used in the Neolithic settlements of Kalavasos-*Tenta* and Khirokitia were either of calcitic or gypsum composition. The petrographic analysis has shown that in most cases lime constituted the binder, but there were also some cases in Kalavasos-*Tenta* where gypsum seems to have been the binder material. In the abovementioned Neolithic settlements the walls were usually coated with a thin whitish plaster layer laid on a base of friable mud plaster. It is noteworthy that in Cyprus during the first periods of antiquity (Neolithic), the knowledge of the technology for plaster production, in relation with the rest of the civilization of the island gives the impression of a very advanced culture.

The microstructural investigation of prehistoric mortars demonstrated that the lime heating process while known since the Neolithic period, was widespread on the island during the Chalcolithic period, as several plaster floors of considerable thickness consisting of one or two dense layers of lime, 0.05–0.10 m thick, were found (Fig. 7). These thick lime floors were quite different from the Neolithic thin final layers of lime. The detailed investigation of various samples under the scanning electron microscope revealed information about the microstructure of these mortars. The very small size of the particles of the binder verified the use of lime and microcrystalline-calcite as a binder material. These plasters were harder and thicker compared to the Neolithic ones suggesting that the combustion process of limestone was widespread during this period.

In the main rooms of important Late Bronze Age public buildings, the coatings of the flooring were thick and consisted of a hard calcitic material [1]. The discovery and use of crushed brick-lime mortars for the first time in the Late Bronze Age constitutes the most important evolution in the manufacture of lime mortars during this period. Their main use was in floors where the capillary rise of water was expected (bathrooms and laboratory areas) and in various water-related structures (water channels). The close examination of thin sections under the petrographic microscope as well as the examination of samples under the scanning electron microscope have shown that these mortars were prepared with the insertion of extremely fine ceramic (brick powder) which has a high pozzolanicity when it is heated in low temperatures. Ancient ceramics seem to have been heated in relatively low temperatures so they were suitable as pozzolanic additives. The appearance of the crushed brick-lime plasters in Cyprus during this period can be associated with the overall prosperity of the era, and particularly with the emergence of urban centres [20].

With the exception of the wide use of gypsum plasters in the Neolithic settlement of Kalavasos-*Tenta*, gypsum plasters are to be found mainly during the Late Bronze Age for special applications (i.e. as a material for the fastening of wooden elements). The limited use of gypsum is surprising, since the island has very notable deposits of gypsum rocks.

From the above it is clear that during prehistoric period there was a great knowledge of the different properties of each material used for mortars and plasters. Thus, the most appropriate material (mud, lime, gypsum etc.) was selected to be used in each case. Mud was mainly used as mortar between stones and adobes due to the easy way of preparation. Lime was mainly used for the construction of floors especially for important buildings due to its better properties. Hydraulic lime was used only in the structures that needed a waterproof material whereas gypsum was used in combination with timber elements due to their compatibility.

2.4 The Use of Timber in the Construction of Roofs and Walls

In prehistoric architecture the use of timber, especially pine and cypress, was observed mainly for the construction of roofs, storey floors, openings, staircases and to a more limited degree for auxiliary walls as well as for reinforcing the structure of main walls in important public buildings. Timber was also used for the construction of the upper part of the staircases, opening panels (shutters), horizontal lintels of the doors and windows and in some cases the lower part of the openings.

The roofs of structures in the prehistoric period (circular and linear) were most probably flat. The main timber beams rested on the walls and sometimes on intermediate stone or timber supports. Smaller beams rested on the main ones, followed by tree trunks, reeds sometimes in intersecting layers and covered by two or more layers of mud. The intermediate posts were placed in a row and sometimes surrounded a central hearth in order to support an elevation of the roof for the extraction of smoke. The construction of storey floors of the Neolithic circular structures as well as of the intermediate floors of the two-storey buildings of the Late Bronze Age was also made of timber and followed the same structural details. The beams also rested on the intermediate stone pillars or timber elements and on the side walls, and they were covered by reeds and soil [1].

Construction of walls with the use of timber elements was rather limited during the prehistoric period. There are a few examples of very primitive light timber structures (utilizing reeds and brush) of the earliest periods of antiquity. Timber elements were either vertical or inclined creating conical structures. Some walls of important buildings of the Late Bronze Age were probably constructed with the use of timber frames, comprising of horizontal and vertical elements and filled with adobes. These frames rested on a stone base made of ashlar and aimed to reinforce the structure of the wall. The timber elements were fastened in small blind holes on the upper face of the *orhtostats* with the help of smaller timber elements and lead (Pb) [21].

3   Discussion and Conclusions

The very high level of the building technology observed during the first periods of antiquity is impressive. The prehistoric people since the earliest periods of antiquity managed to reach a very high level in many aspects of the civilization of the island including the construction industry. The burning of limestone and gypsum rocks in order to prepare lime and gypsum mortars, the addition of ceramics in lime mortars in order to prepare hydraulic lime, the procedure of quarrying and dressing stone as well as the preparation of adobes using moulds constitute some of the most important innovations in the field of construction of this period. It is also remarkable that in the vernacular architecture of the last centuries the same traditional techniques have been followed, underlining the fact that prehistory people managed to solve from very early most of the structural problems following simple and economic solutions.

At the same time there is a chronological development in the use of building materials during prehistoric period. Timber which was the earliest material used in architecture before the Neolithic period, was soon replaced by more durable materials such as stone and clay, as these materials could better meet the needs of a permanent habitation. The irregularly shaped adobes of the Neolithic period were replaced later by the Late Bronze Age rectangular adobes prepared with the use of moulds. The early exclusive use of rubble stone was later (during the Late Bronze Age) enriched with the appearance of ashlar stone. The clay coatings were gradually replaced by lime coatings, which were used since the earliest periods to a limited scale. The discovery of crushed-brick lime mortars during the Late Bronze Age constitutes an important innovation of the period as it constitutes a first form of cement.

In the various structures, the builders proceeded to the best combination of building materials, which has been followed diachronically, even in the vernacular architecture of the last centuries. Stone was used for the lower part of the structures for structural reasons, as well as for the protection of the structures from moisture, adobe which is a more light material was used for the superstructure, and timber for the roofs and sometimes for creating a frame of the superstructure. Generally, there is a selective use of the most suitable materials in each case, which implies a clear understanding of the capabilities and properties of each material. There was always a selection of the most suitable stone for ashlar, the most suitable clay soil for adobes and as a covering material of ceilings, the most suitable calcitic rocks for the preparation of lime as well as the most suitable timber for the construction of the wooden frame of roofs and openings.

In prehistoric architecture, there was always an effort to satisfy the various functional needs and to solve the various constructional problems, with the most economical materials and the simplest techniques without paying particular attention to decoration issues. Thus, preference was given to adobes instead of baked bricks due to the easier and most economical way of preparing them without the use of fuel, clay was used instead of lime as a bonding mortar, timber instead of metals, etc. The same strategies have diachronically been employed in the island, especially for the erection of simple domestic dwellings.

The geophysical resources as well as the climatic conditions of the island played an important role in the creation of the built environment. The existence of appropriate rocks which were easy to quarry, helped towards the widespread use of ashlar all over the island. The abundance of clayey soil in the areas around the settlements resulted in the extensive use of this material both in the manufacture of adobes and as a bonding mortar. In addition to the geological environment, the role of the dry climate of the island which allowed the wide use of adobes (easy drying and preservation), was decisive. The use of adobes as well as the use of timber reinforcing frame in the superstructure may also be connected to their anti-seismic properties.

As aforementioned, the high level of the building technology of the prehistoric period is closely interlinked with other social, economic and political factors and can be attributed to the overall prosperity and cultural development of the era. The simplicity and efficiency of these materials and techniques are evident in the fact that the same traditional building materials and techniques have diachronically been employed in the island for the construction of domestic dwellings, even in the vernacular architecture of the last centuries without major changes. As this paper has indicated, this can be attributed to the high structural level that was reached during the prehistoric period, facilitated by the environmental and climatic conditions of the island, which led to the use of readily available materials and functional solutions that could satisfy the same needs and requirement for shelters diachronically.

References

1. Philokyprou, M.: Building materials and structures in ancient cypriot architecture. Unpublished Ph.D. thesis, University of Cyprus (1998)
2. Dikaios, P.: Enkomi. Exvavations 1948-58. The Architectural Remains, vol. 3. Philipp von Zabern, Mainz am Rheim (1969)
3. Philokyprou, M.: Building materials and methods employed in prehistoric and traditional architecture in Cyprus. In: Ethnography of European Traditional Cultures, Arts, Crafts of Heritage, Athens, pp. 150–164 (1998)
4. Hult, G.: Bronze Age Ashlar Masonry in the Eastern Mediterranean, Cyprus, Ugarit and the Neighbouring Regions. Studies in the Mediterranean Archaeology, vol. LXVI. Paul Åströms Förlag, Göteborg (1983)
5. Philokyprou, M.: The initial appearance of ashlar stone in Cyprus. Issues of provenance and use. Mediterr. Archaeol. Archaeom. Int. Sci. J. **2011**(2), 37–53 (2011)
6. Coulton, J.J.: Lifting in early Greek architecture. J. Hell. Stud. **94**, 1–19 (1974)
7. Karageorghis, V., Demas, M.: Excavations at Kition V. The Pre-Phoenician Levels, I-II. Department of Antiquities, Nicosia (1985)
8. Negbi, O.: The climax of the urban development in Bronze Age Cyprus. Report of the Department of Antiquities, Cyprus, pp. 97–121 (1986)
9. Hadjisavvas, S.: Ashlar buildings and their role in Late Bronze Age Cyprus. In: Acts of the Third International Congress of Cypriot Studies, Nicosia, pp. 387–398 (2000)
10. Philokyprou, M.: The earliest use of adobes in Cyprus: issues of provenance and use. In: Proceedings of Terra 2016: XIIth World Congress on Earthen Architectures, Lyon (2018)
11. Thomas, G.: Prehistoric Cypriot mud buildings and their impact on the formation of archaeological sites. Ph.D. thesis, University of Edinburgh (1995)
12. Costi di Castrillo, M., Philokyprou, M., Ioannou, I.: Comparison of adobes from pre-history to-date. J. Archaeol. Sci.: Rep. **12**, 437–448 (2017). https://doi.org/10.1016/j.jasrep.2017.02.009
13. Leick, G.: A Dictionary on Ancient Near Eastern Architecture. Routledge, London (1988)
14. Guest-Papamanoli, A.: L'Emploi de la Brique Crue dans le Domaine Egéen à l' Époque Néolithique à l'Âge du Bronze. Bull. Corresp. Hell. **1**, 3–24 (1978)
15. Le Brun, A.: Fouilles Récent a Khirokitia (Chypre) 1977-1981. Editions Recherche sur les civilisations, Paris (1984)
16. Todd, I.: Vasilikos Valley Project 6. Excavations at Kalavasos Tenta. Sima, vol. LXXI:6. Åström, Goterberg (1987)
17. Kingery, W.D., Vandiver, P.B., Prickett, M.: The beginnings of pyrotechnology, part ii: production and use of lime and gypsum plaster in the pre-pottery neolithic near east. J. Field Archaeol. **15**, 219–244 (1988)
18. Moropoulou, A., Bakolas, A., Bisbikou, K.: Investigation of the technology of historic mortars. J. Cult. Herit. **1**, 45–58 (2000)
19. Philokyprou, M.: The beginnings of pyrotechnology in Cyprus. Int. J. Arch. Herit. (Taylor Fr.) **6**(2), 172–199 (2012)
20. Theodoridou, M., Ioannou, I., Philokyprou, M.: New evidence of early use of artificial pozzolanic material in mortars. J. Archaeol. Sci. (Elsevier Ed. Syst.) **40**(8), 3263–3269 (2013)
21. Wright, G.R.H.: Ancient Building in Cyprus. E.J. Brill, Leiden (1992)

TLS Survey and FE Modelling of the Vasari's Cupola of the Basilica dell'Umiltà (Italy). An Interdisciplinary Approach for Preservation of CH

Grazia Tucci[ID], Gianni Bartoli[ID], and Michele Betti[(✉)][ID]

Department of Civil and Environmental Engineering,
University of Florence, Florence, Italy
{grazia.tucci,gianni.bartoli,michele.betti}@unifi.it

Abstract. This paper presents an interdisciplinary approach for identification and assessment of historic buildings that combines Terrestrial Laser Scanning (TLS) survey and Finite Element (FE) numerical modeling. The structural analysis of an historic building requires the development of an interconnected series of operations aimed at obtaining a satisfactory knowledge of the building, where usually in-situ investigations are performed together with advanced computational analyses. In this process, the geometric and topographic survey plays a pivotal role and therefore the possibility to rapidly acquire large amounts of spatial data (and to geo-reference any kind of information) allows to provide effective geometric and monitoring data that can be subsequently employed for structural analyses. In this respect, the interchange between Geo-informatics and Engineering sciences can be considered a challenging issue in the field of conservation/preservation of cultural heritage (CH). On the one hand, in fact, the accuracy of measured data directly affects decision-making and analysis process. On the other hand, the merging of digital documentation technologies with innovative computational techniques supports the creation of an inter/trans-disciplinary cooperation model towards sustainable preservation of CH. These issues are herein addressed through the discussion of an emblematic case study: the Cupola of the Basilica dell'Umiltà in Pistoia (Italy) designed and realized by Giorgio Vasari in the middle of the sixteenth century.

Keywords: 3D digitization of cultural heritage · Metric survey ·
Numerical modeling and structural analysis · Dome

1 Introduction

The capacity to rapidly acquire large quantities of spatial data, to geo-reference several type of information, to obtain detailed models that allow even more accurate analyses and simulations, puts Geo-informatics at the center of the attention of many research areas. Among these, particularly interesting is the use of these techniques for the analysis of the cultural heritage (CH). Assess the current stability of a building, identifying its degradation conditions, monitor the evolution over time of a failure,

© Springer Nature Switzerland AG 2019
A. Moropoulou et al. (Eds.): TMM_CH 2018, CCIS 961, pp. 487–499, 2019.
https://doi.org/10.1007/978-3-030-12957-6_34

preventing the potential causes of damage, simulate or reconstruct the behavior of a building under seismic actions, are some of the ways in which the geometric properties of a structure, acquired with the most up-to-date automated surveying systems, are employed to support structural integrity analyses.

From the point of view of the restoration science, a scientific approach that interprets the text of a construction must be based on a systematic observation and measurement of the building. The complexity of the relations between its elements, determined by their position in space, can be systematically approached only by a virtual model that takes into account the reciprocal spatial relations of the collected data. For this reason, well before that this analytic and descriptive approach was rigorously defined, the survey - and its subsequent architectural representation in two and three dimensions - has constituted the main instrument for the study of the CH [1, 2].

Nevertheless, even if the survey boasts a long and solid tradition in the field of restoration, its role has been truly appreciated and understood mainly in recent times. In fact, an examination of the so-called *Carte del Restauro*, reveals that references to the survey are both superficial and fragmentary. The *Carta di Atene* (1931) recommends "precise surveys" only for archeological excavations at sites that will be reburied. The *Carta di Venezia* (1964) discusses "precise documentation in the form of analytical and critical reports, illustrated with drawings and photographs", without specifying the metrological value of the drawings. The *Carta del Restauro* (1972), finally, indicates the content of the restoration project, stating that "it should be based on a complete graphic and photographic survey to be interpreted metrologically." It was not until the *Carta del Rilievo Architettonico* of 1999–2000 [3] that was provided a study explicitly dedicated to the survey. This document recommends that the regulatory criteria for a survey should incorporate a project as well as oversight of the operations and testing. The survey should additionally include a report indicating the implemented criteria, objectives and degree of precision, so as to make qualitative evaluation possible. The survey is further defined as "an open system of knowledge" that brings together all relevant data and whose creation involves multiple professional specializations. This document emphasizes the structural survey among the themes for detailed study, pointing to the dual objective of "illustrating the structural model in its overall configuration" and "documenting the geometric characteristics and the materials necessary for the engineer to conduct the required assessments and tests," in consideration of the need for further experimentation to integrate the survey methods with non-destructive analyses of structures covered by plaster.

The ICOMOS Charter - ISCARSAH Principles (2003) - devotes a section to the structural surveys, beginning with the mapping of visible damage and cracks and of the materials and their state of decay, and moving on to a consideration of geometric and structural irregularities and the relations with the environmental context. Although it highlight the importance of the survey in the restoration project, these documents lack precise specifications, regarding, for example, scale, level of detail, the instruments needed to carry out and verify results. Such shortcomings are probably due to the great variability of situations involved in creating a survey of an architectural structure. Typically, those most familiar with a monument were surveyors armed with tape measures and plumb lines who gauged spaces and noted shapes and dimensions in accordance with their training in the history of architecture; yet the badges of land surveyor or "measurer" have not generally been happily worn by architects or engineers, who consider themselves to be in possession of a different set of skills. Measurements were essentially limited to distances, related to planes whose spatial positioning was not simple to realize. The traditional survey, therefore, was often unsatisfactory above all for measuring buildings with complex floor plans or spaces of great height [4].

For these reasons, researchers have for decades been experimenting with the use of technical instruments which have already been tested for land surveying and cartographical production, such as topography and photogrammetry. In the last decades, with the advances in computer science and the transformation to digital platforms, it have been seen a dramatic change in documentation systems. Terrestrial Laser Scanning (TLD) and Structure from Motion (SfM) digital photogrammetry have for several years now allowed to survey complex geometries at lower costs. If the geospatial market as applied to constructions is considered, attention is no longer directed toward the acquisition itself but rather toward integration and automation of the various technologies:

- Autonomous Technologies - collision avoidance, data visualization, and real-time navigation support are driving commercial adoption of autonomous vehicles in a wide variety of field operations;
- Portable Scanners - handheld scanners, indoor mapping carts, and backpack systems are meeting the needs to extract specific objects and generate 3D models from tight, busy spaces;
- Immersive Data Visualization - moving beyond abstract data sets is enabling professionals to fully manipulate and interact with data in three dimensions;
- All-digital Environments - be it Building Information Modelling or Virtual Design and Construction, AEC firms are rapidly moving to an all-digital design environment to save significant time and money, improve design quality, and increase worker safety;
- Data Processing Automation - organizations are breaking through the bottleneck of overwhelming digital datasets by automating data processing and 3D feature extraction.

All of these techniques aim to define the position of high-density points in space. Their results can be integrated as long as they are expressed in a single reference system defined by a reference network created with a total station or with GPS. Moreover, the permanent materialization of the network and the preservation of all data and metadata allow us to verify the accuracy of the result and its successive integration as well as to monitor the evolution of conditions over time [5–7].

The most interesting aspect of the availability of new digital instruments for reality acquisition - from the point of view of Engineering science - is that they enable to obtain virtual models that are to be understood not only in the architectural or computer-graphic sense as 2- or 3-dimensional images of a building, but from the broader perspective of a conceptual representation that is faithful and objective (apart from measurement uncertainty and inaccuracies in following the adopted procedures, which are in any case verifiable through quality statistical parameters), and thus useful for describing specific phenomena [8–11]. As in the past, proximity to the object, both during acquisition and data processing, places the surveyor in a better position to identify aspects that are less easily detected from a more comprehensive perspective. Three-dimensional acquisition can also be employed as a diagnostic instrument to identify anomalies and irregularities and investigate the deviation between the real geometry and ideal surfaces used as reference points, to monitor the development of instability or of other phenomena, or to automatically characterize decay [12–14].

In summary, surveyors before the digital era, following a principle of economy, acquired data that was sufficient to produce the two-dimensional representations necessary for defining a conceptual model of three-dimensional reality, a model so abstract as to be fully comprehensible only by specialists. Today, on the other hand, digital transformation, driven by technological innovation, allows surveyors not only to acquire 3D geometric information (which can be reduced to 2D data for operational requirements) but also to create 4D models that also take the temporal dimension into account with continuous or repeated measurements. It is therefore evident that the idea of the survey, understood as a synthesis of an open system of knowledge which is always organized and articulated in a spatial form, today finds application in a 3D digital model that is considered the framework for the geolocation of every type of data.

2   Surveyed Structure: The Basilica dell'Umiltà

What has been reported above, delineates a path of organic information acquisition that has been traced with regard to its principle aspects. It must now be shown that in practice its stages are still often applied in a discontinuous way, in accordance with the respective needs of individual projects. It may therefore prove useful to investigate for which aims and in which ways the techniques of geomatics have been used in projects for structural evaluations of significant historical buildings.

An important example of the great potential residing in the adoption of the techniques of geomatics concerns the survey of the great domes [15, 16]. This has always been a challenging work because it involves broad and tall structures which are difficult to measure with precision by direct means. It is further difficult to measure their thicknesses because the extrados is only visible in cramped crawl spaces: if no direct openings are available, creating a topographic network to connect the internal and external surfaces becomes a daunting task. The Vasari's Cupola of the Basilica of the Madonna dell'Umiltà in Pistoia (Italy) (Fig. 1), in this respect, is a broader case for the application of the techniques of geomatics for the study of domes.

2.1   History

The construction of the Basilica della Madonna dell'Umiltà in Pistoia (Fig. 1) started at end of the fourteenth century, following the original design of Giuliano da Sangallo (1445–1516), and required more than seventy years to be completed. After the death of Giuliano da Sangallo the construction was first continued by the architect Ventura Vitoni from Pistoia (1442–1522), and subsequently by Giorgio Vasari (1511–1574) that worked on the Basilica from 1561 to 1567. Vasari designed the last level of the octagonal tambour, together with the Cupola.

Fig. 1. The Basilica della Madonna dell'Umiltà.

The design of the cupola was clearly inspirited to the Brunelleschi's cupola of Santa Maria del Fiore in Florence; in his design Giorgio Vasari reproduced the main key features of the Brunelleschi's dome such as the double masonry shell system, the ribs between the webs and several materials and outer decorative details. Nevertheless, despite such similarities, Vasari did not adopt any of the technical innovation proposed by Brunelleschi in his cupola. Main technical differences can be summarized as follows: (i) the shape of the Cupola of the Basilica della Madonna dell'Umiltà shifted from the gothic to the hemispherical one; (ii) the brick pattern used to build the Cupola was different from that employed by Brunelleschi in the Florentine dome: Vasari laid the brick courses along the parallels disregarding the double curved herringbone courses of the Brunelleschi's cupola; (iii) the connection between the inner and the outer shell was realized with weak joints, disregarding the strong ring and the rib system that held together the inner and outer shell of the Brunelleschi's dome [17, 18].

Due to the three above mentioned constructive drawbacks, and others not discussed here such as the lateral force-carrying capacity of the tambour, a complex and widespread crack pattern formed soon after the completion of the dome (even before the installation of the heavy lantern) that evolved during centuries [19]. Vasari himself carried out the first retrofitting works, through the insertion of steel chains inside and outside the Cupola. Additional chains have been inserted over the centuries due to the evolution of the cracking pattern.

Since 2008, the Cupola of the Basilica dell'Umiltà has been interested by a series of studies, experimental investigations and analyses - promoted by *Soprintendenza* (the Institution in charge of conservation of historical buildings) and financed by the *Cassa di Risparmio di Pistoia e Pescia* Foundation - aimed at the conservation of the structural complex. The project consisted of the restoration of the whole Basilica; the restoration of the Cupola started in 2012 and concluded in 2014 with the renovation of the lantern.

2.2   Geomatic Survey

Recent methodological developments in analyzing masonry structures have led to fresh interest in studying the basilica with the aid of geomatics technologies. The survey is not only the basis on which to create structural models, but especially in such sophisticated and complex cases represents the only tool which allows correlation of the overall geometry, the construction techniques, the characterization of the texture of the materials, the position of the tie-rods, the cracks detection and mapping, etc. Combining these data with historical and archival information enables researchers to gain access to the reasoning behind the initial project and to the modifications carried out over the centuries.

In this case as well, the greatest challenges in performing the survey (extended to the entire basilica) regarded:

- The difficulty in carrying out topographical measurements and laser scans both in the tight crawlspace between the two domes and in surveying the exterior, given that the building is quite high and located in a network of narrow roads;
- The need to conduct a detailed survey of small elements, such as decorative features and the position of cracks, in spite of these limitations.

The survey allowed researchers to more clearly understand the morphology of the work, in particular of elements which are not directly visible. Vasari himself, when the first cracks appeared, modified the system of vertical connections and the crawlspace structures of the dome, adding eight ribs in the crawlspace to sustain the weight of the lantern [20, 21].

The survey (Figs. 2 and 3) highlighted and made it possible to quantify:

- The number of ribs, their correct placing, form and dimension;
- The exact placing and dimension of the internal and external tie-rods;
- The actual thickness of the two calottes; the building materials and techniques;
- The consolidation of the interventions; the internal connections, both horizontal and vertical;
- The cracks present on both calottes, highlighting the most damaged awnings;
- The links among the geometry, structure and deformations shown over time.

2.3   Damage Survey

The Cupola of the Basilica dell'Umiltà is affected by a widespread cracking pattern. The cracks initiated at the springing of the Cupola, and propagated towards the crown, cutting the entire thickness of the inner shell. To avoid the collapse of the cupola, the initial design was modified with the insertion of some tie-rods buried into the masonry during the construction of the dome. Vasari himself inserted the first two iron tie-rods at the base of the Cupola: the first in the corridor between the two shells and the second around the outer dome. In so doing, the construction of the dome was completed. Subsequently the lantern, which was designed by Vasari imitating the one of the Brunelleschi's dome, and particularly heavy, was installed. This increased the lower

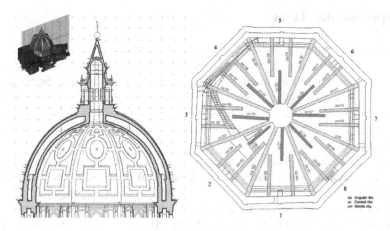

Fig. 2. Madonna dell'Umiltà in Pistoia: cross section of the dome (left) and the ribs between the two calottes (right).

Fig. 3. Madonna dell'Umiltà in Pistoia: 3D model of the space between the two calottes.

thrust of the cupola, and contributed to the enhancement of the cracks on the structure [17]. A third tie-rod was inserted ten years after the completion of the works, around 1585, by Bartolomeo Ammannati (1511–1592) and two more tie-rods were inserted in 1592. Taking into account further interventions during the century, today, eight tie-rods are present on the Vasari's cupola, seven of which disposed at the extrados of the outer masonry shell.

3 Experimental Tests

The experimental campaign included single and double flat-jack tests, and was aimed at assessing the path of internal stresses on the tambour and on the cupola. Even if flat-jack tests are only slightly destructive (since after the test is completed, the flat-jack can usually be easily removed and the mortar layer restored to its original condition), position, number and distribution of tests were selected to minimize obtrusiveness and, at the same time, to obtain the effective structural information needed for the subsequent numerical modeling.

Both single and double flat-jack tests were executed in positions selected in accordance with the *Soprintendenza*. The location of the double flat-jack test takes advantage of previous single one (i.e., after the execution of the single flat-jack test a new flat-jack was inserted and the test was carried out) to reduce the obtrusiveness. Overall four levels (Lev. #1, #3, #4 and #5; Fig. 4) were investigated, alternating the walls in order to obtain a global picture of the state of stress. For the execution of the in situ tests, the following equipment was used: rectangular flat-jack (dimensions: $400 \times 200 \times 4.0$ mm); hydraulic circular saw (ø350 mm and thickness 3.5 mm); hydraulic hand pump [equipped with two manometers, with full scales of 25 and 100 bar (2.5 and 10 MPa)]; and mechanical displacement transducers (with a base length of 200 and 400 mm, with resolution equal to 0.01 mm).

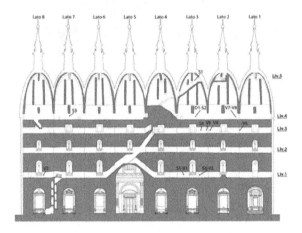

Fig. 4. Position of single (S) and double (D) flat-jack tests. Tests S1, D1s, S3, S4, and S5 were performed on the extrados of the inner tambour/cupola. Tests S2 and S6 were performed on the intrados of the outer tambour.

The measure of pressure applied by the flat-jack, taking into account the corrective factors (Table 1), approximately corresponds to the local pressure in masonry, and

Table 1. Estimated in-place stress and correction coefficients for the single flat-jack tests.

Test#	k_m	k_a	Stress (MPa)
S1	0.743	0.860	0.15
D1s	0.743	0.856	0.15
S2	0.743	0.860	0.18
S3	0.743	0.851	0.15
S4	0.743	0.842	0.28
S5	0.860	0.816	1.28
S6	0.860	0.825	1.28

therefore the average compressive stress in the masonry, σ_m, can be evaluated as follows:

$$\sigma_m = k_m \cdot k_a \cdot p \tag{1}$$

where k_a is the factor accounting for the ratio between the bearing area of the jack in contact with the masonry and the bearing area of the slot; k_m is the factor that accounts for the physical characteristic of the jack; and p is the pressure required to restore the original distance between the gauge points. Position of tests is reported in Fig. 4, while the results, expressed in terms of average compressive stress in the masonry, are summarized in Table 1. It is possible to observe that at the higher levels (Lev. # 3 and #4) the internal stresses are about the same and range between 0.15 MPa (Lev. #4) and 0.28 MPa (Lev. #3), while at Lev. #1 it increases up to 1.28 MPa. The Young's modulus, obtained with the double flat-jack test, was found to be about 800 N/mm^2.

4   Numerical Modelling

Numerical modeling has been proven to be an effective tool to help the comprehension of the structural behavior of ancient structures, and the inherent literature reports a large number of illustrative case studies. Lourenço et al. [22], through the discussion of the case study of the Monastery Jeronimos in Lisbon (Portugal), highlighted the role of advanced numerical simulations for the understanding of the structural behavior of CH. The finite-element technique was used by Taliercio and Binda [23] to analyze the Basilica of San Vitale in Ravenna (Italy), a building which suffers diffused cracking and excessive deformation. Chiorino et al. [24] analyzed the Dome of Vicoforte (Italy), the largest elliptical dome ever built, combining limit analysis and finite-element technique. The dome has been analyzed through models, which are able to provide reliable interpretations of its behavior and damage state. A careful use of numerical analyses dealing with practical engineering problems has been shown by Bartoli and Betti [25], who discussed the cracking pattern on the Prince's Chapel (the Medici's mausoleum) in the Basilica of San Lorenzo (Florence, Italy).

Based on this growing literature several numerical models of increasing complexity were developed to analyze the structural behavior of the Basilica dell'Umiltà. Unknown parameters were assumed by considering reasonable values for historic masonry. These preliminary elastic analyses, even if the hypothesis of linear behavior for masonry is to a large extent an unrealistic assumption, are exclusively aimed at describing the overall behavior of the entire structure, recognizing the role of its structural elements, and evaluating the entity of the principal stresses acting on its main elements from a qualitative point of view.

Results of simplified model were then compared with those obtained from more sophisticated numerical analyses performed using the commercial code ANSYS and adopting solid hexahedral elements to model all the geometrical components [26].

The final three-dimensional solid model was developed based on the geomatic survey, encompassing the entire monument. In addition, taking into account the aim of the research, the modeling strategy used was the macro-modeling technique, which is, moreover, convenient for large-scale models. According to this technique, the masonry units and the mortar elements are assumed to be smeared, and an isotropic (or aniso-tropic) material represents the smeared units and mortars in the masonry. Other strategies, mainly suitable for small-size models, rely on micro-modeling approaches where units and mortar are modeled separately. The main results of simplified linear analyses are reported in Figs. 5 and 6. In Fig. 5 the experimental stress is compared with the one obtained with the numerical model. It is possible to observe a good agreement between the two values. The same can be observed for the remaining test, with the exception of test S6 (Fig. 6) where a difference of about 10% is observed.

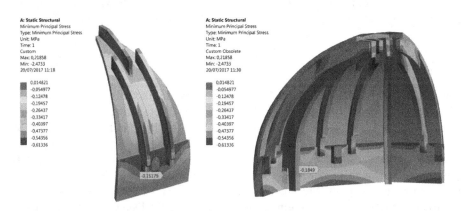

Fig. 5. Test D1s (left) - $\sigma_{m, exp}$ = 0.15 MPa; Test S2 (right) - $\sigma_{m, exp}$ = 0.18 MPa.

Fig. 6. Test S4 (left) - $\sigma_{m,\ exp}$ = 0.28 MPa; Test S6 (right) - $\sigma_{m,\ exp}$ = 1.18 MPa.

5   Conclusive Remarks

The paper showed, discussing the case study of the Basilica della Madonna dell'Umiltà in Pistoia, how the creation of an effective geometric model, through proper instruments of the geomatics, is a mandatory step in the interpretation of the architectural and constructive significance of the various elements that constitute the structure. It must be observed that, after all, this is the underlying approach of the Building Information Modelling that is steadily gaining ground in the design of new construction. The HBIM (Heritage-BIM), the specific application for cultural heritage buildings, is still only used sporadically and experimentally: nonetheless, it is evident that the possibility of gathering all the information about a structure in a single database, organized spatially and capable of progressive integration in light of later investigations, would be particular advantageous for those great complexes, about which knowledge is usually divided among many different specialized fields.

Acknowledgments. The metric survey, coordinated by the authors, has been developed during the Specialization Thesis of the arch. M. Riemma and the Phd Thesis of the arch. A. Nobile: the Figs. 2, 3 and 4 show their drawings. The authors wish also to acknowledge: Arch. V. Tesi of the Soprintendenza Archeologia Belle Arti e Paesaggio per la città metropolitana di Firenze e per le province di Pistoia e Prato; Eng. G. Palchetti of the Cassa di Risparmio di Pistoia e Pescia Foundation, who supported the research; V. Bonora, L. Carosso, F. Panighini and I. Tomei who took part in different survey campaigns, together with the authors; Prof. F. Russo with Zenith Ingegneria Srl, a spinoff of the Engineering Dept. of the University of Ferrara who carried out the topographical framework and E. Fiorati, Canon of the Basilica dell'Umiltà.

References

1. Migliari, R.: Per una teoria del rilievo architettonico. In: Disegno come Modello–Riflessioni sul disegno nell'era informatica. Edizioni Kappa, Roma (2004). (in Italian)
2. Musso, S.: Rilevare - Restaurare: una diade inscindibile. In: ΑΝΑΓΚΗ - Speciale GeoRes2017, pp. 2–7, Altralinea Edizioni, Firenze (2017). (in Italian)
3. Almagro, A., et al.: Verso la Carta del Rilievo Architettonico - Testo di base per la definizione dei Temi. In: Seminario Internazionale di Studio "Gli strumenti di conoscenza per il progetto di restauro", Valmontone, Italy (1999). (in Italian)
4. Tucci, G., Conti, A., Fiorini, L.: Scansione laser per il rilievo dei giardini storici. Geomedia 17(6), 14–18 (2013). (in Italian)
5. Grussenmeyer, P., et al.: Accurate documentation in cultural heritage by merging TLS and high resolution photogrammetric data. In: Proceedings of SPIE XI, Munich, Germany (2011)
6. Voltolini, F., et al.: Digital documentation of complex architectures by integration of multiple techniques - the case study of Valer Castle. In: Proceedings of SPIE IX (2007)
7. Rodríguez-Gonzálvez, P., et al.: 4D reconstruction and visualization of cultural heritage: analyzing our legacy through time. Int. Arch. Photogramm. Remote Sens. Spat. Inf. Sci. 42, 609–616 (2017)
8. Tucci, G., Bonora, V.: Geomatics and management of at-risk cultural heritage. Rendiconti Lincei 26(1), 105–114 (2015)
9. Korumaz, M., et al.: An integrated terrestrial laser scanner (TLS), deviation analysis (DA) and finite element (FE) approach for health assessment of historical structures. A minaret case study. Eng. Struct. 153, 224–238 (2017)
10. Castellazzi, G., et al.: An innovative numerical modeling strategy for the structural analysis of historical monumental buildings. Eng. Struct. 132, 229–248 (2017)
11. Pieraccini, M., et al.: Dynamic identification of historic masonry towers through an expeditious and no-contact approach. Application to the "Torre del Mangia" in Siena (Italy). J. Cult. Herit. 15(3), 275–282 (2014)
12. Sidiropoulos, A.A., Lakakis, K.N., Mouza, V.K.: Localization of pathology on complex architecture building surfaces. Int. Arch. Photogramm. Remote Sens. Spat. Inf. Sci. 42, 617–621 (2017)
13. Nespeca, R., De Luca, L.: Analysis, thematic maps and data mining from point cloud to ontology for software development. Int. Arch. Photogramm. Remote Sens. Spat. Inf. Sci. XLI-B5, 347–354 (2016)
14. Del Pozo, S.J., et al.: Multi-sensor radiometric study to detect pathologies in historical buildings. Int. Arch. Photogramm. Remote Sens. Spat. Inf. Sci. 40(5), 193–200 (2015)
15. Balletti, C., et al.: Ancient structures and new technologies: survey and digital representation of the wooden dome of SS. Giovanni e Paolo in Venice. Int. Arch. Photogramm. Remote Sens. Spat. Inf. Sci. II-5/W1, 25–30 (2013)
16. Bartoli, G., Betti, M., Torelli, G.: Damage assessment of the Baptistery of San Giovanni in Florence by means of numerical modelling. Int. J. Mason. Res. Innov. 2(2–3), 150–168 (2017)
17. Foraboschi, P.: Resisting system and failure modes of masonry domes. Eng. Fail. Anal. 44, 315–337 (2014)
18. Blasi, C., et al.: On the hooping of large masonry domes: discussion on strain measurements and interaction with masonry. In: Proceedings of SAHC 2014, Mexico City, Mexico (2014)
19. Tonietti, U., Ensoli, L., Calonaci, M.: Sulle condizioni statiche della Madonna dell'Umiltà: storia di una costruzione non propriamente brunelleschiana. In: Centenario del miracolo della Madonna dell'Umiltà di Pistoia, Pistoia, Italy (1992). (in Italian)

20. Tucci, G., Nobile, A., Riemma, M.: The Basilica della Madonna dell'Umiltà in Pistoia: survey, analysis and documentation. In: 23rd International CIPA Symposium, 12–16 September, Prague, Czech Republic (2011)
21. Tucci, G., Nobile, A., Riemma, M.: Laser scanner surveys and the study of the geometry and structure of the dome in the Basilica della Madonna dell'Umiltà in Pistoia. In: Proceedings of the International Congress "Domes in the World", Florence, Italy (2012)
22. Lourenço, P.B., et al.: Failure analysis of Monastery of Jeronimos, Lisbon: how to learn from sophisticated numerical models. Eng. Fail. Anal. 14(2), 280–300 (2007)
23. Taliercio, A., Binda, L.: The Basilica of San Vitale in Ravenna: investigation on the current structural faults and their mid-term evolution. J. Cult. Herit. 8(2), 99–118 (2007)
24. Chiorino, M.A., et al.: Modeling strategies for the world's largest elliptical dome at Vicoforte. Int. J. Arch. Herit. 2(3), 274–303 (2008)
25. Bartoli, G., Betti, M.: Cappella dei Principi in Firenze, Italy: experimental analyses and numerical modeling for the investigation of a local failure. J. Perform. Constr. Facil. 27(1), 4–26 (2013)
26. Betti, M., Galano, L., Vignoli, A.: Seismic response of masonry plane walls: a numerical study on spandrel strength. AIP Conf. Proc. 1020(1), 787–794 (2008)

Interdisciplinary Approaches in Cultural Heritage Documentation: The Case of Rodakis House in Aegina

Georgiadou Zoe[2]([✉]), Alexopoulou Athina-Georgia[1],
and Ilias Panagiotis[3]

[1] Department of Interior Architecture, University of West Attica, Athens, Greece
athfrt@uniwa.gr
[2] Department of Conservation of Antiquities and Works of Art,
University of West Attica, Athens, Greece
zoegeo@uniwa.gr
[3] Department of Photography and Audiovisual Arts,
University of West Attica, Athens, Greece
panil@uniwa.gr

Abstract. The surveying and documentation of cultural heritage monuments are widely considered to be the very first steps towards their preservation. In practice, these procedures are undertaken with the use of advanced equipment, which yields the capabilities of detailed recording, representation and digitisation, so as to provide easy access to the material gathered and thus ensure its availability for future research. The expected actions and procedures appear to unfold smoothly when the cultural artifact being documented is public, and is presented as a focal point of cultural interest, ascribing to it an active role in the social, cultural and economic development of the place where it is located. On the contrary, the documentation of private monuments exhibits considerable difficulties in obtaining access permissions, gathering archival material, using specialised technical equipment and coordinating actions in the context of research procedures and interdisciplinary approaches.

The vernacular house of Rodakis is a private monument on the island of Aegina. Despite the fact that it was well known during its creator's lifetime, continued to be a point of attraction and interest for experts and a place of informal education for students, and was officially declared as a monument of cultural heritage, none of the above was able to prevent its destruction, without the required maintenance over time. The changes in its property status established new conditions for the exclusion of experiential visits at the natural monument, which had indeed become dangerous without the necessary restoration works.

The references and studies on the character of Rodakis house as a monument of architectural heritage cover a period of 110 years, during which an extended directory of archival material of various types was compiled. Overall, these elements place the house in a prominent position, but have not yet been assembled in a comprehensive interdisciplinary study. The present paper attempts to outline the conditions and the possibilities for cooperation between different scientific fields, in the interdisciplinary surveying and documentation of private architectural monuments.

© Springer Nature Switzerland AG 2019
A. Moropoulou et al. (Eds.): TMM_CH 2018, CCIS 961, pp. 500–512, 2019.
https://doi.org/10.1007/978-3-030-12957-6_35

Keywords: Cultural heritage · Rodakis house · Documentation ·
Interdisciplinary approach · Private monument · Digitisation

1   Introduction

Alekos Rodakis house in Aegina is a rural house well known as unique case of
vernacular architecture before 1907. It was discovered by the German archaeologist
Adolf Furtwangler, who discerned the value and the authentic character of this local
house of Aegina decorated with lots of ornaments and symbols. In Greece this house
was publicized by Dimitris Pikionis who studied it with sketches and drawings, as it is
shown in the surveying and photographs included in the Modern Greek Architecture
Archives of Benaki Museum which are dated from 1912–18 [1] (Fig. 1). It seems that
these studies were part of his research about contemporary Greek architectural tradition, with references to the local houses in Aegina. In his critical research about Greek
traditional vernacular art [2], pp. 144–158 he refers to the local styles which originate
from the spiritual expression of the necessity and nature, interpret older forms and
create new expressive forms: decoration [3], p. 21. Part of this archival material is
published in 1994 in an anniversary edition for Pikionis, which was edited by his
daughter Agnes, with systematic research and chronicle of his works [3].

It seems that the house soon became familiar to many architects, such as Aristotelis
Zachos [4] (Fig. 1) and George Candilis, who describes his experience in his book
"Building Life, an Architectural testimony of its era" [5]. Particularly, he characterises
the house as exhibiting vivid architecture, describing a process of constant developments and artistic interventions at the building's form and ornaments. Vasiliadis [6, 7],
as well as Michelis [8] refer to this house at their publications on vernacular architecture in Aegina and note its typological relevance to the main formal types of
Aegina's houses which are recorded and analysed.

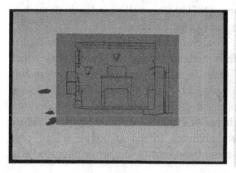

Fig. 1. Pikionis, D., (left): Rodakis House, study 1912. Elevation with fire place.
ANA_67_01_47. Zachos, A., (right): Rodakis House, photograph 1928. ANA_30_35_13. Both
in the Modern Greek Architecture Archives of Benaki Museum. © [1, 4].

As a stand-alone project, Rodakis house was presented with illustrated material granted by Pikionis, in the book "A vernacular house in Aegina" by Klaus Vrieslander and Julio Kaimi (1934). The second edition of Vrieslander's and Kaimi's book (1997) [9] includes a short preface note by Aris Konstantinides [9], p. 9, an introduction by Dimitris Filippidis titled "Between text words and image" [9: 11–24], as well as an addendum by the editor of the publication Michel Faise titled "A vernacular house in Aegina: coincidences and paradoxical" [9: 63–73].

For the first time Rodakis house was presented among other historical monuments of Aegina in a Guide by Gogo Koulikourdi and Spyros Alexiou (around 1950), with a two-page description, which classifies it as a "monument outside the city" and characterises it as "Vernacular Art" [10], pp. 110–111. In a recent publication involving a critique of the aesthetics, the conceptual content and the classification of decorative arts, Philippidis [11], pp. 36–37 comments that Rodakis' house became a landmark of the Greek movement for returning to our roots in tradition and includes a drawing by Pikionis [11], p. 37 of an unknown house with all the morphological influences by Rodakis house: the snake, the birds, the Turk's figure.

However, the significance of Rodakis house is not limited to the above references, but extends deeper, is more influential and continues to be an all-time great, locally as well as internationally. Aldo Van Eyck's Exhibition in the National Gallery (1983) is dedicated to Pikionis and Rodakis from Aegina. In his introductory note for the exhibition under the title of the Greek sigh "Ah-1891-vah," Candilis wrote "I would wish for this small work about construction to be a substantiation of honor for Pikionis and Rodakis, who has left us a legend just before embarking on the ship, which was standing above his door. And he had added without knowing it, this thing that was necessary for a simple construction to became architecture: love and sensibility" [12]. The official declaration of the building as a monument of contemporary architectural heritage in 2001 [13], almost a century after it was first revealed, has found it denuded of its sculptures (since 1995), in bad condition (already in 1989) [14] which deteriorated up to 2009, the year that part of the roof collapsed. Throughout this period, public bodies such as the National Technical University of Athens (NTUA), as well as individuals such as the local teacher Nektarios Koukoulis, have intervened unsuccessfully.

A poetic reference to Rodakis house as an architectural construction connected to an imaginary narration about its creator's life is made in a video by Nikolai [15], which was part of Athens Biennale (2007). The video's material reveals some interesting plans of ornaments that are have been destroyed, such as the graded cabinet. The same is observed in Tritsibidas' video [16]. Important documents are included in individual collections, such as in the photographic archives of Dimitris Moraitis who had visited Rodakis house in 1980. From his visit we obtained fifteen analogue colour photographs, which depict the condition of the monument during the period that Rodakis' daughter Marina lived in it.

In 2011, as the building has been abandoned and one could notice a number of interventions (related to the electricity supply, etc.) that have altered the authentic character of the building, a new detailed surveying is completed by a team of students of the School of Architecture of NTUA, as part of the "building construction" course of their study programme. For this surveying we know from an article published by their supervising professor, Papaioannou [17] who points out that the detriments of the building cannot be reversed, and notices that as early as 1956 George Seferis warned: "How many years this house is going to live? 5 or 10; the strange thing is that now an inscription on the quay suggests to the traveller to visit Rodakis' House. In the past, at least something was wasted and almost nobody knew about it" [18].

The cultural value of Rodakis house was the inspiration for a visual dialogue between photographers Mary Modeen and Panagiotis Ilias, in a group exhibition [19] that took place in cooperation with the Master program of Dundee University in Scotland. In 2016, Rodakis house was transferred to an individual who began its preservation. In a recent article, Papaioannou writes (2017) "And as if this was not enough, a few months later we came to the desecration and the vulgarisation (of the house). We have constructed secretly and without a building permission, a miserable concrete block on the rooftop, with unspeakable skylights, like a permanent tombstone! And I say "we" because we are all in a way responsible for this crime" [20].

2   The Research Out-Line

The references and studies on Rodakis house cover a period of 110 years and during these years a comprehensive body of archival material of different kinds has been constructed: surveying, sketches, photographs—analogue and digital, videos, books, etc. Overall, these elements place the house in a prominent position, although in the original site exists only a simple vernacular, primitive house of the common local type- with a flat roof, main house and subordinate buildings, as most of the contemporary houses of Mesagros in places lacking water. Our research, which took place in 2014–2016 had three objectives: first, the production of primary visual documentation material with digital photographs, in order to represent the present situation of the monument. The collapse of the roof of the main building established the conditions for its rapidly evolving destruction that seemed unpreventable a midst the current economic crisis. In addition, we aimed at collecting, classifying and digitising (in case of analogue photographs) all the known documentary material from different sources: bibliography, classified archival material, collection of unknown elements—mostly unpublished photographs, printed sources of various kinds, online sources, videos, artistic installations, etc., as well as the representation of all the ornaments and sculptures in drawings, using the above informational sources (primary and secondary) in order to record the overall decoration of the house. We should note that the destruction of the roof and the rain water which ruined the surface layer of the plaster, revealed pre-existing murals, that are not tracked in previous surveys. In the recent research they are placed in the interior space of the main house. Secondly, we worked towards the integration of different findings of previous studies in a unified research study and its enrichment with new evidence coming from interdisciplinary approaches, such as the investigation of samples of different rooms and the chemical analysis of the paint layers in order to study the decoration microstructure. Finally, the third objective was to revive a public dialogue, which seemed to have faded, but ultimately led to the participation (through the memorandum of agreement signed between the laboratory of Design, Interior Architecture and Audio-visual Documentation, and the Municipality of

Aegina) and co-ordination of the Project – course of the 2nd class of high school of Aegina. Through this, we have introduced the students to the value and recognition of their local cultural heritage. After several in-situ visits at Rodakis house all the student works were edited in an issue. Additionally, the research material was presented in two events – one that took place in the high school and was organised by the Women Association of Aegina (2018) and the other in the 2nd Greek Conference of Digitisation of Cultural Heritage (2017) [23].

On the one hand, the above-mentioned review of the sources and the archival material presented, highlights the significance of the monument, which exhibits all the primary and secondary characteristics of the common local architectural type, with human scale, analogies, integration in the place, while at the same time reveals its holistic design with unique decorative interventions and ornaments, which were developed and re-casted in a way that characterised it as vivid architecture. On the other hand, the impact of the monument is revealed at all tiers of the public discussion—in research and education at a local, national and international level.

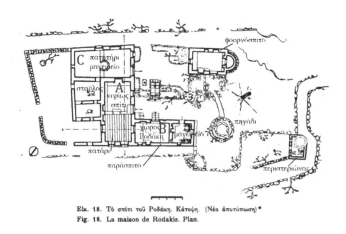

φουρνόσπιτο
C
πατητήρι
μαγειρείο
σταύλος
A
κυρίως σπίτι
χώρος Ροδάκη
B
μαγειρείο
πηγάδι
πατάρι
παράσπιτο
περιστεριώνας
Εἰκ. 18. Τὸ σπίτι τοῦ Ροδάκη. Κάτοψη. (Νέα ἀποτύπωση)*
Fig. 18. La maison de Rodakis. Plan.

Fig. 2. Rodakis house. Ground plan. Source - Michelis, P., A, (1981: 157) [8].

2.1 Architectural Typology and Decorative Interventions in Rodakis House

As an architectural type the building consists of three volumes. The main house, which is a single-space building (A), connected with the subordinate building with three rooms (B) where Alekos Rodakis had chosen to put his personal spaces, and the place for grape stomping (C), which functioned also as a cooking place according to some researchers (Fig. 2). The whole set of the three buildings falls into one of the two main rural house types of Aegina and their later forms, which are recorded in Vasiliadis' studies [6], namely it consists of a single room wide-face house with an above-ground space and a basement. This single-space room extends into two levels with an above-ground space for sleep and fireplace in the upper level, and a basement under the sleeping space, combined with the type of the house covered with flat rooftop in two levels [6]. The third building, which is also accessible independently from the courtyard, has a fireplace like the main house, and is connected to it. It was built later (possibly in 1891) and represents a place for grape stomping and cooking, which seems to be of great importance in the buildings' order, in its size as well as in the decorative interventions in place. These morphological elements commonly feature in subordinate buildings, but also seem to attach to it a more important role with primary functions. This building, although constructed later, seems to have been abandoned unfinished during the period between 1912 and 1917 [1] perhaps due to local grape must production, which is reduced at the time. The buildings described are accompanied by all the auxiliary buildings expected for this rural typology: the oven building, the dove house, the stables and the water well. Conceptually, the floor plan from L-turns to U-shaped, enclosing the oven building as well as the courtyard, which functions as a multi-level extension of the interior space, adapted to the climate conditions of the island.

The buildings follow smoothly the natural ground curves, and integrated to the landscape according to human scale. The courtyard is part of their functional configuration, interchanging the ground levels of the exterior space in a manner similar to the interior space (Fig. 3). The ground plan is oriented towards the south in the courtyard, and benefits from the view of the roof levels of the (B) building. The configuration of the courtyard follows the building standards described by Vasiliadis [6] and is covered with paving stones, in order "not to be covered with mud during rainfall in winter" [7].

Fig. 3. The access to building B, where Alekos Rodakis had chosen to put his personal spaces (left). Rodakis hands curved on the fireplace columns, building B (middle). The front window, building B (right). Photographs by Panagiotis Ilias ©

What, then, makes a farmer's house with flat roof, which has all the primary and secondary characteristics of the local communal architectural type, unique and famous? A first response could be the human scale, the analogies, the integration, the "spirit of the place", characteristics that someone could also find in other local houses. There cognition of the harmonic treatment of the volumes, the patterns and the accepted communal forms, the connection between the social structure and all these character-istics of eternal architecture, are still not the only ones that make Rodakis house unique. Its differentiation lies in the enrichment of this open communal form with exceptional decorative interventions (Fig. 3). The study and recording of these features, concerning the ornaments and the apparent influence of the past prototypes, as well as the visual memories, expressed in Rodakis' sculptures, are these elements which revealed the expressive value of this particular house. But it seems that the building continues to reveal its ornaments, which range on a wide scale of folk art with many decorative applications: sculptures and painted motives, decorated functional and constructive pieces, lyrics and sighs (Fig. 3). All these consist of a decoration with qualities and forms which are produced by Rodakis' creativity. The symbols take their place on a house where everything, even the stone forms and sizes, are combined on the walls and in the courtyard with constructive details that can be characterised as decoration, and highlight an open dialogue between the plastic figurative configuration and the deco-rative ornaments, both sculptured, carved and painted.

2.2   Recording and Documentation

Throughout the literature reviewed above, the exterior façades were emphatically dis-played along with the sculptural decoration that was recorded on the buildings. Yet, the interior spaces, which have been shown only in sketches and not in published pho-tographs, have ornaments that are depicted fragmentarily in a number of sources. Their presentation comprehends the holistic design idea, where, as James Trilling (2001) mentions "neither the innate beauty of the ornament nor the benefits of the functional form can assure success, if both of them are not in complete equilibrium" [22]. The starting point for recording the decoration in Rodakis house is based on the core of Nikos Politis' conception (1909: 14) [23] of folk art, which distinguishes sculpture and graphic two- and three-dimensional representations, from the crafts with functional purpose and the ornaments, and also the aesthetics. Based largely on this distinction, but with the purpose of recording the decorative elements as an organic "whole" composed by all the separate aspects that constitute it[1], the classification is realised in table-form in four categories: sculptures and paintings, curved or repoussé ornaments, and functional artifacts with references to the sources (primary or secondary or a combination both), so that missing or destroyed decoration is recorded (Fig. 4). Because there exist unique sculptures and paintings in the house, these are distinguished from the ornaments and utensil crafts with the assumption that "ornament is one of the fundamental categories of art, along with architecture, sculpture and painting, and yet alike them, ornament has not a recognized place in today's cultural landscape" [22]. This assumption highlights the

---

[1] Philippidis insightfully notices that folk artifacts, in order to be more easily classified as kinds in different research projects, are converted into entities that can be easily collected and studied as subjects of scientific research. The easier they can be isolated and detached for practical reasons, the more prompt is the estimation of their value. Philippidis (1998: 34) [11].

Fig. 4. The connection between the second room and the hold (building B) (left). Flower shelf with engraved motif and raised tassels (building A) (right). Photographs by Panagiotis Ilias ©

aesthetic value of decoration, both in the form of borrowing, self-construction and social prominence of its creator, and in the variety of the expressive means that Rodakis uses by altering several typical characteristics of the vernacular houses of Aegina, and thus creating unique decorative interventions.

We should also note that the information gathered from different sources refer to ornaments and decorated utensils, portable or not, and their position in the interior space, or in curved and repoussé ornaments, such as the boat above the entrance door, the woman on the flower-pot, Rodakis hands on the fireplace columns, the graded cabinet, and others. The painted decoration which is shown on the murals revealed after the destruction of the roof top of the main house, are not recorded in sources such as Pikionis' drawings or other published material (Fig. 5). Therefore, part of the study aims at recording the decorative interventions and also for the vocalization of a schematic hypothesis based on the painted decoration, concerning the motives, the patterns and their repetitions, as well as their classification and relevance with the sculptured and curved patterns. Several sculptures do not exist on the buildings—some of them are missing and some others are in the hands of Rodakis' heirs.

In the process of the in situ photographic documentation, which was completed with a full frame camera, zoom-lens and tripod, a number of significant difficulties were faced. In many rooms, the access was almost impossible because of piles of discarded materials and ruined furniture. The worst conditions were met at the main house (A) with the ruined top-roof, which made for an unstable, flimsy base where the equipment could not be secured. In addition, detached fragments and an excessive amount of dust further deteriorated the existing conditions. In other rooms, the marginal dimensions combined with the declining structure created additional difficulties.

The lighting conditions, which were insufficient, also established inadequate circumstances for the shooting process, as we did not use any technical light (flash, tungsten light, or led). The documentation of the present condition of the monument, as well as the digitization of analog photographs gathered from new private collections, are carried out by Panagiotis Ilias, while the co-ordination of the research project, the recording with sketches and the classification of the decorative elements are carried out by Zoe Georgiadou, within the framework of research interdisciplinary activities of the laboratories of "Design, Interior Architecture and Audiovisual Documentation", and "Advanced Research Technologies for Investigation and Conservation" of the School of Applied Arts and Culture at the University of West Attica.

Fig. 5. The icon-cabinet and part of the revealed murals, main house (building A) (left) and murals (building A) (right). Photographs by Panagiotis Ilias ©

2.3 Study of the Painted Decoration Microstructure

For the purposes of the present study, chemical analysis was carried out on five (5) micro-samples taken from collected detached fragments of interior walls of the house. Four of them (AG01, AG02, AG03 and AG04) came from the rooms of the main house (A) and one (1) sample (AG05) from the wall of building (B). The aim of the followed methodology was to investigate the layers of coatings and colour decorations, and to determine the chemical composition of the pigments used. Microscopy techniques were preferred, which have a prominent position among instrumental chemical analyses and prove to be particularly effective for microstructure examination and material characterization [24, 25]. More precisely, optical microscopy in visible and fluorescence operating modes, as well as Scanning Electron Microscopy (SEM-EDAX), were applied for the study of the paint layer structure and the identification of the inorganic elements of the sample material.

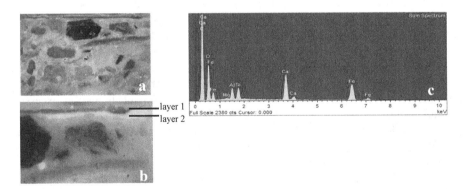

Fig. 6. Sample A1. a: Microphotography in visible, DF, ×50. b: Microphotography of the fluorescence stimulated by blue light, ×100. Layer 1 indicates the superficial red colour of ∼20–50 μm thickness. Layer 2 indicates the underlying preparation lime layer of ∼0.1 mm thickness. c: SEM spectra of the layer 1. The presence of Fe confirms the use of pure red ocher. (Color figure online)

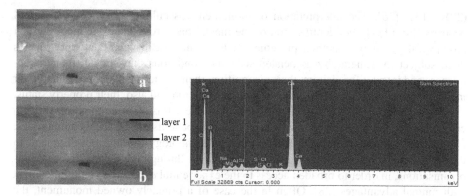

Fig. 7. Sample A2. a: Microphotography in visible, DF, ×50. b: Microphotography of the fluorescence stimulated by blue light, ×50. Blue paint layers of ∼0.1–0.2 mm thickness (layers1 and 3) present a non -homogenous density with vague boundaries. Layer 2 indicates the intermediate lime layer of ∼0.3–0.4 mm thickness. c: SEM spectra of the layer 1 and 3. The presence of Na and S reinforces the assumption of use of the blue pigment ultramarine. (Color figure online)

The analysis was carried out in the Laboratory of Advanced Research Technologies for Investigation and Conservation of the School of Applied Arts and Culture at the University of West Attica. As a result of all the observations and analyses, the fragments studied represent a process of decoration which, in terms of red colour, consists in depositing the pigment layer on a white preparation background over the mortar (Fig. 6a, b). On the contrary, for the blue colour, consecutive white and blue layers are observed, as the walls may have been decorated and then white-washed repeatedly (Fig. 7a, b).

The interior wall mortar appears to be a typical plaster with aggregates according to the geology of the area (Fig. 6a), while for the colour decoration of the rooms, the red ocher (Fig. 6c) and probably the artificial ultramarine (Fig. 7c) has been used, although indigo cannot be ruled out. However, further investigation is needed, regarding the mapping of the decoration and its condition, and for the organic binder identification.

3 Synopsis

Today, beyond the objective of promoting cultural heritage as a way to discover identity in national traditions, as a memorial link to the past, as a means for education or recreation, or even as a method of edification, we cannot ignore the fact that such monuments are also recognized as cultural resources. Georgios Lavas notices (2010: 112) [26] that this perspective places them at the heart of cultural evolution, gives them a direct educational role, and connects them to the social, cultural and economic development of a specific area. Through this process we seek, in a way, their protection and viability, with their disclosure and placement within an informational framework that "indicates, reminds and recalls the natural, historical, artistic or ethnological environment, and gives it its scientific interpretation beyond the epiphemonenon"

(2010: 132) [26]. The interpretation of monuments as cultural resources or cultural deposits (Eco) [27], the identification of the mental background and the ideology that produced them, as well as their placement within a historical social continuity, makes them subject to a hierarchy associated with the conditions they are in, their attractiveness and interest for visitors, their recognition, their scattering over an area and their inclusion into a network of services, their documentation and also, their property status.

According to the UNESCO (2017) objectives, the diffusion of cultural information can be used in the long run as a way to promote the benefits for the local communities, by combining traditional means and information technology for their development. In the short term, it can strengthen online collaboration through the databases of various organisations, in order to broaden access to the experiential exposure to cultural content with mutual advantages [28]. Often in the case of a privately owned monument, the intentions of the local community, the limitations in the experiential sightseeing, the exclusive private use, the protection and sustainability that are at the disposal of the owner, on the one hand obstruct the conceptual continuity of the past into the present, and on the other hand, render impossible the creation of a personal narrative in the monument's understanding, if we associate it with the overall progress of the place.

As we have pointed out, the characteristics of the public visualisation of Rodakis house are focused on the experiential meaning of visits made privately or within the framework of informal educational processes, and are based on fragmented research material, but also on digitised archival collections or educational portals. We could assume that one of the primary objectives of cultural heritage management is to communicate its content to every potential visitor and acquaint him with the features of the monument in order to protect and enhance it. Thus, between the previous and the current situation, there is already a differentiation that has occurred, and lies in the decline in its accessibility. Private initiative has yielded valuable material but, in many cases, this remains unknown. We can mention the photographs taken privately, for which we were informed by chance. Similar photographs are included in files or fragmentary photographic records, unknown to us yet. The comparison between different periods of photographing the monument up to date, demonstrates the evolution of its situation and contributes to the recording of its parts, which have been destroyed or lost. It also outlines chronologically the multiple interventions that the monument has suffered since Rodakis' death. For example, it appears that during the residence of the house by Marina Rodaki—following her father's death, all the murals of the main house (A) were whitewashed and covered, along with the coloured parts of the icon-cabinet. The same happened on the outer walls of the house with multiple repetitions of the process, to the point that the small mermaid engraved in stone, which is depicted in the sketches by Pikionis [1], was re-discovered by Papaioannou [17].

During the investigation and chemical analysis of the micro-samples taken from collected detached fragments, two ways of applying colour were discovered. A very thin layer was applied on the masonry coating of the walls of the main house and then the representation was painted over it. In some of the samples, it appears that more painted layers have been used in the same way using mortars, with rough grains clearly visible to the naked eye. The whitewashed layer with pigments was a frequently used technique on both interior and exterior walls in Aegina. In building B, a different technique was applied, with a typical thick coloured plaster and other additions according to the geology of the area. However, all of these observations are only incidental and fragmentary, as we do not know the exact part of the wall the samples belong to, while further study of the decoration micro-structure is required along with systematic sample research, in order to reach safer conclusions.

The construction of digital representations of cultural information, using the modern means, is an important proposed practice, with multiple benefits for formal research, as well as informal educational processes. In this case, as we have pointed out above, there is a wide range of evidence and information which is available, concerning different periods and material status of the monument, which may perhaps be used in its digitization. This, could be consolidated and form a rich body of literature, with the appropriate management and qualification in order to communicate the monument. The digitisation of the archival material collected should be further enriched within inter-disciplinary approaches, and may represent an alternative proposal for the communication of the culture and the values present. The "electronic culture" and "digital memory" can contribute substantially to research and the dissemination of information, but they cannot substitute the significance of a physical tour with material features. In fact, the digital body of the monument, its virtual representation, is a whole new construction with an independent identity and experiential dimension that leads to new visual perceptions. Its credibility depends on the sufficiency of its material documentation, analysis and presentation, as well as on the assignment of identity to the generated information. The virtual reconstruction of the monument's life has the advantage of an extended audience and, in the case of Rodakis house, it seems to be a unique chance to transfer knowledge to prospective recipients without the limitation of inaccessible physical space. It may also help to halt further dramatic changes in its characteristics. In this monument, and consequently in a series of other private monuments, the interdisciplinary approaches can contribute to the enrichment of the available data and compose a body of information through research collaborations in an environment that can hardly fund this size and type of projects.

References

1. Benaki Museum: Modern Greek Architecture Archives (ANA_67, D. Pikionis) 1912–1917
2. Pikionis, D.: Our Vernacular Art and Us. Filiki Etaireia, Odessa (1925)
3. Pikioni, A. (ed.): Pikionis Dimitris: The Architectural Work, 1912–1934, vol. II. Bastas-Plessas Publications, Athens (1994)
4. Benaki Museum: Modern Greek architecture archives (ANA_30, Aristotelis Zachos)
5. Candilis, G.: Bâtir la Vie: Un Architect Témoin de son Temps. Infolio, Paris (1977)
6. Vasiliadis, D.B.: Vernacular architecture of Aegina. Laographia **16**, 413–512 (1956–1957)
7. Vasiliadis, D.B.: Vernacular architecture of Aegina. Laographia **17**, 197–254 (1957–1958)
8. Michelis, P.A.: Students' Works A. The Greek House, 3rd edn. NTUA, Athens (1981)
9. Vrieslander, K., Kaimi, J.: Rodakis House in Aegina. Akritas, Athens (1997)
10. Koulikourdi, G., Alexiou, S.: Guide for the history and monuments of Aegina (1950)
11. Philippidis, D.: Decorative Arts. Melissa Publications, Athens (1998)
12. Catalogue for Aldo Van Eyck's Exhibition, National Gallery and Museum Alexander Soutsos, September 17–October 23, 1983, Athens (1983)

13. FEK 1252/27-09-2001, B': Number of Ministerial Decision Ministry of Culture/ΔΙΛΑΠ/Γ/ 963/46628 (5)

14. Tsomis, J., Frangoulis, T.: Rodakis house in Aegina. Anti **409**, 50–51 (1989)

15. Nikolai, O. (2007). http://www.eigen-art.com/index.php?article_id=223&clang=1. Accessed 18 Jan 2013

16. Tritsibidas, J. (2012). https://vimeo.com/34739133. Accessed 01 Apr 2016

17. Papaioannou, T.: A monument of traditional architecture is falling in Efimerida syntakton, 10 Jan 2016

18. Seferis, G.: Days Z, 1956–1960. Ikaros Editions, Athens (1996)

19. Ωscillations: Aegina 08–18 of September 2016. https://dura-dundee.org.uk/2016/06/20/oscillations/

20. Papaioannou, T.: Desecrating Rodakis House in Aegina Island. Efimerida syntakton, 24 April 2017

21. Ilias, P., Georgiadou, Z.: Digital memory and private cultural heritage: the case of Rodakis house in Aegina Island. In: 2nd Greek Conference for Digitization of Cultural Heritage Proceedings, Volos, pp. 244–255 (2017)

22. Trilling, J.: The Language of Ornament. Thames and Hudson Ltd., London (2001)

23. Politis, N.G.: Laografia. Laografia **1**, 3–18 (1909)

24. Skoog, D., Holler, J., Nieman, T.: Principles of Instrumental Analysis, 6th edn. Harcourt College Publishers, Boston (2006)

25. Sansonetti, J., Striova, D., Biondelli, E., Castellucci, M.: Anal. Bio-Anal. Chem. **397**(7), 2667–2676 (2010)

26. Lavas, G.P.: Issues About Cultural Management. Melissa Publications, Athens (2010)

27. Eco, U.: Cultural Deposits. Paratiritis, Athens (1992)

28. UNESCO Web Archives. http://webarchive.unesco.org/20170129000029/. Accessed 02 Oct 2017

Assessment of Masonry Structures Based on Analytical Damage Indices

Athanasia Skentou[1], Maria G. Douvika[1], Ioannis Argyropoulos[1],
Maria Apostolopoulou[2], Antonia Moropoulou[2],
and Panagiotis G. Asteris[1(✉)]

[1] Computational Mechanics Laboratory, School of Pedagogical
and Technological Education, 14121 Heraklion, Athens, Greece
panagiotisasteris@gmail.com
[2] Laboratory of Materials Science and Engineering, School of Chemical
Engineering, National Technical University of Athens, Athens, Greece

Abstract. In this paper, a methodology is presented aiming to predict the vulnerability of masonry structures under seismic action. Masonry structures, among which many are cultural heritage assets, present high vulnerability under earthquake. Reliable simulations of their response under seismic stresses are exceedingly difficult because of the complexity of the structural system and the anisotropic and brittle behavior of the masonry materials. Within this framework, a detailed analytical methodological approach for assessing the seismic vulnerability of historical and monumental masonry structures is presented, taking into account the probabilistic nature of the input parameters by means of analytically determining fragility curves. The emerged methodology is presented in detail following an in depth application on both theoretical basis and existing cultural heritage masonry structures.

Keywords: Damage index · Failure criteria · Fragility analysis ·
Masonry structures · Monuments · Restoration mortars · Seismic assessment

1 Introduction

Masonry corresponds to one of the most ancient building materials and hence structure types. This explains the fact that the majority of monuments are masonry structures using stone elements joined together with the use of mortars. The inhomogeneous and anisotropic nature of this particular structure type, as well as that of the materials comprising it, define to a great extent the seismic response of monumental buildings. Notably, a common characteristic of these structures is their high seismic vulnerability when subjected to seismic stresses, which is attributed to the highly brittle behavior and relatively low tensile strength of the materials comprising masonries.

It is imperative that these invaluable historical buildings should be treated with respect and attention to detail, due to their architectural uniqueness and without overlooking the importance of their cultural and social impact. Thus, any measures taken for their protection, aiming to decrease their vulnerability, must comply with the principles of reversibility and compatibility. These limitations create the necessity for

© Springer Nature Switzerland AG 2019
A. Moropoulou et al. (Eds.): TMM_CH 2018, CCIS 961, pp. 513–531, 2019.
https://doi.org/10.1007/978-3-030-12957-6_36

strict compliance of protective measures within the regulatory frameworks which govern this type of structures, both on national and international levels.

Already from the beginning of the previous century [1, 2], regulatory frame-works regarding the protection of historical buildings were formulated and established, such as the guidelines given by the International Scientific Committee of the Analysis and Restoration of Structures of Architectural Heritage of ICOMOS in 2001, and more specifically the ICOMOS Charter regarding the Principles for the Analysis, Conservation and Structural Restoration of Architectural Heritage, ISCARSAH Principle. This particular regulatory framework is based on the principles of research and documentation, authenticity and integrity, aesthetic harmony, least invasive interventions and reversibility, principles which are in accordance with the demands of the Athens and the Venice Charters, as well as with the principles established by The Secretary of the Interior's Standards for Historic Preservation Projects [3]. An additional prerequisite for the successful protection of these structures is related to the interdisciplinary approach that needs to be adopted during the investigation and assessment of any repair scenario.

A prerequisite for the formulation of a reliable methodology to predict the seismic vulnerability of historical/monumental masonry structures is the successful simulation of the structural system, as well as of the materials comprising the masonry, through the formulation of appropriate analytical statutory laws. In this direction however, the complex mechanical behavior of masonries, which is a multiphase material, is a serious obstacle. Furthermore, an additional and at the same time basic difficulty regarding the formulation of such a methodology is related to the probabilistic nature of the parameters influencing the behavior of masonry structures. Among these parameters are the values of the mechanical properties of the materials (due to the wide dispersion of these values regarding the whole of the structure or due to limitation regarding the accuracy of the measurements, related to the lack of sufficient accuracy of methods and instruments used). Additionally, the probabilistic nature of earthquakes, directly connected and influenced by a significant number of parameters, must be taken under consideration. Due to the high uncertainty of the parameters influencing the behavior of masonry structures, the assessment of their vulnerability cannot be conducted in terms of a deterministic approach. To the contrary, a probabilistic approach would be more appropriate, in order to be applied in cases where the response of the structure is evaluated and compared with limit states, such as specific limit values of response directly interlinked with structural damages.

In the framework of the above limitations along with the issues of specific consideration, this study presents an analytical methodology for the evaluation of seismic vulnerability of masonry structures, considering the probabilistic nature of the parameters involved through the development of analytical fragility curves.

2 Proposed Methodology

Taking into account the principles and guidelines of ICOMOS, as well as the results of relative research projects [4–17], a specific methodology has been developed in relation to the restoration of historical masonry structures. The flow chart of the proposed methodology is presented in Fig. 1.

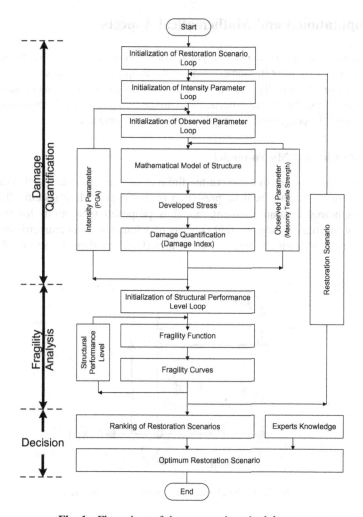

Fig. 1. Flow chart of the proposed methodology.

The proposed methodology is formulated from discrete steps. These steps include the evaluation or/and determination of the mechanical properties of the materials, the simulation of the structural system and of the forces, the analysis of the structure regarding specific stresses, the determination of the failure areas of the structure and the respective damage indices, regarding both the model of the structure in its current condition, as well as for the models of the structure in relation to different repair scenarios. Finally, based on the damage indices, fragility curves are developed, which, as will be illustrated in the following section, contribute to the quantified assessment of the structure's vulnerability in its current condition, and to the assessment of the effectiveness of different repair scenarios.

3   Computational and Mathematical Aspects

In this section, the fundamental analytical constitutive laws and numerical models, required for the successful implementation of the proposed methodology, are presented in detail. More specifically, the finite element model for the macro-modeling of masonry structures, the failure criteria, the damage indices, the performance levels and the mathematical background of fragility curves, are presented.

3.1   Finite Element Macro-model

Since the basic principles that govern the finite element method are well documented, only the essential features will be presented in this paper. More specifically, an anisotropic (orthotropic) finite element model is proposed (and used) for the macro-modeling of masonry structures, that consists of a four-node isoparametric rectangular finite element model with 8 degrees of freedom (DOF) as depicted in Fig. 2.

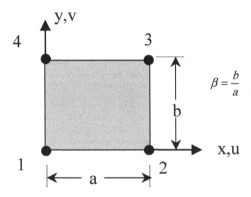

Fig. 2. Finite element macro-model dimensions.

The major assumption of modeling the masonry behavior under plane stress is that the material is homogeneous and anisotropic. In particular, the material shows a different modulus of elasticity (E_x) in the x direction (direction parallel to the bed joints of brick masonry) and a different modulus of elasticity (E_y) in the y direction (perpendicular to the bed joints). For the plane stress, the elasticity matrix is defined by:

$$E = \frac{E^2 G_{xy}}{1 - v_{xy} v_{yx}} \begin{bmatrix} \frac{1}{E_y G_{xy}} & \frac{v_{xy}}{E_y G_{xy}} & 0 \\ \frac{v_{xy}}{E_x G_{xy}} & \frac{1}{E_x G_{xy}} & 0 \\ 0 & 0 & \frac{1 - v_{xy} v_{yx}}{E^2} \end{bmatrix} \quad (1)$$

where v_{xy} and v_{yx} are the Poisson's ratios in the x-y and y-x plane respectively, and G_{xy} is the shear modulus in the y plane.

3.2 Masonry Failure Criterion Under Biaxial Stress State

Masonry is a material that exhibits distinct directional properties due to the fact that the mortar joints act as planes of weakness. To define failure under biaxial stress, the derivation of a 3D surface in terms of the two normal stresses along with shear stress (or the two principal stresses and their orientation to the bed joints) is required. A failure surface of this form has been recently derived by Syrmakezis and Asteris [18] and Asteris [19], where the authors present a method to define a general anisotropic failure surface of masonry under biaxial stress, using a cubic tensor polynomial. In particular, an analytical methodology is proposed in order to describe the masonry failure surface under plane stress via a regular surface, that is, a surface defined by a single equation of the form:

$$
\begin{aligned}
&2.27\sigma_x + 9.87\sigma_y + 0.573\sigma_x^2 + 1.32\sigma_y^2 + 6.25\tau^2 - 0.30\sigma_x\sigma_y \\
&+ 0.009585\sigma_x^2\sigma_y + 0.003135\sigma_x\sigma_y^2 + 0.28398\sigma_x\tau^2 + 0.4689\sigma_y\tau^2 = 1
\end{aligned}
\tag{2}
$$

The main disadvantage of this anisotropic failure criterion (2) is that it applies only to the specific masonry material to which the test data refers to. However, the failure criteria could potentially be generalized if expressed in a dimensionless form, thus enabling it to be applied to a wide range of masonry materials. This can be achieved by dividing and multiplying (at the same time) each term of Eq. 2 by the material uniaxial strength raised in the sum of the exponents of the variables (as appeared in each term). By selecting to use the uniaxial compressive strength across the y-axis, which, in terms of the masonry material corresponds to the uniaxial compressive strength denoted with the symbol $f_{wc}^{90^\circ}$, Eq. 2 obtains the following form:

$$
\begin{aligned}
&17.15\left(\frac{\sigma_x}{f_{wc}^{90^\circ}}\right) + 74.57\left(\frac{\sigma_y}{f_{wc}^{90^\circ}}\right) + 32.71\left(\frac{\sigma_x}{f_{wc}^{90^\circ}}\right)^2 + 75.34\left(\frac{\sigma_y}{f_{wc}^{90^\circ}}\right)^2 + 356.74\left(\frac{\tau}{f_{wc}^{90^\circ}}\right)^2 \\
&- 17.12\left(\frac{\sigma_x}{f_{wc}^{90^\circ}}\right)\left(\frac{\sigma_y}{f_{wc}^{90^\circ}}\right) + 4.13\left(\frac{\sigma_x}{f_{wc}^{90^\circ}}\right)^2\left(\frac{\sigma_y}{f_{wc}^{90^\circ}}\right) + 1.35\left(\frac{\sigma_x}{f_{wc}^{90^\circ}}\right)\left(\frac{\sigma_y}{f_{wc}^{90^\circ}}\right)^2 + 122.46\left(\frac{\sigma_x}{f_{wc}^{90^\circ}}\right)\left(\frac{\tau}{f_{wc}^{90^\circ}}\right)^2 \\
&+ 202.20\left(\frac{\sigma_y}{f_{wc}^{90^\circ}}\right)\left(\frac{\tau}{f_{wc}^{90^\circ}}\right)^2 = 1
\end{aligned}
\tag{3}
$$

or in dimensionless terms $\left(\bar{\sigma}_x = \frac{\sigma_x}{f_{wc}^{90^\circ}}, \bar{\sigma}_y = \frac{\sigma_y}{f_{wc}^{90^\circ}}, \bar{\tau} = \frac{\tau}{f_{wc}^{90^\circ}}\right)$:

$$
\begin{bmatrix}
17.15\,\bar{\sigma}_x + 74.57\,\bar{\sigma}_y + 32.71\,\bar{\sigma}_x^2 + 75.34\,\bar{\sigma}_y^2 + 356.74\,\bar{\tau}^2 \\
- 17.12\,\bar{\sigma}_x\bar{\sigma}_y + 4.13\,\bar{\sigma}_x^2\,\bar{\sigma}_y + 1.35\,\bar{\sigma}_x\,\bar{\sigma}_y^2 + 122.46\,\bar{\sigma}_x\bar{\tau}^2 \\
202.20\,\bar{\sigma}_y\bar{\tau}^2 = 1
\end{bmatrix}
\tag{4}
$$

The following figure depicts the contour map of Eq. 3, that is, the non-dimensional failure surface of masonry in normal stress terms (with values from 0.00 up to 0.45 in steps of 0.05) (Fig. 3).

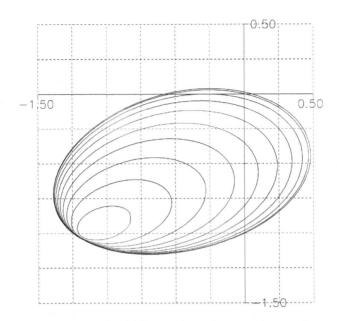

Fig. 3. Non-dimensional failure surface of masonry in normal stress

Furthermore, beyond the above anisotropic failure criterion, isotropic failure criteria such as the ones proposed by Syrmakezis et al. 1995 [6] and Kupfer et al. 1969 [20] have been used in this research.

3.3 Damage Index

Damage control in a building is a complex task, especially under seismic action. There are several response parameters that can be instrumental in determining the level of damage that a particular structure suffers during a ground motion; with the most important ones being: deformation, relative velocity, absolute acceleration, and plastic energy dissipation (viscous or hysteretic). Controlling the level of damage in a structure consists primarily in controlling its maximum response. Damage indices establish analytical relationships between the maximum and/or cumulative response of structural components and the level of damage they exhibit [14, 21, 22]. A performance-based numerical methodology is possible if limits, through the use of damage indices, can be established to the maximum and cumulative response of the structure, as a function of the desired performance of the building for the different levels of the design ground motion. Once the response limits have been established, it is then possible to estimate the mechanical characteristics that need to be supplied to the building so that its response is likely to remain within the limits.

For the case of masonry structures a new damage index is proposed by Asteris et al. 2014 [14], which employs as response parameter the percentage of the damaged area of the structure relatively to the total area of the structure. The proposed damage index (*DI*), for a masonry structure can be estimated by the following equation:

$$[DI] = \frac{A_{fail}}{A_{tot}} \times 100 \tag{5}$$

where A_{fail} is the damaged surface area of the structure and A_{tot} the total surface area of the structure.

3.4   Structural Performance Levels

As it is widely practiced today, performance-based seismic design is initiated with an interplay between demands and appropriate performance objectives while the Engineer then has to develop a design capable of meeting these objectives. Performance objectives are expressed as an acceptable level of damage, typically categorized as one of several performance levels, such as immediate occupancy, life safety or collapse prevention, given that ground shaking of specified severity is experienced.

In the past, the practice of meeting performance-based objectives was already included in design practice yet through a rather informal, simplistic and non-standard approach. Some Engineers would characterize performance as life-safety or not; others would assign ratings that range from "poor" to "good". This qualitative approach adopted for performance prediction was appropriate given the limited capability of seismic-resistant design technology to deliver building designs capable of quantifiable performance.

Three structural performance levels are considered: (a) heavy damage, (b) moderate damage and (c) insignificant damage, in a similar approach to the Federal Emergency Management Agency [23]. The performance levels are defined by the values of *DI*, whereas a value less than 15% can be interpreted as insignificant damage; from 15% to less than 25% as moderate damage; and larger or equal than 25% as heavy damage. Nevertheless, other approaches can be used, according to the recent European Codes [24], based on a more engineered (and more detailed) estimation of damage.

3.5   Fragility Analysis

One of the problems engineers must face and resolve at later stages of the global involves the quantitative vulnerability assessment of the building at its current state (damaged or not), as well as its behavior once it is "modified" after all interventions. Alternatively stated, a method is necessary to assess the seismic vulnerability of the existing structures, as well as to assess the intervention scenarios and categorize them based on the reduction they induce to the structures' seismic vulnerability, thus leading to the selection of the optimal scenario. One of the most important tools is considered to be fragility analysis, which provides a measure of the safety margin of the structural system above specified structural performance/hazard levels.

A number of methodologies for performing fragility analysis have been proposed in the past (used to assess the behaviour of structural systems). Simplified methodologies for fragility evaluation have been proposed by Kircher et al. 1997 [25].

Evaluating seismic fragility information curves for structural systems involves: (a) information on structural capacity and (b) information on the seismic hazard. Due to the fact that both the aforementioned contributing factors are uncertain to a large extent, the fragility evaluation cannot be carried out in a deterministic manner. A probabilistic approach, alternatively, needs to be utilized in the cases in which the structural response is evaluated and compared against "limit states" that is limiting values of response quantities correlated to structural damage.

Fragility, as it is shown in Fig. 4, is the probability of the structural damage to reach or exceed a certain damage threshold *DI* (damage index or performance level) under a given earthquake level (e.g. Peak Ground Acceleration (PGA)). As a rule, it increases as the earthquake intensity level increases. The failure domain is where a *DI* overcomes a specified threshold.

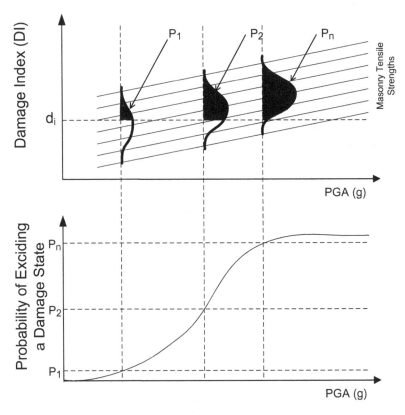

Fig. 4. Development process of analytical fragility curves.

Fragility is evaluated as the total probability of a response parameter R exceeding the allowable response value r_{lim} (limit-state), for various earthquake intensities I.

In mathematical form, this is simply a conditional probability (Barron – Corverra 2000 [26] and Reinhorn et al. 2001 [27]) given by the following equation:

$$Fragility = P[R \geq r_{\lim}|I] = \sum_{j}^{3} P[R \geq r_{\lim}|I, C] P(C = c_j) \tag{6}$$

where $P(C = cj)$ is the probability that capacity cj occurs. In the following examples basic steps of the development of the fragility curves are briefly presented.

3.6   Case Studies

Aiming to assess the proposed methodology, its implementation is presented in this section both on theoretical, as well as on actual structures.

4   2D Masonry Walls

The behavior of three 2D masonry walls, with square openings as presented in Fig. 5, was studied. The values of the percentage of the openings (surface of the opening/wall surface) ranged between 0%, 16%, 36% and 64%. The mechanical characteristics of the masonry material include the properties of Elasticity Modulus and Compressive Strength at 4.4 MPa and 7.6 MPa in X and Y directions respectively while the Tensile Strength is 0.4 MPa in X direction and 0.1 MPa in Y. The Poisson's Ratio is 0.20 in both X-Y and Y-X plane, while the Specific Weight is 20kN/m³. It should be noted, that the data taken under consideration are characterized as typical and are stated in the experimental study of Page, conducted in 1981 [28], the results of which have been widely utilized by the majority of researchers investigating the behaviour of masonries, both on experimental and numerical levels.

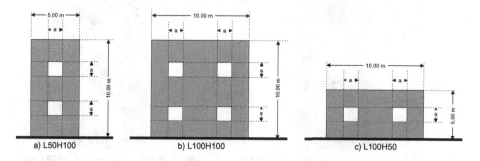

Fig. 5. Cases of masonries examined.

In Fig. 6, the failure areas of 2D masonry wall L50H100 are presented. This wall has two openings reaching a total opening percentage of 16%, while two different failure criteria are utilized, with three different values of maximum ground acceleration (0.24 g, 0.32 g & 0.40 g). These diagrams are especially useful for the determination of failing areas of the structure, as well as for the selection of the optimum repair measures. More specifically, the failure areas are marked with different colors high-lighting in distinct ways the kind of stress underpinning the failure. As it can be easily perceived, other repair measures are demanded when failure occurs under biaxial compression, other measures are demanded in the case of biaxial tension and other measures are demanded in the case of heterosemous biaxial stress state

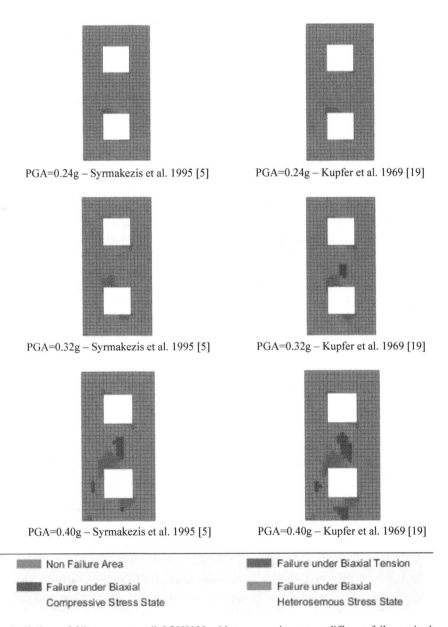

PGA=0.24g – Syrmakezis et al. 1995 [5]   PGA=0.24g – Kupfer et al. 1969 [19]

PGA=0.32g – Syrmakezis et al. 1995 [5]   PGA=0.32g – Kupfer et al. 1969 [19]

PGA=0.40g – Syrmakezis et al. 1995 [5]   PGA=0.40g – Kupfer et al. 1969 [19]

Non Failure Area   Failure under Biaxial Tension

Failure under Biaxial   Failure under Biaxial
Compressive Stress State   Heterosemous Stress State

Fig. 6. Failure of 2D masonry wall *L50H100* with two openings, two different failure criteria and three different values of peak ground acceleration.

(tension/compression). In Fig. 7 respectively, the failure areas of masonry wall L100H100 are presented for three different opening percentages (0%, 16& & 36%) using two different failure criteria and for a peak ground acceleration (PGA) of 0.32 g.

In Fig. 8 the fragility curves of the current condition of masonry wall *L50H100* are presented, using the failure criterion of Syrmakezis et al. 1995 [6] and for an opening

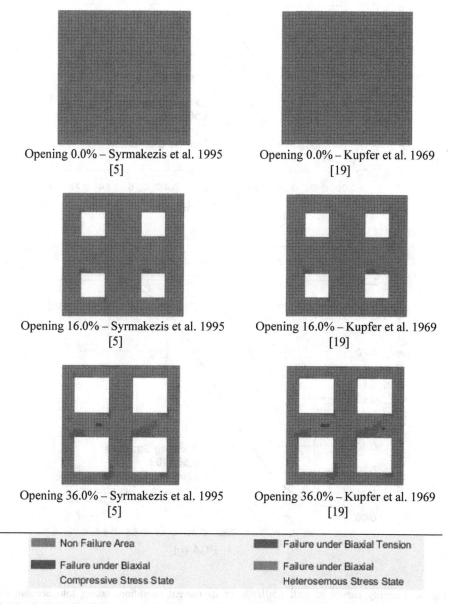

Opening 0.0% – Syrmakezis et al. 1995 [5]   Opening 0.0% – Kupfer et al. 1969 [19]

Opening 16.0% – Syrmakezis et al. 1995 [5]   Opening 16.0% – Kupfer et al. 1969 [19]

Opening 36.0% – Syrmakezis et al. 1995 [5]   Opening 36.0% – Kupfer et al. 1969 [19]

Non Failure Area   Failure under Biaxial Tension

Failure under Biaxial Compressive Stress State   Failure under Biaxial Heterosemous Stress State

Fig. 7. Failure of 2D masonry wall *L100H100* for three different opening percentages, two different failure criteria and peak ground acceleration value PGA = 0.32 g.

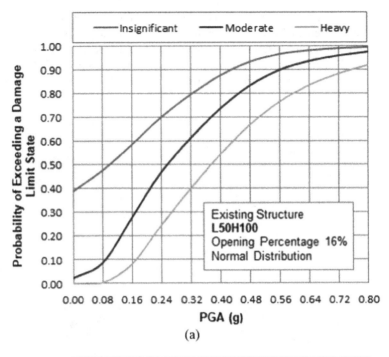

(a)

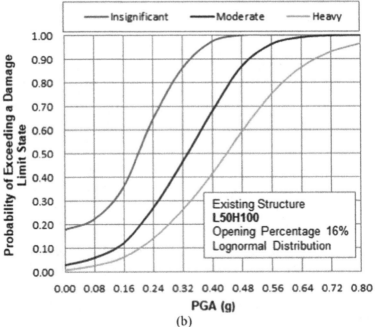

(b)

Fig. 8. Fragility curves of wall *L50H100* in its current condition, taking into account the Syrmakezis et al. 1995 failure criterion [5] and for an opening percentage of 16% with (a) normal distribution and (b) lognormal distribution.

percentage of 16% for normal and lognormal distribution. In Fig. 9 the fragility curves of the current condition of wall *L50H100* are presented, using the Kupfer et al. 1969 failure criterion [20], four different opening percentages (0%, 16%, 36% & 64%), lognormal distribution and moderate damage performance level.

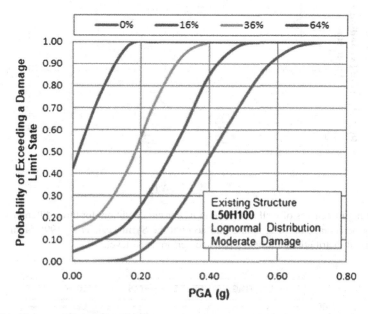

Fig. 9. Fragility curves of wall *L50H100* in its current condition, taking into account the Kupfer et al. 1969 failure criterion [19], four different opening percentages (0%, 16%, 36% & 64%), lognormal distribution and moderate damage performance level.

In Fig. 10 the fragility curves of the current condition of three different masonry walls with percentage openings of 16% are presented, using the Syrmakezis et al. 1995 failure criterion [6], lognormal distribution and heavy damage performance level. This particular figure is indicative of the potential of fragility curves in quantifying the vulnerability of structures and especially in classifying structures according to their vulnerability.

In Fig. 11 the fragility curves for the case of wall *L50H100* are presented, using the Syrmakezis et al. 1995 failure criterion [6], regarding the current state of the wall, as well as the repaired states in the cases of using three different repair mortars M5, M10 and M15, of compressive strength 5.0 MPa, 10.0 MPa and 15.0 MPa respectively, using lognormal distribution and moderate damage performance level. These three types of restoration mortars were evaluated in order to cover the wide range of mechanical properties, which have been assessed to be compatible for use in monuments and historical buildings (Moropoulou et al. 2005 [29]).

Based on Figs. 10 and 11, it is evident that the proposed approach offers a classification method which can greatly assist authorities to optimize their decision-making

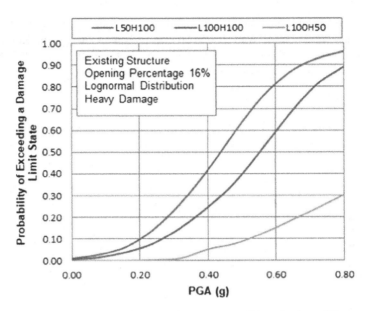

Fig. 10. Fragility curves of wall *L50H100* in its current condition, for three different walls with an opening percentage of 16%, taking into account the Syrmakezis et al. 1995 failure criterion [5], lognormal distribution and heavy damage performance level.

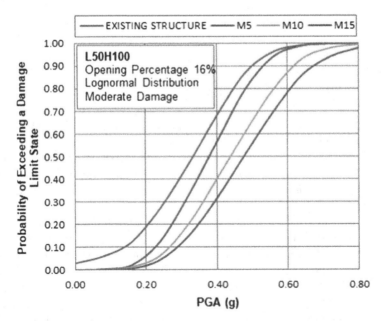

Fig. 11. Fragility curves of wall *L50H100*, taking into account the Syrmakezis et al. 1995 failure criterion [5], for its current condition, as well as its behavior with the use of three different repair mortars of 5.0 MPa, 10.0 MPa and 15.0 MPa compressive strength (M5, M10, M15), for lognormal distribution and moderate damage performance level.

mechanisms in relation to the selection of the structure most in need of repair, as well as provide the criteria to select the optimum repair scenario.

5 Monumental Masonry Structures

In this section the reliability of the proposed methodology is presented, through its application on two real case monumental masonry structures. More specifically, the presented methodology was employed in order to investigate the vulnerability of two different real case monumental masonry structures. The first structure is the Catholicon of the Kaisariani Monastery (Fig. 12), which was built in Athens in the late 11th – early 12th century AD, while in the second case study the structure is the Palace of Chania, Crete (Fig. 13), which was built in the Chalepa area of Crete in 1882. In 1898, when Crete became autonomous and acquired a ruler and a constitution, Prince George arrived in Crete as high commissioner of Crete. The mansion in question, the Palace, is selected as the most appropriate structure in order to serve as his dwelling.

Fig. 12. Front view of the Catholicon of the Kaisariani Monastery.

In Fig. 14 the fragility curves for the current situation for both monumental structures are presented, in lognormal distribution and heavy damage performance level. This figure depicts the potential of fragility curves to quantify the vulnerability of structures and furthermore to classify them, classifying them by descending degree

Fig. 13. Front view of the Palace of Chania.

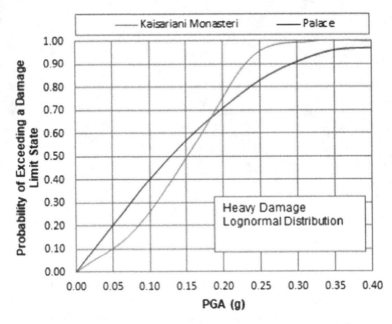

Fig. 14. Fragility curves of the current condition for two real case monumental structures using lognormal distribution.

vulnerability. Based on this figure it is also evident that for peak ground acceleration values lower than 0.18 g the Catholicon of Kaisariani Monastery presents lower vulnerability compared to the Palace of Chania, while the opposite results are observed for values greater than 0.18 g.

The results for these two monumental structures, regarding the evaluation of their seismic vulnerability and the selection of the optimum repair scenarios, are presented in detail in the Masters' theses of Douvika 2017 [30], Skentou 2018 [31] and in the research papers of Asteris et al. 2014, 2016, 2017 [14, 16, 32], Moropoulou et al. 2016 [33], Douvika et al. 2016 [34] and Apostolopoulou et al. 2017 [17].

6 Conclusions

In this study a methodology is presented for the evaluation of the seismic vulnerability of masonry structures, taking into account the probabilistic nature of the parameters involved in the simulation of the structure as well as in the mechanical characteristics of the materials and seismic forces. The evaluation of the vulnerability of the structures, as well as the evaluation of different repair scenarios and the selection of the optimum scenario are amongst the main conclusions of this paper. Furthermore, the proposed approach can act as a valuable tool for authorities involved in monument protection, as it offers a classification methodology, which can greatly facilitate their decision-making mechanisms, not only in relation to the selection of the structure most in need of protection and/or repair, amongst a vast variety of monuments, but also towards the selection of the optimum repair scenario for each individual monument.

References

1. ICOMOS1931. The Athens charter for the restoration of historic monuments. In: Paper Presented at the First International Congress of Architects and Technicians of Historic Monuments, Athens, Greece (1931). http://www.icomos.org/athens_charter.html
2. ICOMOS1964. The Venice charter for the restoration of historic monuments. In: Paper Presented at the Second International Congress of Architects and Technicians of Historic Monuments, Venice, Italy (1964). http://www.international.icomos.org/charters/venice_e.htm
3. Morton, B., Hume, G.L.: The Secretary of the Interior's standards for historic preservation projects with guidelines for applying the standards, U.S. Dept. of the Interior, Heritage Conservation and Recreation Service (1979)
4. Syrmakezis, C.A., Asteris, P.G., Sophocleous, A.A.: Earthquake resistant design of masonry tower structures. In: Proceedings, 5th STREMA Conference on Structural Studies, vol. 1, pp. 377–386 (1997)
5. Syrmakezis, C., Sophocleous, A., Asteris, P., Liolios, A.A.: Earthquake resistant design of masonry structural systems. Adv. Earthq. Eng. 2, 717–726 (1995)
6. Syrmakezis, C.A., Chronopoulos, M.P., Sophocleous, A.A., Asteris, P.G.: Structural analysis methodology for historical buildings. Architectural Stud. Mater. Anal. 1, 373–382 (1995)
7. Binda, L., Saisi, A., Tiraboschi, C.: Investigation procedures for the diagnosis of historic masonries. Constr. Build. Mater. 14(4), 199–233 (2000)

8. Lourenço, P.B.: Recommendations for restoration of ancient buildings and the survival of a masonry chimney. Constr. Build. Mater. **20**(4), 239–251 (2006). https://doi.org/10.1016/j.conbuildmat.2005.08.026

9. Asteris, P.G.: On the structural analysis and seismic protection of historical masonry structures. Open Constr. Build. Technol. J. **2**(1), 124–133 (2008)

10. Onaka, T.: A study of the documentation process for conservation of architectural heritage sites: illustrated by examples from Egypt and Belgium. In: 22nd CIPA Symposium, 11–15 October 2009, Kyoto, Japan (2009)

11. Tassios, T.P.: Seismic engineering of monuments. Bull. Earthq. Eng. **8**(6), 1231–1265 (2010)

12. Chronopoulos, P.M., Zigouris, N., Asteris, P.G.: Investigation/documentation and aspects of seismic assesment and redesign of traditional masonry buildings in Greece. In: 5th European Conference on Structural Control (EACS 2012), Genoa, Italy, 18–20 June 2012 (2012)

13. Asteris, P.G., Tzamtzis, A.D., Vouthouni, P.P., Sophianopoulos, D.S.: Earthquake resistant design and rehabilitation of masonry historical structures. Pract. Periodical Struct. Des. Constr. (ASCE) **10**(1), 49–55 (2005)

14. Asteris, P.G., et al.: Seismic vulnerability assessment of historical masonry structural systems. Eng. Struct. **62–63**, 118–134 (2014). https://doi.org/10.1016/j.engstruct.2014.01.031

15. Asteris, P.G., et al.: Numerical Modeling of Historic Masonry Structures. Handbook of Research on Seismic Assessment and Rehabilitation of Historic Structures (2015). (Asteris, P.G., Plevris, V. eds.)

16. Asteris, P.G., Douvika, M., Karakitsios, P., Moundoulas, P., Apostolopoulou, M., Moropoulou, A.: A stochastic computational framework for the seismic assessment of monumental masonry structures. In: 5th International Conference on Integrity, Reliability and Failure, Faculty of Engineering, U. Porto (2016)

17. Apostolopoulou, M., et al.: A methodological approach for the selection of compatible and performable restoration mortars in seismic hazard areas. Constr. Build. Mater. **155**, 1–14 (2017)

18. Syrmakezis, C.A., Asteris, P.G.: Masonry failure criterion under biaxial stress state. J. Mater. Civil Eng. Am. Soc. Civil Eng. (ASCE) **13**(1), 58–64 (2001)

19. Asteris, P.G.: Unified yield surface for the nonlinear analysis of brittle anisotropic materials. Nonlinear Sci. Lett. A **4**(2), 46–56 (2013)

20. Kupfer, H., Hilsdorf, H.K., Rusch, H.: Behavior of concrete under biaxial stresses. J. Am. Concrete Inst. **66**(8) (1969)

21. Park, Y.J., Ang, H., Wen, Y.K.: Damage-limiting a seismic design of buildings. Earthq. Spectra **3**(1), 1–26 (1987)

22. D'Ayala, D.: Assessing the seismic vulnerability of masonry buildings. In: Tesfamariam, S., Goda, K. (eds.) Handbook of Seismic Risk Analysis and Management of Civil Infrastructure Systems. Woodhead Publishing, Cambridge, pp. 334–365 (2013). https://doi.org/10.1533/978057098986.3.334

23. FEMA-273. NEHRP guidelines for the seismic rehabilitation of buildings. Federal Emergency Management Agency (1997)

24. CEN. Eurocode 8: design of structures for earthquake resistance. Part 3: assessment and retrofitting of buildings. EN 1998-3, March 2005 (2005)

25. Kircher, C.A., Nasser, A.A., Kutsu, O., Holmes, W.T.: Developing of building damage functions for earthquake loss estimation. Earthq. Spectra **13**(4), 664–681 (1997)

26. Barron-Corverra, R.: Spectral evaluation of seismic fragility in structures. Ph.D dissertation. Department of Civil, Structural & Environmental Engineering, University at Buffalo, The State University of New York, Buffalo, NY (2000)

27. Reinhorn, A.M., Barron-Corverra, R., Ayala, A.G.: Spectral evaluation of seismic fragility of structures. In: Proceedings ICOSSAR 2001, Newport Beach CA, June 2001
28. Page, A.W.: The biaxial compressive strength of brick masonry. Proc. Instn. Civ. Engrs. **71** (2), 893–906 (1981)
29. Moropoulou, A., Bakolas, A., Anagnostopoulou, S.: Composite materials in ancient structures. Cement Concrete Comp. **27**(2), 295–300 (2005)
30. Douvika, M.G.: Seismic Vulnerability Assessment of Monumental Masonry Structures, Master thesis, School of Pedagogical & Technological Education, Athens, Greece (2017). https://www.researchgate.net/publication/318259801_Seismic_Vulnerability_Assessment_of_Monumental_Masonry_Structures
31. Skentou, A.D.: Seismic Vulnerability Assessment of Masonry Structures, Master thesis, School of Pedagogical & Technological Education, Athens, Greece (2018). https://www.researchgate.net/publication/324684962_Seismic_Vulnerability_Assessment_of_Masonry_Structures
32. Asteris, P.G., Douvika, M.G., Apostolopoulou, M., Moropoulou, A.: Seismic and restoration assessment of monumental masonry structures. Materials **10**(8), 895 (2017)
33. Moropoulou, A., et al.: The combination of NDTS for the diagnostic study of historical buildings: the case study of Kaisariani Monastery. In: COMPDYN 2015 5th ECCOMAS Thematic Conference on Computational Methods in Structural Dynamics and Earthquake Engineering Crete Island, Greece, 25–27 May 2015, pp. 2321–2336 (2015)
34. Douvika, M.G., Apostolopoulou, M., Moropoulou, A., Asteris, P.G.: Seismic Vulnerability Assessment of Monumental Masonry Structures (in greek). In: 17th Panhellenic Concrete Conference, Thessaloniki, Greece, November 2016 (2016)

A Historical Mortars Study Assisted by GIS Technologies

Panagiotis Vryonis[(✉)], George Malaperdas, Eleni Palamara,
and Nikos Zacharias

Laboratory of Archaeometry, University of the Peloponnese, Tripoli, Greece
vrionpan@yahoo.gr, envcart@yahoo.gr,
el.palamara@gmail.com, zacharias@uop.gr

Abstract. Geographic Information Systems (GIS) have a vital role on broadening the understanding of the relationship of space, place, and culture. In recent years, a steady increase in the use of GIS in the fields of Archaeology and Cultural Heritage Management can be attested [1]. GIS has nowadays become one of the most versatile and comprehensive analytical tools in Archaeology in terms of handling archaeological data and exploring human space [2].

In this paper we will present the contribution of GIS in the study of the historical mortars of one of the most important castles in the Peloponnese, the castle of Androusa in Messenia. GIS was used in the documentation of the fortification of the castle and in the organization and archiving of the analyses of mortars. An interactive database was created, including the fortification ground plan, photographs and the results of the analytical study of eleven mortar samples. This database offers an easy access platform of the archiving data, with the potential of continuous update, while maintaining the historical and archaeological data in the same time.

Keywords: Castle of Androusa · Mortars · Construction phases · GIS · Interactive database

1 Introduction

1.1 Historical Information About the Castle

The castle of Androusa is located in the homonymous local community of the Municipality of Messenia in the Peloponnese. It occupies the northeastern edge of a low flat plateau (128 m. height) at the western slopes of the plain of Messenia. According to the Aragonese version of the Chronicle of Morea[1], the erection of the castle dates back to the middle of the 13th century and is attributed to the renowned Frankish ruler William Villehardouin [3][2]. During its history, the castle came under the

---

[1] The Chronicle of Morea is a long history text that contains a great amount of information regarding the Frankish conquest of mainland Greece. It was probably first written in French, but there are also versions in Greek, Italian and Aragonese, all of which have slight differences [15].

[2] The Aragonese version of the Chronicle of Morea mentions in paragraph 216: «…et en la castelania de Calamata fizo fer el castiello de Druges…» (meaning the castle of Androusa) [3].

© Springer Nature Switzerland AG 2019
A. Moropoulou et al. (Eds.): TMM_CH 2018, CCIS 961, pp. 532–540, 2019.
https://doi.org/10.1007/978-3-030-12957-6_37

control of different conquerors (Franks, Byzantines, Venetians, Ottomans), resulting to several additions and alterations to its initial fortification design. More specifically three building phases can be distinguished: 1st building phase during the Frankish period (southeastern tower), 2nd building phase during the Byzantine period (eastern and southern fortification), and 3rd building phase during Ottoman period (northern fortification) (Fig. 1) [4].

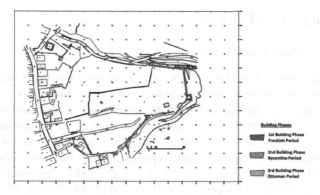

Fig. 1. Plan of the castle of Androusa depicting the building phases of its fortification

As most medieval castles, the ground plan of the curtain wall is trapezoidal and generally follows the contours of the terrain. The walls are reinforced by towers of various shapes placed in unequal intervals. The inner face of the walls has a series of blind relieving arches on which the wall-walk of the castle is housed. Unfortunately, only a small part of the castle remains today, mainly due to the modern village of Androusa, built in the northern and western sides of the castle. The only parts of the castle that are preserved are the eastern wall, parts of the southern and northern wall and six towers, whereas the western part of the fortification is completely missing [5].

During recent years, the castle underwent a series of restoration works aiming to its preservation. The most recent restoration project lasted from 2012 until 2015 and included the restoration of the eastern section of its fortification.

1.2   The Use of GIS

The study of the castle of Androusa was implemented as part of a research project, involving the use of new technologies and archaeometric analytical techniques and focusing mainly on the fortification of the castle, the mortars used for its construction and the relation of the castle with its landscape [6].

An essential tool in the study of the castle was GIS (Geographic Information Systems). GIS is a computer-based technology that is used to produce, organize and analyze spatial information. The capabilities of GIS include database management, mapping, image processing and statistical analysis [7]. GIS provides the potential to update geographical information index in a continuous and interactive way, to process and store large volume of different source data and to create thematic maps based on specific inquiries [8]. In our case, the use of GIS served a twofold purpose: (1) to represent the ground plan of the castle and (2) to aid the organization, documentation and archiving of the analyses of the historic mortar samples that were collected from the masonries of the monument. More specifically, GIS helped in the identification of parts of the castle that were missing and in the examination of the historic mortars that had been used in each building phase, establishing some chronological relationships between the different fortification areas.

2   Methodology and Results

2.1   The GIS Map of the Castle

As already mentioned before, the castle of Androusa is not preserved intact. Large parts of the northern and southern sides of its fortification are missing or lying in ruins. However, the biggest problem is located on the western side of the castle, where there are no visible traces of the fortification, except from a single tower. As a result, the whole image of the castle looks quite fragmented, providing limited information about it.

Initially, an effort was made to create a representation of the castle by employing the GIS ArcMap10 program and various maps, as well as, satellite and aerial photos of the region. The idea was to use the collected data and the geomorphology of the landscape as guidelines to make a hypothetical representation of the actual form and shape of its fortification ground plan. The datasets containing different type of information had to be digitized in different layers in the GIS program in order to be used in the process.

Georeferencing, the initial step when using GIS, is the procedure of transforming a map image to a reference coordinate system in order to be used by GIS software [9]. All the maps and the photos of the castle were georeferenced on the EGSA'87 coordinate system that is used in Greece and were inserted as layers in the GIS program.

An excerpt map of the region of Androusa was obtained from the LandSat (2007) and was used as the basic reference map. Two extra maps of the castle were also used: (1) a topographical map provided from the Ephorate of Antiquities of Messenia, and (2) an old hand made map[3] made by the French expeditor A. Bon during his visit in Androusa published in 1969 (Fig. 2) [10]. The topographical map depicted the parts of the castle that are preserved today and provided an accurately coordinated ground plan of the fortification, since it had been created with the use of new GPS systems. The old map of A. Bon proved to be an excellent source of information about the missing parts of the fortification of the castle, since it depicted remnants of the western, southern and northern sides that today are not visible or preserved. Both maps were digitized in different layers and then they were overlaid and correlated, resulting to a more complete image of the castle[4].

---

[3] Old maps constitute a valuable source of information regarding the historical landscape. In recent years, they are widely used in archaeology, mainly due to the rapid increase of the use of GIS and the potentials it offers [9].

[4] The correlation of the maps was based on the coordinates of the digitized fortification plan of the topographical map, provided by the Ephorate of Antiquities of Messenia since it was more accurate.

Fig. 2. Maps of the castle of Androusa. From left to right: Excerpt map of the region of Androusa (LandSat 2007); Depiction of the castle according to A. Bon [10]; Topographical map of the castle [4]

The final step was to connect the parts of the castle using polylines, thus completing the missing sections. This process was based on the geomorphology of the terrain and the orientation of the existing fortification parts. The result was a hypothetical representation of the castle, which seems to cover an area of approximately $20.000m^2$ (Fig. 3).

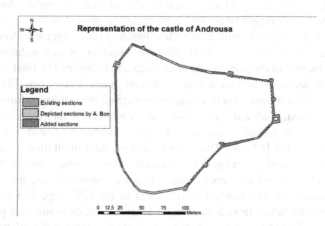

Fig. 3. Representation of the castle of Androusa based on the use of GIS

Based on the digitized map which was produced a landscape and viewshed analyses were carried out, showing that the location of the castle had been carefully selected in order to be: (1) well protected, (2) have proximity to fertile lands and natural water sources, (3) provide a good oversight of the wider area. However, the details of these analyses are beyond the scope of this paper (Fig. 4).

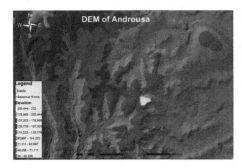

Fig. 4. Maps of Androusa in GIS: On the left the Digital Elevation Model (DEM) made for the landscape analysis; On the right the Viewshed analysis of the castle [6]

2.2 The GIS Database

Having acquired a clearer image about the fortification of the castle, the study focused on the analysis of the historical mortars of the castle. Eleven mortar samples were collected from different areas of the fortification covering all 3 building phases (Frankish, Byzantine and Ottoman). As shown in the Fig. 5, the examination of the mortar samples was conducted using a multi-technique approach: Optical Digital Microscopy LED, Electronic Scanning Microscopy (SEM) equipped with X-ray Microanalyzer (EDX), X-ray Fluorescence (XRF) and Granulometric Analysis. The analyzed mortars were lime mortars and were mainly consisted of fine aggregates. The samples collected from areas built in the same building phase appear to have the same chemical composition. Samples from different building phases exhibited relative heterogeneity mainly in the concentration of silica and calcium. The final results of the aforementioned analytical techniques are presented in detail elsewhere [6].

GIS was used for the creation of a database where all the resulted information could be stored, processed, analyzed and updated in an interactive and dynamic way. The application of GIS in a similar way is well known and has been widely employed in archaeological research [11–13]. The geographical location of all mortar samples were recorded on the digitized ground plan of the castle and were joined with their respective datasets. The exact location of each sample was measured with the use of a Differential GPS device and was marked with a point in the GIS map. The datasets were inserted in attribute tables in GIS in form of texts, tables, diagrams and photographs (raster graphics). The user can have access to information about: the different areas of the fortification of the castle (naming, building phase/period, photos), the sampling process of mortars (date of sampling, sampling area, building phase/period, photo of the area), the macroscopic characteristics of the samples (weight, color, preservation state, OM-LED photos), the results of the XRF analysis [type of device and settings, major oxides (wt% normalized to 100%), trace elements (ppm)], the results of the SEM-EDS analysis (chemical composition of the major oxides, photos under SEM microscopy) and the results of the Granulometric Analysis (grain size distribution,

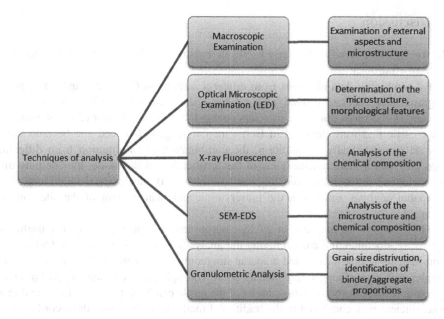

Fig. 5. Flowchart illustrating the analytical techniques used for the characterization of mortar samples

diagrams). All the above information is available just by clicking on the areas of interest on the map or on the table of contents in the GIS program. The user can also select the desired type of information to be displayed and have a quick overview of them (Fig. 6).

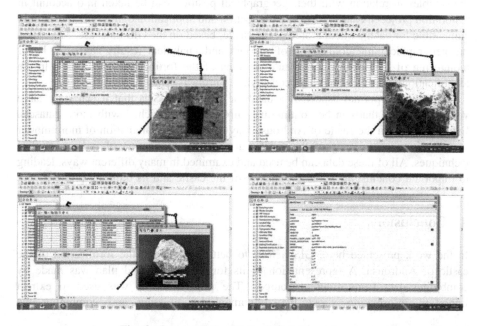

Fig. 6. Demonstration of the GIS database in use

3 Discussion

The use of GIS proved particularly useful in the study of the mortar samples from the castle of Androusa.

By combining data elements from different sources in GIS, we managed to get a geographically accurate ground plan of the castle. The use of GIS can assess the accuracy of old maps drawn by travelers during the past [14]. In our case, the sections of the castle that were depicted in Bon's map were georeferenced and became more accurate geographically. In addition, the combination of information from different maps provided a more holistic view of the fortification of the castle, thus helping the archaeological research by indicating areas of interest. By achieving detailed mapping, the spatial analysis and hence the preservation and management of the site can be enhanced.

The creation of a database in a GIS environment was proved to be very useful in organizing, documenting and archiving the analyses of the mortar samples. One of the main benefits is that it provided a spatial interconnection of the analytical data of the samples with their exact geographical location. Through this, the examination of mortars from different areas and building phases of the castle can be done easier, faster and in a more efficient way compared to the traditional methods. In our case, the recording and classification of the mortar samples in a GIS database helped us to compare their composition and to highlight the homogeneity or heterogeneity they presented with each other, simplifying and facilitating the whole process. In addition, the visualization of data helped the identification of spatial patterns associated with the composition of mortars, the different areas of the fortification and the likely building phases.

This GIS database can offer more possibilities in the future. For example, it can be used in restoration and conservation works at the castle of Androusa. The analysis of the samples in relation with their geographical position can be taken into account in restoration projects, contributing to the selection and application of a proper mortar mix compatible for each particular area of the castle, based on the GIS database. Moreover, the geographical interconnection of mortars and analytical results can offer better monitoring of their preservation state per area of the fortification and thus help the identification and organization of conservation projects when needed.

The most important aspect of a database made in GIS is that it can store large volumes of data that can be continuously updated and enriched with more datasets. Future projects in the castle of Androusa may include the examination of more mortar samples from other areas of the fortification or the application of more analytical techniques. All of these data can be used and examined in many different ways, leading to important conclusions about the history of the castle and its building phases.

4 Conclusions

In the work presented here, GIS helped to better understand the fortification of the castle of Androusa. A representation of its fortification ground plan was made by combining data from different sources. The produced map was used to extract important information about the monument by utilizing and combining the archaeological knowledge with spatial information and the visualization techniques offered by GIS.

In addition, GIS was implemented in the management of the information collected from the analytical study of the historic mortar samples. The positions of the samples were documented in GIS and were linked with their respective datasets. As a result, a database was created where all the acquired information can be stored and processed in a dynamic and interactive way. This database offers easy access and the potential of continuous update of the archiving data, while protecting and preserving the historical and archaeological data in the same time. In the future the database can be further enriched with more information about the castle.

The joining of GIS with the archaeological and archaeometrical studies allows a better approach of the history, archaeology and the cultural heritage in general. In our case, GIS has proved to be an excellent tool for processing, synthesizing, analyzing and interpreting archaeological information that can lead to the creation of new knowledge. The flexibility of the GIS applications is what makes them such a valuable tool for research and knowledge in the field of archaeology.

Acknowledgements. The authors of the paper are thankful to the personnel of the Ephorate of Antiquities of Messenia for providing access to the archaeological and bibliographical material.

References

1. Petrescu, F.: The use of GIS technology in cultural heritage. In: Georgopoulos, A. (ed.) XXI International CIPA Symposium, Athens (2007)
2. Tsiafakis, D., Evagelidis, V.: GIS as an Interpretative Tool in Greek Archaeological Research. Nottingham (2006)
3. Fernández de Heredia, J., Morel-Fatio, A.: Chapitre VI: Citation de G. de La Roche. In: Morel-Fatio, A. Libro de los Derechos et Conquistas del Principado de la Morea. Imprimerie Jules -Guillaume Fick. Geneva (1885)
4. 26th Ephorate of Byzantine Antiquities: Κάστρο Ανδρούσας - Τροποποιητική Μελέτη Νότιου Τείχους. Ministry of Culture and Sports. Kalamata (2013)
5. Ephorate of Antiquities of Messenia: Restoration of the eastern section of the Androusa castle. Information brochure. Ministry of Culture and Sports. Kalamata (2016)
6. Vryonis, P.: The castle of Androusa: Characterization of mortars and application of the Geographical Information Systems (GIS). Master Thesis. University of the Peloponnese, Kalamata (2018)
7. Box, P.: GIS and Cultural Resource Management: Manual for Heritage Managers. UNESCO, Bangkok (1999)
8. Hadjimitsis, D.G., Themistokleous, K., Ioannides, M., Clayton, C.: The registration and monitoring of cultural heritage sites in the Cyprus landscape using GIS and satellite remote sensing. In: The 7th International Symposium on Virtual Reality, Archaeology and Cultural Heritage. VAST (2006)
9. Cajthaml, J., Pacina, J.: Old maps as source of landscape changes, georeferencing, accuracy and error detections. In: 15th International Multidisciplinary Scientific Geo-Conference SGEM 2015, pp. 1004–1009 (2015)
10. Bon, A.: La Moree Franque recherches historiques, topographiques et archaeologiques sur la principalite d' Achaie (1205–1430). Eddtions E De Boccard, Paris (1969)

11. Malaric, I., Gasparovic, M.: GIS of the Crikvenica - "Igraliste" Archaeological Site. Kartografija i Geoinformacije, Zagreb (2009)
12. Tantillo, M.D.: GIS application in archaelogical site of Solunto. In: XXI International CIPA Symposium, Athens (2007)
13. Khan, Z.L., et al.: Geoinformatics for cultural heritage mapping - A case study of Srinagar city, Jammu and Kashmir. Int. J. Technol. Res. Eng. **2**, 1234–1241 (2015)
14. Sarris, A., Déderix, S.: GIS for archaeology and cultural heritage management in Greece. Quo Vadis? In: Zacharias, N. (ed.) 3rd Symposium Archaeological Research and New Technologies ARCH_RNT. University of the Peloponnese, Kalamata (2012)
15. Hetherington, P.: Byzantine and Medieval Greece – Churches, Castles and Art of the Mainland Peloponnese. John Murray Ltd., London (1991)

Towards a Blockchain Architecture for Cultural Heritage Tokens

Aristidis G. Anagnostakis$^{(\boxtimes)}$ (ID)

University of Ioannina, 45110 Ioannina, Greece
arian@teiep.gr

Abstract. Disputes over Cultural Heritage tokens and collections claims among collectors, organized social groups, countries, ethnicities, even civilizations, are but uncommon over time. Universal ontologies such as the CIDOC Conceptual Reference Model (ISO 21127:2014) have emerged as global standards over the past years, to allow for seamless structuring and interchange Cultural Heritage artifacts documentation information, in spite of their actual nature (physical, intangible, digital, etc.).

Yet, no objective universal procedure exists to safeguard the originality of the records and the archives; the immunity of essential sensitive data of the documentation tokens (dates, places, owners, etc.) is still questionable.

Aiming toward a coherent, effective Blockchain architecture to establish an immune, objective, collective archive of the documented Cultural Heritage tokens, the present proposes an implementation based on a comparative analysis of the prominent Blockchain architectures.

Keywords: Blockchain use cases · Blockchain for cultural heritage · Hyperledger · Cultural Heritage Documentation · CIDOC blockchain

1 Introduction

Currently, Cultural Heritage data are being structured, processed and archived in libraries and services such as the Archaeology Data Service (ADS) [5]. Even the best structured and formatted data suffer from serious deficiencies; there is no objective way to verify that the data has not been altered over time, neither intentionally, nor as the result of an attack. There is no objective mechanism to safeguard the ownership of the data, leading in many times to misconceptions upon the ownership of the heritage token itself. Different "documentalists" may describe the same artifact subjectively. There is no objective way to merely verify the succession of the recordings in time!

Yet, the historian of the future should have no doubt that the "Physical, man-made thing" of subtype "building" [1] entitled "Acropolis", was situated at the referred location known as "Athens" at the time of existence or at the time of the first recorded documentation, (*even this has been altered/vanished in the meantime).

At a minimum, an assimilated hash fingerprint of the XML structured document describing "Acropolis" (which has the form of a graph), has to be stored in an immune public index, allowing for verification of the original documentation. The actual content of the documentation may still relay under the copyright of the "documentalist".

© Springer Nature Switzerland AG 2019
A. Moropoulou et al. (Eds.): TMM_CH 2018, CCIS 961, pp. 541–551, 2019.
https://doi.org/10.1007/978-3-030-12957-6_38

Blockchain can offer such an immutable public record for cultural heritage archives, upon which everyone will be able to verify at low to no cost the validity of the documentation process.

2   Methods

Blockchain is a series of public records, the sequence and the content of which is extremely hard to question, thus forming a "universally" accepted ledger. It may be stored in a central point or be distributed over the network. For a number of reasons, i.e. to robustly operate among non-trusting third parties, to avoid "single point of failure" vulnerability, etc. the distributed ledger has evolved rapidly over the past years [8–25].

One of the key operations of a distributed ledger mechanism is to ensure that the entire network collectively agrees on the contents of the ledger; the mechanism is known as "consensus" mechanism. A consensus mechanism assures that each next block added to the chain, is actually representing the most recent link, thus preventing arbitrary infinite "forking".

The state of arbitrary forking suggests that different copies of the initial Blockchain evolve over time; the issue is known as "*double spending*" or "*nothing at stake*" under the context of the *Proof of Work (PoW)* and *Proof of Stake (PoS)* methodologies respectively.

The most popular and validated Blockchain consensus method is the Proof of Work (PoW). The Proof of Stake (PoS) is emerging to tackle specific inefficiencies of the PoW [3–8].

2.1   Consensus via Transparent Proof of Work

Transparent Proof of Work (PoW) as a consensus mechanism was first published by Dwork and Naor in 1993 [9], however, it wasn't until 1999 the actual term "Proof of Work" was coined by Jakobsson et al. [10]. Irrespective of the implementation variances, the PoW has some major properties:

- There is a Prover (i.e. the "minter-miner") and a Verifier; the Prover solves a mathematical puzzle at the work-cost w, and the Verifier verifies the puzzle is correctly solved at the cost of z.
- The puzzles are asymmetric; it is difficult for Prover to solve but the solution is easily verified by a Verifier (one-trap-door). "*A POW may be regarded as efficient if the Verifier performs substantially less computation than the Prover*" Cai et al. [12], Such a proof is considered to have "*large advantage*" if $\frac{z}{w} \ll 1$, or if z is asymptotically lower than w (as in Jakobsson et al. [10]).
- The puzzles are statistically independent; no skills are involved, they require brute force. This ensures certain Provers do not gain an unfair advantage over others. The only way for a Prover to improve his odds of solving a puzzle is to acquire additional computational power (more w at the solution time interval [ts, tc]); something that is very energy and capital-demanding. The only way to overpower the network strength of Blockchain networks is through a "51% attack" [11].

- The puzzle parameters are periodically updated in order to keep the next-block time consistent (lower bound the [ts,tc] interval, thus limiting the occurrence of inflation. As an example, the Bitcoin protocol [11] sets the desired *"block generation time"* to 10 min. If the average *"block generation time"* gets below this interval during a specific timeframe, the network automatically increases the difficulty of the puzzle (Fig. 1).

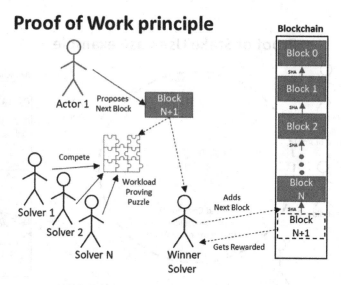

Fig. 1. Proof of work blockchain principle.

PoW has proved its' validity; yet it suffers from deficiencies (i.e. 51% type of attack vulnerability, increased latency, high energy consumption) that makes it hard to apply generally.

2.2   Consensus via Proof of Stake

Proof of Stake, is currently gaining popularity [8, 14, 24, 25] as a consensus mechanism for validating Blockchain links and achieving consensus, bearing a number of comparative advantages over the PoW, notably security, reduced risk of centralization, and energy efficiency [13].

Basic properties of the PoS include:

- Consensus finality: *"if a valid block appends to a Blockchain at some point in time, it can never after be removed"* [17]. PoW Blockchains does not satisfy consensus finality (see the "double spending" issue).
- Energy efficiency: In PoS, there is no mathematical puzzle, instead, the creator of a new block is chosen based on his stake, or even randomly.

- Reduced centralization risk: The "stake" policy on a PoS blockchain can be designed to provide "blind" equality to all nodes irrespective of their power.
- Penalization: *Greedy & malicious* behaviors and various forms of 51% attacks can be made unaffordably expensive; *"it's as though your ASIC farm burned down if you participated in a 51% attack"* Vlad Zamfir [22].
- Node Identity: The BFT PoS approach to consensus typically requires each node to know the entire set of the nodes participating in the consensus process. PoS is in many cases closely bound to the notion of "Smart-contracts" [17] (Fig. 2).

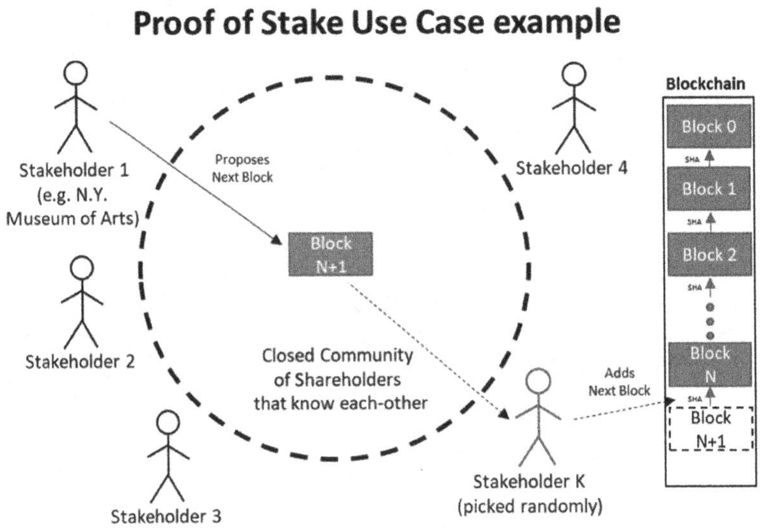

Proof of Stake Use Case example

Fig. 2. An example of Proof of Stake consensus Blockchain.

2.3 Smart Contracts

Smart-contract, introduced in 1994 by Szabo [15] is a *"computerized transaction protocol that executes the terms of a real-life contract"*. The idea behind the Smart-contract is to establish an automated way to fulfill the demands of real-life contracts (e.g. payment terms, confidentiality, conditional execution), to minimize deliberate and accidental *"exceptions"* and consequently to minimize the need for *trusted intermediaries*.

"Smart-contract use cases take the Blockchain well beyond its original cryptocurrency orientation, back to the domain of database replication protocols, notably, the classical state-machine replication" [20]. A Smart-contract can be modeled as a state machine, executing consistently across multiple nodes in an interconnected environment, using *state-machine replication*. A family of such protocols of high interest for Blockchain is the Byzantine Fault Tolerant (BFT) *state-machine replication* protocols, which provide consensus regardless of the existence of malicious (Byzantine) nodes in the network.

Smart-contracts bound to a Blockchain are used as general purpose computations that take place on a Hyperledger. Robust, nearly Turing complete languages have been proposed (i.e. Ethereum Foundation [16], IBM [17], the Hyperledger Fabric [26]), facilitating effective real-life Smart-contract implementations.

2.4   Transparent "Proof of Work" vs BFT "Proof of Stake"

A high-level comparison PoW and the BFT PoS for some important Blockchain properties are summarized after [20] on Table 1:

Table 1. Comparison of important PoW and PoS BFT properties.

	PoW consensus	PoS BFT consensus
Node Identity	Unnecessary, open to anyone without the need of identity awareness	Necessary for every node in the network to know the identity of all nodes
Scalability/distribution	High. Supported and tested on thousands of nodes and millions of clients, e.g. in the Bitcoin Blockchain	Node scalability not extensively tested; proved in practice for thousands of clients
Performance	Poor response time and power consumption. PoW uses power as the valued asset, and in some implementations, specific time duration to down-limit it	High performance in response time and power consumption. PoS BFT introduces minimal latency and computational power need
Attack power vulnerability	25% of the total network power is needed to gain control of the chain Eyal and Sirer [23]	Varies on implementation; Dwork Lynch and Stockmeyer [24] suggest that generally n/3 malicious nodes are adequate to gain control
Network synchronization	Physical timestamps required to validate nodes	Not required; temporal synching is needed only to keep the network alive

Perhaps the most important architectural difference between the two summarizes in the fact that in PoS, each node has to be aware of all the node id's participating in the network. This imposes a "controlled", centralized group of nodes (i.e. documenting entities in the domain of discourse); to add a new node in the network an additional round of voting may be necessary, to limit the risk of introducing non-reliable members.

In addition, a mechanism to restrict the function of non-reliable nodes -should such exist over time- has to be foreseen; the nodes already in the Blockchain network may act as a central trusted party and re-configure the rules at any future time.

3   Choosing the Appropriate Consensus Model

Building a Hyperledger for Cultural Heritage documents poses some major domain-specific considerations:

1. Administrators are in majority public and non-profit organizations. This calls for transactions of low to no cost and proclaims PoS consensus.
2. Administrators bear a significant know-how in documenting Cultural Heritage artifacts. This suggests a finite, controlled network of nodes and pinpoints the need for membership rules.
3. The members safeguard the validity of the process. They thus may accept or discard a newcomer, acting together as an internal trusted party. The confidence of the network on each has to be constantly re-evaluated.
4. All members have the same "rights" in terms of functionality, yet the relative significance of each vary vastly among them.
5. Documentation data may become significantly large in volume; this is especially true when hi-resolution images, videos, sounds and 3D scannings are part of the documentation. Importing the master documentation in the Hyperledger is both impractical and of limited value. An assimilated fingerprint of the document (a hashed SHA version of the document) is proposed to be used instead.
6. Yet, for the Hyperledger to be of practical use to third parties, a minimal set of reference information to the initial artifact has to be contained to the Blocks in plain, (un-hashed) text.

Considerations 1, 2 and 4 indicate PoS BFT as the most appropriate consensus model (see Table 1). Considerations 3 and 4 stress the need for functionality agreement among all members; this is efficiently compromised by existing Smart-contracts frameworks, (i.e. the Hyperledger Fabric, Ethereum, etc.). Considerations 5 and 6 pinpoint the need for customized software agents to bridge current documentations DBMS's to the Blockchain.

In addition, for a Blockchain architecture to be suitable for the domain of discourse:

- It has to be distributed; cultural heritage being public in notion and essence, its' documentation should be decentralized as well, to address the "single point of failure" issues and distribute trust and liability among all involving actors.
- It has to be fair; no member should be able to gain control of the process over the rest, in spite of his computational power or stake, especially without the explicit approval of the rest.
- It has to be documentation method-and-content agnostic; minimal id data of the imported tokens, (e.g. "Title" or "Id") are desired but not mandatory, and shall exist solely to facilitate third parties' quests in the future. The documenting authority shall fully preserve content and rights and prove these undoubtedly via the chain.

4 Proposed Architecture

4.1 Structuring the Content

Cultural artifacts documentation, irrespective of their actual storage format (flat file, E-R/OO DBMS, etc.) may, in the generic case be represented as structured XML [1, 2] (Fig. 3).

```
<CRM Entity (E1)>
 - <Number (E60)>: #32142
 - <Temporal Entity (E2)>

       ...

        - <Beginning of existence (E63)>:
            - <Time primitive(E59)>: "130 BC"

       ...

 - <Persistent Item (77)>

       ...

        - <Physical Man-Made Thing (E24)>:
"Aphrodite of Milos"

       ...

 - <Place (E53)>: "Milos Island, Greece"
        - <Spatial Coordinate(E47)>: "36° 44' 35"
N, 24° 25' 28' W"
 ...
```

Fig. 3. CIDOC XML documentation excerpt example.

The volume of the documentation being considerably high, block chaining the master documentation becomes evidently impractical; instead, a signed, assimilated fingerprint is used. The fingerprint bears and verifies the validity of the original, which may by no means be altered over time without being noticed [18] (Fig. 4).

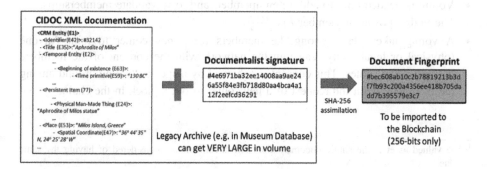

Fig. 4. Fingerprinting the Cultural Heritage artifact documentation.

The Blockchain will act as an immune distributed ledger of cultural heritage tokens' documentation. Each items' *Identifier* and *Title* (CIDOC E42 and E35 respectively) is also proposed to be added to the Block un-hashed, to allow for back-reference by third clients (Fig. 5).

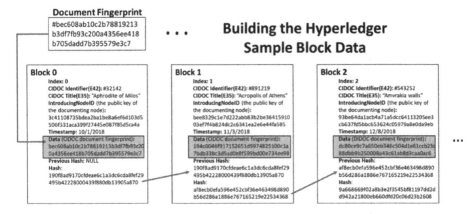

Fig. 5. Building the chain: putting essential data to blocks.

4.2 Forming the Contract

Let the universe of all cultural heritage artifacts $\{1, N\}$, and the universe of all documented artifacts $\{1, M\}, (M, N \in \mathbb{N})$. The Documentation may be modeled as a one-way function $\{1, M\} \xrightarrow{D_A} \{1, N\}$ carried out by member $i \in \{1, K\}$, where K is the number of active members in the Blockchain network.

The following contract is proposed:

- Each document has a specific Document Value of 1 DV, in a way analogous to Ethereum Wei [25][1].
- The first member to put records in the Blockchain is assigned a stake value of TDV_M (*Total Documents Value*) representing the value of all documents in the chain[2].
- Voting is needed only to add a new member, and to re-validate memberships.
- The voting power of member i is $\frac{DV_i}{TDV_M}$.
- A voting takes place among the members for a new-comer to join. Anyone (physical or legal entity) can join the network with the consent of the members holding $\frac{2}{3} TDV$ votes. This consent is achieved by a special "voting" round among the members, upon the request of a newcomer to put a block in the chain.

---

[1] The valued asset is the single documentation. All documents are considered of having the same value.

[2] Each document corresponds to a single Block. The voting power of each member is proportional to the number of Blocks it has proposed and have been successfully imported to the Hyperledger.

- All members are considered *trusted trustees* and *consent* to add a Block to the chain upon the *request* of any other member.
- A special round of "re-validation" between the members is executed, should any member tend to reach the critical limit (1/3 of the votes) i.e. $DV_i \leq \frac{1}{3}TDV_M - l, \forall i \in [1, K]$, where $l(>0)$ is the "safety" limit, preventing a member to acquire potential control of the chain without the explicit consent of the rest. The member questioned each time does not participate in the voting.
- The "re-validation" cycle also takes place if a member is marked as *"unavailable"*[3].
- Upon exclusion, the stakes of the rejected member DV_r are assigned to the rest proportionally to their stakes (i.e. $DV_r * \frac{DV_i}{TDV_M}$).
- Each member i is chosen to add a block to the chain explicitly, with probability $\frac{DV_i}{TDV_M}$ in a pseudorandom way. Upon unavailability, another node is chosen following the same rule.

The notion of "rewarding" in this type of Hyperledger is unnecessary. Increasing the length of the chain over time, $(M \rightarrow N)$ the real value of it is expected to increase proportionally; voting power is introduced as a "rewarding" mechanism on the long term[4].

5   Challenges

The proposed Blockchain architecture presents attractive features in terms of fairness, efficiency, computational load and response time.

New Blocks are imported at $O(1)$ cost, while new members are added at the cost of $O(K)$. Periodic re-validation of the memberships requires $O\left((K-1)^2\right)$, distributed over the network.

Even though recent PoS implementations suggest they easily overwhelm the criteria set, choosing the most suitable is a major challenge. *Hyperledger Fabric* [26] and *Ethereum* [14] are the most prominent, yet their attributes remains to be proved in practice. The proposed architecture is subject to further testing and development.

6   Conclusion

This study illustrates the prospective of introducing Hyperledger to the vast repository of Cultural Heritage documentation tokens. It reveals the potential, identifies the core features, and sets the requirements for a universal Hyperledger in the domain of discourse. It proposes a solid architecture and elaborates an initial performance evaluation.

---

[3] Has failed to add a block to the chain, even if he was chosen to, in more than $n\%$ *of the* calls.

[4] It also safeguards the smooth operation of the chain, since the largest members are expected to present higher availability.

Blockchain shall not resolve disputes by itself. The validity of the documentation shall always rely on the validity of the "documentalist", yet the community is empowered with the means to constantly re-evaluate the trustworthiness of its' members.

Introducing Blockchain in Cultural Heritage documentation shall provide a priceless, universal, immune time sequence of the Cultural Heritage documentation records (i.e. ISO 21127:2014), safeguarding the validity of the data, the value of the tokens, and facilitating in the most effective way the resolution of ownership and valorization issues for the generations to come.

References

1. International council of museums, International committee for documentation official site, definition of the CIDOC Conceptual Reference Model. http://www.cidoc-crm.org/. Accessed 12 May 2018
2. ISO standards official homepage - ISO 21127:2014 A reference ontology for the interchange of cultural heritage information. https://www.iso.org/standard/57832.html. Accessed 28 Aug 2018
3. Web archives homepage, Hal Finney Reusable Proofs of Work. https://web.archive.org/web/20071222072154/http://rpow.net/. Accessed 28 Mar 2010
4. Hackernoon tech BlogSpot, Decentralized Objective Consensus without Proof-of-Work, C.V. Alkan. https://hackernoon.com/decentralized-objective-consensus-without-proof-of-work-a983a0489f0a. Accessed 5 Feb 2017
5. Archaeology Data Service UK official site. http://archaeologydataservice.ac.uk/. Accessed 14 Aug 2018
6. Proof of stake vs Proof of work, Github Software Archive. https://github.com/ethereum/wiki/wiki/Proof-of-Stake-FAQs#what-is-proof-of-stake. Accessed 10 June 2018
7. Miller, A., LaViola Jr., J.J.: Anonymous Byzantine consensus from moderately-hard puzzles: a model for Bitcoin. Technical report, University of Central Florida, CS-TR-14-01, Florida (2014)
8. Proof of work vs proof of stake Turner Schumann, Hackernoon tech blogspot. https://hackernoon.com/consensus-mechanisms-explained-pow-vs-pos-89951c66ae10. Accessed 5 Apr 2018
9. Dwork, C., Naor, M.: Pricing via processing or combatting junk mail. In: Brickell, E.F. (ed.) CRYPTO 1992. LNCS, vol. 740, pp. 139–147. Springer, Heidelberg (1993). https://doi.org/10.1007/3-540-48071-4_10
10. Jakobsson, M., Juels, A.: Proofs of work and bread pudding protocols (Extended abstract). In: Preneel, B. (ed.) Secure Information Networks. IFIP AICT, vol. 23, pp. 258–272. Springer, Boston, MA (1999). https://doi.org/10.1007/978-0-387-35568-9_18
11. Bitcoin official homepage, A Peer-to-Peer Electronic Cash System white paper. https://bitcoin.org/bitcoin.pdf. Accessed 14 Aug 2018
12. Cai, J., Lipton, R., Sedgewick, R., Yao, A.: Towards uncheatable benchmarks. In: IEEE Structures, pp. 2–11 (1993)
13. Ethereum official blog, Vitalik Buterin p-epsilon attack. https://blog.ethereum.org/2015/01/28/p-epsilon-attack/. Accessed 10 Aug 2018
14. Ethereum's Casper protocol explained in simple terms. https://www.finder.com/ethereum-casper. Accessed 12 July 2018

15. University of Amsterdam on-line courses, Smart contracts definition (Nick Szabo 1994). http://www.fon.hum.uva.nl/rob/Courses/InformationInSpeech/CDROM/Literature/ LOTwinterschool2006/szabo.best.vwh.net/smart.contracts.html. Accessed 4 July 2018
16. The Ethereum official homepage. https://www.ethereum.org/. Buterin, Vitalik. "Ethereum Whitepaper". Accessed 1 June 2017
17. IBM Research – Zurich official homepage, Christian Cachin "Architecture of the Hyperledger Blockchain Fabric". https://www.zurich.ibm.com/dccl/papers/cachin_dccl.pdf. Accessed 1 Oct 2016
18. Vukolić, M.: The quest for scalable blockchain fabric: proof-of-work vs. BFT replication. In: Camenisch, J., Kesdoğan, D. (eds.) iNetSec 2015. LNCS, vol. 9591, pp. 112–125. Springer, Cham (2016). https://doi.org/10.1007/978-3-319-39028-4_9
19. Goldreich, O.: Foundations of Cryptography I: Basic Tools. Cambridge University Press, Cambridge (2001). ISBN 978-0-511-54689-1
20. Schneider, F.B.: Implementing fault-tolerant services using the state machine approach: a tutorial. ACM Comput. Surv. 22(4), 299–319 (1990)
21. Lamport, L., Shostak, R., Pease, M.: The Byzantine generals problem. ACM Trans. Program. Lang. Syst. 4, 382–401 (1982)
22. Medium Blogspot Vlad Zamfir "Simple model of an internal PoW attacker". https://medium. com/@Vlad_Zamfir/simple-model-of-an-internal-pow-attacker-1a713cf00672. Accessed 2 May 2017
23. Eyal, I., Sirer, E.G.: Majority is not enough: Bitcoin mining is vulnerable. In: Christin, N., Safavi-Naini, R. (eds.) FC 2014. LNCS, vol. 8437, pp. 436–454. Springer, Heidelberg (2014). https://doi.org/10.1007/978-3-662-45472-5_28
24. Dwork, C., Lynch, N., Stockmeyer, L.: Consensus in the presence of partial synchrony. J. ACM 35, 288–323 (1988)
25. Wood, G.: Ethereum: A Secure Decentralised Generalised Transaction Ledger Byzantium Version e94ebda - 2018-06-05. The Ethereum yellow paper homepage. https://ethereum. github.io/yellowpaper/paper.pdf. Accessed 8 July 2018
26. The Hyperledger Fabric platform official homepage. https://www.hyperledger.org/projects/ fabric. Accessed 14 Aug 2018

Protection and Highlighting of a Waterfront Zone Disposing Strong Cultural Characteristics

Dimitris Psychogyios[1]([⊠]) and Helen Maistrou[2]

[1] U.A.S. of Thessaly, Thessaly, Greece
dcyxos@gmail.com
[2] School of Architecture N.T.U.A., Athens, Greece
elmais@central.ntua.gr

Abstract. The scope of this intervention is to present a project concerning the protection and the functional and aesthetic upgrade of a coastal zone which lies at the tip of the Argolic gulf in the Eastern Peloponnese. The coastal zone connecting two very well-known historic cities Nafplio and Nea Kios runs for approximately 14 km and it disposes some interesting features. It constitutes an important seaside wetland which concentrates rare flora and fauna species, mainly avifauna, having a particular financial importance for the production cycle of food, as well as for the broader balance of the region's ecology. Important historic and archaeological sites which can and should be highlighted lie in immediate vicinity to that part of the coastal zone, as well as historic rivers which flow into the Argolic gulf, all of which offer the occasion to take a cultural route in the outdoors as it combines elements having environmental, historic and regional value. The protection and highlighting of this zone will contribute on one hand to the sensitization of the local population in matters of cultural heritage and environmental protection, and on the other to the reinforcement of cultural tourism. The study was interdisciplinary, as it required historical and archaeological documentation of the intervention site and had to address urban and development issues, issues of protection of historic sites, natural environment and landscape, architectural issues and issues pertaining to electrical installations, big data and digital enhancement issues and, of course, specialized wetland protection issues.

Keywords: Historical urban landscapes · Digital mapping · Network

1 Introduction

1.1 Brief Description

Firstly, there is a brief description of the research project at the level of analysis and at the proposal level. Subsequently, emphasis is placed on the large volume and types of information for the study area. The successful management of this data constitutes a critical parameter for the sustainability of the area. With this in view, afterwards, there is an outline of the technological directions that are suggested in the work and are based on 'digital mapping' and on 'Networked Distributed micro-Architectural Systems'.

© Springer Nature Switzerland AG 2019
A. Moropoulou et al. (Eds.): TMM_CH 2018, CCIS 961, pp. 552–566, 2019.
https://doi.org/10.1007/978-3-030-12957-6_39

Next, follows the analytical description of specific technological applications that were suggested and whose main aim is innovative ways of educating the young and creating citizen awareness to issues of environmental protection and cultural heritage. Finally, there is a description of the general context of the technological suggestions of the study, which is based on modern learning processes and mobility practices, on the concepts of 'historical urban landscape' and 'allocation' as well as the prevalence of 'the network example' in all sectors of human activities.

The sections of the proposal are the following:

A. Brief description of the overall project. Analysis and Suggestions.
B. Management of big data. A critical parameter of design
C. Digital Mapping and Networked Distributed micro-Architectural Systems
D. Description of suggestion of technological educational application for the environment and cultural heritage.
E. Summary of basic terms
F. Epilogue

Subsequent paragraphs, however, are indented.

1.2   Analysis and Suggestions

Nafplio is a very well-known historic city, whose existence dates to the Prehistoric era. Nea Kios, which is a small city adjacent to Nafplio, was established in 1933 by refugees who had arrived from Kios in Asia Minor. The waterfront zone connecting the two cities runs for approximately 14 km. and it disposes some interesting features, as well as substantial problems. It offers an exceptional view towards Palamidi, Akronafplia, Bourtzi and the broader Argolic gulf and it presents the features of a seaside wetland which concentrates rare flora and fauna species, mainly avifauna. Important historic and archaeological sites lie in immediate vicinity to that part of the coastal zone, as well as historic rivers which flow into the Argolic gulf, all of which offer the occasion to take a cultural route in the outdoors. Nonetheless, this coastal zone is also characterized by problems which were examined within the context of the present study, in order to formulate proposals for the protection of the natural environment, for the promotion of the region's historic features, for the connection of the coastal zone with the functions of the urban space of the adjacent areas and for the region's sustainable economic development.

A multidisciplinary team [1] was set up to analyze not only the problems but also the strong points of the study area and to formulate relative proposals. Architects, environmental scientists, historians, a hydraulic and an electrical engineer, as well as experts from the Greek Ornithological Society, collaborated to complete the study. The study has drawn up proposals to highlight the relationship between the coastal zone and the sites of historical and archaeological interest of the immediate area, to formulate a regulatory framework for the protection and enhancement of the environment and the wetland and to design light constructions protecting the area as well as information signs and mobile digital applications that inform about the importance of the site.

At the level of analysis, the historical importance of the area and the mythological references were documented. The geological and hydrological characteristics of the area and the environmental data (water quality, soil quality, characteristics of flora and fauna) were studied. Then came the recording and the assessment of the environmental and landscape characteristics of the coastal zone and the wider environment, as well as the commitments resulting from the overlying design. The characteristics and problems of the wetland, where 160 bird species live were studied in particular. It is a zone with unique flora, which constitutes a concentration area of avifauna largely migratory, which often offers the observer ornithological 'treasures', of rare winged visitors. The research team then documented the uses and activities of the coastal zone, the existing technical infrastructure, the institutional framework concerning environmental issues and the organisation of urban and peri-urban space as well as the existing studies and opinions of local bodies.

The outcome of this analysis was the realisation that there was insufficient protection and promotion of the important wetland and lack of connection between the archaeological site of Ancient Tirintha with the beach. In addition, there was lack of promotion of the rural culture of the area and the powerful natural element of the river Inahos as well as the unexploited today abandoned old industrial complex 'Nafplian Can Company Pelargos'. Another observation was that pedestrians couldn't move along the coastland, approach the beach and the wetland and observe the avifauna without causing disturbance (Managing human access to protect wetland birds from disturbance).

At a proposal level, suggestions at an institutional level were composed for the protection of nature, for the environmental improvement of the area and the promotion of the natural elements and the wetland. A comprehensive network for visitation and tour of the coastal zone with interventions in order to promote the area was designed. Constructions were built in order to facilitate (a) the movement of pedestrians, (b) the movement of cyclists and (c) accessibility of persons with restricted mobility in the coastal zone. Also, equipment was designed and specified that includes:

- Signposts that show information or educational material
- Theme information kiosks
- Fauna observation posts
- Seaside and waterfront recreation facilities
- Constructions for child recreation
- Urban equipment: benches, bins, lighting, shelters
- Technological facilities

The design of the routes – footpaths, was organised in such a way that the human presence (visitors) would not constitute a means of disturbance to the whole of the wetland of the seazone. Along the routes and at key points, information signs linked to the area have been added, but also benches, designed to offer rest to the walkers.

2 Management of Big Data. A Critical Parameter of Design

As mentioned above, the study area has a length of about 14 km, connects 8 Districts (Nafplion, Tirintha, Midea, Dalamanara, N. Kios, Argos, Lerna and Kiveri), belongs to 2 Capodistrian Municipalities, has a variety of places of environmental, historical and touristic interest and hosts a large number of urban activities, many times with conflicting interests. The area constitutes a dynamic and under continuous negotiation area of 'habitation'. It is a complex 'historical urban landscape', an environment that creates the culture, the spiritual culture and generally the identity traits of the collectivities that 'inhabit' it. The results of the analysis phase of the research reveal as main characteristics of the area of study its 'complexity' and 'change over time'. The suggestions for the promotion and exploitation of the area of study require a 'dynamic' model of management that includes the concept of complexity and change. The successful management of digital information and the resulting data from the area constitute a critical factor of sustainable development of the area.

In view of the above, it is essential to develop a management model that on an ongoing basis will allow the collection of data from the area, the suggestion of solutions, the evaluation and the redefining of those solutions. It is also important that the management model be based on open data, on non-central control and take into consideration the collectivities of the area. Next are briefly mentioned the categories of data for the study area. There are spatial, time, thematic, management data, scale data, data of use and project data. Combinations of all of the above constitute either recording data or results of proposals. For the study area the following have been identified and categorized:

- Spatial data (points, axes, areas)
- Time data (direct, short term, long term, continuous, periodical)
- Thematic data (environment, culture, tourism, legislation, society, architecture, technology, sport, property)
- Management data (consultations, decisions, approvals, studies, implementations, competitions, evaluations, participations, proposals)
- Scale data (environment, spatial planning, urban planning, village, building, equipment, building detail)
- Data of use (residence, work, mobility, public service, property, cultivation, storage)
- Data of technical projects and research

Indicatively, we can mention (see Table 1) some of the specific data of the area that comes from private bodies, from government agencies and independent collectivities.

Special mention must be made to the activities of the Environmental Education Centre [23] which was founded in 2005 and is the main body of education and awareness on environmental issues. The centre aims to create awareness on the necessity to protect nature through experience learning. It is the development of a spirit of co-operation, friendship, teamwork and understanding. The Environmental Education Centre has developed a set of tools such as the use of the Internet, field actions, actions at its premises in N Kios and actions in the town. Especially on the Internet, it has created a reliable website for the environment. At the same time, personal pages or groups have been developed at social media from 'friends' and 'associates' of the Environmental Education Center. Additionally, it has developed actions for students as, for instance, the program 'on the wings of the seagull'.

Another important digital infrastructure for recording and information both for the study area and other areas is the website of the Hellenic Ornithological Society [35],

Table 1. Table of the specific data of the area

Digital data management entity	Data
Nafplio and Argos Municipal Water and Sewerage Companies	• *Drinking Water analysis* • *Seawater analysis* • *River water analysis* • *Soil analysis*
Hellenic Statistical Authority	• *Distribution of land and cultivation usage in the Drainage Basin of the Argolic Gulf (GR31)* • *Car accidents*
National Statistical Service	• *Agricultural water supply needs* • *Average annual mixed water runoff at rivers and water systems* • *Condition of coastal water systems* • *Qualitative condition of groundwater systems* • *Annual feed and yield from groundwater systems of the drainage basin of the Argolic Gulf* • *Quantitative – chemical condition of groundwater bodies – Drainage Basin of the Argolic Gulf (GR31)*
Institute of Geology and Mineral Exploration	• *Average monthly precipitation* • *Meteorological data* • *T = Average annual temperature* • *Tmax = Absolute maximum temperature* • *Tmin = Absolute minimum temperature* • *U = Average relative humidity* • *η = Cloudiness* • *P = Precipitation* • *np = Average annual number of days of precipitation*
Greek land registry	• *Property ownership status*
Ministry of the environment	• *Traditional settlement and listed building archive* • *National information system of energy* • *Natura 2000* • *National network for information about the environment* • *Database for the monitoring of wastewater treatment plants*
National Technical University of Athens	• *'Filotis' Database for the Natural Environment of Greece*
Hellenic Ornithological Society	• *Recording of birds*

which includes general and specific information. Also, through the website, the members, supporters and friends of birdwatching organise daily bird-watching excursions to the Moustos wetland in Astros and to the seaside wetland of N Kios. Finally, we would like to mention all the autonomous blogs of friends and fans of bird-watching in which they upload information, pictures and experiences from their recordings.

The above indicative reports on digital data about the area do not cover the total of data, nor the total of scientific fields. The list is long and the data constantly change and get updated. In order for the study team to be able to create a complete plan for the area, it needs to take into account and manage the total amount of information. As a result, there is a pressing need to create a Local Database and to connect it to the central public database catalogue Data.gov [24]. As mentioned in the relevant website, the aim of data.gov is the provision of access to high value machine-readable data as well as internet services (cataloguing, indexing, storing, searching etc.) to the public and other computer systems.

Open data is crucial for good governance, as it supports fundamental democratic values such as transparency, responsibility of governments and citizen participation, and at the same time constitutes an important element for development and a basic requirement for digital politics. Some of the sectors where open data has been of great value are the following:

- transparency and democratic control,
- participation, innovation,
- production of new products and services,
- improvement of productivity of government services as well as
- production of new knowledge from the combination and the use of big data.

According to the research [3] of European Data Portal [25] on the benefits from open data for the period 2016–2020, the cumulative size of the market is estimated at 325 billion Euros. Also, until 2020 the jobs related to open data will be 100.000. In addition, cost saving for the 28+ members of the EE is expected to be around 1.7 billion Euros. Always according to the research, open data is capable of saving 7.000 lives per year and 629 million hours of unnecessary waiting in the streets of the EE can be avoided.

Consequently, the study area provides valuable information data on an ongoing basis, the use of digital technologies can contribute in two basic directions. The first is at the level of environment management and participation of collectivities with the use of open geo-spatial data, the second at the level of digital accessibility and promotion of the environment. We can, for instance, observe, record and evaluate environmental data (such as quality of water, temperature, humidity, flora status, fauna status, pollution, rubbish etc.) and subsequently suggest solutions to the problems that will arise. Also, there is the opportunity to develop open educational programs and to support touristic and educational activities with the creation of guides that will exploit mobile technologies. The methodology and the tools that will allow us to develop successful applications are based on "digital mapping" and "networked distributed micro-Architectural Systems".

2.1 Digital Mapping/Networked Distributed micro-Architectural Systems (NdmAS)

In this section, there is an outline of the technological directions that are suggested for the project and that are based on 'digital mapping' and 'Networked Distributed micro-Architectural Systems' (NdmAS).

Digital mapping concerns the development of a digital platform based on geographical data systems. The data content may originate either from official bodies like

'Koitida' [8], 'Map of traditional settlements' [26] and 'Filotis' [27], or from self-organised collectives like 'Mapping the commons' [28], or from automated systems that have been installed or act in the natural environment within some specific program like 'UIROOMS simulator' [15], "Meteorological data" [29], marine traffic [30] and planefinder [31]. The digital mapping may offer an integrated platform for the recording and use of the data of the area. In that way, the open data can yield multiple benefits at the level of management, research, economy etc.

The Networked Distributed micro-Architectural Systems concern networks of natural areas of a small scale that can be combined with technological applications and mobile activities and services. This involves the organisation of small spatial units of basic organisation, limited range of action and influence. However, the relationships that the units develop allow much greater results. A highly complicated reality is created which requires the integration into the design of action and time. The design of Networked Distributed micro-Architectural Systems is based on 'the triple nature of the object', the 'real', the 'narrative' and the 'collective'. Consequently, the characteristics of Networked Distributed micro-Architectural Systems are objective, collective and narrative. The design must define small scale constructions, the online applications, the mobile applications. In addition, it must define the ways of coexistence, exchange of information and personalised information. Also, it must design the material evidence, the digital content and the usage scenarios (see Table 2). There are two types of Networked Distributed micro-Architectural Systems: the ones with hierarchical structure like Hybrid city [32] and the ones that have non-hierarchical structure like vrnafplio [33], Leafsnap [34] and kew garden app [4, 21].

Table 2. Table of characteristics for Networked Distributed micro-Architectural Systems (NDmAS).

Objective characteristics	Collective characteristics	Narrative characteristics
Small scale constructions	Coexistence	Material Evidence
Online applications	Information exchange	Digital content
Mobile applications	Personalised information	Usage scenarios

At this point it is important to emphasize that although computer systems have a relative autonomy, they cannot operate successfully if they aren't combined with the other suggestions for the wetland, such as the network of routes, the points of interest, the visitor support construction network etc. Computer systems constitute one more layer of equipment aiming to manage and promote the area as well as the information, entertainment and education of the visitors. For example, the functionality of computer systems directly depends on the functionality of light constructions of a small scale and on equipment components of the outdoor area; the visitor that searches for information through his mobile phone or tablet must be in a comfortable and friendly environment. Access to specific content of geospatial information needs to be combined with the appropriate configuration of route or with the placement of signposts with information about the area.

3 Description of Suggestion of Technological Educational Application for the Environment and Cultural Heritage

Next comes the analytical description of proposed specific technological applications whose central objective is innovative ways of educating the young and creating citizen awareness to issues of protection of the environment and cultural heritage. The research team recommends the systematic completion, organisation and free distribution of information material of the wider study area, through an expanded online digital platform that will be addressed both to the on-site visitor of the physical location and to the distant internet visitor.

The organisation of the functionality of the applications will be based on:

- Promotion, which will include issues addressed to the public and local bodies such as tourist associations, nature clubs etc.
- Management, which will include issues addressed to limited public, to local regions, municipalities etc.
- Research, which will include issues addressed to limited public, to universities, research centres etc.

This expanded online digital platform will include many mutually supportive applications of free and open access. Indicatively we mention the following:

- Website
- Augmented reality smartphone app for on-site visit
- Connection to social networks
- Thematic list of articles on Wikipedia
- Image list on Flickr
- Google Earth, Google maps and Bing maps
- Inclusion of the produced material in the National Documentation Centre and Europeana.
- Digital educational actions making use of existing programs and producing new ones.

The informational material included in the above platforms will contain:

- General Information
- Tourist Information
- Interactive map with places of interest in the study area
- Visitor navigation applications
- Special information about the flora and fauna of the area
- Informative documents about issues involving the management of wetland areas
- Educational programs
- Information about events, scientific meetings
- Thematic bibliography with the possibility to enrich and update the information
- Bibliography of scientific projects on the area with the possibility to update the information
- Bibliography of legislations, decisions about the area with the possibility to update the information

- Connection to databases for the continuous flow of specific information such as, for example, water quality analysis, soil and subsoil quality analysis, land use allocations, condition of coastal water systems, qualitative condition of subsoil water systems, quantitative – chemical condition of underground bodies, meteorological data, property ownership, flora and fauna recordings etc.

More specifically, the study recommends the development of a pilot mobile educational application (Apps4Argolida) which aims to promote the important historical, environmental and landscape elements of the area. It also aims the connection through promotion of neighbouring areas as well as to increase the awareness of the local population on issues of cultural heritage and protection of the environment. In particular, the following groups have been selected:

- Agricultural land
- Local produce
- 'Tirintha' and fortifications
- Humans and the environment
- Environment and development
- 'Timenio' and ancient seaports
- 'Stork'
- Mythology and water

Each of the above subject headings was placed at specific points in the specific study area which will be a pilot guide for interventions in the surrounding area. The selection of the points and the placement of the subject headings was a result of research that took into consideration a series of parameters, environmental, cultural, landscape and managerial. 8 points of interests, E2 through E9 were created and two more, E1 and E10 will be added in the town of Nafplio and the village of N Kios that are not included in the pilot area though. At the same time, a route is created, one that will connect all the points of interests mentioned above, while along all the route, the visitor may discover the wealth of the natural landscape and make use of facilities such as benches, taps and information signposts. The digital information network works based on this network of light constructions, routes, seats and information signposts.

Especially the educational application "Apps4Argolida" consists of 5 basic sections Home/Map/Flora/Fauna/Games (see Fig. 1). The application has a special home page depending on the position of the user, so different content appears if, for example, the user is at the Roumani wetland, different if they are in the touristic zone of N. Kios and different if they are within the limits of the town of Nafplio. In the Home Page there is also an application that informs the visitor about the events that are taking place in the area during his visit.

The subpage Map shows a map of the area and finds the exact location of the visitor in it. At the same time, it helps the orientation of the visitor, giving them information such as points of interest, entrances, cafeterias and points of rest, playgrounds and exhibition areas. When the visitor approaches one of the points of interest, if they wish, the application gives them analytical information about the point where they are. Under the Home page, the visitor can see the subpage Fauna, in which there is information and pictures mainly about the birds that one can meet at the wetland. The application

Fig. 1. Snapshots from the educational application "Apps4Argolida"

detects the time period of the visit and makes a special note about the species that are hosted during that season. Also, under the Home page, the visitor can see the subpage Flora in which there is information and pictures about the plants and trees that one can meet at the wetland. The application also allows the user to identify the species of plant by taking a photo of the leaf. Finally, the subpage games gives the chance to many users to simultaneously organise and play games of hidden treasure, discovering information by walking in the wetland.

4 Summary of Basic Terms

This section describes the general context of the technological suggestions of the study, based on modern learning processes and the new mobility practices, on the concepts of 'historical urban landscape' and 'allocation' and the prevalence of the 'network example' in all aspects of human activity.

'Historical urban landscapes' as defined by official Unesco texts [19, 20] and elsewhere [9, 10] are the tangible or intangible elements that shape the character of the landscape and more specifically concern topography and terrain, space organisation, land usage, visual relations, buildings, vegetation, the element of water but also technical infrastructure and urban equipment. Also included in historical urban landscapes are tangible and intangible social, cultural, economic and ecological factors, which form its evolution.

The concept of the network in the light of the technologies of information formed a new model for the perception of space. van Dijk [6] in the 90s introduces the concept of 'the network society' defining it as 'a society where a combination of networks of societies and media shapes the dominant way of organisation and the most important structures at all levels (personal, organisational and communal)'. He compares that type of society to 'a mass society that is shaped by teams, organisations and communities ('masses') that are organised through physical presence'. On the other hand, Manuel Castells claims that 'networks constitute the new social morphology of our societies' [5], while he gives an alternative definition: '... a society where the main social structures and activities are organised around tne networks that digitally process information. As a result, it isn't just about networks or social networks, because social networks are very old forms of organisation. It's about networks that process and manage information and use technologies based on microelectronics'. Papalexopoulos [13] finds that 'the network example is predominant in today's state of affairs; it signifies the transition from linear and clearly defined organisation of production of products and services, to the gradual prevalence and acceptance of organisation through distributed networks with unclear boundaries and identity... The network example determines our way of thinking, the knowledge structures, defines what is normal and acceptable and what isn't.'

The concept of distribution and 'distributed museum' [14] was recently introduced to the research field for museums by researchers Bautista and Balsamo [2] describing the modern institutions that act as main hubs at 'distributed learning networks', which characterise today's electronic age. The term is used to describe the new museum format as a hub of a network of knowledge environment. The museum is not considered a building that belongs exclusively to natural space, but expands its presence through new digital applications and mobility technologies. According to the researchers, museum experience can be found in a specific natural space or can be approached through the routes of the visitors in a series of successive spaces. They note in that way the move for the validation and design of the museum 'from the concept of space to the concept of place' [13].

Bautista and Balsamo, with reference to the definition of space according to Michel de Certeau as 'a practiced place' [7], described this process as a transition from a 'cultural institution based on the place to a dispersed (meta) modern place' and they separated it to three categories of efforts that modern museums make in order to overcome their natural space and move into the sphere of virtual, mobile and open learning (real/virtual, fixed/mobile, closed/open).

Specifically, as a sociologist, Michel de Certeau claims that the concept of place has been used from the dominant classes for the organisation and control of society through urban planning and architecture. The concept of space, however, constitutes the result of living and movement. Place implies stability, 'an instant formation of various positions', while space includes 'direction, speed and time variables'. As a result, space constitutes 'junctions of mobile elements' [7], and therefore can be characterised as 'a practiced place'.

New practices of mobility contributed to the dispersion of place in the modern museum. Recently, the speed of cultural change of what Williams [22] named 'mobile privatization' is increased, as new technologies, such as laptops, PDA and mobile telephones are largely used. Based on this transformation, Spigel [17] redefines Williams' phrase as 'privatized mobility', a concept that confirms the cultural importance of portability (now in connection with mobility), which brings activities of the private sector to public spaces.

Modern museology displayed an important innovation when mobile communication was widely applied in new practices that brought the museum experience outside the natural place of the museum. Today, the 'mobile museum' consists of satellite museum spaces, museum programs that are conducted by the personnel of the museum at schools, libraries and communal places, and special vehicles designed to provide learning experience through multimedia that is based on museum collections that travel all around the town. During the last decade, the 'mobile museum' has been transformed into what we call 'distributed museum', as a postmodern form through which the modern museum perfectly adapts its traditional functions and spaces to the new cultural environment of the digital age.

Especially for the Internet technologies related to space, Komninos, Schaffers and Pallot [18], claim that from 2009 onwards, we are in the third phase, in which interconnected services covering a wide range and meeting the requirements of the management of a large mass of data on knowledge, services, 'things' (Internet of Things) are provided. It could be considered the space of real time cooperative editing on the Internet. In this space of flows or networked space, Bautista and Balsamo place the modern museum as one of its main hubs. They claim that the space of new museology is similarly dispersed, personalised and non-linear, similar to the digital age. It isn't any more committed to a specific place, but new museology can be described through changes in practice, relationships and emerging experiences. Internet technology gives the chance of instant access to information from distant places, expanding the limits of the current existing space [14]. As a result, the museum will be a number of hubs and links which will develop collective intelligence.

Researchers Susana Bautista and Anne Balsamo claim that as the learning process nowadays is shifting from teacher guidance, or in the specific case expert guidance, to 'communal exchange of knowledge and information', museums are obliged to review their role in the 'distributed, dispersed and decentralised space of the digital age', while at the same time they maintain their traditional functions of guarding, exhibiting and maintenance of exhibits, education and creation of community awareness.

5 Conclusions

The design and installation of computer systems for interconnection and access to the Internet are a subject that has great research and commercial interest. At the same time, public bodies have invested and are going to invest large amounts of money in the development of digital applications, in the context of open government aiming to provide optimal services to the citizen. The analysis of the study area through the prism of new technologies led us to the conclusion that in the area there is a large and valuable amount of information data the management of which constitutes a critical factor for sustainable development. At the same time, through the documentation and evaluation of the existing information infrastructure it appeared that any actions need to also include the existing information infrastructure of the study area.

The modern research directions through which new goals and priorities must be drawn are the promotion and management of historical urban landscapes, the networks, the new mobility practices and modern trends in learning processes. The available technologies and the relevant projects are constantly on the rise and the experience from use as well as the study of the research results must be guides of good practices. The present study recommends the systematic completion, organisation and open provision of information material concerning the study area, through an expanded digital platform, both to the on-site visitor and to the distant Internet browser.

References

1. Project title: Highlighting and development of Nafplio - N Kios coastaline. Research team: Eleni Maistrou (team leader) Costas Moraitis, Dimitris Psychogyios (main researcher), Elena Konstantinidou, Stavros Gyftopoulos, Valentini Karvountzi, Maria Marlanti, Dimitris Sagonas (architects engineers and urban designers) Fran Vargas (Hellenic ornithological Society) Nikos Tsiopelas (program coordinator LIFE+, Hellenic ornith. Soc) Minos Maistros (mechanical and electric engineer), Maria Papaioannou (architect, agronomist)
2. Bautista, S., Balsamo, A.: Understanding the distributed museum: mapping the spaces of museology in contemporary culture. In: Trant, J., Bearman, D. (eds.) Museums and the Web 2011: Proceedings. http://www.museumsandtheweb.com/mw2011/papers/understanding_the_distributed_museum_mapping_t.html. Accessed 7 Sep 2018
3. Carrara, W., San Chan, W., Fischer, S., Van Steenbergen, E.: Creating Value through Open Data. Study on the impact of Re-use of Public Data Resources, 1st edn. Publication Office of the European Union, Luxembourg (2015). https://www.europeandataportal.eu/sites/default/files/edp_creating_value_through_open_data_0.pdf. Accessed 7 Sep 2018
4. Mann, C.: A study of the iPhone app at Kew Gardens: improving the visitor experience. https://ewic.bcs.org/content/ConWebDoc/46082. Accessed 7 Sep 2018
5. Castells, M.: The Rise of the Network Society. Blackwell, London (1996)
6. Van Jan, D.: The Network Society. Sage, London (1999)

7. De Certeau, M.: The Practice of Everyday Life (S. Rendall, Trans.). University of California Press, Los Angeles (1984)
8. Maistrou, H., Psychogios, D.: Presentation of an integrated system for the recording and documentation of the cultural heritage of a historic city. Digital registry for the historic center of the city of Nafplio. In: Ioannides, M., Arnold, D., Niccolucci, F., Mania, K. (eds.) The E-volution of Information Communication Technology in Cultural Heritage, Where Hi-Tech Touches the Past: Risks and Challenges for the 21st Century, pp. 114–116. EPOCH Publication, Cyprus (2006). ISBN-10 963 8046 73 2 Ö, ISBN-10 963 8046 74 0
9. Maistrou, E.: Analysis of urban patterns in historic settlements, as basis for their conservation and planning. In: European Research on Cultural Heritage, vol. 4. ARCCHIP (2006)
10. Maistrou, E.: The role of the integrated conservation of cultural heritage for a creative, resilient and sustainable city. In: Colletta, T. (ed.) ACTA of the ICOMOS - CIVVIH Symposium, FRANCOANGELI/Urbanistica (2012)
11. Minker, W., Weber, M., Hagras, H., Callagan, V., Kameas, A.D. (eds.): Advanced Intelligent Environments. Springer, US (2009). https://doi.org/10.1007/978-0-387-76485-6. ISBN 978-0-387-76484-9
12. Kumar, N., et al.: Leafsnap: a computer vision system for automatic plant species identification. In: Fitzgibbon, A., Lazebnik, S., Perona, P., Sato, Y., Schmid, C. (eds.) ECCV 2012, Part II. LNCS, pp. 502–516. Springer, Heidelberg (2012). https://doi.org/10.1007/978-3-642-33709-3_36
13. Papalexopoulos, D.: Digital Regionalism. Libro, Athens (2008)
14. Papalexopoulos, D.: Urban Hybrid Networks, The Commons Dimension, In Hybrid City II: Subtle rEvolutions, Athens (2013). http://archtech.arch.ntua.gr/forum/Hybrid_City_papalexopoulos%20final%20002.pdf. Accessed 7 Sep 2018
15. Psychogyios, D., Dimakis, N.: Ubiquitous information rooms simulator version 1.0. In: Mutations and Discontinuities. Practices, Policies and Knowledge About Urban Space on Proceedings, Alexandria, Athens, pp. 150–158 (2008). ISBN 978-960-221-471-8
16. Psychogios, D., Kechrinioiti, M., Stolidou, R.: Design of ubiquitous information environments. Tour of Nafplio project. In: Minker, W., Weber, M., Hagras, H., Callagan, V., Kameas, A.D. (eds.) Proceedings of the 3rd International Conference on Intelligent Environments, IE 2007. Advanced Intelligent Environments, Ulm (2007). ISBN 978-0-387-76484-9
17. Spigel, L.: Portable TV: studies in domestic space travels. In: Sturken, M., Thomas, D., Ball-Rokeach, S. (eds.) Technological Visions: The Hopes and Fears That Shape New Technologies, pp. 110–144. Temple University Press, Philadelphia (2004)
18. Schaffers, H., Komninos, N., Pallot, M., Aguas, M., Almirall, E., et al.: Smart Cities as Innovation Ecosystems sustained by the Future Internet [Technical report] (2012). https://hal.inria.fr/hal-00769635/document. Accessed 7 Sep 2018
19. UNESCO: UNESCO Recommendation concerning the Safeguarding and Contemporary Role of Historic Areas (1976)
20. UNESCO: Vienna Memorandu Mon World Heritage and Contemporary Architecture Managing the Historic Urban Landscape (2005)
21. Waterson, N., Saunders, M.: Delightfully Lost: A New Kind of Wayfinding at Kew Royal Botanic Gardens. In Museum and the Web (2012). http://www.museumsandtheweb.com/mw2012/papers/delightfully_lost_a_new_kind_of_wayfinding_at_.html. Accessed 7 Sep 2018
22. Williams, R.: Culture and Society: 1780–1950. Columbia University Press, New York (1983)

23. Environmental Education Center of N Kios. https://blogs.sch.gr/kpearg/about. Accessed 7 Sep 2018
24. Data.gov. http://data.gov.gr/. Accessed 7 Sep 2018
25. European Data Portal. https://www.europeandataportal.eu/. Accessed 7 Sep 2018
26. Architectural analysis of traditional buildings and complexes. 5th semester cross-sectoral digitization program. http://5a.arch.ntua.gr/villages_map. Accessed 7 Sep 2018
27. Filotis – Database for the natural Environment of Greece. https://filotis.itia.ntua.gr/. Accessed 7 Sep 2018
28. Mapping the Commons, research project on Urban Commons. http://mappingthecommons. net/en/about/. Accessed 7 Sep 2018
29. Hydrological and Meteorological Stations of the Decentralized Administration of Crete. https://www.apdkritis.gov.gr/en/node/1305. Accessed 7 Sep 2018
30. Real time marine traffic. https://www.marinetraffic.com/el/ais/home/centerx:24.2/centery:38. 4/zoom:7. Accessed 7 Sep 2018
31. Real time plane finder. https://planefinder.net/. Accessed 7 Sep 2018
32. Hybrid city events, Codes of Disobedience & Disfunctionality. http://empedia.info/maps/41. Accessed 7 Sep 2018
33. Virtual tour of Nafplio. https://www.nafplio-tour.gr/. Accessed 7 Sep 2018
34. Leafsnap an electronic leaf guide. http://leafsnap.com/. Accessed 7 Sep 2018
35. Hellenic orntithological society. http://www.ornithologiki.gr/. Accessed 7 Sep 2018

Author Index

Printed in the United States
by Bookmasters

Printed in the United States
By Bookmasters